CONVECTIVE HEAT TRANSFER

CONVECTIVE HEAT TRANSFER

Second Edition

LOUIS C. BURMEISTER
Department of Mechanical Engineering
University of Kansas
Lawrence, Kansas

A Wiley-Interscience Publication
JOHN WILEY & SONS, INC.
New York · Chichester · Brisbane · Toronto · Singapore

This text is printed on acid-free paper.

Copyright © 1993 by John Wiley & Sons, Inc.

All rights reserved. Published simultaneously in Canada.

Reproduction or translation of any part of this work beyond
that permitted by Section 107 or 108 of the 1976 United
States Copyright Act without the permission of the copyright
owner is unlawful. Requests for permission or further
information should be addressed to the Permissions Department,
John Wiley & Sons, Inc., 605 Third Avenue, New York, NY
10158-0012.

Library of Congress Cataloging in Publication Data:

Burmeister, Louis C.
    Convective heat transfer/Louis C. Burmeister.—2nd ed.
        p.   cm.
    Includes index.
    ISBN 0-471-57709-X (alk. paper)
        1. Heat—Transmission.   2. Heat–Convection.   I. Title.
    QC320.B87 1993
    536′.25—dc20   93-1032

Printed in the United States of America

10 9 8 7 6 5 4 3 2

To Rosalyn, Elise, Amanda, and my parents

CONTENTS

PREFACE

Happily, a decade of use showed the organization of the first edition to be good; but, it showed the topic of transport properties to be expendable in favor of new material in the confines of limited space. Accordingly, the fundamental topics other than transport properties and the sequence of their presentation have been maintained. Valuable new material, primarily pertaining to flow and heat transfer in porous media and computational fluid dynamics, has been added in a complementary manner. Excision of material, trimming of descriptions of the solution processes for end-of-chapter problems, and terser exposition enabled additions while reducing the length of the book. Yearly teaching of a course with this selection of material confirmed that coverage of the essence of each chapter can be accomplished in one semester. Of course, three-dimensional fluid flow and heat transfer is so rich in physical phenomena that a one-semester course of study, for which this textbook was intended, provides only a fundamental base on which further exploration can be built.

Uncounted man-years have gone into developing computer-aided methods for solving the equations that describe velocity and temperature distributions in flowing fluids. Upon learning to use the power of these methods to predict details of heat, mass, and momentum transport, a student finds the theory and the correlations more juicy and meaningful; use of a computer program, either self-developed or commercially available, should be included in a course on convective heat transfer. Not having space for exposition of more than one, the SIMPLE method was selected because its basis is the control-volume method of analysis introduced in early chapters. Diverse problems for application of numerical methods are usually suggested in the manuals that accompany commercially available computational fluid dynamics programs, and it is left to the students and instructors to indulge their own interests through problem selection in that area.

Exposition of flow and heat transfer through porous media extends the coverage of natural convection. Ranging from applications in earthen environments to packaging of electronic equipment, porous media applications have become important and were added on that basis. Some information on generalized correlations and thermal plumes has been added, as well, for its pertinence to these applications. The derivation of the equations describing flow and heat transfer through porous media again illustrates that progress is sometimes best made with simple approximate equations rather than with complex exact, Navier–Stokes and energy in this case, equations.

Other additions are less prominent. Onto the end of each chapter have been added new problems, for one. During annual teaching of a required senior-level capstone thermal-fluid design course, it became evident that its coverage of broad design applications need not be extended into one on the subject of convective heat transfer. An emphasis on end-of-chapter problems that illuminate the basics was retained for that reason. Leaving illustration of applications to the instructors enables the richness of their experiences to season the pedagogical stew, so to speak. Last, the stability of a horizontal layer of fluid heated from below was added to the appendix on instability.

The names of the people who produced the results discussed in this textbook are cited to bring the importance of the individual into view. Historical vignettes are extant, but space limitations require that such surveys as those cited in Chapter 1 be consulted for detail. *International Journal of Heat and Mass Transfer* biographical sketches will also be of interest in this regard, for example.

Several anonymous reviewers gave valuable suggestions. Much, too, is owed to those who used the first edition and shared their ideas for improvements. As would be expected, keyboard operators played a key role. This time they were Mrs. Georgia Porter and Mrs. Tammy Barta.

The citations of literature provide a means of checking the accuracy of reporting and, just as important, show the sorts of publications in which to seek pertinent information since an idea of the sources of information is a large part of a successful search for it. The assignment of some problems that require library work is recommended, and these literature leads are helpful to such assignments.

Heavy time requirements for homework often make in-class examinations unneeded, experience has shown. In many instances, computer-aided exploration of problem solution is the consumer of time. Realistically, cooperative work on such problems has to be accepted since unproductive expenditure of time often accompanies the initial stages of work on an unfamiliar computer system. This cooperative effort can be kept productive by assigning different problems to groups in the class, sometimes with results presented to the class by each group, as the experienced instructor doubtless knows.

Errors can plague any new publication. Errors in the material of the first edition were found and eliminated. Naturally, preventing them in the second edition was a high priority. Finding them in the second edition is unlikely, as a result. In a few places, though, the careful reader might find some that escaped repeated searches. Financial reward cannot be offered to the reporting readers, but gratitude certainly is.

Textbook balance between brevity of discourse that speeds delivery of the large message and completeness of development that speeds comprehension of small messages has been sought. Younger readers appreciate the details. Older ones prefer discussion of concepts and insights. Needed is the book that gratifies both. Each who passes from the first to the second role can judge that balance.

LOUIS C. BURMEISTER

Lawrence, Kansas
July 1992

PREFACE TO THE FIRST EDITION

This book evolved from notes prepared over a period of years and was used in a graduate course in convective heat and mass transfer. Heat transfer by convection is heavily emphasized, with mass transfer by convection included primarily by analogy. Expositions of events at the molecular level first give insight into the physical origin of the transport properties. Proceeding from this sound base, the equations describing convective transport on the continuum level are derived next. Examples of physical situations described by one-dimensional formulations and their solutions follow, giving insight into important conclusions and solution techniques. Only after these basics have been presented are more complex applications, such as laminar and turbulent duct and boundary-layer flows, treated. Problems, many with answers, are given at the end of each chapter. Learning is best accomplished by practice, as every teacher knows.

Every author has an intended audience. This book was written primarily for the beginning engineering graduate student. Having had prior application courses at the undergraduate level, such a student is ready for a more sharply defined course. Applications, although always the ultimate motivation, need not be emphasized as a result. The resurgence of interest in energy and its transport in thermal form, most effectively done by convection, maintains the importance of detailed knowledge of convective heat transfer and the courses offered in that subject in most schools of engineering.

When presenting a subject as rich in physical phenomena as convective transport, a class schedule must be judiciously balanced between breadth and depth. Almost all transport properties can be viewed as tabulated quantities. Little classroom time need be devoted to Chapters 1–3 if that view is adopted, although the presence of the first three chapters gives the reader a choice. Knowledge in depth of the equations describing convective transport in Chapter 4 is important since their use without understanding is dangerous. Exposition of one-dimensional problems in Chapter 5 is thorough, so little classroom time is needed. Duct and boundary-layer flows are usually given scant attention in undergraduate courses, so Chapters 6–11 should be given full consideration. Integral methods covered in Chapter 8 need only brief class time since the student can read them unaided. Numerical methods for boundary layers at the ends of Chapters 7 and 10 can be given special attention; however, detailed discussion of numerical methods is usually best reversed for a following course (e.g., in computational fluid mechanics). Driving forces for convection not externally imposed are the common denominators of Chapters 12–14. A substantial coverage of them is recommended since most undergraduate courses do not treat natural convection, boiling, and condensation in detail despite their technological importance. Realizing that only a guide is possible, I suggest the following schedule for a 14-week semester.

Chapter	1–3	4	5	6	7–8	9	10	11	12–14	Examinations and Special Topics
Class periods	5	4	2	3	6	3	4	3	8	4

Key topics are covered and, especially if examinations are of the take-home variety, there is time for special topics.

Numerous journals and reference books exist. End-of-chapter references suggest many of them as sources of additional information. Seldom is it possible to satisfy every interest in a single book. Students and practitioners seeking grounding in the basics will find the present book helpful and, it is hoped, they will appreciate the benefits of a clear separation between textbooks such as this one and the archival and reference literature. Huge amounts of relevant information exist. A coherent assemblage of this information with sufficient, but not overwhelming, detail is a formidable undertaking. Very often many sources of differing levels of complexity and detail must be consulted. Eliciting the coherent base of the subject is the first and most difficult step, and it is that base that the present book is intended to provide.

Systems of units used are the Systéme International primarily and the English system secondarily. Even though the SI system has been officially adopted almost everywhere, much information is in the English system and facility in conversion of units should be maintained. Eventually the SI system will be in common use, but that time has not yet come. Numerical conversion factors are provided to expedite conversions.

Acknowledgment of influential preceding works is appropriate here. Gifted writers serve us better than is commonly realized. Realistically, few writings do more than report and organize information developed by others. Exceptional, though, are the earlier textbooks entitled *Transport Phenomena* by Bird, Stewart, and Lightfoot and *Heat and Mass Transfer* by Eckert and Drake. A reading of the present book will show their subtle influence.

There are many who assisted in the preparation of this book. Little exaggeration is possible in saying that the students in the graduate classes, who were taught from the notes out of which this book evolved, were immensely helpful. It is with pleasure that the contribution of Mrs. Georgia Porter in typing the manuscript is recognized. Great support was provided by my family. Help of indispensable nature was provided by the publisher through careful editing and preparation of final drawings. Thanks are also due to anonymous reviewers who made valuable suggestions for improvements.

Inevitably, a few errors will have escaped repeated proofreading. Should they be found, please communicate them to me. No amount of care is sufficient, it seems, to ensure perfect copy. It does help to eliminate those that are found, though. No bounty is offered.

Effort is usually accompanied by gain. That is what is hoped for the reader. When this book is finished, a working understanding of convective heat transfer should have been acquired. Only the reading and studying remain for the reader to supply.

Louis C. Burmeister

Lawrence, Kansas
July 1982

CONVECTIVE HEAT TRANSFER

1

INTRODUCTION

The area of principal interest in this book is the convective transport of heat, mass, and momentum. It is recognized that heat can be transferred by the three modes of conduction, convection, and radiation. Usually all three modes act simultaneously, but cases in which conduction and convection are predominant are considered to the near exclusion of radiation.

In any discussion of heat transfer it is appropriate to recall that heat is energy flowing across a boundary as a result of a temperature difference across that boundary. A container of hot coffee moved by hand is a convective energy flow not directly caused by a temperature difference, and is not in the strictest sense a heat-transfer mechanism, for example.

Since convective transport requires fluid motion, it is necessary to describe the manner in which the velocity distributions affect temperature and concentration distributions. It will be seen from these descriptions that there are many similarities between heat, mass, and momentum transfer, although differences will also be apparent.

Recent surveys of the history of heat transfer by Brush [1], Cheng and Fujii [2], and Cheng [3, 12] are engrossing sources of information concerning important workers and the dates of their contributions. An outline of the history of fluid mechanics is provided by White [4], based on the history of hydraulics by Rouse and Ince [5]. Carvill [6] and the *International Journal of Heat and Mass Transfer* provide accounts of the lives of many of these contributors.

1.1 CONSERVATION PRINCIPLES

The basic relationships on which an analysis of convective transport phenomena can be based are few in view of the complexity of the problems to be solved. The basic relationships may be categorized as conservation principles and rate equations.

Physical quantities that are conserved are mass, energy, species of mass (in the absence of chemical reactions), and electrical charge. In most instances it is not necessary to apply the conservation of electrical charge principle. However, in problems that involve the interaction of electric and magnetic fields with fluids, relationships involving electromagnetic quantities (e.g., the conservation of electrical charge) are needed.

Energy can be converted into different forms without violating the energy conservation principle, credited to Mayer (1842 [2]), and one kind (species) of mass can be converted into another kind (species); the only requirement of the mass conservation principle is that the total amount of mass remain constant. In any such conversion the mass of each chemical element remains constant, disregarding relativistic effects. For mass to be converted from one species to a different species, a chemical reaction is required. For example, a container filled with a stoichiometric mixture of gaseous hydrogen and oxygen contains two species of mass. If no chemical reaction occurs, the mass of each of the two original species, the hydrogen and the oxygen, remains constant. A spark would cause the hydrogen and oxygen to combine into a third species of mass, water, destroying the original two species of hydrogen and oxygen, although the masses of hydrogen and oxygen would be unchanged.

Newton's laws of motion are needed for a description of the manner in which a fluid can move. Although they are not conservation laws in the strictest sense of the term, they can still be usefully considered as such. For example, Newton's second law of motion for a particle of constant mass m is

$$\mathbf{F} = \frac{m}{g_c} \frac{d\mathbf{V}}{dt} \tag{1-1}$$

where $\mathbf{F}$ is the net vector force acting on the particle, $\mathbf{V}$ is the vector velocity of the center of mass, and $g_c = 32.174 \ \text{lb}_m \ \text{ft}/\text{lb}_f \ \text{sec}^2 = 1 \ \text{kg m}/\text{N s}^2$ is a constant of proportionality. This may be rewritten, since m is constant, as

$$\mathbf{F} = \frac{d(m\mathbf{V}/g_c)}{dt} \tag{1-2}$$

and interpreted as a conservation law if $\mathbf{F}$ is considered as representing the rate at which the momentum of the solid particle is generated. Then, since $m\mathbf{V}/g_c$ is the momentum of the particle, in the absence of a net external force the momentum of the particle is conserved.

1.2 RATE EQUATIONS

It is necessary to be able to predict the rate at which a quantity can diffuse relative to the medium through which it is passing. The predictive equations are appropriately termed rate equations; they are sometimes also called *phenomenological relationships* because one pertains to the phenomenon of heat conduction, one pertains to the phenomenon of electrical conduction, and so forth. A basic recounting of these relations follows; more details are given in Appendix A.

Fourier's Law for Heat Conduction

Consider a large slab of homogeneous material as depicted in Fig. 1-1 without motion of one part of the slab relative to another part. A small temperature difference is steadily imposed across the slab, and it is found that the steady-state heat flow rate Q_x by conduction in the x direction is predictable by

$$Q_x = kA\frac{T(x) - T(x + \Delta x)}{\Delta x} \qquad (1\text{-}3)$$

where the thermal conductivity k is a property of the material, A is the area across which heat flows, and T is temperature. The heat flows perpendicularly to the faces of the slab (in the x direction) from the high to the low temperature. In the limit as $\Delta x \to 0$ (it may be helpful to envision a small slice being taken from the original slab, but the slice thickness cannot approach the distance between molecules), Eq. (1-3) becomes

$$q_x = \frac{Q_x}{A} = -k\frac{dT}{dx} \qquad (1\text{-}4)$$

which is *Fourier's law*, credited to him on the basis of an 1822 paper, relating the heat flux q_x to the temperature gradient dT/dx.

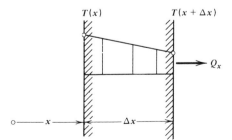

Figure 1-1 One-dimensional steady heat conduction.

Fick's Law for Binary Mass Diffusion

Consider the large slab of homogeneous material 2 in Fig. 1-2. Through this slab a different material 1 diffuses steadily as a consequence of steadily maintaining the amount of material 1 at a slightly higher level on one side of the slab than on the other. The slab could consist of a stagnant layer of oxygen through which hydrogen gas passes by diffusion, for example, but it could be solid; it is required here only that the parts of the slab not move relative to one another. It is found that the steady diffusion rate $\dot{M}_{1x}$ of species 1 in the x direction is predictable by

$$\dot{M}_{1x} = \rho D_{12} A \frac{\omega_1(x) - \omega_1(x + \Delta x)}{\Delta x} \tag{1-5}$$

where $\rho = \rho_1 + \rho_2$ is the total density of the slab and D_{12} is the mass diffusivity of species 1 through species 2. The cross-sectional area across which flow occurs is A, and $\omega_1 = \rho_1/\rho$ is the mass fraction of species 1. The diffusion of species 1 is perpendicular to the slab faces and from the slab face that has the amount of species 1 maintained at a high level to the slab face that has a low level of species 1. Because Eq. (1-5) applies to slabs of all thicknesses (so long as the slab thickness does not decrease to such an extent as to approach the dimensions of the spacing between molecules), the limit of Eq. (1-5) as $\Delta x \to 0$ gives the flux $\dot{m}_{1x}$ as

$$\dot{m}_{1x} = \frac{\dot{M}_{1x}}{A} = -\rho D_{12} \frac{d\omega_1}{dx} \tag{1-6}$$

which is commonly called *Fick's law for binary diffusion* on the basis of his 1855 paper [7]. Fick's law in the form of Eq. (1-5) is restricted to cases in which the total density ρ of the slab does not vary much in the distance Δx. Also, it should not be applied across an interface between two materials or two phases. When the density ρ is nearly constant, Eq. (1-5) can be rewritten as

$$\frac{\dot{M}_{1x}}{A} = D_{12} \frac{\rho_1(x) - \rho_1(x + \Delta x)}{\Delta x}$$

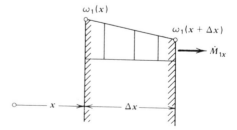

Figure 1-2   One-dimensional steady binary mass diffusion.

To illustrate the difficulty encountered in applying Eq. (1-6) across an interface, consider a pool of liquid water in equilibrium with a mixture of its vapor and air. No net mass transfer occurs; yet there is a difference in mass fraction across the liquid–vapor interface in the amount of

$$\frac{\rho_{liq}}{\rho_{liq}} - \frac{\rho_{vap}}{\rho_{vap} + \rho_{air}} \neq 0$$

Substitution of this difference into Eq. (1-6) would erroneously predict a new mass flow across the interface.

Newton's Viscosity Law

Consider a slab of fluid contained between two plates as shown in Fig. 1-3*a*. The top plate steadily moves to the right, relative to the bottom plate and parallel to it in response to the steady net force F_{ext} imposed on the top plate. The relative motion of the two plates demonstrates the fact that most fluids are unable to support shear forces without undergoing continuous displacement. The resultant flow can be smooth with a particle of fluid moving steadily in a smooth line parallel to the plates—since a thin layer of fluid then moves as a lamination, such a flow is called *laminar flow*. Or, the flow can be erratic and chaotic with a particle of fluid moving unsteadily in an unpredictable path—such a flow is called *turbulent flow*. Laminar flow is generally expected to occur when velocities are low; turbulent flow is expected to occur when velocities are high. As velocities increase from low values, the fluid flow undergoes a transition from laminar to turbulent flow, the velocity at which transition occurs depending on the geometry and the pressure gradient.

For the parallel laminar flow case shown in Fig. 1-3, it is found that the net external force that must be steadily applied to the top plate is predictable

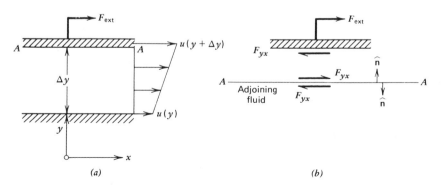

Figure 1-3 One-dimensional steady laminar flow.

(provided the slab thickness substantially exceeds the molecular spacing) by

$$F_{ext} = \mu A \frac{u(y + \Delta y) - u(y)}{\Delta y} \qquad (1\text{-}7)$$

where the dynamic viscosity μ is a property of the fluid and A is the surface area of the plate. The free-body diagram of the top plate shown in Fig. 1-3b illustrates that the fluid must exert an opposing leftward force on the plate of magnitude $F_{yx} = F_{ext}$ in order for the plate to move at constant velocity; by Newton's third law of equal action and reaction, the plate exerts a rightward force of magnitude $F_{yx} = F_{ext}$ on the adjoining fluid surface.

As shown in Fig. 1-3b for surface A–A, Newton's third law of equal action and reaction requires forces of equal magnitude but opposite direction on each side of a fluid surface. At this point it is necessary to stipulate the side of a fluid surface on which the acting shear force is to be calculated. This choice did not arise in the heat conduction or binary mass diffusion cases; the stipulation is arbitrary, dictated by both the convenience of the results obtained from it and the desirability of adhering to conventions.

The fluid mechanics convention is to identify the force acting on a surface by two subscripts, the first one designating the face on which the force acts and the second one designating the direction in which the force acts, since complete description of a force requires specification of both its point of application and direction of action. If the force acts on a positive face (a face whose surface normal $\hat{n}$ points in the positive direction denoted by the first subscript), the magnitude of the force is positive if the force is directed in the positive direction denoted by the second subscript. If the force acts on a negative face (a face whose surface normal $\hat{n}$ points in the negative direction denoted by the first subscript), the magnitude of the force is positive if the force is directed in the negative direction denoted by the second subscript.

The force F_{yx} on the positive y face of surface A–A is shown in Fig 1-3 to act in the positive x direction. In the limit as $\Delta y \to 0$ the shear stress τ_{yx} on the upper face of surface A–A is obtained from Eq. (1-7) as

$$\tau_{yx} = \frac{F_{yx}}{A} = \mu \frac{du}{dx} \qquad (1\text{-}8)$$

which is *Newton's law of viscosity* [6, p. 52; 8] on the basis of Newton's analyses of fluid mechanics problems.

Newton's Law of Cooling

Consider a hot solid wall at temperature T_w that is exposed to a cool flowing fluid at temperature T_f as shown in Fig. 1-4. It is observed that heat flows from the wall into the cooler fluid and that the rate of heat transfer Q is

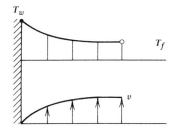

Figure 1-4 Convective heat transfer from a solid surface to a flowing fluid.

proportional to both the surface area A and to the temperature difference $T_w - T_f$. These observations are incorporated in the predictive equation for the convective heat flux q

$$q = \frac{Q}{A} = h(T_w - T_f) \tag{1-9}$$

which is referred to as *Newton's law of cooling*. The proportionality factor h is a *heat-transfer coefficient*. Under normal conditions, at the wall there is no relative velocity between the fluid and the wall and the fluid temperature equals the wall temperature. In other words, at a solid wall a fluid normally experiences no slip and no temperature jump.

It is experimentally observed that the heat-transfer coefficient h in Eq. (1-9) is not a constant; it is sensitive to the flow conditions in the ambient fluid. On physical grounds it can be understood how this might be so. The thickness of the fluid film, or boundary layer (since it is also a layer at a boundary), near the wall is observed to decrease as the fluid flows by more rapidly. Since the constant imposed temperature difference occurs over the largely stagnant film that is of diminished thickness, the heat flux should increase, thus requiring an increased heat transfer coefficient h. It is apparent that convection is a combination of several fundamental transport mechanisms.

It is evident that the magnitude of the heat-transfer coefficient is dependent on the rate at which the fluid can convect a particle away from the wall after it has been heated. Since the fluid experiences no slip at the wall and thermal radiation is often negligible at the moderate temperatures envisioned here, heat can flow from the wall into the fluid only by conduction. As the heat diffuses further into the fluid where velocity is high, convection of heated particles becomes significant. Finally, near the outer edge of the boundary layer, almost all the energy transport occurs by convection. The simple form of Eq. (1-9) tends to obscure the basic phenomena that are active, but its convenience is responsible for its continued use—see Adiutori [9] for an alternative discussion of the origin and utility of the heat-transfer coefficient h.

Thermal Radiation

Thermal radiation is sometimes an important mechanism of heat transfer. Examples of processes in which radiative transport plays a major role are heat transfer from spacecraft, in the fireboxes of boilers of central electrical generating plants, and warming of solar collectors by the sun.

The rate at which a perfect emitter, called a *blackbody* since it also absorbs all incident thermal radiation, emits energy by the mechanism of thermal radiation is given by

$$Q = A\sigma T^4 \tag{1-10}$$

where $\sigma = 0.1714 \times 10^{-8}$ Btu/hr ft^2 R^4 = 5.66961 × 10^{-8} W/m^2 K^4 is the Stefan–Boltzmann constant in honor of Stefan's 1879 experimentally based and Boltzmann's analytically based formulations, T is the emitter absolute temperature, Q is the heat flow rate, and A is the surface area. A nonblackbody emits at a lesser rate as expressed by insertion of a multiplicative emittance ϵ in Eq. (1-10) to give

$$Q = \epsilon A\sigma T^4 \tag{1-11}$$

where $0 \leqslant \epsilon \leqslant 1$. The emitted energy can be regarded as propagating at the speed of light either as electromagnetic waves or as packets of photons.

The fraction of thermal radiation incident on a body that is absorbed is denoted by the absorptance α where $0 \leqslant \alpha \leqslant 1$. The transmittance τ and reflectance ρ denote the fraction of incident radiation that is transmitted and reflected, respectively, with $0 \leqslant \tau \leqslant 1$ and $0 \leqslant \rho \leqslant 1$. Conservation of energy requires that the relationship $\tau + \rho + \alpha = 1$ always be satisfied. It can be additionally shown that at a specific wavelength $\epsilon = \alpha$, a relationship referred to as *Kirchhoff's law*.

When the space between two solid surfaces is transparent (because of either the absence of intervening matter as in a vacuum or the inability of the intervening matter to be substantially affected by the frequencies present in the electromagnetic wave), the net rate of heat transfer is given by

$$Q_{1-2} = \mathscr{F}_{1-2}\sigma\left(T_1^4 - T_2^4\right)A_1 \tag{1-12}$$

in which the factor $\mathscr{F}_{1-2}$ contains the combined effects of the radiative properties and geometric orientation of the surfaces. The geometric orientation is accounted for by a geometric shape factor F_{1-2} that can be interpreted as being the ratio of the diffuse radiation leaving surface 1 that strikes surface 2.

The rate of radiative heat transfer is seen in Eqs. (1-11) and (1-12) to be proportional to a difference of the fourth power of absolute temperatures, a relationship that is accurate when the space between the absorber and the emitter is transparent. When matter that is capable of strongly absorbing and emitting radiation intervenes between the two surfaces, the rate of heat transfer is characterized more by diffusion as a transport process than by thermal radiation. This fact is used in the Rosseland diffusion approximation [10, 11] of radiative energy transport.

1.3   ANALOGIES FOR ONE-DIMENSIONAL DIFFUSION

For convenience and ease of perception, the four rate equations are displayed in Table 1-1 under the assumption of constant properties. There it is apparent that all the rate equations are of the same form—flux = constant × potential gradient. The proportionality constant is a function of the material involved in the transport process and is called a *transport property*. The transport properties in Table 1-1 are k (thermal conductivity), D_{12} (mass diffusivity), and μ (viscosity). The presence of a positive sign in Newton's law of viscosity is solely due to the choice of sides of a fluid surface on which to evaluate the acting stress; if the other side had been chosen, a negative sign would appear in the rate equation.

The left-hand side of each rate equation is a flux, having the units of the diffusing substance/area per time. This point is less clear for the case of viscous fluid shear for which the shear stress τ_{xy} has units of N/m^2 in SI units (lb_f/ft^2 in English units). Conversion of units by multiplication by g_c then gives the units of shear stress as $kg\ m\ s^{-1}/s\ m^2$ in SI units (lb_m ft $sec^{-1}/sec\ ft^2$ in English units), more clearly signifying that shear stress has units of momentum/area per time and showing that the diffusing quantity is momentum (horizontal momentum for the situation shown in Fig. 1-3 and vertical momentum for the formulation in Table 1-1).

TABLE 1-1   Recast One-Dimensional Rate Equations

Process	Recast Rate Equation	Diffusivity	Rate Equation in SI Units	
Heat conduction	$q_x = -\alpha \dfrac{d(\rho C_p T)}{dx}$	$\alpha = \dfrac{k}{\rho C_p}$	$\dfrac{J}{s\ m^2}$	$= -\dfrac{m^2}{s}\dfrac{d(J/m^3)}{d(m)}$
Binary mass diffusion	$\dot{m}_{1,x} = -D_{12}\rho \dfrac{d\omega_1}{dx}$	D_{12}	$\dfrac{kg_1}{s\ m^2}$	$= -\dfrac{m^2}{s}\dfrac{d(kg_1/m^3)}{d(m)}$
Viscous fluid shear	$\tau_{xy} = \nu \dfrac{d(\rho v)}{dx}$	$\nu = \dfrac{\mu}{\rho}$	$\dfrac{kg\ m\ s^{-1}}{s\ m^2}$	$= \dfrac{m^2}{s}\dfrac{d(kg\ m\ s^{-1}/m^3)}{d(m)}$

Consideration of the second column of Table 1-1 shows the common form

$$\frac{\text{diffusing quantity}}{\text{time area}} = \pm(\text{diffusivity})\,\frac{d(\text{diffusing quantity/volume})}{d(\text{distance})}$$

suggesting that the flux of a diffusing quantity is proportional to the concentration gradient of that quantity. The property of diffusivity that is a coefficient has the units $(\text{distance})^2/\text{time}$ regardless of the physical phenomenon considered. The ratio of diffusivities is a dimensionless number that indicates the relative rate of diffusion of the corresponding physical quantities; if the ratio is near unity, the physical mechanisms of the two diffusions are likely to be the same. The Prandtl number is $\text{Pr} = \nu/\alpha$, the Schmidt number is $\text{Sc} = \nu/D_{12}$, and the Lewis number is $\text{Le} = D_{12}/\alpha$.

The common functional form of the rate equations and the similar nature of the conservation principles suggested that analogies between heat, mass, and momentum transport can be found. Of course, final proof that analogies exist must be deferred until the full mathematical descriptions for a specific physical situation are shown to be of the same form. When an analogy does occur, and it does not always, information gained with an experimentally convenient physical phenomenon can be applied to the analogous physical phenomena whose experimental treatment is more difficult.

1.4 OVERVIEW OF FOLLOWING CHAPTERS

In Chapter 2 fluids are considered to be continuous, their molecular makeup tacitly acknowledged by the use of transport properties. The partial differential equations that describe convection are derived, showing how the effects of nonuniformities in temperature, velocity, and mass fraction can be taken into account, and including the case in which a continuous fluid saturates and flows through a porous solid medium.

In Chapter 3 the essential features of convective transport by forced convection are ascertained by relatively easy solutions to a few illustrative one-dimensional problems. In Chapters 4–6 these insights are extended by detailed consideration of laminar flow in ducts and boundary layers. Technically important conclusions and analytical methods are presented. In Chapters 7–9 the insights acquired from study of laminar flow, together with additional empirical information, are applied to forced turbulent flow in ducts and boundary layers.

In Chapter 10–12 situations are considered in which flow is not externally forced. Natural convection, boiling, and condensation are treated in a manner that merges previously developed insights and techniques with pertinent new information.

REFERENCES

1. S. G. Brush, in E. T. Layton and J. H. Lienhard, Eds., *History of Heat Transfer*, ASME, 1988, pp. 25–51.
2. K. C. Cheng and T. Fujii, in E. T. Layton and J. H. Lienhard, Eds., *History of Heat Transfer*, ASME, 1988, pp. 213–260.
3. K. C. Cheng, in G.-J. Hwang, Ed., *Transport Phenomena in Thermal Control*, Hemisphere, 1989, pp. 3–35.
4. F. M. White, *Viscous Fluid Flow*, 2nd ed., McGraw-Hill, 1991, pp. 1–4.
5. H. Rouse and S. Ince, *History of Hydraulics*, State University of Iowa, Institute of Hydraulic Research, Iowa City, Iowa.
6. J. Carvill, *Famous Names in Engineering*, Butterworths, 1981.
7. A. Fick, *Ann. Phys. Chem.* **94**, 59 (1855).
8. H. Lamb, *Hydrodynamics*, 6th ed., Dover, 1945, p. 588.
9. E. F. Adiutori, *Mech. Eng.* **112**, 46–50 (1990).
10. R. Siegel and J. R. Howell, *Thermal Radiation Heat Transfer*, McGraw-Hill, 1972, pp. 469–474.
11. S. Rosseland, *Theoretical Astrophysics; Atomic Theory and the Analysis of Stellar Atmospheres and Envelopes*, Clarendon, 1936.
12. K. C. Cheng, *Heat Transfer Engineering* **13**, 19–37 (1992).

2

EQUATIONS OF CONTINUITY, MOTION, ENERGY, AND MASS DIFFUSION

Often transport properties can be taken to be known quantities since the most common desire is to determine the distributions of bulk velocity, temperature, or mass fraction. These, when inserted into the appropriate rate equation (e.g., Fourier's law), yield the quantity of major interest (surface shear stress, heat flux, or mass flux). The determination of every detail of bulk velocity distribution that comes from accounting for random molecular motion in gases, for example, leads to the Boltzmann integrodifferential equation. The Boltzmann equation represents the most detailed description of transport, but it is so difficult that it has been solved neither exactly nor numerically for many cases [1–3]. The first terms of a series solution of the Boltzmann equation verify that fluid bulk velocity and so forth can usually be accurately obtained from simpler equations that consider the fluid to be a continuous substance and, as long as extremely steep gradients and rapid changes are not encountered, the rate equations of Chapter 1 are accurate. The Burnett equations [4, 5, 53] provide a description intermediate between that of the Boltzmann equation and a continuous description.

Usually, the equations describing bulk velocity, temperature, and mass fraction distributions for a continuous fluid provide sufficient accuracy, and they are derived in this chapter. Conservation principles are applied to a convenient control volume, and the rate equations of Chapter 1 are used to relate fluxes to gradients.

2.1 GENERAL ASSUMPTIONS

Property values are assumed to be available; see Liley [6] and others [7–15] for discussions of the theory and correlations pertinent to such transport

12

properties as thermal conductivity, viscosity, and mass diffusivity. When large temperature excursions are encountered, it is presumed that the variation of properties with temperature can be evaluated. Equations of state are assumed to be known to relate density to temperature and pressure; the equation of state $\rho = pM/RT$ for a perfect gas is a simple example. It is usually assumed that the materials encountered are isotropic—no property depends on direction.

The continuous nature of matter is assumed because the derivation of differential equations involves taking the limit as a control volume becomes of infinitesimal size. Although rarefied gases may violate this assumption when the mean free path of molecules is of the same order of magnitude as the system dimensions, at room conditions even a gas can be considered to be a continuous medium. Liquids and solids, which are substantially denser than gases, are even better regarded as continuous materials.

It is assumed that pure substances in the thermodynamic sense are under consideration. A pure substance [16] is homogeneous in composition and invariable in chemical aggregation. For example, water existing as a vapor, a liquid, or a solid or a combination of these states is a pure substance. A mixture of hydrogen and oxygen gases is a pure substance as long as no liquid or solid phase appears since then one phase will be richer in hydrogen than the other. If part of the system is combined to form water, this system is not a pure substance because it is not homogeneous in chemical composition. This assumption, when taken together with the local equilibrium assumption, allows such quantities as enthalpy and internal energy to be related to temperature and pressure.

Local equilibrium is assumed. The essence of this assumption is that the departure from equilibrium is sufficiently slight that the rate equations are accurate. Without equilibrium, the concept of temperature loses much of its meaning. The temperature (1.5×10^{12} K is the highest possible temperature; above that, added energy is used to convert nuclear matter into hadron matter [17]) of a gas, for instance, is strictly defined only if all energy storage modes are in a state of thermal equilibrium. Since individual gas molecules store energy in translational, rotational, and electronic modes of motion, complete equilibrium is attained only when all the modes of energy storage have population distributions corresponding to the same temperature [18]. A rapid change such as might be experienced in a shock wave, rapid decompression, or rapid chemical reaction may result in a delay of the order of several microseconds [19] before equilibrium is again attained. In a monatomic gas only three translation modes exist for energy storage, with rotation seemingly unimportant at normal temperatures, and equilibrium is achieved rapidly. A diatomic or polyatomic gas, on the other hand, first stores energy in translational motion. Then, after a number of collisions with other molecules, energy is stored in the rotational mode of motion; after a larger number of collisions, vibrational modes of motion store energy. The number of collisions required for a gas to reachieve equilibrium ranges from a few in the case of air to several thousand in the case of carbon dioxide [20, 21]. The new equilibrium condition is asymptotically approached [1, p. 125] with a relaxation time proportional to μ/p for the translational mode [7].

2.2 EQUATION OF CONTINUITY

The conservation of mass principle is applied first to the general control volume shown in Fig. 2-1. This principle states that the mass of a system of fixed identity is constant. Let the system of fixed identity be bounded by the dotted line; as the fluid flows and distorts the system, the dotted line will also move and distort in such a manner as to always enclose the same fluid particles. The solid line denotes an arbitrary control volume within which the system is initially contained. At initial time t the system occupies regions 1 and 3; at a later time $t + \Delta t$ it occupies regions 1 and 2. It is presumed that the elapsed time Δt is sufficiently small that region 1 occupies some part of the control volume. Since the system mass is constant,

$$m_1(t) + m_3(t) = m_1(t + \Delta t) + m_2(t + \Delta t)$$

where m is the mass in the identified region of space. Rearrangement gives

$$m_1(t + \Delta t) - m_1(t) = m_3(t) - m_2(t + \Delta t)$$

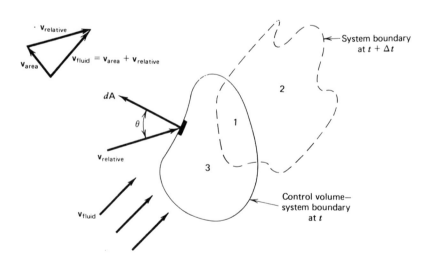

Figure 2-1   General control volume through which fluid flows.

Division by Δt then yields

$$\frac{m_1(t + \Delta t) - m_1(t)}{\Delta t} = \frac{m_3(t)}{\Delta t} - \frac{m_2(t + t)}{\Delta t} \qquad (2\text{-}1)$$

As $\Delta t \to 0$, region 1 approaches coincidence with the control volume. Then, the left-hand side represents the rate at which mass is stored in the control volume, the first term of the right-hand side represents the rate at which mass enters the control volume since the mass $m_3(t)$ initially occupying region 3 must be replaced if the fluid is continuous, and the second term on the right-hand side represents the rate at which mass leaves the control volume since the mass $m_2(t + \Delta t)$ in region 2 was initially inside the control volume. The conservation of mass principle applied to a control volume is then stated as

$$\sum \dot{m}_{stored} = \sum \dot{m}_{in} - \sum \dot{m}_{out} \qquad (2\text{-}2)$$

The dot notation signifies a rate with respect to time. For many problems involving finite control volumes, Eq. (2-2) is the most useful statement of the conservation of mass principle.

Equation (2-2) can be generally reformulated as

$$\frac{\partial(\int_V \rho \, dV)}{\partial t} = -\int_A \rho(\mathbf{V}_{rel} \cdot d\mathbf{A}) \qquad (2\text{-}3)$$

since the mass in any infinitesimal volume dV is $\rho \, dV$ and the net rate at which mass flows across an infinitesimal area $d\mathbf{A}$ is

$$\rho(V_{fluid} - V_{area}) \cos \theta \, dA$$

or

$$\rho \mathbf{V}_{rel} \cdot d\mathbf{A}$$

in which it is assumed that the surface normal of an area element points outward. The divergence theorem relates volume integrals to area integrals according to

$$\int_A \mathbf{X} \cdot d\mathbf{A} = \int_V \text{div}(\mathbf{X}) \, dV$$

in which A is the area that encloses the volume V. Recognizing that $\mathbf{X} = \rho \mathbf{V}_{rel}$ in this case allows Eq. (2-3) to be rewritten more conveniently as

$$\frac{\partial(\int_V \rho \, dV)}{\partial t} + \int_V \text{div}(\rho \mathbf{V}_{rel}) \, dV = 0 \qquad (2\text{-}4)$$

Equation (2-4) applies to a control volume that can distort as time passes. When the control volume is of fixed shape, the time derivative can be taken inside the integral to obtain

$$\int_V \left[\frac{\partial \rho}{\partial t} + \text{div}(\rho \mathbf{V}_{\text{rel}}) \right] dV = 0 \qquad (2\text{-}5)$$

Focusing attention on an infinitesimal control volume over which no property experiences appreciable variation, Eq. (2-5) is well approximated by

$$\left[\frac{\partial \rho}{\partial t} + \text{div}(\rho \mathbf{V}_{\text{rel}}) \right] dV = 0$$

which is satisfied only by

$$\frac{\partial \rho}{\partial t} + \text{div}(\rho \mathbf{V}_{\text{rel}}) = 0 \qquad (2\text{-}6)$$

which is a *conservation-law form*. If the control volume is additionally constrained to be motionless, $\mathbf{V}_{\text{area}} = 0$ and so $\mathbf{V}_{\text{rel}} = \mathbf{V}_{\text{fluid}}$. If subscripts are dropped and $\mathbf{V}_{\text{fluid}}$ is denoted by $\mathbf{V}$, Eq. (2-6) then is

$$\frac{\partial \rho}{\partial t} + \text{div}(\rho \mathbf{V}) = 0 \qquad (2\text{-}7)$$

Equation (2-7) is the *continuity equation* (because the divergence theorem employed at one step in its derivation requires density and velocity to be continuous functions) and is the form most useful for infinitesimal control volumes.

Equation (2-7) has been derived without specification of a coordinate system and is applicable to any coordinate system. In rectangular coordinates, for which $\mathbf{V} = u\hat{i} + v\hat{j} + w\hat{k}$, the continuity equation takes the form

$$\frac{\partial \rho}{\partial t} + \frac{\partial(\rho u)}{\partial x} + \frac{\partial(\rho v)}{\partial y} + \frac{\partial(\rho w)}{\partial z} = 0 \qquad (2\text{-}8)$$

In steady state $\partial \rho / \partial t = 0$; and for a constant-density fluid, $\rho = \text{const.}$, it is always the case that

$$\frac{\partial u}{\partial x} + \frac{\partial v}{\partial y} + \frac{\partial w}{\partial z} = 0 \qquad (2\text{-}9)$$

The continuity equation has been derived in a general way, following which specific assumptions were made to arrive at Eq. (2-8). It is instructive to first make the specific assumptions and then to apply the conservation

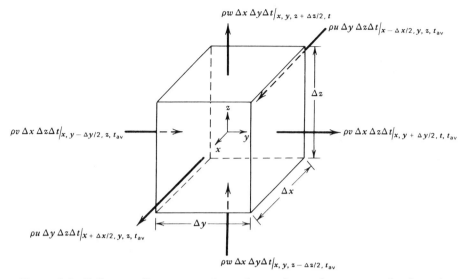

Figure 2-2 Bulk mass flows across the surfaces of a stationary control volume in rectangular coordinates.

principle. To this end, it is assumed from the beginning that the control volume is a cube fixed in space as shown in Fig. 2-2. A fluid, having velocity components u, v, and w, flows through the space occupied by the cube from time t to time $t + \Delta t$. The cube faces are of length Δx in the x direction, Δy in the y direction, and Δz in the z direction. As shown in Fig. 2-2, the rate at which total mass flows through a face is evaluated at the average time t_{av}, $t < t_{av} < t + \Delta t$, as density $\times$ velocity perpendicular to area at area center and average time $\times$ area. The conservation principle expressed by Eq. (2-2) is, in terms of the mathematical quantities shown in Fig. 2-2,

$$\frac{\rho\,\Delta x\,\Delta y\,\Delta z|_{x,y,z,t+\Delta t} - \rho\,\Delta x\,\Delta y\,\Delta z|_{x,y,z,t}}{\Delta t}$$

$$= \rho u\,\Delta y\,\Delta z|_{x-\Delta x/2,y,z,t_{av}} - \rho u\,\Delta y\,\Delta z|_{x+\Delta x/2,y,z,t_{av}}$$

$$+ \rho v\,\Delta x\,\Delta z|_{x,y-\Delta y/2,z,t_{av}} - \rho v\,\Delta x\,\Delta z|_{x,y+\Delta y/2,z,t_{av}}$$

$$+ \rho w\,\Delta x\,\Delta y|_{x,y,z-\Delta z/2,t_{av}} - \rho w\,\Delta x\,\Delta y|_{x,y,z+\Delta z/2,t_{av}}$$

Dividing this equation by $\Delta x\,\Delta y$, Δz and taking the limit as all differential quantities approach zero, one has, by the definition of a partial derivative,

$$\frac{\partial \rho}{\partial t} + \frac{\partial(\rho u)}{\partial x} + \frac{\partial(\rho v)}{\partial y} + \frac{\partial(\rho w)}{\partial z} = 0$$

which is Eq. (2-8) as it should be. It is important to fully comprehend this second manner of derivation.

The density is the bulk (average) density of the possibly multicomponent fluid, and the velocity is the bulk (average) macroscopic velocity that gives the bulk mass flow rate—$\mathbf{V} = \Sigma_i \rho_i \mathbf{V}_i / \rho$, where $\rho = \Sigma_i \rho_i$. The continuity equation for cylindrical and spherical coordinates is given in Appendix C.

2.3   SUBSTANTIAL DERIVATIVE

The continuity equation, Eq. (2-7) in general or Eq. (2-8) for rectangular coordinates, can be manipulated into an alternative form that will be convenient later. To see the nature of the manipulation, Eq. (2-8) is considered first. Taking the derivatives of products in Eq. (4-8), one obtains

$$\left(\frac{\partial \rho}{\partial t} + u \frac{\partial \rho}{\partial x} + v \frac{\partial \rho}{\partial y} + w \frac{\partial \rho}{\partial z} \right) + \rho \left(\frac{\partial u}{\partial x} + \frac{\partial v}{\partial y} + \frac{\partial w}{\partial z} \right) = 0$$

This result can be more compactly written as

$$\frac{D\rho}{Dt} + \rho \left(\frac{\partial u}{\partial x} + \frac{\partial v}{\partial y} + \frac{\partial w}{\partial z} \right) = 0 \qquad (2\text{-}10)$$

where the operator

$$\frac{D(\ )}{Dt} = \frac{\partial(\ )}{\partial t} + u \frac{\partial(\ )}{\partial x} + v \frac{\partial(\ )}{\partial y} + w \frac{\partial(\ )}{\partial z} \qquad (2\text{-}11)$$

is the *substantial derivative* in rectangular coordinates.

The general form of the substantial derivative D/Dt can be established by manipulations on Eq. (2-6) whose second term can be separated into two parts as

$$\left(\frac{\partial \rho}{\partial t} + \mathbf{V}_{\mathrm{rel}} \cdot \nabla \rho \right) + \rho \nabla \cdot \mathbf{V}_{\mathrm{rel}} = 0$$

This result is more compactly written as

$$\frac{D\rho}{Dt} + \rho \, \mathrm{div}(\mathbf{V}_{\mathrm{rel}}) = 0 \qquad (2\text{-}12)$$

Here, the general form of the substantial derivative is seen to be

$$\frac{D(\ )}{Dt} = \frac{\partial(\ )}{\partial t} + \mathbf{V}_{rel} \cdot \nabla(\ ) \tag{2-13}$$

whose compactness explains its frequent use in convective heat transfer and fluid mechanics. The words of Davis [22] are recommended for an alternative discussion that includes historical comments.

2.4 EQUATIONS OF MOTION

The motion of a fluid particle is influenced by the external forces acting on it. This fact can be taken into account by use of Newton's second law which, for an inertial coordinate system and a system of fixed identity, is

$$\mathbf{F} = \frac{m}{g_c} \frac{d\mathbf{V}}{dt} \tag{2-14}$$

where $\mathbf{F}$ is the vector sum of all external forces, m is the total mass of the system, and $\mathbf{V}$ is the center of mass velocity referred to inertial coordinates. The constant g_c has been explicitly written in Eq. (2-14) as a reminder that it must always be included at some stage to make the equation dimensionally consistent. In the English system of units $g_c = 32.2$ lb_m ft/lb_f sec^2, and serious numerical errors (in addition to dimensional inconsistency) can result from its omission. In the SI system of units $g_c = 1$ kg m/N s^2, and no numerical error results from its omission. To emphasize the importance of momentum changes, Newton's second law is rewritten as

$$\mathbf{F} = \frac{d(m\mathbf{V}/g_c)}{dt} \tag{2-15}$$

The momentum $m\mathbf{V}/g_c$ of the system is shown by Eq. (2-15) to change with time by the action of the external force. Inasmuch as the change in momentum is stored in the system, it could be said that, for a system of fixed identity,

$$\mathbf{F} = \mathbf{momentum}_{stored}$$

and that $\mathbf{F}$ acts like a momentum generation term.

Because it is impractical in a moving fluid to permanently identify a system of fixed identity, it is desired to have a form that is appropriate for a control volume through whose surfaces fluid flows. To this end, a system of fixed identity is identified and followed for only a short time Δt. Referring to

Fig. 2-1, it is seen that the system momentum initially equals

$$\frac{m_1 V_1(t)}{g_c} + \frac{m_3 V_3(t)}{g_c}$$

and finally equals

$$\frac{m_1 V_1(t + \Delta t)}{g_c} + \frac{m_2 V_2(t + \Delta t)}{g_c}$$

The change in system momentum is, accordingly,

$$\Delta \frac{mV}{g_c} = \frac{[m_1 V_1(t + \Delta t) - m_1 V_1(t)]}{g_c} + \frac{m_2 V_2(t + \Delta t)}{g_c} - \frac{m_3 V_3(t)}{g_c}$$

Division by the elapsed time Δt—sufficiently small that region 1 lies within the control volume—followed by substitution into Eq. (2-15) shows that

$$F = \frac{m_1 V_1(t + \Delta t) - m_1 V(t)}{g_c \Delta t} + \frac{m_2 V(t + \Delta t)}{g_c \Delta t} - \frac{m_3 V_3(t)}{g_c \Delta t} \quad (2\text{-}16)$$

As $\Delta t \rightarrow 0$, region 1 coincides with the control volume. The terms on the right-hand side of Eq. (2-16) are identical in form with those of Eq. (2-1) in the continuity equation derivation and can be similarly interpreted. The only difference is that momentum takes the place of mass. Accordingly, Eq. (2-16) can be stated as

$$\textbf{Force} + \textbf{momentum}_{\text{in}} - \textbf{momentum}_{\text{out}} = \textbf{momentum}_{\text{stored}} \quad (2\text{-}17)$$

Again, the dot notation signifies a rate with respect to time. Equation (2-17) is a convenient form of Newton's second law for problems involving finite control volumes. It is a vector equation, having three components.

Equation (2-17) can be generally stated as

$$F - \int_A \frac{V}{g_c} \rho V_{\text{rel}} \cdot dA = \frac{d\left[\int_V (\rho V/g_c)\, dV \right]}{dt} \quad (2\text{-}18)$$

The second term on the left-hand side is the net efflux of momentum, which is seen to be momentum per unit mass × mass flow rate. The vector identity

$$\int_A y(x \cdot dA) = \int_V [y\, \text{div}(x) + (x \cdot \nabla)y]\, dV$$

allows a surface integral to be expressed as an integral over the volume enclosed by the surface. Its use allows Eq. (2-18) to be rewritten as

$$\mathbf{F} = \frac{d\left[\int_V (\rho\mathbf{V}/g_c)\,dV\right]}{dt} + \int_V \frac{(\mathbf{V}/g_c)\,\mathrm{div}(\rho\mathbf{V}_{\mathrm{rel}}) + (\rho\mathbf{V}_{\mathrm{rel}} \cdot \nabla)\mathbf{V}}{g_c}\,dV \quad (2\text{-}19)$$

If the control volume is of fixed shape, the order of differentiation and integration can be interchanged to give

$$\mathbf{F} = \int_V \left[\partial(\rho\mathbf{V})\,\partial t + \mathbf{V}\,\mathrm{div}(\rho\mathbf{V}_{\mathrm{rel}}) + (\rho\mathbf{V}_{\mathrm{rel}} \cdot \nabla)\mathbf{V}\right]\frac{dV}{g_c}$$

If the control volume is of infinitesimal size, experiencing no appreciable property variation over its extent, this relationship is well approximated by

$$\frac{g_c\mathbf{F}}{dV} = \frac{\partial(\rho\mathbf{V})}{\partial t} + \mathbf{V}\,\mathrm{div}(\rho\mathbf{V}_{\mathrm{rel}}) + (\rho\mathbf{V}_{\mathrm{rel}} \cdot \nabla)\mathbf{V}$$

which is a conservation-law form. This equation can be rearranged as

$$\frac{g_c\mathbf{F}}{dV} = \mathbf{V}\left[\frac{\partial\rho}{\partial t} + \mathrm{div}(\rho\mathbf{V}_{\mathrm{rel}})\right] + \rho\left[\frac{\partial\mathbf{V}}{\partial t} + (\mathbf{V}_{\mathrm{rel}} \cdot \nabla)\mathbf{V}\right]$$

The first bracketed term on the right-hand side is zero from the continuity equation [Eq. (2-6)]. Thus Newton's second law applied to an infinitesimal control volume of fixed shape is

$$\frac{\mathbf{F}}{dV} = \rho\left[\frac{\partial(\mathbf{V}/g_c)}{\partial t} + (\mathbf{V}_{\mathrm{rel}} \cdot \nabla)\frac{\mathbf{V}}{g_c}\right] \quad (2\text{-}20)$$

A more compact form of Eq. (2-20) is

$$\frac{\mathbf{F}}{dV} = \rho\frac{D(\mathbf{V}/g_c)}{Dt} \quad (2\text{-}21)$$

Equation (2-21) is identical in form with Newton's second law for a system of fixed identity, as comparison with Eq. (2-14) shows. Some derivations of the equations of motion merely adopt Eq. (2-14) and identify the derivative as the "derivative following the motion," termed the *substantial derivative* here. Such a derivation, although it has speed and brevity to commend it, is not really correct since a particle is followed for a distance (a Lagrangian point of view) before reverting to a coordinate system fixed in the laboratory (an Eulerian point of view). Having been derived without specification of a coordinate system, Eq. (2-20) is applicable to any coordinate system.

In the case for which the control volume is stationary, $V_{area} = 0$; thus $V_{rel} = V_{fluid}$. If subscripts are dropped so that $\mathbf{V}$ represents only the fluid velocity, Eq. (2-20) becomes

$$\frac{\mathbf{F}}{dV} = \rho \frac{\partial \mathbf{V}/\partial t + (\mathbf{V} \cdot \nabla)\mathbf{V}}{g_c} \tag{2-22}$$

In rectangular coordinates for which $\mathbf{F} = F_x\hat{\mathbf{i}} + F_y\hat{\mathbf{j}} + F_z\hat{\mathbf{k}}$ and $\mathbf{V} = u\hat{\mathbf{i}} + v\hat{\mathbf{j}} + w\hat{\mathbf{k}}$, Eq. (2-22) becomes

$$\frac{F_x}{dV} = \rho \frac{\partial u/\partial t + u\,\partial u/\partial x + v\,\partial u/\partial y + w\,\partial u/\partial z}{g_c} \tag{2-23a}$$

$$\frac{F_y}{dV} = \rho \frac{\partial v/\partial t + u\,\partial v/\partial x + v\,\partial v/\partial y + w\,\partial v/\partial z}{g_c} \tag{2-23b}$$

$$\frac{F_z}{dV} = \rho \frac{\partial w/\partial t + u\,\partial w/\partial x + v\,\partial w/\partial y + w\,\partial w/\partial z}{g_c} \tag{2-23c}$$

It is instructive to first make all assumptions and then to apply the conservation principle. Therefore, assume that the control volume is a cube fixed in space with rectangular coordinates as shown in Fig. 2-3. The rate at which a momentum component flows through one of the cube faces is

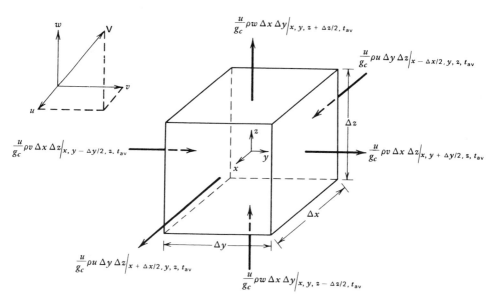

Figure 2-3 x-Momentum flows across the surfaces of a stationary control volume in rectangular coordinates.

evaluated at the average time t_{av}, $t \leqslant t_{av} \leqslant t + \Delta t$, as

Momentum component/mass

$\times$ mass flux at area center and average time $\times$ area

where the flow is observed for the time interval between t and $t + \Delta t$. With the x, y, and z velocity components being u, v, and w, respectively, and focusing on the x component of momentum whose flow through each face is shown in Fig. 2-3, application of the momentum principle expressed by Eq. (2-17) gives

$$F_x = \frac{\left[(u/g_c)\rho\,\Delta x\,\Delta y\,\Delta z|_{x,y,z,t+\Delta t} - (u/g_c)\rho\,\Delta x\,\Delta y\,\Delta z|_{x,y,z,t}\right]}{\Delta t}$$

$$+ \frac{u}{g_c}\rho u\,\Delta y\,\Delta z|_{x+\Delta x/2,\,y,\,z,\,t_{av}} - \frac{u}{g_c}\rho u\,\Delta y\,\Delta z|_{x-\Delta x/2,\,y,\,z,\,t_{av}}$$

$$+ \frac{u}{g_c}\rho v\,\Delta x\,\Delta z|_{x,\,y+\Delta y/2,\,z,\,t_{av}} - \frac{u}{g_c}\rho v\,\Delta x\,\Delta z|_{x,\,y-\Delta y/2,\,z,\,t_{av}}$$

$$+ \frac{u}{g_c}\rho w\,\Delta x\,\Delta y|_{x,\,y,\,z+\Delta z/2,\,t_{av}} - \frac{u}{g_c}\rho w\,\Delta x\,\Delta y|_{x,\,y,\,z-\Delta z/2,\,t_{av}}$$

Dividing by $\Delta x\,\Delta y\,\Delta z$ and taking the limit as all differential quantities approach zero, one obtains by the definition of a partial derivative,

$$\frac{g_c F_x}{dx\,dy\,dz} = \frac{\partial(\rho u)}{\partial t} + \frac{\partial(\rho uu)}{\partial x} + \frac{\partial(\rho uv)}{\partial y} + \frac{\partial(\rho uw)}{\partial z}$$

Taking the derivatives of the products, one then has

$$\frac{g_c F_x}{dx\,dy\,dz} = u\left[\frac{\partial\rho}{\partial t} + \frac{\partial(\rho u)}{\partial x} + \frac{\partial(\rho v)}{\partial y} + \frac{\partial(\rho w)}{\partial z}\right]$$

$$+ \rho\left[\frac{\partial u}{\partial t} + u\frac{\partial u}{\partial x} + v\frac{\partial u}{\partial y} + w\frac{\partial u}{\partial z}\right]$$

The first bracketed term on the right-hand side of this equation is zero according to the continuity equation [Eq. (2-8)]. Hence the result is

$$\frac{F_x}{dx\,dy\,dz} = \frac{\rho(\partial u/\partial t + u\,\partial u/\partial x + v\,\partial u/\partial y + w\,\partial u/\partial z)}{g_c}$$

which is the same as Eq. (2-23a), as it should be.

The external forces either act on the surface of the control volume and are called *surface forces*, or they act on the distributed mass in the control volume and are called *body forces*. Shear and pressure give rise to surface forces, for example, whereas gravity and electromagnetic fields (see Viskanta [23]) give rise to a body force. Attention is directed to a stationary cubical control volume in rectangular coordinates. As illustrated in Fig. 2-4, the surface force acting on a face is composed of three components. A surface force is given by the product of the force/area, denoted by τ, and area. The stress on a face is denoted by τ_{ij} in which the convention is adopted that the first subscript i designates the face on which the stress acts and the second subscript j designates the direction in which the stress acts. On a face whose outward directed normal points in a positive coordinate direction, the positive stress points in a positive coordinate direction; if the outward directed surface normal points in a negative coordinate direction, so does the positive stress. The body force is given by the product of the force/volume, denoted by B, and volume. The net x component of the external force is then

$$
\begin{aligned}
F_x = {} & \left(\tau_{xx}|_{x+\Delta x/2,\, y,\, z,\, t_{av}} - \tau_{xx}|_{x-\Delta x/2,\, y,\, z,\, t_{av}} \right) \Delta y\, \Delta z \\
& + \left(\tau_{yx}|_{x,\, y+\Delta y/2,\, z,\, t_{av}} - \tau_{yx}|_{x,\, y-\Delta y/2,\, z,\, t_{av}} \right) \Delta x\, \Delta z \\
& + \left(\tau_{zx}|_{x,\, y,\, z+\Delta z/2,\, t_{av}} - \tau_{zx}|_{x,\, y,\, z-\Delta z/2,\, t_{av}} \right) \Delta x\, \Delta y \\
& + B_x|_{x,\, y,\, z,\, t_{av}} \Delta x\, \Delta y\, \Delta z
\end{aligned}
$$

in which time dependence is taken into account by evaluating quantities at an average time t_{av} between the start t of the observation period and its end $t + \Delta t$. Dividing by $\Delta x\, \Delta y\, \Delta z$ and taking the limit as all differential quantities approach zero, one obtains

$$
\frac{F_x}{dV} = B_x + \frac{\partial \tau_{xx}}{\partial x} + \frac{\partial \tau_{yx}}{\partial y} + \frac{\partial \tau_{zx}}{\partial z} \tag{2-24}
$$

where $dV = dx\, dy\, dz$. Substitution of Eq. (2-24) into the x equation of motion [Eq. (2-23a)] gives

$$
\rho \frac{D(u/g_c)}{Dt} = B_x + \frac{\partial \tau_{xx}}{\partial x} + \frac{\partial \tau_{yx}}{\partial y} + \frac{\partial \tau_{zx}}{\partial z} \tag{2-25a}
$$

In similar fashion it is found for the y and z equations of motion in rectangular coordinates that

$$
\rho \frac{D(v/g_c)}{Dt} = B_y + \frac{\partial \tau_{xy}}{\partial x} + \frac{\partial \tau_{yy}}{\partial y} + \frac{\partial \tau_{zy}}{\partial z} \tag{2-25b}
$$

$$
\rho \frac{D(w/g_c)}{Dt} = B_z + \frac{\partial \tau_{xz}}{\partial x} + \frac{\partial \tau_{yz}}{\partial y} + \frac{\partial \tau_{zz}}{\partial z} \tag{2-25c}
$$

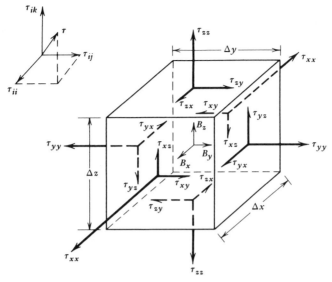

Figure 2-4 Surface and body-force components acting on a stationary control volume in rectangular coordinates.

The number of unknowns in Eqs. (2-25) must be reduced if a solution is to be achieved, and information from an independent source must be provided for that purpose.

For most important fluids, such as air and water, there is a linear relationship between stresses and velocities; such fluids are called *Newtonian fluids*. Fluids having a nonlinear relationship between stresses and velocities are called *non-Newtonian fluids*. The major types of relationships between stresses and velocities are illustrated in Fig. 2-5 for a simple flow situation. Additional discussion is given in Appendix B.

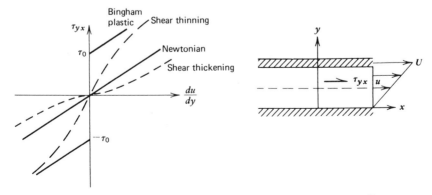

Figure 2-5 Relationships between shear stress and velocity gradient.

For a Newtonian fluid, the linear relationship between stresses and velocities (shown in detail in Appendix B) in rectangular coordinates is

$$\tau_{xy} = \mu\left(\frac{\partial u}{\partial y} + \frac{\partial v}{\partial x}\right) = \tau_{yx} \tag{2-26a}$$

$$\tau_{yz} = \mu\left(\frac{\partial v}{\partial z} + \frac{\partial w}{\partial y}\right) = \tau_{zy} \tag{2-26b}$$

$$\tau_{xz} = \mu\left(\frac{\partial w}{\partial x} + \frac{\partial u}{\partial z}\right) = \tau_{zx} \tag{2-26c}$$

$$\tau_{xx} = -p + \mu\left(2\frac{\partial u}{\partial x} - \frac{2}{3}\,\text{div V}\right) \tag{2-26d}$$

$$\tau_{yy} = -p + \mu\left(2\frac{\partial v}{\partial y} - \frac{2}{3}\,\text{div V}\right) \tag{2-26e}$$

$$\tau_{zz} = -p + \mu\left(2\frac{\partial w}{\partial z} - \frac{2}{3}\,\text{div V}\right) \tag{2-26f}$$

The viscosity μ is a property of the fluid and originates in random molecular motion. The stress normal to a surface τ_{ii} is composed of the sum of pressure stress p and viscous stress (which is not always a shear stress) as

$$\tau_{ii} = -p + \tau_{ii_{\text{viscous}}}$$

If the fluid is inviscid, $\mu \equiv 0$, so that there cannot be any viscous stress,

$$\tau_{ii} = -p$$

Equations (2-26d–f) added show that for a viscous fluid pressure can be defined as

$$p = -\frac{(\tau_{xx} + \tau_{yy} + \tau_{zz})}{3}$$

which conforms to the expectation one has for inviscid fluids.

Introducing Eqs. (2-26) into the equations of motion [Eqs. (2-25)] that follow from Newton's second law, [Eqs. (2-25)] one obtains

$$\frac{\rho}{g_c}\frac{Du}{Dt} = B_x - \frac{\partial p}{\partial x} + \frac{\partial}{\partial x}\left[\mu\left(2\frac{\partial u}{\partial x} - \frac{2}{3}\operatorname{div}\mathbf{V}\right)\right]$$

$$+ \frac{\partial}{\partial y}\left[\mu\left(\frac{\partial u}{\partial y} + \frac{\partial v}{\partial x}\right)\right] + \frac{\partial}{\partial z}\left[\mu\left(\frac{\partial w}{\partial x} + \frac{\partial u}{\partial z}\right)\right] \qquad (2\text{-}27a)$$

$$\frac{\rho}{g_c}\frac{Dv}{Dt} = B_y - \frac{\partial p}{\partial y} + \frac{\partial}{\partial x}\left[\mu\left(\frac{\partial u}{\partial y} + \frac{\partial v}{\partial x}\right)\right] + \frac{\partial}{\partial y}\left[\mu\left(2\frac{\partial v}{\partial y} - \frac{2}{3}\operatorname{div}\mathbf{V}\right)\right]$$

$$+ \frac{\partial}{\partial z}\left[\mu\left(\frac{\partial v}{\partial z} + \frac{\partial w}{\partial y}\right)\right] \qquad (2\text{-}27b)$$

$$\frac{\rho}{g_c}\frac{Dw}{Dt} = B_z - \frac{\partial p}{\partial z} + \frac{\partial}{\partial x}\left[\mu\left(\frac{\partial w}{\partial x} + \frac{\partial u}{\partial z}\right)\right] + \frac{\partial}{\partial y}\left[\mu\left(\frac{\partial v}{\partial z} + \frac{\partial w}{\partial y}\right)\right]$$

$$+ \frac{\partial}{\partial z}\left[\mu\left(2\frac{\partial w}{\partial z} - \frac{2}{3}\operatorname{div}\mathbf{V}\right)\right] \qquad (2\text{-}27c)$$

for rectangular coordinates.

When visocity is constant; Eqs. (2-27) assume a simpler form as can be shown through the example of the x-direction equation of motion. Expansion of the derivatives in Eq. (2-27a) with μ constant gives

$$\frac{\rho}{g_c}\frac{Du}{Dt} = B_x - \frac{\partial p}{\partial x} + \mu\left\{\left[\frac{\partial(\partial u/\partial x)}{\partial x} + \frac{\partial(\partial u/\partial y)}{\partial y} + \frac{\partial(\partial u/\partial z)}{\partial z}\right]\right.$$

$$+ \left.\left[\frac{\partial(\partial u/\partial x)}{\partial x} + \frac{\partial(\partial v/\partial x)}{\partial y} + \frac{\partial(\partial w/\partial x)}{\partial z}\right] - \frac{2}{3}\frac{\partial\operatorname{div}(\mathbf{V})}{\partial x}\right\}$$

The second bracketed term inside the braces is div(**V**) if the order of differentiation is interchanged for each term. Then the x-direction equation of motion is

$$\frac{\rho}{g_c}\frac{Du}{Dt} = B_x - \frac{\partial p}{\partial x} + \mu\left[\frac{\partial^2 u}{\partial x^2} + \frac{\partial^2 u}{\partial y^2} + \frac{\partial^2 u}{\partial z^2} + \frac{1}{3}\frac{\partial\operatorname{div}(\mathbf{V})}{\partial x}\right] \qquad (2\text{-}28)$$

The equations of motion for the other two directions are of the same form. In general the constant viscosity case can be expressed as

$$\frac{\rho}{g_c} \frac{D\mathbf{V}}{Dt} = \mathbf{B} - \nabla p + \mu \left[\nabla^2 \mathbf{V} + \frac{1}{3} \nabla (\nabla \cdot \mathbf{V}) \right]$$

If density is also constant, the continuity equation [Eq. (2-7)] shows that div($\mathbf{V}$) = 0. Then the equations of motion in rectangular coordinates are

$$\frac{\rho}{g_c} \frac{Du}{Dt} = B_x - \frac{\partial p}{\partial x} + \mu \left(\frac{\partial^2 u}{\partial x^2} + \frac{\partial^2 u}{\partial y^2} + \frac{\partial^2 u}{\partial z^2} \right) \tag{2-29a}$$

$$\frac{\rho}{g_c} \frac{Dv}{Dt} = B_y - \frac{\partial p}{\partial y} + \mu \left(\frac{\partial^2 v}{\partial x^2} + \frac{\partial^2 v}{\partial y^2} + \frac{\partial^2 v}{\partial z^2} \right) \tag{2-29b}$$

$$\frac{\rho}{g_c} \frac{Dw}{Dt} = B_z - \frac{\partial p}{\partial z} + \mu \left(\frac{\partial^2 w}{\partial x^2} + \frac{\partial^2 w}{\partial y^2} + \frac{\partial^2 w}{\partial z^2} \right) \tag{2-29c}$$

This set of equations describing the motion of a constant density and viscosity fluid can be more compactly written in vector notation as

$$\frac{\rho}{g_c} \frac{D\mathbf{V}}{Dt} = \mathbf{B} - \nabla p + \mu \nabla^2 \mathbf{V} \tag{2-30}$$

Equations of motion for cylindrical and spherical coordinate systems are given in Appendix C. It is usually convenient to obtain the equations of motion by a coordinate transformation of the equations for rectangular coordinates, rather than by applying conservation principles directly to an often awkwardly shaped control volume in nonrectangular coordinates.

2.5 ENERGY EQUATION

The conservation of energy principle in the form that applies to a system always composed of the same particles of mass, also known as the *first law of thermodynamics*, is expressed in mathematical terms as

$$Q_{in} = \Delta E + W_{out} \tag{2-31}$$

where Q_{in} is the heat transferred into the system from its surroundings, W_{out} is the work done by the system on its surroundings, and ΔE is the change in the system internal energy. The mass of the system can store energy internally in a number of forms; thus the internal energy is

$$E = E_{\text{kinetic}} + E_{\substack{\text{potential} \\ \text{due to body force}}} + E_{\text{surface tension}} + E_{\text{electromagnetic}} + E_{\text{thermal}} + \cdots$$

For the developments of this chapter, the contributions of surface tension, electromagnetic, and other energy forms are taken as unchanged by the physical processes encountered; since they do not change, they can be neglected. Also, the potential energy due to location in a body force field is considered to be the result of work done by the body force, a permissible though arbitrary decision. With these choices in mind, $E = E_{\text{kinetic}} + E_{\text{thermal}}$.

To cast the conservation of energy principle into the form needed for application to a control volume through which fluid flows, Eq. (2-31) is applied to the system of fixed identity illustrated in Fig. 2-1 for a brief time Δt. During this time interval the internal energy change of the system is

$$\Delta E = \left[E_1(t + \Delta t) + E_2(t + \Delta t) \right] - \left[E_1(t) + E_3(t) \right]$$

Insertion of this result into Eq. (2-31) and division by Δt gives

$$\frac{Q_{\text{in}}}{\Delta t} - \frac{W_{\text{out}}}{\Delta t} = \frac{E_1(t + \Delta t) - E_1(t)}{\Delta t} + \frac{E_2(t + \Delta t)}{\Delta t} - \frac{E_3(t)}{\Delta t}$$

As Δt approaches zero the two rightmost terms on the right-hand side of this equation can be interpreted as the rate of internal energy leaving and entering, respectively. Thus the conservation of energy principle for a control volume is

$$\dot{Q}_{\text{in}} - \dot{W}_{\text{out}} = \text{internal energy}_{\text{stored}} + \text{internal energy}_{\text{out}} - \text{internal energy}_{\text{in}}$$
$$(2\text{-}32)$$

Equation (2-32) is a useful form of the conservation of energy principle for problems involving finite control volumes.

Equation (2-32) can be generally stated in mathematical terms as

$$\dot{Q}_{\text{in}} - \dot{W}_{\text{out}} = \frac{d\left(\int_V e\rho \, dV \right)}{dt} + \int_A e(\rho \mathbf{V}_{\text{rel}} \cdot d\mathbf{A})$$

in which e represents internal energy per unit mass. The divergence theorem permits the area integral to be cast as a volume integral, giving

$$\dot{Q}_{\text{in}} - \dot{W}_{\text{out}} = \frac{d\left(\int_V e\rho \, dV \right)}{dt} + \int_V \text{div}(e\rho \mathbf{V}_{\text{rel}}) \, dV$$

For a control volume of fixed shape, the order of integration and differentiation can be interchanged in the preceding equation to obtain

$$\dot{Q}_{in} - \dot{W}_{out} = \int_V \left[\frac{\partial(e\rho)}{\partial t} + \text{div}(e\rho\mathbf{V}_{rel}) \right] dV$$

For an infinitesimal control volume, this becomes

$$\frac{\partial(e\rho)}{\partial t} + \text{div}(e\rho\mathbf{V}_{rel}) = \frac{\dot{Q}_{in}}{dV} - \frac{\dot{W}_{out}}{dV} \tag{2-33}$$

which is a conservation-law form. Equation (2-33) can be rearranged into

$$e\left[\frac{\partial \rho}{\partial t} + \text{div}(\rho\mathbf{V}_{rel}) \right] + \rho\left[\frac{\partial e}{\partial t} + \mathbf{V}_{rel} \cdot \nabla e \right] = \frac{\dot{Q}_{in}}{dV} - \frac{\dot{W}_{out}}{dV}$$

The first bracketed term equals zero by the continuity equation [Eq. (2-6)], and the second bracketed term can be written as De/Dt by use of Eq. (2-13). Then Eq. (2-33) becomes

$$\rho\frac{De}{Dt} = \frac{\dot{Q}_{in}}{dV} - \frac{\dot{W}_{out}}{dV} \tag{2-34}$$

The heat term $\dot{Q}_{in}$ is treated first without specification of a coordinate system. Heat enters the control volume by diffusion relative to the bulk flow due to random molecular motion. It is possible for some energy that entered the control volume in chemical, atomic, or electromagnetic forms to be converted into a thermal form of energy, leading to an internally distributed heat source that accounts for things not specifically enumerated. Hence

$$\dot{Q}_{in} = -\int_A \mathbf{q} \cdot d\mathbf{A} + \int_V q''' \, dV$$

in which $\mathbf{q}$ is the diffusive flux of heat and q''' is the rate at which thermal energy is liberated per unit volume. Appeal to the divergence theorem gives

$$\dot{Q}_{in} = \int_V [q''' - \text{div}(\mathbf{q})] \, dV$$

For an infinitesimal control volume, this relationship becomes

$$\frac{\dot{Q}_{in}}{dV} = q''' - \text{div}(\mathbf{q})$$

Insertion of this result into Eq. (2-34) gives the energy equation as

$$\rho \frac{De}{Dt} + \text{div}(\mathbf{q}) - q''' = \frac{-\dot{W}_{out}}{dV} \tag{2-35}$$

In rectangular coordinates, Eq. (2-35) would be written in full as

$$\rho \left(\frac{\partial e}{\partial t} + u \frac{\partial e}{\partial x} + v \frac{\partial e}{\partial y} + w \frac{\partial e}{\partial z} \right) + \left(\frac{\partial q_x}{\partial x} + \frac{\partial q_y}{\partial y} + \frac{\partial q_z}{\partial z} \right) - q''' = \frac{-\dot{W}_{out}}{dx\, dy\, dz} \tag{2-36}$$

Although the foregoing derivation is general and independent of any coordinate system, it tends to obscure the point that

Total energy flux = bulk convection + diffusion relative to bulk motion

To amplify this point, an alternative development is presented for a stationary control volume in rectangular coordinates. The energy flows across surfaces of the control volume are shown in Fig. 2-6. The energy conservation

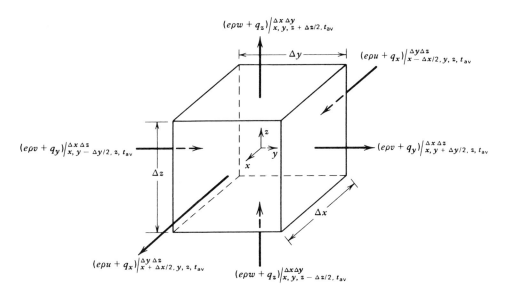

Figure 2-6 Energy flows across surfaces of a stationary control volume in rectangular coordinates.

principle of Eq. (2-32) accepts the terms partially enumerated in Fig. 2-6 to show

$$
\left(e\rho|_{x,y,z,t+\Delta t} - e\rho|_{x,y,z,t}\right)\frac{\Delta x\,\Delta y\,\Delta z}{\Delta t}
$$
$$
+\left[(e\rho u + q_x)|_{x+\Delta x/2,y,z,t_{av}} - (e\rho u + q_x)|_{x-\Delta x/2,y,z,t_{av}}\right]\Delta y\,\Delta z
$$
$$
+\left[(e\rho v + q_y)|_{x,y+\Delta y/2,z,t_{av}} - (e\rho v + q_y)|_{x,y-\Delta y/2,z,t_{av}}\right]\Delta x\,\Delta z
$$
$$
+\left[(e\rho w + q_z)|_{x,y,z+\Delta z/2,t_{av}} - (e\rho w + q_z)|_{x,y,z-\Delta z/2,t_{av}}\right]\Delta x\,\Delta y
$$
$$
- q'''|_{x,y,z,t_{av}}\,\Delta x\,\Delta y\,\Delta z = -\dot{W}_{out}
$$

in which $t \leqslant t_{av} \leqslant t + \Delta t$. With division by $\Delta x\,\Delta y\,\Delta z$, and in the limit as all differential quantities approach zero, this relationship becomes

$$
\frac{\partial(e\rho)}{\partial t} + \frac{\partial(e\rho u + q_x)}{\partial x} + \frac{\partial(e\rho v + q_y)}{\partial y} + \frac{\partial(e\rho w + q_z)}{\partial z} - q''' = \frac{-\dot{W}_{out}}{dx\,dy\,dz}
$$

This result can be rearranged, by taking derivatives of products and sums, into

$$
\rho\left(\frac{\partial e}{\partial t} + u\frac{\partial e}{\partial x} + v\frac{\partial e}{\partial x} + w\frac{\partial e}{\partial z}\right) + \left(\frac{\partial q_x}{\partial x} + \frac{\partial q_y}{\partial y} + \frac{\partial q_z}{\partial z}\right)
$$
$$
+ e\left(\frac{\partial\rho}{\partial t} + \frac{\partial\rho u}{\partial x} + \frac{\partial\rho v}{\partial y} + \frac{\partial\rho w}{\partial z}\right) - q''' = \frac{-\dot{W}_{out}}{dx\,dy\,dz}
$$

The third term in parentheses on the left-hand side is zero, according to the continuity equation [Eq. (2-8)], so that Eq. (2-36) is duplicated as it should be.

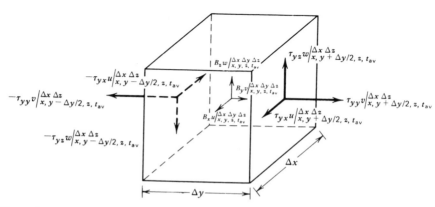

Figure 2-7 Work done by surface and body forces on a stationary control volume in rectangular coordinates.

The work term $-\dot{W}_{out}$ on the right-hand side of Eq. (2-35) is evaluated next. Insight into the physical processes involved is achieved by considering a stationary control volume in rectangular coordinates. The rate at which work $\dot{W}_{in}$ ($= -\dot{W}_{out}$) is done by the environment on the mass inside the control volume is the product of the force exerted by the environment and the parallel component of the velocity of the particle exerting the force. Since the particle exerting the force on the control volume touches the control volume, its velocity is the fluid velocity at the control volume surface. $\dot{W}_{in}$ at two control volume surfaces (note that the rate of work input is negative when force and velocity are oppositely directed) is shown in Fig. 2-7; similar terms appear for the other four faces. The body force is shown as doing work, also; the potential energy due to position in the force field cannot then also be considered to be a part of internal energy of the mass in the control volume. Collecting terms, dividing by $\Delta x\,\Delta y\,\Delta z$, and taking the limit as all differential quantities approach zero (again, $t \leqslant t_{av} \leqslant t + \Delta t$), one obtains

$$
\begin{aligned}
-\dot{W}_{out}/dV = {}&+ \frac{\partial(\tau_{xx}u)}{\partial x} + \frac{\partial(\tau_{yx}u)}{\partial y} + \frac{\partial(\tau_{zx}u)}{\partial z} + B_x u \\[2mm]
&+ \frac{\partial(\tau_{xy}v)}{\partial x} + \frac{\partial(\tau_{yy}v)}{\partial y} + \frac{\partial(\tau_{zy}v)}{\partial z} + B_y v \\[2mm]
&+ \frac{\partial(\tau_{xz}w)}{\partial x} + \frac{\partial(\tau_{yz}w)}{\partial y} + \frac{\partial(\tau_{zz}w)}{\partial z} + B_z w \qquad (2\text{-}37)
\end{aligned}
$$

which is a conservation-law form. Expansion and collection of terms gives, with Eq. (2-25) taken into account,

$$
\begin{aligned}
\frac{-\dot{W}_{out}}{dV} = {}&u\left(\frac{\partial\tau_{xx}}{\partial x} + \frac{\partial\tau_{yx}}{\partial y} + \frac{\partial\tau_{zx}}{\partial z} + B_x\right) \frac{\rho}{g_c}\frac{Du}{Dt} \\[2mm]
&+ v\left(\frac{\partial\tau_{xy}}{\partial x} + \frac{\partial\tau_{yy}}{\partial y} + \frac{\partial\tau_{zy}}{\partial z} + B_y\right) \frac{\rho}{g_c}\frac{Dv}{Dt} \\[2mm]
&+ w\left(\frac{\partial\tau_{xz}}{\partial x} + \frac{\partial\tau_{yz}}{\partial y} + \frac{\partial\tau_{zz}}{\partial z} + B_z\right) \frac{\rho}{g_c}\frac{Dw}{Dt} \\[2mm]
&+ \tau_{xx}\frac{\partial u}{\partial x} + \tau_{yx}\frac{\partial u}{\partial y} + \tau_{zx}\frac{\partial u}{\partial z} \\[2mm]
&+ \tau_{xy}\frac{\partial v}{\partial x} + \tau_{yy}\frac{\partial v}{\partial y} + \tau_{zy}\frac{\partial v}{\partial z} \\[2mm]
&+ \tau_{xz}\frac{\partial w}{\partial x} + \tau_{yz}\frac{\partial w}{\partial y} + \tau_{zz}\frac{\partial w}{\partial z} \qquad (2\text{-}38)
\end{aligned}
$$

The first three terms of Eq. (2-38) constitute ρD (kinetic energy/mass)$/Dt$ as can be seen by considering

$$u\frac{Du}{Dt} + v\frac{Dv}{Dt} + w\frac{Dw}{Dt} = \frac{Du^2/2}{Dt} + \frac{Dv^2/2}{Dt} + \frac{Dw^2/2}{Dt} = \frac{D}{Dt}\left(\frac{|\mathbf{V}|^2}{2}\right)$$

Introducing the preceding expression for $-\dot{W}_{out}/dV$ into Eq. (2-35) and letting the internal energy I per unit mass due to temperature be $I = e -$ kinetic energy per unit mass $= e - |\mathbf{V}|^2/2g_c$, one finds that

$$\rho\frac{DI}{Dt} + \text{div}(\mathbf{q}) - q''' = \tau_{xx}\frac{\partial u}{\partial x} + \tau_{yx}\frac{\partial u}{\partial y} + \tau_{zx}\frac{\partial u}{\partial z}$$

$$+ \tau_{xy}\frac{\partial v}{\partial x} + \tau_{yy}\frac{\partial v}{\partial y} + \tau_{zy}\frac{\partial v}{\partial z}$$

$$+ \tau_{xz}\frac{\partial w}{\partial x} + \tau_{yz}\frac{\partial w}{\partial y} + \tau_{zz}\frac{\partial w}{\partial z} \qquad (2\text{-}39)$$

Note that both kinetic and potential forms of energy are absent from the "energy" equation [Eq. (2-39)].

Surface stresses τ_{ij} are rarely specified directly (they might be set to zero at a free surface), so it is necessary to relate them to the velocities that are more commonly specified. For a Newtonian fluid, introduction of Eqs. (2-26) into Eq. (2-39) results in

$$\rho\frac{DI}{Dt} + \text{div}(\mathbf{q}) - q''' = -p\left(\frac{\partial u}{\partial x} + \frac{\partial v}{\partial y} + \frac{\partial w}{\partial z}\right) + \mu\left\{\left[2\left(\frac{\partial u}{\partial x}\right)^2 - \frac{2}{3}\frac{\partial u}{\partial x}\text{div}(\mathbf{V})\right]\right.$$

$$+ \left[\frac{\partial u}{\partial y} + \frac{\partial v}{\partial x}\right]\frac{\partial u}{\partial y} + \left[\frac{\partial w}{\partial x} + \frac{\partial u}{\partial z}\right]\frac{\partial u}{\partial z}$$

$$+ \left[\frac{\partial u}{\partial y} + \frac{\partial v}{\partial x}\right]\frac{\partial v}{\partial x} + \left[2\left(\frac{\partial v}{\partial y}\right)^2 - \frac{2}{3}\frac{\partial v}{\partial y}\text{div}(\mathbf{V})\right]$$

$$+ \left[\frac{\partial v}{\partial z} + \frac{\partial w}{\partial y}\right]\frac{\partial v}{\partial z} + \left[\frac{\partial w}{\partial x} + \frac{\partial u}{\partial z}\right]\frac{\partial w}{\partial x} + \left[\frac{\partial v}{\partial z} + \frac{\partial w}{\partial y}\right]\frac{\partial w}{\partial y}$$

$$+ \left.\left[2\left(\frac{\partial w}{\partial z}\right)^2 - \frac{2}{3}\frac{\partial w}{\partial z}\text{div}(\mathbf{V})\right]\right\}$$

This equation can be more compactly written in the general form

$$\rho \frac{DI}{Dt} + \text{div}(\mathbf{q}) - q''' = -p \, \text{div}(\mathbf{V}) + \mu \Phi \qquad (2\text{-}40)$$

Here, Φ is defined in rectangular coordinates as

$$\Phi = 2\left[\left(\frac{\partial u}{\partial x}\right)^2 + \left(\frac{\partial v}{\partial y}\right)^2 + \left(\frac{\partial w}{\partial z}\right)^2\right] - \frac{2}{3}[\text{div}(\mathbf{V})]^2$$

$$+ \left(\frac{\partial u}{\partial y} + \frac{\partial v}{\partial x}\right)^2 + \left(\frac{\partial v}{\partial z} + \frac{\partial w}{\partial y}\right)^2 + \left(\frac{\partial w}{\partial x} + \frac{\partial u}{\partial z}\right)^2 \qquad (2\text{-}41)$$

and is called the *dissipation function* since it represents the irreversible conversion of mechanical forms of energy to a thermal form. The squared terms show the irreversibility since $\Phi \geqslant 0$ and *always* acts as a *source* of thermal energy. In contrast, $p \, \text{div}(\mathbf{V})$ represents the reversible work done on the environment by the expanding mass inside the control volume.

The enthalpy H per unit mass is introduced next, in a search for a more convenient form of the "energy" equation, as a replacement for thermal internal energy I per unit mass. By definition,

$$H = I + \frac{p}{\rho}$$

Thus

$$\frac{DI}{Dt} = \frac{DH}{Dt} - \rho^{-1}\frac{Dp}{Dt} + p\rho^{-2}\frac{D\rho}{Dt} \qquad (2\text{-}42)$$

The continuity equation [Eq. (2-12)] shows that the last term on the right-hand side of this relation equals $-p\rho^{-1}\,\text{div}(\mathbf{V})$. Incorporation of this fact into Eq. (2-42) and substitution of that result into Eq. (2-40) gives

$$\rho \frac{DH}{Dt} + \text{div}(\mathbf{q}) - q''' = \frac{Dp}{Dt} + \mu \Phi \qquad (2\text{-}43)$$

The Dp/Dt term in Eq. (2-43) is often preferred over the $p \, \text{div}(\mathbf{V})$ term in Eq. (2-40) because it is more common for a fluid to undergo a nearly constant pressure process (for which $Dp/Dt \sim 0$) than a nearly constant volume process (for which $\text{div}\,\mathbf{V} \sim 0$). It is apparent that the "energy" equation can be manipulated into a variety of forms, some of which are more convenient for a given purpose than others.

Equation (2-43) is occasionally a convenient form for applications with phase change [24–26] as encountered in manufacturing processes and thermal energy storage as well as gases that undergo dissociation or ionization [27]. Usually, however, a further manipulation is required so that boundary conditions given in terms of temperature can be met by the solution. Accordingly, recourse to thermodynamics is next.

For a pure substance in the absence of motion, surface tension, and electromagnetic effects, there are only two independent properties [28]—$H = H(p, T)$, where T is temperature. Thus

$$DH = \frac{\partial H}{\partial T}\bigg|_{p=\text{const}} DT + \frac{\partial H}{\partial p}\bigg|_{T=\text{const}} Dp \tag{2-44}$$

From thermodynamics, with $s = $ entropy,

$$DH = T\,Ds + \frac{Dp}{\rho}$$

which can be rephrased as

$$\frac{DH}{Dp} = T\frac{Ds}{Dp} + \frac{1}{\rho} \tag{2-45}$$

Applying Eq. (2-45) to an infinitesimal process at thermodynamic equilibrium, since thermodynamic equilibrium implies constant temperature, gives

$$\frac{\partial H}{\partial p}\bigg|_{T=\text{const}} = T\left(\frac{\partial s}{\partial p}\right)\bigg|_{T=\text{const}} + \frac{1}{\rho} \tag{2-46}$$

The Maxwell relations of thermodynamics [28] reveal that*

$$\frac{\partial s}{\partial p}\bigg|_{T=\text{const}} = \rho^{-2}\frac{\partial \rho}{\partial T}\bigg|_{p=\text{const}}$$

---

*Gibb's free energy is $G = I + p/\rho - Ts$. Its change is given by

$$dG = \left[dI + p(d\rho^{-1}) - T\,ds\right] + \rho^{-1}\,dp - s\,dT$$

The bracketed term is zero according to the first law of thermodynamics. Thus

$$dG = \rho^{-1}\,dp - s\,dT$$

Comparison with the chain rule $dG = (\partial G/\partial p)\,dp + (\partial G/\partial T)\,dT$ shows

$$\frac{\partial G}{\partial p} = \rho^{-1} \quad \text{and} \quad \frac{\partial G}{\partial T} = -s$$

Taking second derivatives yields

$$\frac{\partial G^2}{\partial T\,\partial p} = -\rho^{-2}\frac{\partial \rho}{\partial T}\bigg|_{p=\text{const}} \quad \text{and} \quad \frac{\partial^2 G}{\partial p\,\partial T} = -\frac{\partial s}{\partial p}\bigg|_{T=\text{const}}$$

This relationship, when substituted into Eq. (2-56), gives

$$\left(\frac{\partial H}{\partial p}\right)\Bigg|_{T=\text{const}} = T\rho^{-2}\frac{\partial\rho}{\partial T}\Bigg|_{p=\text{const}} + \frac{1}{\rho} \tag{2-47}$$

Recall that the specific heat at constant pressure is defined as

$$C_p = \frac{\partial H}{\partial T}\Bigg|_{p=\text{const}} \tag{2-48}$$

and the coefficient of thermal expansion is defined as

$$\beta = -\rho^{-1}\left(\frac{\partial\rho}{\partial T}\right)_{p=\text{const}} \tag{2-49}$$

Insertion of Eqs. (2-47)–(2-49) into Eq. (2-44) shows that

$$DH = C_p\,DT + \rho^{-1}(1 - \beta T)\,Dp \tag{2-50}$$

The energy equation, on substitution of Eq. (2-50) into Eq. (2-43), then is

$$\rho C_p\frac{DT}{Dt} = -\text{div}(\mathbf{q}) + q''' + \beta T\frac{Dp}{Dt} + \mu\Phi \tag{2-51}$$

Relation of conductive heat flux to temperature is accomplished by Fourier's law $\mathbf{q} = -k\nabla T$. The energy equation then is

$$\rho C_p\frac{DT}{Dt} = \text{div}(k\nabla T) + q''' + \beta T\frac{Dp}{Dt} + \mu\Phi \tag{2-52a}$$

For a constant-density fluid, $\beta = 0$. If viscous dissipation is neglected, $\Phi = 0$. Then, for constant thermal conductivity, Eq. (2-52a) in rectangular coordinates for a stationary control volume is

$$\rho C_p\left(\frac{\partial T}{\partial t} + u\frac{\partial T}{\partial x} + v\frac{\partial T}{\partial y} + w\frac{\partial T}{\partial z}\right) = k\left(\frac{\partial^2 T}{\partial x^2} + \frac{\partial^2 T}{\partial y^2} + \frac{\partial^2 T}{\partial z^2}\right) + q''' \tag{2-52b}$$

2.6 BINARY MASS DIFFUSION EQUATION

The conservation of mass principle that led to the continuity equation makes no distinction among the different kinds (species) of mass that can comprise a fluid. If a fluid is made up of several kinds of mass and if one is unusually abundant in a particular region, it will tend to diffuse in such a way as to render its abundance uniform. To describe this behavior, a single species of mass must be tracked, using the conservation of a species of mass principle.

The conservation of a species of mass principle, differing slightly from the conservation of mass principle in that it is possible to create or destroy a species of mass by a chemical reaction, is given for species i as

$$\Sigma \text{mass}_{i_{in}} - \Sigma \text{mass}_{i_{out}} + \Sigma \text{mass}_{i_{generated}} = \Sigma \text{mass}_{i_{stored}} \qquad (2\text{-}53)$$

This principle can be applied to the general control volume illustrated in Fig. 2-1. Each species moves with its own velocity V_i, has its own mass density ρ_i, and might be created at the rate r_i per unit volume. The procedure gives

$$\frac{\partial \rho_i}{\partial t} + \text{div}\left(\rho_i V_{i_{rel}}\right) = r_i''' \qquad (2\text{-}54)$$

For each component of the fluid, there will be one equation of the preceding form. These equations can be summed to give

$$\frac{\partial(\Sigma_i \rho_i)}{\partial t} + \text{div}\left(\sum_i \rho_i V_{i_{rel}}\right) = \sum_i r_i''' \qquad (2\text{-}55)$$

It is reasonable to define the bulk density as

$$\rho = \sum_i \rho_i$$

and the bulk mass flow across the control volume surface as

$$\rho V_{rel} = \sum_i \rho_i V_{i_{rel}}$$

from which it is seen that the bulk velocity is a mass-averaged velocity

$$V_{rel} = \rho^{-1} \sum_i \rho_i V_{i_{rel}}$$

Because $\Sigma_i r_i''' = 0$ —as much species mass must be destroyed as is created if bulk mass is to be conserved—Eq. (2-55) is recognized to be the continuity equation [Eq. (2-6)] for the bulk fluid. Thus only $n - 1$ components of an n-component fluid need be tracked if the bulk continuity equation is included.

Two new unknowns have been introduced—the species mass density ρ_i and the species mass velocity $V_{i_{rel}}$. To reduce the number of new unknowns to just one, an idea used for the derivation of the energy equation is employed. The flux of species i mass across a surface $\rho_i V_{i_{rel}}$ is set equal to the sum of that $\rho_i \mathbf{V}_{\mathrm{rel}}$ convected by the bulk flow and that $\dot{\mathbf{m}}$ which diffuses relative to the bulk convection. Thus

$$\rho_i \mathbf{V}_{i_{\mathrm{rel}}} = \rho_i \mathbf{V}_{\mathrm{rel}} + \dot{\mathbf{m}}_i \tag{2-56}$$

Introduction of Eq. (2-56) into Eq. (2-54) gives

$$\frac{\partial \rho_i}{\partial t} + \mathrm{div}(\rho_i \mathbf{V}_{\mathrm{rel}}) = r_i''' - \mathrm{div}(\dot{\mathbf{m}}_i) \tag{2-57}$$

Let the mass fraction ω_i of species i be

$$\omega_i = \frac{\rho_i}{\rho} \tag{2-58}$$

Then Eq. (2-57) becomes, after expansion,

$$\omega_i \left[\frac{\partial \rho}{\partial t} + \mathrm{div}(\rho \mathbf{V}_{\mathrm{rel}}) \right] + \rho \left[\frac{\partial \omega_i}{\partial t} + (\bar{\mathbf{V}}_{\mathrm{rel}} \cdot \nabla) \omega_i \right] = r_i''' - \mathrm{div}(\dot{\mathbf{m}}_i)$$

The first bracketed term equals zero by the bulk continuity equation [Eq. (2-6)]. The second bracketed term can be written as $D\omega_i/Dt$ by use of the substantial derivation defined by Eq. (2-13). Thus

$$\rho \frac{D\omega_i}{Dt} = r_i''' - \mathrm{div}(\dot{\mathbf{m}}_i) \tag{2-59}$$

For a binary fluid, Fick's law relates the mass fraction to the diffusive flux of species 1 relative to the bulk convection as

$$\dot{\mathbf{m}}_1 = -\rho D_{12} \nabla \omega_1 \tag{2-60}$$

Although Eq. (2-59) applies to a multicomponent fluid, Eq. (2-60) applies only to a binary fluid, and its use in Eq. (2-59) results in the form for a binary fluid

$$\rho \frac{D\omega_1}{Dt} = r_1''' + \mathrm{div}(\rho D_{12} \nabla \omega_1) \tag{2-61}$$

which is the *binary mass-diffusion equation*. For a binary fluid, no corresponding equation need be written for the second component if the bulk continuity equation is used to obtain the bulk density since then

$$\frac{\rho_2}{\rho} = \omega_2 = 1 - \omega_1 \tag{2-61a}$$

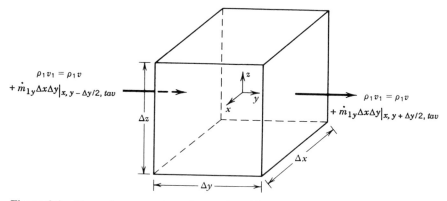

Figure 2-8   Flow of one species of mass in a binary fluid across the x surface of a stationary control volume in rectangular coordinates.

For a constant-property fluid such as would result if there were vanishingly small amounts of species 1 in the fluid, Eq. (2-61) becomes

$$\frac{D\omega_1}{Dt} = D_{12}\nabla^2\omega_1 + \frac{r_1'''}{\rho} \tag{2-62}$$

In rectangular coordinates for a stationary control volume Eq. (2-62) is

$$\frac{\partial\omega_1}{\partial t} + u\frac{\partial\omega_1}{\partial x} + v\frac{\partial\omega_1}{\partial y} + w\frac{\partial\omega_1}{\partial z} = D_{12}\left(\frac{\partial^2\omega_1}{\partial x^2} + \frac{\partial^2\omega_1}{\partial y^2} + \frac{\partial^2\omega_1}{\partial z^2}\right) + \frac{r_1'''}{\rho} \tag{2-63}$$

These ideas can be more concretely developed by applying the conservation of species of mass principle to a control volume fixed in a rectangular coordinate system. It is assumed that a binary fluid flows. The rate at which species 1 mass flows across two control volume surfaces is shown in Fig. 2-8. The flux of species 1 mass across a surface in an x–z plane, for example, is expressible as either $\rho_1 v_1$ or $\rho_1 v + \dot{m}_{1_y}$, where v_1 is the velocity of species 1 and v is the bulk mass-averaged fluid velocity. In other words,

Species 1 mass flux = bulk convection + diffusion relative to bulk convection

or

$$\rho_1 v_1 = \rho_1 v + \dot{m}_{1_y} \tag{2-64}$$

Application of the conservation of species principle expressed by Eq. (2-53) then results in

$$(\rho_1 u_1|_{x-\Delta x/2, y, z, t_{av}} - \rho_1 u_1|_{x+\Delta x/2, y, z, t_{av}}) \Delta y \Delta z$$

$$+ (\rho_1 v_1|_{x, y-\Delta y/2, z, t_{av}} - \rho_1 v_1|_{x, y+\Delta y/2, z, t_{av}}) \Delta x \Delta z$$

$$+ (\rho_1 w_1|_{x, y, z-\Delta z/2, t_{av}} - \rho_1 w_1|_{x, y, z+\Delta z/2, t_{av}}) \Delta x \Delta y + r_1'''|_{x, y, z, t_{av}} \Delta x \Delta y \Delta z$$

$$= (\rho_1|_{x, y, z, t+\Delta t} - \rho_1|_{x, y, z, t}) \frac{\Delta x \Delta y \Delta z}{\Delta t}$$

With division by $\Delta x \Delta y \Delta z$ and in the limit as all differential quantities approach zero, this equation becomes

$$\frac{\partial \rho_1}{\partial t} + \frac{\partial (\rho_1 u_1)}{\partial x} + \frac{\partial (\rho_1 v_1)}{\partial y} + \frac{\partial (\rho_1 w_1)}{\partial z} = r_1''' \qquad (2\text{-}65)$$

For the second fluid, one finds by similar means that

$$\frac{\partial \rho_2}{\partial t} + \frac{\partial (\rho_2 u_2)}{\partial x} + \frac{\partial (\rho_2 v_2)}{\partial y} + \frac{\partial (\rho_2 w_2)}{\partial z} = r_2''' \qquad (2\text{-}66)$$

When Eqs. (2-65) and (2-66) are added, the result is

$$\frac{\partial (\rho_1 + \rho_2)}{\partial t} + \frac{\partial (\rho_1 u_1 + \rho_2 u_2)}{\partial x} + \frac{\partial (\rho_1 v_1 + \rho_2 v_2)}{\partial y} + \frac{\partial (\rho_1 w_1 + \rho_2 w_2)}{\partial z}$$

$$= r_1''' + r_2'''$$

which is the bulk continuity equation

$$\frac{\partial \rho}{\partial t} + \frac{\partial (\rho u)}{\partial x} + \frac{\partial (\rho v)}{\partial y} + \frac{\partial (\rho w)}{\partial z} = 0 \qquad (2\text{-}8)$$

since $\rho = \rho_1 + \rho_2$, $r_1''' + r_2''' = 0$, $\rho u = \rho_1 u_1 + \rho_2 u_2$, and so forth.

Focusing on species 1 with use of the idea in Eq. (2-64), Eq. (2-65) becomes

$$\frac{\partial \rho_1}{\partial t} + \frac{\partial (\rho_1 u)}{\partial x} + \frac{\partial (\rho_1 v)}{\partial y} + \frac{\partial (\rho_1 w)}{\partial z} = r_1''' - \left(\frac{\partial \dot{m}_{1x}}{\partial x} + \frac{\partial \dot{m}_{1y}}{\partial y} + \frac{\partial \dot{m}_{1z}}{\partial z} \right)$$

$$(2\text{-}67)$$

The mass fraction ω_1 of species 1 is defined as

$$\omega_1 = \frac{\rho_1}{\rho} = \frac{\rho_1}{\rho_1 + \rho_2}$$

With ω_1 employed in Eq. (2-67), one has, after rearrangement,

$$\omega_1 \left[\frac{\partial \rho}{\partial t} + \frac{\partial(\rho u)}{\partial x} + \frac{\partial(\rho v)}{\partial y} + \frac{\partial(\rho w)}{\partial z} \right]$$

$$+ \rho \left[\frac{\partial \omega_1}{\partial t} + u\frac{\partial \omega_1}{\partial x} + v\frac{\partial \omega_1}{\partial y} + w\frac{\partial \omega_1}{\partial z} \right] = r_1''' - \left(\frac{\partial \dot{m}_{1x}}{\partial x} + \frac{\partial \dot{m}_{1y}}{\partial y} + \frac{\partial \dot{m}_{1z}}{\partial z} \right)$$

The first bracketed term is zero by the bulk continuity equation [Eq. (2-8)].
Use of Fick's law as

$$\dot{m}_{1x} = -\rho D_{12} \frac{\partial \omega_1}{\partial x}$$

with similar expressions for the y and z fluxes then results in, assuming
constant ρ,

$$\frac{\partial \omega_1}{\partial t} + u\frac{\partial \omega_1}{\partial x} + v\frac{\partial \omega_1}{\partial y} + w\frac{\partial \omega_1}{\partial z} = D_{12} \left(\frac{\partial^2 \omega_1}{\partial x^2} + \frac{\partial^2 \omega_1}{\partial y^2} + \frac{\partial^2 \omega_1}{\partial z^2} \right) + \frac{r_1'''}{\rho}$$

which is identical to Eq. (2-63), as it should be.

A summary of multicomponent diffusion is provided by Cussler [29], which
can be consulted if other than binary diffusion is encountered.

2.7 ENTROPY EQUATION

The irreversible diffusion of heat and momentum that is caused by nonuniformities of temperature and velocity in a fluid is accompanied by diffusion
and volumetric generation of entropy. The equations that describe the
entropy production associated with heat conduction and viscous friction are
developed next.

A review of the generalized development of the mass diffusion equation
[Eq. (2-59)] shows that if P is a property per unit mass of the bulk fluid, $\dot{P}$ is
its diffusive flux relative to bulk convection, and P''' is its volumetric rate of
generation, the describing equation is

$$\rho \frac{DP}{Dt} = -\text{div}(\dot{P}) + P'''$$

Let the property per unit mass of the bulk fluid be entropy s, the diffusive
flux of entropy relative to bulk convection be $\dot{s}$, and the volumetric rate of

generation of entropy per unit mass be s'''. Then

$$\rho \frac{Ds}{Dt} = -\text{div}(\dot{\mathbf{s}}) + s''' \tag{2-68}$$

For a reversible process, it is known from thermodynamics that

$$T Ds = DI + p D\rho^{-1} \tag{2-69}$$

It is assumed that local equilibrium is present, so that Eq. (2-69) can be applied with substantial accuracy. Use of the energy equation [Eq. (2-40)] in Eq. (2-69) gives

$$T\frac{Ds}{Dt} = \frac{-\text{div}(\mathbf{q}) + q''' - p\,\text{div}(\mathbf{V}) + \mu\Phi}{\rho} - \frac{p}{\rho^2}\frac{D\rho}{Dt}$$

The continuity equation [Eq. (2-12)] shows the last term to be

$$\frac{p}{\rho}\,\text{div}(\mathbf{V})$$

Combination of terms then gives

$$\frac{Ds}{Dt} = \frac{-\text{div}(\mathbf{q})}{T} + \frac{q''' + \mu\Phi}{T}$$

Recall that the divergence of a scalar times a vector is

$$\text{div}\left(\frac{\mathbf{q}}{T}\right) = \frac{\text{div}(\mathbf{q})}{T} + \mathbf{q}\cdot\nabla\left(\frac{1}{T}\right)$$

$$= \frac{\text{div}(\mathbf{q})}{T} - \frac{\mathbf{q}\cdot\nabla T}{T^2}$$

Therefore, the substantial derivative of entropy is

$$\rho\frac{Ds}{Dt} = -\text{div}\left(\frac{\mathbf{q}}{T}\right) + \left(-\frac{\mathbf{q}\cdot\nabla T}{T^2} + \frac{q''' + \mu\Phi}{T}\right) \tag{2-70}$$

Comparison of Eqs. (2-70) and (2-68) shows that the diffusive flux of entropy relative to bulk convection is given by

$$\dot{\mathbf{s}} = \frac{\mathbf{q}}{T} \tag{2-71}$$

and the volumetric rate of generation of entropy per unit mass is given by

$$s''' = -\frac{\mathbf{q} \cdot \nabla T}{T^2} + \frac{q''' + \mu \Phi}{T} \tag{2-72}$$

where the dissipation function Φ is defined in Eq. (2-41).

A fuller discussion and treatment is given by Hirschfelder et al. [21, p. 700]. The viscous dissipation function Φ is always positive, and so viscous friction always generates entropy as befits an irreversible process. The flow of heat is also seen to always generate entropy.

Bejan [30, 31] and Witte [32] discuss application of relations similar to Eq. (2-72), to the design of heat exchangers that minimize the production of entropy as have Ranasinghe and Reistad [33] with consideration of the available energy utilized to form the apparatus.

2.8   EQUATIONS FOR POROUS MEDIA

When a fluid flows through a porous medium, such as a packed-bed regenerative heat exchanger or bed of gravel, solution of the preceding equations for the fluid becomes difficult, primarily due to the tortuous geometry of the interstices between the solid particles through which the fluid flows. An approximating set of equations to describe conditions in the fluid is obtained by averaging quantities over a control volume that is large compared to the solid particles and interstices of the porous medium.

Application of the conservation of mass principle to the control volume shown in Fig. 2-9 yields, with u_f being the local x component of fluid velocity in the pores,

$$\frac{\phi \varrho \, \Delta x \, \Delta y \, \Delta z|_{x,y,z,t+\Delta t} - \phi \varrho \, \Delta x \, \Delta y \, \Delta z|_{x,y,z,t}}{\Delta t}$$

$$= \int_{y-\Delta y/2}^{y+\Delta y/2} \int_{z-\Delta z/2}^{z+\Delta z/2} \varrho u_f \, dy \, dz|_{x-\Delta x/2, \, t_{av}}$$

$$- \int_{y-\Delta y/2}^{y+\Delta y/2} \int_{z-\Delta z/2}^{z+\Delta z/2} \varrho u_f \, dy \, dz|_{x+\Delta x/2, \, t_{av}}$$

$$+ \int_{x-\Delta x/2}^{x+\Delta x/2} \int_{z-\Delta z/2}^{z+\Delta z/2} \varrho v_f \, dx \, dz|_{y-\Delta y/2, \, t_{av}}$$

$$- \int_{x-\Delta x/2}^{x+\Delta x/2} \int_{z-\Delta z/2}^{z+\Delta z/2} \varrho v_f \, dx \, dz|_{y+\Delta y/2, \, t_{av}}$$

$$+ \int_{x-\Delta x/2}^{x+\Delta x/2} \int_{y-\Delta y/2}^{y+\Delta y/2} \varrho w_f \, dx \, dy|_{z-\Delta z/2, \, t_{av}}$$

$$- \int_{x-\Delta x/2}^{x+\Delta x/2} \int_{y-\Delta y/2}^{y+\Delta y/2} \varrho w_f \, dx \, dy|_{z+\Delta z/2, \, t_{av}}$$

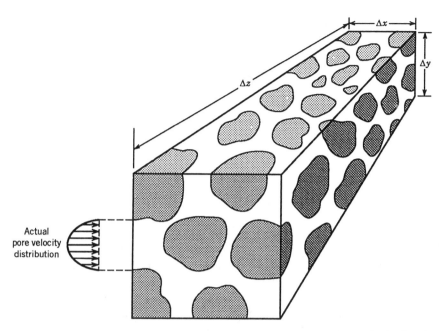

Figure 2-9 Fluid flow through a porous medium, velocity averaging. ϕ is the fraction of the volume that is occupied by the fluid. (Reprinted by permission from A. Bejan, *Convective Heat Transfer*, Copyright © John Wiley & Sons, 1984 [31b].)

Dividing by $\Delta x \, \Delta y \, \Delta z$ and taking the limit as all differential quantities approach a small value, but one larger than the particle or interstice size, gives the continuity equation for flow through porous media as

$$\frac{\partial \phi \varrho}{\partial t} + \frac{\partial \varrho u}{\partial x} + \frac{\partial \varrho v}{\partial y} + \frac{\partial \varrho w}{\partial z} = 0 \qquad (2\text{-}73)$$

or, in vector equation form,

$$\frac{\partial \phi \varrho}{\partial t} + \mathrm{div}(\varrho \mathbf{V}) = 0 \qquad (2\text{-}74)$$

The velocity components have the area-averaged definitions of

$$u(x, y, z, t) = \int_{z-\Delta z/2}^{z+\Delta z/2} \int_{y-\Delta y/2}^{y+\Delta y/2} \varrho(x, y, z, t)$$
$$\times u_f(x, y, z, t) \, dy \, dz / \varrho(x, y, z, t) \, \Delta y \, \Delta z \quad (2\text{-}75)$$

or

$$\varrho u A = \int_A \varrho u_f \, dA \qquad (2\text{-}76)$$

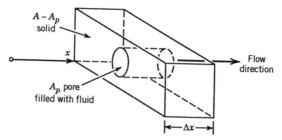

Figure 2-10 Simplistic parallel-pore view of a porous medium.

and so forth. Note that the local fluid velocity in the pores can be much larger than the space-averaged velocity defined by Eq. (2-76). The porosity ϕ is the fraction of a control volume that is occupied by the fluid, the remaining fraction $1 - \phi$ being occupied by the solid of density ρ_s. In a water resources problem, porosity might vary with both position, as different compositions are encountered in different portions of the aquifer, and time, as withdrawal of water allows the porous medium to compact (often irreversibly).

Derivation of the equation of motion for a fluid flowing through a porous medium begins with a view of the pores as being a system of parallel, straight cylindrical ducts, one of which is shown in Fig. 2-10. The x-direction equation of motion for the fluid in the pore is obtained in cylindrical coordinates from Appendix C as

$$\frac{\partial \rho u_f}{\partial t} + \frac{\partial \rho u_f u_f}{\partial x} = -\frac{\partial p}{\partial x} + \frac{\mu}{r}\frac{\partial}{\partial r}\left(r\frac{\partial u_f}{\partial r}\right) + \mu\frac{\partial^2 u_f}{\partial x^2} + \rho g_x$$

for fully developed flow. Multiplication by $dA_p = 2\pi r\, dr$ followed by integration over the pore cross-section area A_p leads to

$$\frac{\partial\left(\int_{A_p}\rho u_f\, dA_p\right)}{\partial t} + \frac{\partial\left(\int_{A_p}\rho u_f u_f\, dA_p\right)^{\approx\left(\int_{A_p}\rho u_f\, dA_p\right)^2}}{\partial x}$$

$$= -A_p\frac{\partial p}{\partial x} + \pi D_p\left(\mu\frac{\partial u_f}{\partial r}\bigg|_{D_p/2}\right) + \mu\frac{\partial^2\left(\int_{A_p}u_f\, dA_p\right)}{\partial x^2} + \rho g_x A_p \quad (2\text{-}77)$$

At this point it is recognized that the wall stress τ_w is given by

$$-\tau_w = \mu\frac{\partial u_f}{\partial r}\bigg|_{D_p/2}$$

and is related to the average pore velocity $\int_{A_p} u_f \, dA_p / A_p$ by

$$\tau_w = \frac{8\mu}{A_p D_p} \int_{A_p} u_f \, dA_p \tag{2-78}$$

as can be shown from the solutions for Problem 3-4. In addition, note that

$$\int_{A_p} \varrho u_f \, dA_p = \varrho u A \tag{2-79}$$

Introduction of Eqs. (2-78) and (2-79) into Eq. (2-77) gives, after division by A_p and rearrangement,

$$\varrho \left[\frac{1}{\phi} \frac{\partial u}{\partial t} + \frac{u}{\phi^2} \frac{\partial u}{\partial x} \right] = -\frac{\partial p}{\partial x} + \varrho g_x - \frac{\mu}{(\phi A_p / 8\pi)} u + \frac{\mu}{\phi} \frac{\partial^2 u}{\partial x^2} \tag{2-80}$$

Because the pores are not really isolated straight cylindrical ducts, the permeability K is used to represent the actual geometric effects on wall shear stress as

$$K \sim \phi A_p / 8\pi$$

It is seen that permeability has the dimension of length squared; a common unit of permeability is the darcy, 1 darcy $= 9.87 \times 10^{-13}$ m^2. In general, it must be measured; representative values of permeability and porosity are given in Table 2-1.

The general vector form of Eq. (2-80) is then

$$\varrho \left[\frac{1}{\phi} \frac{\partial \mathbf{V}}{\partial t} + \frac{1}{\phi^2} (\mathbf{V} \cdot \nabla) \mathbf{V} \right] = \underbrace{-\nabla p + \varrho \mathbf{g} - \frac{\mu}{K} \mathbf{V}}_{\text{Darcy's law}} + \underbrace{\frac{\mu}{\phi} \nabla^2 \mathbf{V}}_{\substack{\text{Brinkman} \\ \text{extension}}}$$

Because the local velocity in a pore can be high enough for the effects of turbulence to be appreciable, the wall friction effects in the third term on the right-hand side are modified by addition of the Forchheimer extension

$$\frac{\varrho C}{K^{1/2}} |\mathbf{V}| \mathbf{V}$$

where the inertia coefficient C is a parameter that depends on the porous

TABLE 2-1   Representative Values of Permeability for Various Substances (Adapted from [34])

Substance	Permeability Range (Permeability in cm^2)	Porosity Range (Porosity in %)
Berl saddles	1.3×10^{-3}–3.9×10^{-3}	68–83
Wire crimps	3.8×10^{-5}–1.0×10^{-4}	68–76
Black slate powder	4.9×10^{-10}–1.2×10^{-9}	57–66
Silica powder	1.3×10^{-10}–5.1×10^{-10}	37–49
Sand (loose beds)	2.0×10^{-7}–1.8×10^{-6}	37–50
Soils	2.9×10^{-9}–1.4×10^{-7}	43–54
Sandstone ("oil sand")	5.0×10^{-12}–3.0×10^{-8}	8–38
Limestone, dolomite	2.0×10^{-11}–4.5×10^{-10}	4–10
Brick	4.8×10^{-11}–2.2×10^{-9}	12–34
Bituminous concrete	1.0×10^{-9}–2.3×10^{-7}	2–7
Leather	9.5×10^{-10}–1.2×10^{-9}	56–59
Fiberglass	2.4×10^{-7}–5.1×10^{-7}	88–93
Cigarette	1.1×10^{-5}	17–49

medium to give the final form of the equation of motion

$$\varrho\left[\frac{1}{\phi}\frac{\partial \mathbf{V}}{\partial t} + \frac{1}{\phi^2}(\mathbf{V}\cdot\nabla)\mathbf{V}\right] = -\nabla p + \varrho\mathbf{g}$$
$$-\left(\frac{\mu}{K} + \frac{\varrho C}{K^{1/2}}|\mathbf{V}|\right)\mathbf{V} + \frac{\mu_e}{\phi}\nabla^2\mathbf{V} \quad (2\text{-}81)$$

The first term on the right-hand side of Eq. (2-81) is most accurately written as $-(1/\phi)\nabla(\phi p)$ for variable porosity, according to David et al. [35].

For packed beds of spheres of diameter d the permeability is given by

$$K = \frac{d^2\phi^3}{150(1-\phi)^2}$$

and the inertia coefficient is given by

$$C = \frac{1.75}{150^{1/2}\phi^{3/2}}$$

according to Ergun [36], although Beckermann et al. [37] have used 175 in place of 150 and Heggs [38] recommends the correlation of MacDonald et al. [39] with 180 in place of 150 and 1.8 in place of 1.75 in these expressions for K and C. MacDonald's correlation covers the widest range of Reynolds number, but is restricted to randomly packed spheres; for other geometries consult Heggs. Hunt and Tien [40] represent the approach of porosity to unity near a solid wall as a consequence of the reduced packing fraction there by

$$\phi = \phi_\infty + (1 - \phi_\infty)\exp(-b\Delta/d)$$

where ϕ_∞ is the porosity far from the wall, Δ is the distance from the wall, d is the particle effective diameter, and b is a parameter between 2 and 8 with the latter a good estimate for irregular solid particles (6 is a commonly used value). The effective viscosity μ_e may be taken to be the fluid viscosity μ, despite the effects of mixing local fluid streams that give rise to the dispersion conductivity discussed later.

The energy equation for flow through a porous medium is derived by first integrating the energy equation for the solid

$$\varrho_s C_{p,s} \frac{\partial T}{\partial t} = k_s \frac{\partial^2 T}{\partial x^2} + q_s'''$$

over the area, $(A - A_p)$ in Fig. 2-10, occupied by the solid to obtain

$$(A - A_p)\varrho_s C_{p,s} \frac{\partial T}{\partial t} = (A - A_p)k_s \frac{\partial^2 T}{\partial x^2} + (A - A_p)q_s''' \quad (2\text{-}82)$$

Second, the energy equation for the fluid, obtained from Appendix C as

$$\varrho C_p \left(\frac{\partial T}{\partial t} + u_f \frac{\partial T}{\partial x} \right) = k \frac{\partial^2 T}{\partial x^2} + \mu\Phi + q'''$$

is integrated over the pore area A_p to obtain

$$\varrho C_p \left[A_p \frac{\partial T}{\partial t} + \left(\int_{A_p} u_f \, dA_p \right) \frac{\partial T}{\partial x} \right] = A_p k \frac{\partial^2 T}{\partial x^2} + \int_{A_p} \mu\Phi \, dA_p + A_p q''' \quad (2\text{-}83)$$

Simplification comes from the realization that $uA = \int_{A_p} u_f \, dA_p$. Viscous dissipation, the last term on the right-hand side of the equation, can be approximated as the work done by the surface and body forces acting on the pore shown in Fig. 2-10. Thus

$$\int_{A_p} \mu\phi \, dA_p \approx u\left(-A\frac{\partial p}{\partial x} + \mu A_p \frac{\partial^2 u}{\partial x^2} + \varrho g_x A \right)$$

Addition of Eqs. (2-82) and (2-83) followed by division by A then gives, subject to the assumption (see Problem 2-22) that the solid and fluid are at the same temperature,

$$\left[\varrho C_p \phi + \varrho_s C_{p,s}(1 - \phi)\right]\frac{\partial T}{\partial t} + \varrho C_p u \frac{\partial T}{\partial x}$$

$$= \left[k\phi + k_s(1 - \phi)\right]\frac{\partial^2 T}{\partial x^2}$$

$$+ \left[\phi q''' + (1 - \phi)q_s'''\right] + u\left(-\frac{\partial p}{\partial x} + \frac{\mu}{\phi}\frac{\partial^2 u}{\partial x^2} + \varrho q_x\right)$$

In vector form the result is

$$\varrho C_p\left(\sigma\frac{\partial T}{\partial t} + (\mathbf{V} \cdot \nabla)T\right) = \nabla \cdot (k_e \nabla T) + q_e''' + \left(\frac{\mu}{K} + \frac{\varrho C}{K^{1/2}}|\mathbf{V}|\right)\mathbf{V} \cdot \mathbf{V}$$

$$(2\text{-}84)$$

Here, $\sigma = \phi + (\varrho_s C_{p,s}/\varrho C_p)(1 - \phi)$, $q_e''' = \phi q''' + (1 - \phi)q_s'''$, and the equation of motion Eq. (2-81) with the convective terms on the left-hand side neglected has been used to relate $-\nabla p + \mu\nabla^2 \mathbf{V}/\phi + \varrho\mathbf{g}$ to velocity. The effective conductivity k_e cannot very often be accurately obtained in the simplistic manner suggested by the one-dimensional result preceding Eq. (2-84) since the pores are tortuous and interconnected. Instead, k_e is the sum of the stagnant conductivity k_0 and the dispersion conductivity k_d. The stagnant conductivity is given by

$$\frac{k_0}{k} + \phi\frac{k_s - k}{k}\left(\frac{k_0}{k}\right)^{1/3} = \frac{k_s}{k}$$

due to Veinberg [41] although more complex relationships such as the one used by David et al. [35] exist.

As for porosity, the stagnant conductivity varies with distance from a solid wall as

$$k_0 = k_{0,\infty} + (k - k_{0,\infty})\exp(-b\Delta/d)$$

following Zehnder and Schlunder [42]. The dispersion conductivity results from the mixing of local fluid streams as the fluid winds its tortuous way around solid particles and is given by

$$k_d = \varrho C_p|\mathbf{V}|l/\gamma$$

where l is a mixing length usually equal to the particle diameter d with near-wall variation described by

$$\frac{l}{d} = \frac{\Delta}{d}, \qquad \frac{\Delta}{d} \lesssim 1$$

and γ is a dispersion coefficient usually equal to about 0.1 and, according to Kuo and Tien [43], related to porosity by

$$\gamma_\infty = 0.75\left[\tfrac{1}{2}(1 - \phi)^{1/3} - \tfrac{2}{3}(1 - \phi)^{2/3} + \tfrac{1}{4}(1 - \phi)\right]$$

far from a wall with wall proximity variation described by

$$\gamma = \gamma_\infty\left[1 - \exp(-b\Delta^2/d^2)\right]$$

For powders and fluidized beds in which the solid particles are movable, additional considerations apply [35, 44, 45]. Kaviany [46] provides additional detail, especially with regard to permeability and porosity of drag models for periodic structures.

PROBLEMS

2-1 Derive the equation of continuity for cylindrical coordinates by:

   (a) Applying the conservation of mass principle to the stationary cylindrical control volume indicated in Fig. 2P-1 for which $dV = r\,\Delta\theta\,\Delta r\,\Delta z$.

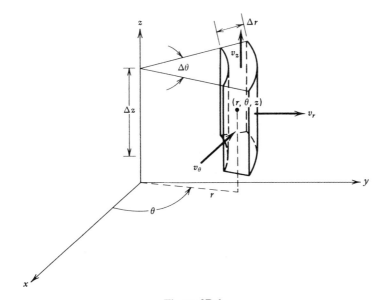

Figure 2P-1

(b) Mathematically transforming the continuity equation from rectangular coordinates through the use of the relationships that

$$x = r \cos \theta \qquad r = \left(x^2 + y^2\right)^{1/2} \qquad v_x = v_r \cos \theta - v_\theta \sin \theta$$

$$y = r \sin \theta \qquad \theta = \arctan \frac{y}{x} \qquad v_y = v_r \sin \theta + v_\theta \cos \theta$$

$$z = z \qquad z = z \qquad v_z = v_z$$

2-2 Derive the equation of continuity for spherical coordinates by applying the conservation of mass principle to the stationary spherical control volume indicated in Fig. 2P-2 for which $dV = r^2 \sin \theta \, \Delta\theta \, \Delta\phi \, \Delta r$.

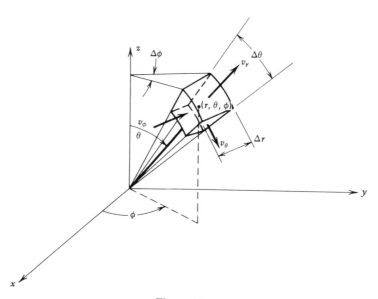

Figure 2P-2

2-3 Consider the constant-property, steady laminar flow illustrated in Fig. 2P-3. A fluid is between a large horizontal stationary bottom plate and large top plate that moves parallel to the bottom plate at a constant velocity U in response to a constant external force per unit area F/A applied to the upper plate. Because of the infinite extent, it can be assumed that no quantity varies in the x and z directions.

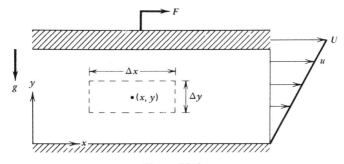

Figure 2P-3

(a) Apply the conservation of mass principle to the dotted control volume (after first making all assumptions) to derive the form of the continuity equation appropriate to this physical situation.

(b) Apply Newton's second law for control volumes to the dotted control volume (after first making all assumptions) to derive the forms of the x- and y-direction equations of motion in terms of pressures and stresses appropriate to this physical situation. Show all nonzero surface and body forces acting on the control volume; note that since $v = 0$, it is reasonable to assume $\tau_{yy} = -p$ is the only stress acting on the control volume top and bottom other than τ_{yx}.

(c) Relate τ_{yx} to du/dy in the equations of motion of part b by applying Newton's viscosity law.

(d) Draw a free-body diagram of the upper plate and relate the external force per unit area F/A to the shear stress τ_{yx} in the fluid.

2-4 Consider the constant-property, steady laminar flow illustrated in Fig. 2P-4. A fluid flows through a very long tube in response to an externally imposed pressure gradient along the tube length. Because of the infinite length, it can be assumed that no quantity (other than pressure) varies in the θ and z directions. Also, under these conditions a fluid particle moves solely in the z direction (i.e., $v_\theta = 0 = v_r$).

(a) Apply the conservation of mass principle to the dotted control volume (after first making all assumptions) to derive the form of the continuity equation appropriate to this physical situation.

(b) Apply Newton's second law for control volumes to the dotted control volume (after first making all assumptions) to derive the specialized z-direction equation of motion in terms of pressures and stresses. Show all nonzero surface and body forces acting on the control volume—note that, since $v_\theta = 0 = v_r$, only τ_{rz} is

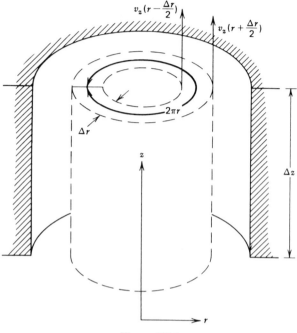

Figure 2P-4

acting on the sides of the annular control volume and, since v_z does not vary in the z direction, $\tau_{zz} = -p$ is the only stress acting on the ends of the annular control volume.

(c) Relate τ_{rz} to dv_z/dr in the equation of motion of part b by considering the annulus to be so thin that it can be flattened to a flat plate of thickness Δr for which Newton's viscosity law is directly applicable.

2-5 Show that a Newtonian fluid of constant viscosity but variable density has y- and z-direction equations of motion of the same form as Eq. (2-28).

2-6 On the three positive faces of the cylindrical and spherical elemental control volumes illustrated in Figs. 2P-1 and 2P-2, draw all the stresses acting and label them with appropriate subscripts.

2-7 Manipulate the energy equation given by Eq. (2-40) into a form that involves T and C_v.

2-8 Show that $\beta = 1/T$ for a perfect gas.

2-9 Show that for a reversible adiabatic process (viscous dissipation and thermal conduction are negligible), the solution to the energy equation

[Eq. (2-52)] for a perfect gas is $p/\rho^\gamma = \text{const}$, where γ is the ratio of specific heats $\gamma = C_p/C_v$.

2-10 Plot the coefficient of thermal expansion–absolute temperature product βT against absolute temperature for water and for air. Comment on the magnitude of βT at high and low temperatures.

2-11 Derive the energy equation in rectangular coordinates for the situation of Problem 2-3 by first making all assumptions and then applying the conservation of energy principle to the indicated control volume.

2-12 Derive the energy equation in cylindrical coordinates for the situation of Problem 2-4 by first making all assumptions and then applying the conservation of energy principle to the indicated control volume.

2-13 To explore the different technique used to derive boundary conditions, consider a stationary plane interface between two phases of a fluid. As illustrated in Fig. 2P-13, the fluid flows from left to right in a one-dimensional manner. Properties are constant in each phase.

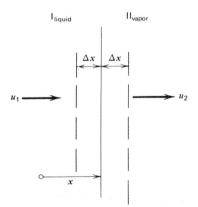

Figure 2P-13

(a) Apply the conservation of mass principle to the control volume shown with dashed lines to show that, in the limit as $\Delta x \to 0$,

$$\rho_1 u_1 = \rho_2 u_2$$

(b) Apply Newton's second law for a control volume to show that, in the limit as $\Delta x \to 0$,

$$p_1 - p_2 \approx \frac{\rho_2 u_2^2 (1 - \rho_2/\rho_1)}{g_c}$$

(c) Apply the conservation of energy principle for control volumes to show that, in the limit as $\Delta x \to 0$,

$$k_2 \frac{dT}{dx} - k_1 \frac{dT}{dx} = \rho_2 u_2 \left[\Lambda + \left(1 - \frac{\rho_2^2}{\rho_1^2} \right) \frac{u_2^2}{2g_c} \right]$$

where Λ is the latent heat of vaporization $\Lambda = H_2 - H_1$.

2-14   Show for a binary fluid that, if $\rho_i \mathbf{V}_i = \rho_i \mathbf{V} + \dot{\mathbf{m}}_i$ for mass diffusion, the definition of mass-averaged velocity as

$$\rho \mathbf{V} = \sum_i \rho_i \mathbf{V}_i$$

requires that

$$\dot{\mathbf{m}}_1 = -\dot{\mathbf{m}}_2 \quad \text{and} \quad \dot{\mathbf{m}}_i = \rho_i (\mathbf{V}_i - \mathbf{V})$$

In other words, for a binary fluid, the two species of mass diffuse at equal rates in opposite directions.

2-15   Consider the one-dimensional form of the energy equation for a constant-property fluid without dissipation or volume sources in stationary coordinates (see Fig. 2P-15)

$$\frac{\partial T}{\partial t} + u \frac{\partial T}{\partial x} = \alpha \frac{\partial^2 T}{\partial x^2} \tag{2-52a}$$

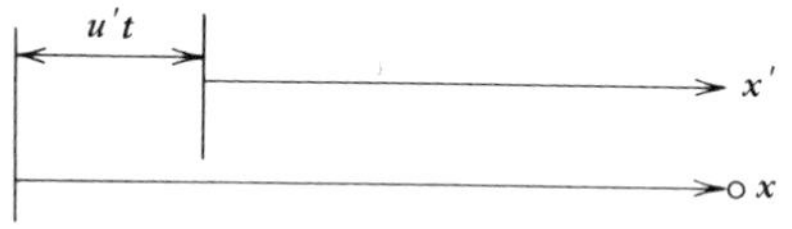

Figure 2P-15

This equation is to be transformed to a coordinate system moving to the right at steady velocity u' according to

$$t' = t \quad \text{and} \quad x' = x - u't$$

Use the results obtained from the chain rule

$$\frac{\partial T}{\partial t} = \frac{\partial T}{\partial t'} \frac{\partial t'}{\partial t} + \frac{\partial T}{\partial x'} \frac{\partial x'}{\partial t} = \frac{\partial T}{\partial t'} - u' \frac{\partial T}{\partial x'}$$

$$\frac{\partial T}{\partial x} = \frac{\partial T}{\partial t'} \frac{\partial t'}{\partial x} + \frac{\partial T}{\partial x'} \frac{\partial x'}{\partial x} = \frac{\partial T}{\partial x'}$$

and

$$\frac{\partial^2 T}{\partial x^2} = \frac{\partial T}{\partial t'}\frac{\partial^2 t'}{\partial x^2} + \frac{\partial t'}{\partial x}\left(\frac{\partial^2 T}{\partial t'^2}\frac{\partial t'}{\partial x} + \frac{\partial^2 T}{\partial x'\,\partial t'}\frac{\partial x'}{\partial x}\right)$$

$$+ \frac{\partial T}{\partial x'}\frac{\partial^2 x'}{\partial x^2} + \frac{\partial x'}{\partial x}\left(\frac{\partial^2 T}{\partial t'\,\partial x'}\frac{\partial t'}{\partial x} + \frac{\partial^2 T}{\partial x^2}\frac{\partial x'}{\partial x}\right)$$

$$= \frac{\partial^2 T}{\partial x'^2}$$

in Eq. (2-52b) to show that in x', t' coordinates the energy equation is

$$\frac{\partial T}{\partial t'} + (u - u')\frac{\partial T}{\partial x'} = \alpha\frac{\partial^2 T}{\partial x'^2}$$

Note: Details of coordinate transformations are available [47].

2-16 The boundary conditions at the moving interface between two phases of a fluid can be obtained by applying conservation principles to a control volume that moves with the interface. Consider the spherical vapor bubble in an infinite liquid as sketched in Fig. 2P-16.

(a) Apply the conservation of mass principle to the spherical annulus control volume suggested in Fig. 2P-16b and show that, since the mass flux across a surface is ρv_{rel},

$$\rho_1\left(v_1 - \dot{R}\right) = \rho_2\left(v_2 - \dot{R}\right) \qquad \text{at} \quad r = R(t)$$

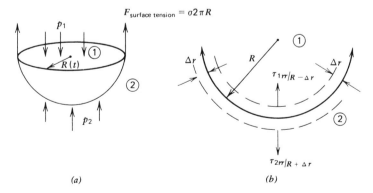

Figure 2P-16

(b) Apply Newton's second law to the hemispherical annulus and show that

$$P_1 - P_2 = \frac{2\sigma}{R} + \rho_1(u_1 - \dot{R})(u_2 - u_1) + \frac{4}{3}\mu_1\left(\frac{\partial u_1}{\partial r} - \frac{u_1}{r}\right)$$
$$- \frac{4}{3}\mu_2\left(\frac{\partial u_2}{\partial r} - \frac{u_2}{r}\right)$$

In view of the continuity equation for an incompressible fluid, the $\frac{4}{3}\mu(\partial u/\partial r - u/r)$ term can also be written as $2\mu\,\partial u/\partial r$.

(c) Apply the conservation of energy principle to the spherical annulus and show that

$$k_2\frac{\partial T_2}{\partial r} - k_1\frac{\partial T_1}{\partial r} = \rho_1(\dot{R} - u_1)\left[\Lambda + \frac{(u_1 - \dot{R})^2}{2g_c} - \frac{(u_2 - \dot{R})^2}{2g_c}\right.$$
$$\left. - \frac{4}{3}\frac{\mu_2}{\rho_2}\left(\frac{\partial u_2}{\partial r} - \frac{u_2}{r}\right) + \frac{4}{3}\frac{\mu_1}{\rho_1}\left(\frac{\partial u_1}{\partial r} - \frac{u_1}{r}\right)\right] \quad \text{at} \quad r = R(t)$$

Detailed discussion is available [48, 49].

2-17 Consider a solid that contacts a fluid with heat exchange occurring by convection. The fluid temperature away from the solid is T_f, and the heat-transfer coefficient is h. Apply the conservation of energy to the control volume shown in Fig. 2P-17 to show that at the interface (a) $k_s\,\partial T_s/\partial x = h(T_s - T_f)$ and (b) $-k_s\,\partial T_s/\partial x = h(T_s - T_f)$.

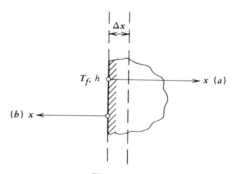

Figure 2P-17

2-18 Consider a fluid flowing vertically upward through a bed of thickness L composed of spherical rocks of diameter d.

(a) Solve the one-dimensional form of Eq. (2-81) to obtain

$$f = \frac{(-\Delta p + \varrho g_x L)d}{L\varrho u^2/2}$$

$$= 2(150)^{1/2}\frac{(1-\phi)}{\phi^{3/2}}\left[150^{1/2}\frac{1-\phi}{\phi^{3/2}}\frac{1}{\mathrm{Re}} + \frac{1.75}{150^{1/2}\phi^{3/2}}\right]$$

where $\mathrm{Re} = \rho u d/\mu$.

(b) Compare numerical values of f from part a with the suggestion by Duffie and Beckman [50] that

$$f = 2\alpha\frac{1-\phi}{\phi^{3/2}}\left[166\frac{1-\phi}{\phi^{3/2}}\frac{1}{\mathrm{Re}} + 4.74\right]$$

in which α is a surface area shape factor that decreases linearly from 2.5 at very small particle sizes to 1.5 for 5-cm-diameter particles. Briefly comment on the factors responsible for the differences between the two results. For explicitness, the case of air flowing through spherical rocks of 1 cm diameter can be used. The porosity ϕ is equal to 0.26 for face-centered cubic, 0.32 for body-centered cubic, and 0.476 for simple cubic-packed conditions for spheres according to Van Vlack [51].

2-19 Consider air flowing through a 2-m-thick bed of 1-cm-diameter granite rocks. The air flow rate is 1500 cfm and the bed cross-sectional area is 16 m^2.

(a) Based on the porosity given in Problem 2-18, determine the numerical values of the stagnant conductivity k_0, the dispersion conductivity k_d, and the effective conductivity k_e.

(b) Determine the numerical value of the viscous dissipation in the energy equation Eq. (2-84).

(c) Determine the numerical value of the temperature rise caused by the viscous dissipation.

(d) On the basis of the foregoing results, state whether or not viscous dissipation is likely to be important in an application. *Answers:* (a) $k_0 = 0.17$ Btu/hr ft °F, $k_d = 0.031$ Btu/hr ft °F; (b) $\mu\Phi = 1.8 \times 10^{-2}$ Btu/hr ft^3; (c) 2×10^{-4} °F.

2-20 Examine the case of fluid flow through a porous medium in which the fluid temperature T and solid temperature T_s are not equal, as in a packed-bed regenerative heat exchanger [52] discussed by Wilmott and Kulakowski. To do this, repeat the derivation of the one-dimensional energy equation in Section 2.8 with the rate of heat transfer between the fluid and solid in a volume $A\,dx$ occurring at a rate given by

$h_v(T - T_s)A\,dx$ with h_v, of dimensions W/m³ °C, being the heat-transfer coefficient per unit volume. *Answer:*

$$\varrho C_p\left(\frac{\partial T}{\partial t}\phi + u\frac{\partial T}{\partial x}\right) = k\frac{\partial^2 T}{\partial x^2}\phi - h_v(T - T_s)$$

$$\varrho_s C_{p,s}\frac{\partial T_s}{\partial t}(1 - \phi) = k_s\frac{\partial^2 T_s}{\partial x^2}(1 - \phi) + h_v(T - T_s)$$

2-21 Show that for fluid flow through a porous medium the heat-transfer coefficient h_v per unit volume is related to the heat-transfer coefficient h_a per unit area as

$$h_v = 6\frac{1 - \phi}{d}\alpha h_a$$

Here ϕ is the porosity, d is the effective particle spherical diameter, and α is the ratio of solid particle surface area A_s to effective particle spherical surface area ($\alpha = A_s/\pi d^2$) in which α varies linearly from 2.5 for crushed gravel of very small size to 1.5 for 50-mm-diameter crushed gravel, according to Duffie and Beckman [50]. *Hint:* Establish that the number N of spherical particles in volume V is $N = 6(1 - \phi)V/\pi d^3$.

2-22 To establish a criterion for judging whether or not the temperature difference between fluid and solid can be ignored in fluid flow through a porous medium, consider a solid of mass m_s and specific heat C_s and zero initial temperature suddenly immersed in a mass m of fluid of specific heat C of initial temperature T_i.

(a) Sketch this physical situation.

(b) Derive, assuming that the solid is at uniform temperature as judged by the criterion that the Biot number $\text{Bi} = h_a d/k_s \leq \frac{1}{10}$, the equations

$$\frac{m_s C_s}{h_a A_s}\frac{dT_s}{dt} = T - T_s \quad \text{and} \quad \frac{mC}{h_a A_s}\frac{dT_s}{dt} = T_s - T$$

$$T(0) = T_i \quad \text{and} \quad T_s(0) = 0$$

that describe the solid temperature T_s and fluid temperature T in terms of the solid surface area A_s and heat-transfer coefficient h_a based on solid surface area.

(c) Show that the solution to the equations of part b diminishes exponentially with elapsed time as

$$\frac{T - T_s}{T_i} = \exp\left(-\frac{t}{\tau}\right)$$

where the time constant τ is

$$\frac{1}{\tau} = h_a A_s \left(\frac{1}{m_s C_s} + \frac{1}{mC}\right)$$

(d) With $m_s = (1 - \phi)V\rho_s$, where V is the total volume considered, $m = \phi V\rho$, and $A_s = N\pi d^2\alpha$, where d is the equivalent spherical diameter of the solid and N and α are described in Problem 2-21, show that a criterion for the temperature difference to be of little importance is $Pe_L < 10^3$. Here, $Pe_L = (uL/\nu)(\nu/\alpha) = Re_L Pr$ with L being the thickness of the porous medium, or packed bed.

REFERENCES

1. S. M. Yen, *Ann. Rev. Fluid Mech.* **16**, 67–97 (1984).

2. C. Truesdell and R. G. Muncaster, *Fundamentals of Maxwell's Kinetic Theory of a Simple Monotonic Gas, Treated as a Branch of Rational Mechanics*, Academic, 1980.

3. N. Corngold, in J. L. Potter, Ed., *Rarefied Gas Dynamics*, Vol. 51, Part II, Progress in Astronautics and Aeronautics, American Institute of Aeronautics and Astronautics, 1977, pp. 651–628.

4. D. Burnett, *Proc. London Math. Soc.* **40**, 382 (1935).

5. C. Simon and J. Foch, in J. L. Potter, Ed., *Rarefied Gas Dynamics*, Vol. 51, Part I, Progress in Astronautics and Aeronautics, American Institute of Aeronautics and Astronautics, 1977, pp. 493–500.

6. P. E. Liley, in M. Kutz, Ed., *Mechanical Engineers' Handbook*, Wiley, 1986, pp. 1449–1490; in W. Rohsenow, J. Hartnett, and E. Ganic, Eds., *Handbook of Heat Transfer Fundamentals*, 2nd ed., McGraw-Hill, 1985, pp. 3-1–3-135; in S. Kakac, R. Shah, and W. Aung, Eds., *Handbook of Single-Phase Convective Heat Transfer*. Wiley-Interscience, 1987, pp. 22-1–22-41.

7. L. C. Burmeister, *Convective Heat Transfer*, Wiley-Interscience, 1983.

8. *Heat Exchanger Design Handbook*, Hemisphere, Vol. 5, 1986.

9. D. W. Green and J. O. Maloney, Eds., *Perry's Chemical Engineers' Handbook*, 6th ed., McGraw-Hill, 1984, pp. 3-279–3-291.

10. C. A. Depew and T. J. Kramer, *Adv. Heat Transfer* **9**, 113–180 (1973).

11. Y. S. Touloukian, Ed., *Thermophysical Properties of Matter*, IFI/Plenum, 1975.

12. *International Critical Tables*, McGraw-Hill, 1933.

13. K. Stephan and T. Heckenberger, *Thermal Conductivity and Viscosity Data of Fluid Mixtures*, Deutsche Gesellschaft fur Chemisches Apparatwesen, Chemische Technik und Biotechnologie e.V., 6000 Frankfurt/Main, 1988.

14. C. F. Beaton and G. F. Hewitt, Eds., *Physical Property Data for the Design Engineer*, Hemisphere, 1989.

15. R. C. Reid, J. M. Prausnitz, and B. E. Poling, *The Properties of Gases and Liquids*, 4th ed., McGraw-Hill, 1987.

16. K. H. Keenan, *Thermodynamics*, Wiley, 1941, p. 18.

17. W. Greiner and H. Stocker, *Scientific American* **252**(1), 76–87 (1985).

18. D. A. Russell, *Astronaut. Aeronautic.* **13**, 50–55 (1975).

19. J. C. F. Wang and G. S. Springer, in J. L. Potter, Ed., *Rarefied Gas Dynamics*, Vol. 51, Part II, Progress in Astronautics and Aeronautics, American Institute of Aeronautics and Astronautics, 1977, pp. 849–858.

20. E. R. G. Eckert and R. M. Drake, *Analysis of Heat and Mass Transfer*, McGraw-Hill, 1972, p. 479.

21. J. O. Hirschfelder, C. F. Curtis, and R. B. Bird, *Molecular Theory of Gases and Liquids*, Wiley, 1966, p. 21.

22. P. K. Davis, *Mech. Eng. News* **2**, 29 (1965).

23. R. Viskanta, in W. Rohsenow, J. Hartnett, and E. Ganic, Eds., *Handbook of Heat Transfer Fundamentals*, McGraw-Hill, 1985, pp. 10-1–10-45.

24. V. R. Voller, *Int. J. Heat Mass Transfer* **30**, 604–607 (1987).

25. K. S. Kim and B. Yimer, *Numerical Heat Transfer* **14**, 483–498 (1988).

26. N. Shamsunder and E. M. Sparrow, *ASME J. Heat Transfer* **97**, 333–340 (1975).

27. E. R. G. Eckert and R. M. Drake, *Analysis of Heat and Mass Transfer*, McGraw-Hill, 1972, p. 433.

28. E. F. Obert, *Concepts of Thermodynamics*, McGraw-Hill, 1960, pp. 266, 368.

29. E. L. Cussler, *Multicomponent Diffusion*, American Elsevier, 1976 (a three-page erratum is available from the publisher).

30. A. Bejan, *Adv. Heat Transfer* **15**, 1–58 (1982).

31. (a) A. Bejan, *Advanced Engineering Thermodynamics*, Wiley-Interscience, 1988, pp. 594–669; (b) A. Bejan, *Convection Heat Transfer*, Wiley-Interscience, 1984.

32. L. C. Witte, in R. Shah, A. Kraus, and D. Metzger, Eds., *Compact Heat Exchangers*, Hemisphere, 1990, pp. 381–393.

33. J. Ranasinghe and G. M. Reistad, in R. Shaw, A. Kraus, and D. Metzger, Eds., *Compact Heat Exchangers*, Hemisphere, 1990, pp. 357–380.

34. A. E. Scheidegger, *The Physics of Flow Through Porous Media*, 3rd ed., University of Toronto Press, 1974.

35. E. David, G. Lauriat, and P. Cheng, *Proc. 1988 National Heat Transfer Conference*, ASME HTD-96, Vol. 1, 1988, pp. 605–612.

36. S. Ergun, *Chem. Eng. Prog.* **48**, 89–94 (1952).

37. C. Beckermann, R. Viskanta, and S. Ramadhyani, *J. Fluid Mech.* **186**, 257–284 (1988).

38. P. J. Heggs, *Heat Exchanger Design Handbook*, Hemisphere, 1986, pp. 2.2.5-1–2.2.5-5.

39. I. F. MacDonald, M. S. El-Sayed, K. Mow, and F. A. L. Dullen, *Ind. Eng. Chem. Fund.* **18**, 198–208 (1979).

40. M. L. Hunt and C. L. Tien, *ASME J. Heat Transfer* **110**, 378–384 (1988).

41. A. K. Veinberg, *Sov. Phys. Dokl.* **11**, 593–595 (1967).

42. P. Zehnder and E. U. Schlunder, *Chem. Eng. Sci.* **42**, 933–940 (1970).

43. S. M. Kuo and C. L. Tien, *Proc. 1988 National Heat Transfer Conference*, ASME HTD-96, Vol. 1, pp. 629–634.

44. D. Green and J. Maloney, Eds., *Perry's Chemical Engineers' Handbook*, 6th ed., McGraw-Hill, 1984, pp. 5-53–5-56.

45. N. Cheremisinoff and R. Gupta, Eds., *Handbook of Fluids in Motion*, Ann Arbor Science, 1983, pp. 623–969.

46. M. Kaviany, *Principles of Heat Transfer in Porous Media*, Springer-Verlag, 1991.

47. L. M. K. Boelter, V. H. Cherry, H. A. Johnson, and R. C. Martinelli, *Heat Transfer Notes*, McGraw-Hill, 1965; D. A. Anderson, J. C. Tannehill, and R. H. Pletcher, *Computational Fluid Mechanics and Heat Transfer*, McGraw-Hill, 1984, pp. 247–255, 519–548; J. F. Thompson, in W. Minkowycz, E. Sparrow, G. Schneider, and R. Pletcher, Eds., *Handbook of Numerical Heat Transfer*, Wiley-Interscience, 1988, pp. 905–947.

48. G. M. Cho and R. A. Seban, *Trans. ASME J. Heat Transfer* **91**, 537–542 (1969).

49. D. Y. Hsieh, *Trans. ASME J. Basic Eng.* **87**, 991–1005 (1965).

50. J. Duffie and W. Beckman, *Solar Energy Thermal Processes*, 2nd ed., Wiley-Interscience, 1980, p. 176.

51. L. H. Van Vlack, *Elements of Materials Science and Engineering*, 5th ed., Addison-Wesley, 1985, p. 70.

52. A. J. Wilmott, pp. 3.15.3-1–3.15.10-7, and B. Kulakowski, pp. 3.15-1–3.15.11-9, *Heat Exchanger Design Handbook*, Vol. 3, Hemisphere, 1986.

53. F. E. Lumpkin and O. R. Chapman, *J. Thermophys. and Heat Transfer* **6**, 419–425 (1992).

3

ONE-DIMENSIONAL SOLUTIONS

Major points discernable from the describing equations in Chapter 2 can be emphasized by solution of well-chosen one-dimensional problems. Although the solutions themselves are of restricted applicability, the conclusions drawn from them are broad.

As a preliminary, note that if properties are constant, the energy and diffusion equation do not affect either the equations of motion or the continuity equation. In other words, the velocity distribution can be determined first, although by way of nonlinear equations. With velocities known, the temperature and mass fraction distributions can subsequently be determined. The energy and diffusion equations are then linear so that superposition may be employed if desired; the only complication introduced by velocity distributions then is that the energy and diffusion equations have variable coefficients. This simplification is characteristic of the problems discussed in this chapter.

Before solving specific problems, let us first determine whether solutions are possible on basic grounds. To be solvable, there should be as many equations as there are unknowns. The listing of Table 3-1 shows that to be so.

3.1 COUETTE FLOW

Couette flow consists of a fluid contained between two parallel plates that are in relative parallel motion. Here, the upper one moves with velocity U while the lower one is stationary as illustrated in Fig. 3-1. The simplicity of this flow situation allows exact solutions of the describing equations. The effect of the imposed flow on the heat transfer resulting from different temperatures at the top and bottom plates is to be ascertained. Assumptions

TABLE 3-1 Equations and Unknowns

Equations	Unknowns
1 continuity (plus $n - 1$ diffusion equations for an n-component fluid)	1 density (plus $n - 1$ mass fractions for an n-component fluid)
3 equations of motion	3 velocity components
1 energy equation	1 temperature
1 $\mu = \mu(p, T)$	1 viscosity
1 $k = k(p, T)$	1 thermal conductivity
$[n - 1 D_{ab} = D_{ab}(p, T)$ for n-component fluid]	$(n - 1$ mass diffusivities for n-component fluid)
1 equation of state (n for an n-component fluid)	1 pressure (plus $n - 1$ partial pressures for an n-component fluid)
8 (5 + 3n for n-component fluid)	8 (5 + 3n for n-component fluid)

for the following developments are (1) steady conditions, (2) laminar flow, (3) constant properties, (4) no pressure gradient in the x direction, (5) no edge effects ($\partial/\partial z = 0$), (6) no end effects ($\partial/\partial x = 0$), (7) Newtonian fluid, and (8) accuracy of Fourier law. The equations of continuity, motion, and energy then reduce to

Continuity $$\frac{dv}{dy} = 0 \qquad (3\text{-}1a)$$

x Motion $$\frac{d^2u}{dy^2} = 0 \qquad (3\text{-}1b)$$

y Motion $$\frac{\rho g}{g_c} + \frac{dp}{dy} = 0 \qquad (3\text{-}1c)$$

z Motion $$\frac{d^2w}{dy^2} = 0 \qquad (3\text{-}1d)$$

Energy $$\frac{d^2T}{dy^2} + \frac{\mu}{k}\left(\frac{du}{dy}\right)^2 = 0 \qquad (3\text{-}1e)$$

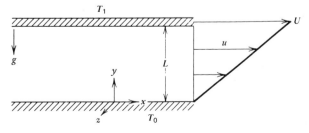

Figure 3-1 Geometry and coordinate system for Couette flow.

Boundary conditions to be imposed on the solutions to these equations are

$$\text{At } y = 0 \qquad u = 0 = w \qquad\qquad \text{no slip} \qquad\qquad (3\text{-}2a)$$
$$v = 0 \qquad\qquad \text{impermeable wall} \qquad (3\text{-}2b)$$
$$T = T_0 \qquad\qquad \text{no temperature jump} \quad (3\text{-}2c)$$
$$\text{At } y = L \qquad u = U, \qquad w = 0 \qquad \text{no slip} \qquad\qquad (3\text{-}2d)$$
$$v = 0 \qquad\qquad \text{impermeable wall} \qquad (3\text{-}2e)$$
$$T = T_1 \qquad\qquad \text{no temperature jump} \quad (3\text{-}2f)$$

Note that the boundary conditions and the basic assumptions have been used to simplify the continuity, motion, and energy equations. The full continuity equation, for example, for constant properties is simplified as

$$\frac{\partial u}{\partial x} + \frac{\partial v}{\partial y} + \frac{\partial w}{\partial z} = 0$$

$$\text{Assumption} \qquad\qquad \underset{(6)}{0} \qquad \underset{(5)}{0}$$

The first and last terms vanish by assumptions 6 and 5, giving Eq. (3-1a), which in turn shows that $v = \text{const}$. But since $v = 0$ at the two plates, $v = 0$ everywhere. The constant-property z-motion equation is simplified as

$$\underset{\text{0 continuity}}{\frac{\partial w}{\partial t}} + u\frac{\partial w}{\partial x} + v\frac{\partial w}{\partial y} + w\frac{\partial w}{\partial z} = \frac{g_c}{\rho}\left(B_z - \underset{\text{0 none imposed}}{\frac{\partial p}{\partial z}}\right)$$

$$\text{Assumption} \qquad \underset{(1)}{0} \quad \underset{(6)}{0} \qquad\qquad \underset{(5)}{0} \qquad\qquad \underset{(5)}{0}$$

$$+ g_c \nu\left(\frac{\partial^2 w}{\partial x^2} + \frac{\partial^2 w}{\partial y^2} + \frac{\partial^2 w}{\partial z^2}\right)$$

$$\underset{(6)}{0} \qquad\qquad \underset{(5)}{0}$$

giving Eq. (3-1d). The constant-property y-motion equation is easily simplified, since it has already been shown that $v = 0$, to Eq. (3-1c) as

$$\underbrace{\frac{\partial v}{\partial t} + u\frac{\partial v}{\partial x} + v\frac{\partial v}{\partial y} + w\frac{\partial v}{\partial z}}_{\text{0 continuity}} = \frac{g_c}{\rho}\left(B_y - \frac{\partial p}{\partial y}\right) + g_c\nu\left(\frac{\partial^2 v}{\partial x^2} + \underbrace{\frac{\partial^2 v}{\partial y^2} + \frac{\partial^2 v}{\partial z^2}}_{\text{0 continuity}}\right)$$

$$-\frac{\rho g}{g_c}$$

The constant-property x-motion equation is simplified to Eq. (3-1b) as

0 continuity   0 none imposed

$$\frac{\partial u}{\partial t} + u\frac{\partial u}{\partial x} + v\frac{\partial u}{\partial y} + w\frac{\partial u}{\partial z} = \frac{g_c(B_x - \partial p/\partial x)}{\rho}$$

Assumption   0   0   0   0

         (1)   (6)   (5)   (6)

$$+ g_c\nu\left(\frac{\partial^2 u}{\partial x^2} + \frac{\partial^2 u}{\partial y^2} + \frac{\partial u^2}{\partial z^2}\right)$$

                       0   0
                    (6)   (5)

The constant-property energy equation is simplified to Eq. (3-1e) as

0 continuity

$$\frac{\partial T}{\partial t} + u\frac{\partial T}{\partial x} + v\frac{\partial T}{\partial y} + w\frac{\partial T}{\partial z} = \alpha\left(\frac{\partial^2 T}{\partial x^2} + \frac{\partial^2 T}{\partial y^2} + \frac{\partial^2 T}{\partial z^2}\right)$$

Assumption   0   0   0   0   0

         (1)   (6)   (5)   (6)   (5)

                         0 none imposed

$$+ \frac{q'''}{\rho C_p} + \frac{\mu}{\rho C_p}\Phi$$

           0 continuity   0 continuity

$$\Phi = 2\left[\left(\frac{\partial u}{\partial x}\right)^2 + \left(\frac{\partial v}{\partial y}\right)^2 + \left(\frac{\partial w}{\partial z}\right)^2\right] - \frac{2}{3}(\text{div } \mathbf{V})^2 + \left(\frac{\partial u}{\partial y} + \frac{\partial v}{\partial x}\right)^2$$

                 0   0   0

Assumption   (6)   (5)   (6)

                           0 z-motion and boundary
                                  conditions

$$+ \left(\frac{\partial v}{\partial z} + \frac{\partial w}{\partial y}\right)^2 + \left(\frac{\partial w}{\partial x} + \frac{\partial u}{\partial z}\right)^2$$

                         0   0   0

                                 (6)   (5)

The y-motion equation [Eq. (3-1c)] shows that pressure varies as

$$p = p_0 - \rho\left(\frac{g}{g_c}\right)y$$

The z-motion equation [Eq. (3-1d)] and its boundary conditions show that $w = 0$. From the x-motion equation the linear variation

$$\frac{u}{U} = \frac{y}{L} \tag{3-3}$$

is found.

The energy equation can be solved now that velocity is known. With the use of Eq. (3-3), the energy equation is

$$\frac{d^2T}{dy^2} = -\frac{\mu}{k}\frac{U^2}{L^2} \tag{3-4}$$

$$\frac{T - T_0}{T_1 - T_0} = \frac{y}{L} + \frac{E\,Pr}{2}\frac{y}{L}\left(1 - \frac{y}{L}\right) \tag{3-5}$$

where E = Eckert number = $U^2/C_p(T_1 - T_0)$, representing a measure of the importance of viscous dissipation relative to the imposed temperature difference, and Pr = Prandtl number = $\mu C_p/k$. The heat flow rate q_w at the lower surface is obtained from the Fourier law applied to Eq. (3-5) as

$$q_w = -k\frac{dT(y = 0)}{dy} = -\frac{k(T_1 - T_0)(1 + E\,Pr/2)}{L}$$

which can be rearranged into

$$q_w = \frac{k\left[T_0 - \left(T_1 + Pr\,U^2/2C_p\right)\right]}{L} \tag{3-6}$$

This Couette flow is an approximation to conditions in a hydrodynamically lubricated bearing, and resembles boundary-layer flow over a flat plate far from the leading edge as shown in Fig. 3-2. The outer edge of the boundary layer is represented by the moving upper plate in the Couette flow model. The disturbing aspect of the heat flow found in Eq. (3-6) is that more than a temperature difference drives the heat flow, and it is expected that the same conclusion applies to boundary-layer flow. Of course, the physical mechanism responsible is that some energy is dissipated near the solid surface into a thermal form.

To recover the idea that heat flows only in response to a temperature difference, consider $Pr\,U^2/2C_p$ to be a recovery or adiabatic wall tempera-

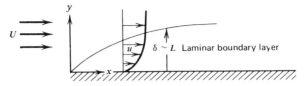

Figure 3-2   Laminar boundary-layer flow resembles Couette flow.

ture. To show that this is a reasonable thing to do, the case where the lower surface is insulated is solved next and the temperature it attains under the influence of viscous dissipation determined. The velocity distribution is unchanged from Eq. (3-3), so that specification of an insulated lower surface requires as the only change that $dT(y = 0)/dy = 0$. As a result, the solution to Eq. (3-4) is

$$T - T_1 = \frac{\Pr U^2}{2C_p}\left(1 - \frac{y^2}{L^2}\right) \tag{3-7}$$

At the lower surface it is then found that

$$T(y = 0) = T_{aw} = \frac{\Pr U^2}{2C_p} + T_1 \tag{3-8}$$

where T_{aw} is the temperature an adiabatic surface assumes under the influence of viscous dissipation. In Eq. (3-6) for heat flow at the lower surface, one can view T_0 as T_{wall} and $T_1 + \Pr(U^2/2Cp)$ as T_{aw}. Then the interpretation of

$$q_w = \frac{k(T_{\text{wall}} - T_{aw})}{L} \tag{3-9}$$

holds true. The conclusion is that the driving temperature difference for convection is really $T_{\text{wall}} - T_{aw}$, with the possibility yet remaining that T_{aw} may differ for different geometries and flow conditions. Heat-transfer coefficients measured or calculated at low speeds, where viscous dissipation is negligible, can then be used in high-speed situations where dissipation is large, provided that fluid properties are evaluated at a reference temperature, and no further experiments are necessary.

The convenience of basing the wall heat flux on the difference between the wall temperature and the adiabatic wall temperature as suggested by Eq. (3-9) can be appreciated by considering the temperature distribution given by Eq. (3-5) in more detail. In Fig. 3-3 it is shown that at $E \Pr > 2$ the wall heat flux actually is into the upper plate even though the imposed temperature difference would initially have been considered to cause the wall heat flux to

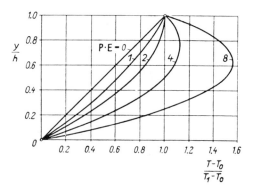

Figure 3-3 Dimensionless temperature variation in Couette flow.

be out of the upper plate. The heat-transfer coefficient h at the lower plate based on the imposed temperature difference $T_0 - T_1$ is

$$h = \frac{q_w}{T_0 - T_1} = \frac{k}{L}\left(1 + \frac{U^2\mu/k}{T_1 - T_0}\right)$$

according to Eq. (3-6). If the heat-transfer coefficient is based on $T_0 - T_1$, the result is a coefficient that depends on both the imposed velocity and the imposed temperature difference; occasionally, this will yield a negative h since the driving potential difference for local heat transfer is incorrect—the imposed temperature difference is *not* proportional to the temperature gradient at the wall. If the heat-transfer coefficient is based on the difference between the imposed and the adiabatic wall temperatures, $T_w - T_{aw}$, then

$$h = \frac{q_w}{T_w - T_{aw}} = \frac{k}{L} \tag{3-10}$$

according to Eq. (3-9), which leads to a Nusselt number Nu of Nu $= hL/k = 1$. This result is more convenient due to its constancy at the value for negligible viscous dissipation for which $T_{aw} \approx T_1$. Because the driving potential difference for local heat transfer is correctly ascertained, h will never be negative.

Another item of information that is particularly useful in the boundary-layer flows can be extracted from Eq. (3-8). In Eq. (3-8) the kinetic energy $U^2/2$ of the fluid at the upper plate is partially converted into thermal form. If the lower plate is insulated, some of the converted energy shows up there and causes a temperature rise $T_{aw} - T_1$. The ratio of this "recovered" thermal energy at the lower wall to the fluid kinetic energy at the upper plate is called the *recovery factor r*. Thus, for Couette flow,

$$r = \frac{C_p(T_{aw} - T_1)}{U^2/2} = \mathrm{Pr} \qquad (3\text{-}11)$$

For gases such as air, $\mathrm{Pr} < 1$ and r is less than unity. For liquids such as water, $\mathrm{Pr} > 1$ and r exceeds unity.

Temperature-dependent viscosity solutions are available [1, 2].

Recovery factors for other situations, including a jet impinging on a plate, are discussed by Eckert [3] who includes the swirling flow in a Hilch–Ranque tube in a survey of energy separation in fluid flows.

3.2 POISEUILLE FLOW

Poiseuille flow is much like Couette flow, except that both surfaces are usually considered to be stationary. Flow is caused by maintaining a pressure differential down the flow channel as illustrated in Fig. 3-4. Assumptions for the following developments are (1) steady conditions, (2) laminar flow, (3) constant properties, (4) constant x-direction pressure gradient, (5) no edge effects ($\partial/\partial z = 0$), (6) no end effects ($\partial/\partial x = 0$ except for pressure), (7) Newtonian fluid, and (8) accuracy of Fourier law.

The equations of continuity, motion, and energy then reduce to

Continuity	$\dfrac{dv}{dy} = 0$	(3-12a)
x Motion	$\dfrac{d^2u}{dy^2} = \dfrac{1}{\mu}\dfrac{dp}{dx}$	(3-12b)
y Motion	$\dfrac{dp}{dy} = -\dfrac{\rho g}{g_c}$	(3-12c)
z Motion	$\dfrac{d^2w}{dy^2} = 0$	(3-12d)
Energy	$\dfrac{d^2T}{dy^2} + \dfrac{\mu}{k}\left(\dfrac{du}{dy}\right)^2 = 0$	(3-12e)

Boundary conditions to be imposed are

At $y = -L$	$u = 0 = w$	no slip	(3-13a)
	$v = 0$	impermeable wall	(3-13b)
	$T = T_0$	no temperature jump	(3-13c)
At $y = L$	$u = 0 = w$	no slip	(3-13d)
	$v = 0$	impermeable wall	(3-13e)
	$T = T_1$	no temperature jump	(3-13f)

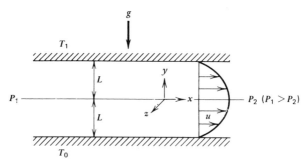

Figure 3-4   Geometry and coordinate system for Poiseuille flow.

The assumptions and boundary conditions have been used to simplify the continuity, motion, and energy equations in the same manner as for Couette flow.

The continuity equation [Eq. (3-12a)] and boundary conditions [Eqs. (3-13b) and (3-13e)] yield $v = 0$. Similarly, the z-motion equation [Eq. (3-13a)] and boundary conditions [Eqs. (3-13a) and (3-13d)] give $w = 0$. The vertical pressure variation is found from the y-motion equation to be hydrostatic as

$$p = p_0 - \rho \frac{g}{g_c} y$$

From the x-motion equation and its boundary conditions, it is found that

$$u = \frac{L^2}{2\mu}\left(-\frac{dp}{dx}\right)\left(1 - \frac{y^2}{L^2}\right) \tag{3-14}$$

The maximum velocity occurs at the center line where $y = 0$ and is

$$u_m = \frac{L^2}{2\mu}\left(-\frac{dp}{dx}\right) = \frac{3}{2}u_{av} \tag{3-15}$$

The average velocity is found from its definition to be

$$u_{av} = \frac{1}{2L}\int_{-L}^{L} u \, dy$$

$$= \frac{L^2}{3\mu}\left(-\frac{dp}{dx}\right)$$

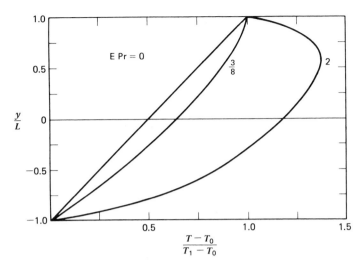

Figure 3-5   Dimensionless temperature variation in Poiseuille flow.

Thus

$$u = u_m\left(1 - \frac{y^2}{L^2}\right) = \frac{3}{2}u_{av}\left(1 - \frac{y^2}{L^2}\right) \qquad (3\text{-}16)$$

Note that it was again possible to solve the equations of motion without consideration of temperature distributions because of the constancy of properties. If velocities are known, the energy equation [Eq. (3-12e)] can next be attacked to find temperature as

$$T = T_0 + \frac{T_1 - T_0}{2}\left(1 + \frac{y}{L}\right) + \frac{\mu}{3k}\mu_m^2\left(1 - \frac{y^4}{L^4}\right) \qquad (3\text{-}17)$$

The temperature distribution predicted by Eq. (3-17) is shown in Fig. 3-5, where it is seen that for $E\,Pr > \frac{3}{8}$, heat flows into the upper plate even though the upper plate might be warmer than the bottom one. The heat flow at the lower surface ($y = -L$) is obtained from the Fourier law as

$$q_w = -k\frac{dT(y = -L)}{dy} = \frac{k}{2L}\left(T_0 - T_1 - \frac{8}{3}\frac{Pr\,u_m^2}{C_p}\right) \qquad (3\text{-}18)$$

where E = Eckert number = $u_m^2/C_p(T_1 - T_0)$ and Pr = Prandtl number = $\mu C_p/k$.

To test the presumption that Eq. (3-18) for lower surface heat flow can be interpreted in the same manner as for the Couette flow case,

$T_1 + \frac{8}{3}\text{Pr}\, u_m^2/C_p = T_{aw}$, let the lower surface be insulated and adiabatic. The boundary condition at the lower surface is restated as

$$\frac{dT(y = -L)}{dy} = 0$$

The velocity distribution being unchanged, the resultant temperature is

$$T - T_1 = \frac{\mu}{3k} u_m^2 \left(5 - \frac{y^4}{L^4} - 4\frac{y}{L} \right)$$

The adiabatic temperature of the lower surface is then

$$T_{aw} = T(y = -L) = T_1 + 6\frac{\text{Pr}\, u_{av}^2}{C_p} \tag{3-19}$$

The truth of the interpretation

$$q_w = \frac{k}{2L}(T_{\text{wall}} - T_{aw})$$

is again seen as is the convenience of having the corresponding heat-transfer coefficient $h = k/2L$ apply to both low-speed and high-speed flow. The Nusselt number Nu is seen to be Nu $= 2hL/k = 1$. For this slightly different flow condition, T_{aw} is shown in Eq. (3-19) to be computed from a formula slightly different from Eq. (3-8) for Couette flow. Accurate computation of T_{aw} requires that properties be evaluated at a reference temperature to account for their temperature dependence.

Inasmuch as the major assumption is the linearity of the energy equation with respect to T, the conclusion that the proper driving temperature difference for convective heat transfer is $T_{\text{wall}} - T_{aw}$ is expected to hold for more complicated flow conditions. The only difficulty is accurate computation of T_{aw}.

The Poiseuille flow case just treated resembles flow down a duct, where the duct is long and wide. In contrast to Couette flow in which the imposed velocity is a natural choice for a reference velocity, the Poiseuille flow has several plausible reference velocities (e.g., u_m or u_{av}), but none is obvious by reason of being imposed. In duct flow it is usually the average velocity that is known. Accordingly, the recovery factor r for this Poiseuille flow is

$$r = \frac{C_p(T_{aw} - T_1)}{u_{av}^2/2} = 12\,\text{Pr}$$

which is of the same form as Eq. (3-11) for Couette flow, but with a different coefficient.

The cases of temperature-dependent viscosity in Poiseuille flow between two planes and in a circular duct are reported by Schlichting [4], as is flow between convergent planes.

3.3 STEFAN'S DIFFUSION PROBLEM

In the two foregoing examples the fluid had a single component. The consequences of allowing the fluid to have more than one component is explored by examining a two-component (binary) convection problem referred to as *Stefan's diffusion problem*. The problem is of sufficient physical simplicity that a one-dimensional solution can be taken as exact.

In Stefan's problem a container partially filled with a volatile liquid is exposed to a stagnant atmosphere poor in vapor. For the physical situation shown in Fig. 3-6, the rate of evaporation is to be related to the saturation vapor pressure specified by the liquid temperature. As the liquid evaporates, its vapor moves from a region of high concentration at the bottom to a region of low concentration at the top where it leaves the container. In like fashion the air tries to move from the top to the bottom of the container; but the liquid is impermeable to air and does not absorb air, so the air is stagnant. Therefore, the downward diffusion of air must be countered by an upward bulk flow. Assumptions for this situation are (1) steady state, (2) stagnant conditions (laminar flow), (3) liquid impermeable to air, (4) constant total pressure and temperature, (5) liquid and vapor in equilibrium at their interface, (6) air and vapor as ideal gases, and (7) variations in the y direction only (this can be inaccurate [5]).

Since it is assumed that $u = 0 = w$, the x- and z-motion equations need not be considered. The assumed constancy of temperature renders the energy equation irrelevant. Likewise, the y-motion equation is unneeded because of the assumed constancy of total pressure. The equations of continuity and diffusion, all that are needed for determination of the vertical

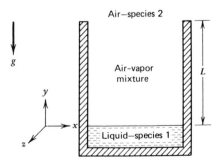

Figure 3-6 Geometry and coordinate system for Stefan's diffusion problem.

bulk velocity v and the mass fraction ω_1 of species 1, reduce to

Continuity $\qquad \dfrac{d(\rho v)}{dy} = 0$ $\hfill$ (3-20a)

Diffusion $\qquad \rho v \dfrac{d\omega_1}{dy} = D_{12} \dfrac{d(\rho \, d\omega_1/dy)}{dy}$ $\hfill$ (3-20b)

The applicable boundary conditions are

At $y = 0^+$ $\qquad \omega_1 = \omega_{10}$ or $p_1 = p_{10}$ $\qquad$ vapor in equilibrium with liquid $\hfill$ (3-21a)

$\qquad\qquad\qquad\quad p_2 v_2 = 0$ $\qquad\qquad\qquad\qquad$ liquid impermeable to air $\hfill$ (3-21b)

At $y = L$ $\qquad \omega_1 = \omega_{1L}$ or $p_1 = p_{1L}$ $\qquad$ some vapor in ambient air $\hfill$ (3-21c)

Note that bulk density ρ has not been assumed to be constant.

An important conclusion can be reached without solving either the diffusion or the continuity equation. The mass flow rate of air (species 2) is given by

$$\rho_2 v_2 = \rho_2 v - \rho D_{12} \frac{d\omega_2}{dy} \hfill (3-22)$$

Since the liquid is impermeable to air, it is everywhere true that $\rho_2 v_2 = 0$. This fact gives the bulk velocity v from Eq. (3-22) as

$$v = \frac{\rho}{\rho_2} D_{12} \frac{d\omega_2}{dy}$$

By definition $\omega_2 = \rho_2/\rho$, so this relationship becomes

$$v = \frac{D_{12}}{\omega_2} \frac{d\omega_2}{dy}$$

Recall also that by definition $\rho = \rho_1 + \rho_2$, so that $1 = \omega_1 + \omega_2$. Incorporation of these relationships into the equation for v gives

$$v = -\frac{D_{12}}{1 - \omega_1} \frac{d\omega_1}{dy} \hfill (3-23)$$

Surprisingly, there is an upward bulk flow, a "blowing," to exactly counteract the downward diffusion of air. The mass flow rate $\dot{m}_1$ of vapor (species 1) must be constant so that

$$\dot{m}_1 = \rho_1 v - \rho D_{12}\frac{d\omega_1}{dy} = \text{const} \tag{3-24}$$

If Eq. (3-23) is introduced into Eq. (3-24), the vapor mass flow rate becomes

$$\dot{m}_1 = -\frac{\rho}{1-\omega_1}D_{12}\frac{d\omega_1}{dy} \tag{3-25}$$

Because $\dot{m}_1$ is constant, comparison of Eq. (3-23) with Eq. (3-25) reveals that ρv is also constant. Evaluation of all quantities at the liquid–vapor interface puts Eq. (3-25) in the form

$$\dot{m}_1 = -\frac{(\rho D_{12}\,d\omega_1/dy)|_{y=0}}{1-\omega_1|_{y=0}} \tag{3-26}$$

Here Fick's law, $\dot{m}_1 = -[\rho D_{12}\,d\omega_1/dy]|_{y=0}$, represents diffusive flow, which is also given in terms of a mass-transfer coefficient h_D as $\dot{m}_1 = \rho h_D \Delta\omega_1$; the effect of bulk convection induced by the mass transfer is incorporated in Eq. (3-26) by $(1-\omega|_{y=0})$ in the denominator. In general, then, the effect of bulk convection can be approximately taken into account in mass transfer in which one surface is impermeable to species 2 of a binary mixture by use of

$$\dot{m}_1 = \frac{\rho h_D \Delta\omega_1}{1-\omega_{10}} \tag{3-27}$$

That the continuity and diffusion equations yield the same result will be shown next. The continuity equation [Eq. (3-20a)] gives, after one integration,

$$\rho v = C$$

which agrees with the earlier deduction. The diffusion equation [Eq. (3-20b)] gives, after one integration,

$$C_1 + \rho v \omega_1 = \rho D_{12}\frac{d\omega_1}{dy} \tag{3-28}$$

Insertion of the expression for ρv from Eq. (3-23) into Eq. (3-28) gives

$$C_1 = \frac{\rho D_{12}}{1-\omega_1}\frac{d\omega_1}{dy} \tag{3-29}$$

Comparison of Eq. (3-26) with Eq. (3-29) reveals that $\dot{m}_1 = -C_1$, as previously argued on physical grounds.

The distribution of species 1 mass fraction ω_1 or its partial pressure p_1 can be determined by a second integration of the diffusion equation [Eq. 3-29)]. Since an exact solution is sought, variable total density is considered. Relations pertinent to a binary mixture of ideal gases are developed next.

Because the equilibrium vapor pressure of a liquid at a specified temperature is commonly given, Eq. (3-29) is recast into a form that displays partial pressures. Recall that for a binary mixture of isothermal perfect gases, the total pressure is the sum of the partial pressures

$$p = p_1 + p_2 \tag{3-30}$$

Also, the equation of state for a perfect gas requires $p = \rho RT/M$ for the mixture and $p_i = \rho_i RT/M_i$ for each component. Insertion of these equations of state into Eq. (3-30) gives

$$\frac{1}{M} = \frac{\rho_1}{\rho}\frac{1}{M_1} + \frac{\rho_2}{\rho}\frac{1}{M_2} \tag{3-31}$$

Use of the equation of state in the definition of mass fraction gives

$$\omega_1 = \frac{\rho_1}{\rho} = \frac{P_1 M_1}{RT}\frac{RT}{pM} = \frac{p_1}{p}\frac{M_1}{M}$$

Substitution of Eq. (3-31) into this relationship for ω_1 gives

$$\omega_1 = \frac{p_1}{p}\left(\frac{\rho_1}{\rho} + \frac{\rho_2}{\rho}\frac{M_1}{M_2}\right)$$

which can be rearranged, since $\omega_i = \rho_i/\rho$ and $1 = \omega_1 + \omega_2$, to give

$$\omega_1 = \frac{p_1(M_1/M_2)}{p - p_1(1 - M_1/M_2)} \tag{3-32}$$

It is further found that the bulk density is given by

$$\rho = \frac{pM}{RT} = \frac{M_2}{RT}\left[p - p_1\left(1 - \frac{M_1}{M_2}\right)\right] \tag{3-33}$$

Equation (3-29) becomes, after use of Eqs. (3-32) and (3-33),

$$C_1 = \frac{pM_1}{RT} \frac{D_{12}}{p - p_1} \frac{dp_1}{dy}$$

whose solution is

$$C_2 - C_1 \frac{RT}{pM_1 D_{12}} y = \ln(p - p_1) \tag{3-34}$$

Evaluation of C_1 and C_2 is accomplished by consideration of the boundary conditions expressed by Eqs. (3-21a) and (3-21c) with the result that

$$\ln\left(\frac{p - p_1}{p - p_{10}}\right) = \frac{y}{L} \ln\left(\frac{p - p_{1L}}{p - p_{10}}\right)$$

which can be rewritten as

$$\frac{p_2}{p_{20}} = \left(\frac{p_{2L}}{p_{20}}\right)^{y/L} \tag{3-35}$$

As illustrated in Fig. 3-7, the partial pressure distribution of Eq. (3-35) is nonlinear. Additionally, it is recalled that $C_1 = -\dot{m}_1$ so that the evaporation

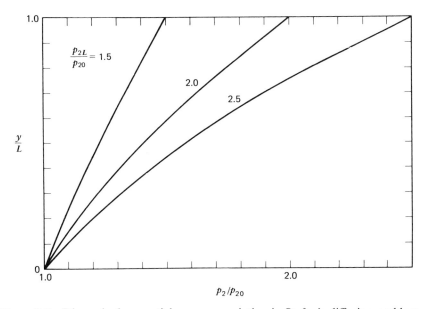

Figure 3-7 Dimensionless partial pressure variation in Stefan's diffusion problem.

rate of species 1 is given by

$$\dot{m}_1 = \frac{pM_1 D_{12}}{RTL} \ln\left(\frac{p - p_{1L}}{p - p_{10}}\right) \tag{3-36}$$

To use the exact solution [Eq. (3-36)] to test the suggestion of Eq. (3-27), Eq. (3-36) is first recast to employ mass fractions with the aid of Eq. (3-32) to obtain

$$\dot{m}_1 = \frac{pM_1 D_{12}}{RTL} \ln\left\{1 + \frac{\omega_{10} - \omega_{1L}}{(1 - \omega_{10})\left[\frac{M_1}{M_2} + \omega_{1L}\left(1 - \frac{M_1}{M_2}\right)\right]}\right\} \tag{3-37}$$

or

$$\dot{m}_1 = \frac{pM_1 D_{12}}{RTL} \ln\left(1 + \frac{p_{10} - p_{1L}}{p - p_{10}}\right)$$

The series expansion of $\ln(1 + x) = x + \cdots$, accurate for small values of x, in Eq. (3-37) shows that for small values of ω_1 (low rates of species 1 evaporation)

$$\dot{m}_1 \approx \frac{pM_2 D_{12}}{RTL} \frac{\omega_{10} - \omega_{1L}}{1 - \omega_{10}} \tag{3-38}$$

or, recognizing that $\rho_2 = pM_2/RT$,

$$\dot{m}_1 \approx \frac{\rho_2 D_{12}}{L} \frac{p_{10} - p_{1L}}{p - p_{10}}$$

Comparison with Eq. (3-27) shows that the mass-transfer coefficient h_D at low transport rates is

$$\dot{m}_1 \frac{1 - \omega_{10}}{\rho_2(\omega_{10} - \omega_{10})} = h_D \approx \frac{pM_2 D_{12}}{\rho_2 RTL} = \frac{D_{12}}{L}$$

which is a constant. From this result, the Sherwood number Sh is found to be

$$Sh = \frac{h_D L}{D_{12}} = 1$$

The "blowing" associated with mass transfer in which one surface is impermeable is accounted for by the $1 - \omega_{10}$ term in the denominator of Eq.

(3-27). Also, $\dot{m}_1$ is more simply expressed in terms of partial pressure than in terms of mass fraction. At large mass-transfer rates, the full solutions to the describing differential equations [e.g., Eqs. (3-34) or (3-35) for Stefan's diffusion problems] must be used for accuracy—the alternative is to consider the mass-transfer coefficient h_D to be dependent on mass fraction, which would be awkward since greatest utility is as a constant.

The rate at which a volatile liquid evaporates into an infinitely long tube determined by Arnold [6] is useful in estimating the time required to achieve steady-state conditions in a tube of finite length.

Not every binary diffusion problem has one surface impermeable to the second species. Another important simple case is that of equimolar counter-diffusion in which the molar transport rates of the two species are equal in magnitude and oppositely directed. This case is the subject of Problem 3-11.

Vapor pressure–temperature data can be obtained from such sources as the *International Critical Tables* [7], as well as the books by Nesmeyanov [8] and Jordan [9].

3.4 TRANSIENT CHANGE OF PHASE

Up to this point no change of phase of the fluid has been taken into account. Applications such as quenching during heat treatment, boiling heat exchangers, and ablative heat shields for reentry vehicles rely on the fact that a large amount of energy must be absorbed at nearly constant temperature by a fluid in order to change phase. Transient vaporization in which a large pool of stagnant liquid is initially at the boiling temperature corresponding to the imposed system pressure is such a situation. A large plate rests horizontally in the pool. Suddenly, the plate attains a high temperature that is substantially above the liquid boiling temperature. As a result, a film of vapor forms between the plate and the liquid. As time passes, the vapor film increases in thickness as a result of heat transfer through the film into the liquid–vapor interface, where more liquid is vaporized (see Fig. 3-8). This is also referred to as *Stefan's problem*, classically discussed by Carslaw and Jaeger [10].

Assumptions applied to this problem are (1) only x-direction variations are important; (2) pressure is constant; (3) the plate is at constant temperature; (4) the liquid is everywhere at its saturation temperature; (5) the density–thermal conductivity product is constant; (6) viscous dissipation is negligible; and (7) thermal radiation is negligible. With these simplifications, the describing equations are, for $0 \leqslant x \leqslant \delta$,

Continuity $$\frac{\partial \rho}{\partial t} + \frac{\partial \rho u}{\partial x} = 0 \qquad (3\text{-}39)$$

Energy $$\rho C_p \left(\frac{\partial T}{\partial t} + u \frac{\partial T}{\partial x} \right) = \frac{\partial}{\partial x} \left(k \frac{\partial T}{\partial x} \right) \qquad (3\text{-}40)$$

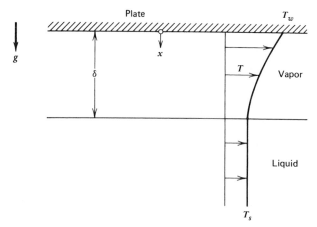

Figure 3-8 Geometry and coordinate system for the problem of one-dimensional transient change of phase.

subject to the boundary conditions that

$$\text{At } x=0 \qquad T = T_w \qquad \text{no temperature jump} \qquad (3\text{-}41a)$$

$$u = 0 \qquad \text{wall impermeable} \qquad (3\text{-}41b)$$

$$\text{At } x=\delta \qquad T = T_s \qquad \text{no temperature jump} \qquad (3\text{-}41c)$$

$$-k\frac{\partial T}{\partial x} = \Lambda\frac{dM}{dt} \qquad \begin{array}{l}\text{heat conducted into}\\ \text{liquid–vapor interface}\\ \text{all used to vaporize liquid}\end{array} \qquad (3\text{-}41d)$$

where δ is vapor film thickness, Λ is latent heat of vaporization, and M is the vapor mass in the film per unit area.

A coordinate transformation is undertaken that will place the continuity and energy equations of Eqs. (3-37) and (3-38) in a Lagrangian form. This transformation, referred to as a *Lagrangian transformation* because it removes convective terms, providing a description from the point of view of an observer on the moving particle, is useful in what are basically one-dimensional problems. To begin, integrate the continuity equation of Eq. (3-39) from $x = 0$ to obtain

$$\int_0^x \frac{\partial \rho}{\partial t}\, dx + \int_0^x \frac{\partial \rho u}{\partial x}\, dx = 0$$

$$\int_0^x \frac{\partial \rho}{\partial t}\, dx + (\rho u)|_{x,t} - (\rho u)\big|_{x=0,t}^{\;0} = 0 \qquad \text{by Eq. (3-41b)}$$

Since x is independent of t, allowing the order of integration and differentiation to be interchanged, one has

$$\frac{\partial}{\partial t}\left(\int_0^x \rho\,dx\right) = -\rho u$$

or

$$\frac{\partial m}{\partial t} = -\rho u \tag{3-42}$$

where $m = \int_0^x \rho(x, t)\,dx$ is the mass of vapor per unit area contained between the plate and a parallel plane a distance x away. Although m depends on both t and x, m can be used as a replacement for x as a spatial position indicator in the energy equation since at any fixed time m is directly related to x. Next, the energy equation is transformed from x, t to m, t' as independent variables. By the chain rule

$$\frac{\partial}{\partial t} = \frac{\partial}{\partial t'}\frac{\partial t'}{\partial t} + \frac{\partial}{\partial m}\frac{\partial m}{\partial t}$$

$$\frac{\partial}{\partial x} = \frac{\partial}{\partial t'}\frac{\partial t'}{\partial x} + \frac{\partial}{\partial m}\frac{\partial m}{\partial x}$$

Arbitrarily set $t' = t$. As a result,

$$\frac{\partial}{\partial t} = \frac{\partial}{\partial t'}\frac{\partial t'}{\partial t}^{\ 1} + \frac{\partial}{\partial m}\frac{\partial m}{\partial t}^{\ -\rho u} \quad\text{and}\quad \frac{\partial}{\partial x} = \frac{\partial}{\partial t'}\frac{\partial t'}{\partial x}^{\ 0} + \frac{\partial}{\partial m}\frac{\partial m}{\partial x}^{\ \rho x}$$

The energy equation in terms of m and t is then

$$\rho C_p\left[\left(\frac{\partial T}{\partial t} - \rho u\frac{\partial T}{\partial m}\right) + \rho u\frac{\partial T}{\partial m}\right] = \frac{\partial}{\partial x}\left(k\rho\frac{\partial T}{\partial m}\right)$$

Note on the right-hand side that $k\rho$ can be assumed to be constant. The energy equation reduces to, for $0 \leqslant m \leqslant M$,

$$C_p\frac{\partial T}{\partial t} = (k\rho)\frac{\partial^2 T}{\partial m^2} \tag{3-43}$$

with the boundary conditions of $T(m = 0, t) = T_w$, $T(m = M, t) = T_s$, and

$$-(k\rho)\frac{\partial T(m = M, t)}{\partial m} = \Lambda\frac{dM}{dt} \tag{3-44}$$

Equation (3-43) has a Lagrangian aspect—all convective terms have disappeared, and it has the form of a conduction problem. Eq. (3-43) is linear, the boundary condition of Eq. (3-44) is not. This is characteristic of problems in which the location of a surface is initially unknown and must be determined in the course of the problem solution. To bring this point into sharper focus, consider that at the liquid–vapor interface $dT = (\partial T/\partial m)\, dm + (\partial T/\partial t)\, dt = 0$ since there $T = T_s = $ const; or

$$\frac{\partial T}{\partial m}\, dm + \frac{k\rho}{C_p}\frac{\partial^2 T}{\partial m^2}\, dt = 0$$

which gives the nonlinear result

$$\frac{dM}{dt} = \frac{-k\rho}{C_p}\frac{\partial^2 T/\partial m^2}{\partial T/\partial m}$$

As things now stand M, the unknown maximum value of m, is bound up in a boundary condition. This awkwardness can be partially overcome by another coordinate transformation which fixes the liquid–vapor interface at unity distance from the plate by using m/M as a measure of spatial position. Let $t'' = t$ and $y = m/M$. By the chain rule

$$\frac{\partial}{\partial t} = \frac{\partial}{\partial t''}\frac{\partial t''}{\partial t} + \frac{\partial}{\partial y}\frac{\partial y}{\partial t}$$

$$\underset{1}{} \qquad \underset{-\dfrac{m}{M^2}\dfrac{dM}{dt} = \dfrac{-y}{M}\dfrac{dM}{dt} = \dfrac{-y}{M^2}M\dfrac{dM}{dt} = -\dfrac{y}{2M^2}\dfrac{dM^2}{dt}}{}$$

$$\frac{\partial}{\partial m} = \frac{\partial}{\partial t''}\frac{\partial t''}{\partial m} + \frac{\partial}{\partial y}\frac{\partial y}{\partial m}$$

$$\underset{0}{} \qquad \underset{\dfrac{1}{M}}{}$$

The new form of the energy equation is then, for $0 \leqslant y \leqslant 1$,

$$M^2\frac{\partial T}{\partial t} = \frac{(k\rho)}{C_p}\frac{\partial^2 T}{\partial y^2} + \frac{y}{2}\frac{dM^2}{dt}\frac{\partial T}{\partial y} \tag{3-45}$$

with boundary conditions of

$$\text{At } y = 0 \quad T = T_w \qquad \text{at } y = 1 \quad T = T_s$$

$$-2\frac{(k\rho)}{C_p}\frac{C_p}{\Lambda}\frac{\partial T}{\partial y} = \frac{dM^2}{dt} \tag{3-46}$$

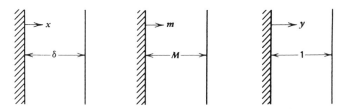

Figure 3-9   Effects of coordinate transformations.

The effects of the transformations are depicted in Fig. 3-9. The idea to be pursued next is that it is possible that all temperature profiles will be of similar appearance in y, t coordinates as time progresses. In other words, it is possible that T depends only on y and not at all on t. For this "similarity solution" to be valid, it must be that $dM^2/dt = 4B^2(k\rho/C_p) = \text{const.}$ With a dimensionless temperature defined as $\theta = (T - T_s)/(T_w - T_s)$, the dimensionless energy equation and boundary conditions are, for $0 \leqslant y \leqslant 1$,

$$0 = \frac{d^2\theta}{dy^2} + 2B^2 y \frac{d\theta}{dy} \tag{3-47}$$

with boundary conditions of

$$\theta(y = 0) = 1 \tag{3-48a}$$

$$\theta(y = 1) = 0 \tag{3-48b}$$

$$\frac{-C_p(T_w - T_s)}{\Lambda} \frac{d\theta(y = 1)}{dy} = 2B^2 \tag{3-48c}$$

Two integrations of the dimensionless energy equation [Eq. (3-47)] result in

$$\frac{T - T_s}{T_w - T_s} = \theta = 1 - \frac{\text{erf}(By)}{\text{erf}(B)} \tag{3-49}$$

where the error function $\text{erf}(x)$ illustrated in Fig. 3-10 is defined as

$$\text{erf}(x) = 2\pi^{-1/2} \int_0^x e^{-u^2} d\eta$$

The constant B is evaluated from use of Eq. (3-49) in Eq. (3-48c) as

$$\pi^{1/2} B e^{B^2} \text{erf}(B) = \frac{C_p(T_w - T_s)}{\Lambda} \tag{3-50}$$

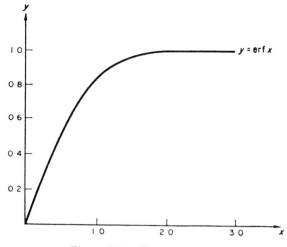

Figure 3-10   Error function.

For small values of $C_p(T_w - T_s)/\Lambda$, Eq. (3-50) is closely approximated by

$$2B^2 \approx \frac{C_p(T_w - T_s)}{\Lambda} \tag{3-51}$$

as shown in Fig. 3-11. The "blowing" directed toward the plate caused by vaporization at the liquid–vapor interface is responsible for the nonlinear film temperature distribution given by Eq. (3-49) and illustrated in Fig. 3-11.

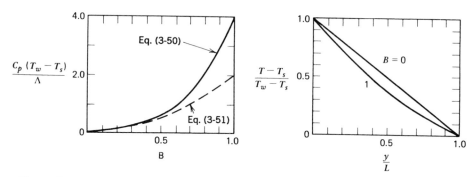

Figure 3-11   Parameter B dependence on $C_p(T_w - T_0)/\Lambda$ and dimensionless temperature variation.

If the dimensionless parameter $C_p(T_w - T_s)/\Lambda$ is known, the rate of evaporation is available since

$$\frac{dM^2}{dt} = \frac{4B^2k\rho}{C_p}$$

One integration gives

$$M^2 = \frac{4B^2k\rho}{C_p}t + \text{const} \tag{3-52}$$

If there is initially no vapor film, the constant of integration must vanish. Provided vapor density is constant, $M \approx \rho\delta$, and the film thickness varies as the square root of elapsed time as

$$\delta = 2B\alpha^{1/2}t^{1/2} \tag{3-53}$$

Also, the Fourier number Fo is found from Eq. (3-53) to be the constant

$$\text{Fo} = \frac{\alpha t}{\delta^2} = \frac{1}{4B^2}$$

The infinitely large heat flux this solution predicts at the beginning when $\delta = 0$ cannot be entirely correct as is discussed in Appendix A.

The vaporization problem that has been solved is in some ways a boundary-layer problem. The vapor film is a layer of thermally affected fluid near a boundary. Although the boundary layer in this case has a definite thickness that is marked by the liquid–vapor interface, it is similar to the boundary layer of less physically definite thickness that forms as air flows over an airfoil. It would be expected, then, that the boundary-layer thickness on an airfoil immersed in an airstream flowing at velocity V would increase in the manner suggested in Eq. (3-53). In other words, since the elapsed time for which a fluid particle moving with the free stream has been exposed to disturbance is $t = x/V$, where x is the distance from the leading edge of the airfoil, the boundary-layer thickness should vary with distance as

$$\delta \sim x^{1/2}$$

The effect of thermal radiation on transient vaporization of a saturated liquid at a constant temperature plate has been studied [11]. In problems of this type there must be a pressure excursion since some time must elapse before the liquid can be displaced to make room for the newly formed vapor. Studies [12–14] of this effect show that the pressure excursion is usually moderate and does not markedly affect the rate of vapor formation, although the film thickness is often substantially affected.

Transient heat transfer in a fluid flowing through a porous medium is treated in Problem 3-27.

PROBLEMS

3-1 (a) Show from Eq. (3-5) that in Couette flow with both plate temperatures equal, the maximum temperature occurs at $y/L = \frac{1}{2}$ and
equals

$$T_{max} - T_0 = \frac{\mu U^2}{8k}$$

which is independent of the plate separation. Sketch the temperature distribution.

(b) Show from the energy equation for Couette flow that the viscous
dissipation rate is a uniformly distributed heat source at the value
$\mu U^2/L^2$ regardless of the boundary conditions imposed.

(c) Show from Eq. (3-3) that the shear stress τ_{yx} is constant as

$$\tau_{yx} = \frac{\mu U}{L}$$

3-2 (a) The maximum oil temperature in a nominally 60°F journal bearing
that rotates at 3000 rpm and has an inner diameter of 2 in. and a
radial clearance of 0.002 in. is to be estimated. The oil properties
are $\mu = 5.82 \times 10^{-2}$ lb$_m$/ft sec, $k = 0.077$ Btu/hr ft °F. Show
that the maximum temperature rise in the oil is 10°F, assuming
that both bearing surfaces are maintained at 60°F. *Note*: At 100°F,
$\mu = 1.53 \times 10^{-2}$ lb$_m$/ft sec so that $A = 9726$°R in the
viscosity–temperature relation $\mu/\mu_0 = \exp[A(1/T - 1/T_0)]$.

(b) Show that the shear stress in the fluid is 284 lb$_f$/ft^2 and that the
torque per unit length required to turn the bearing is 12.3 ft lb$_f$/ft
which yields a power to overcome bearing friction of 7 hp.

3-3 The steady laminar flow of a constant-property Newtonian fluid down
a circular pipe of radius R and constant temperature T_0 in response to
an imposed constant pressure drop dp/dz is to be determined.

(a) Show that the motion and energy equations reduce to

$$z \text{ Motion} \quad \frac{d(r\,dv_z/dr)}{dr} = \frac{r(dp/dz)}{\mu}$$

$$\text{Energy} \quad \frac{d(r\,dT/dr)}{dr} = -\frac{\mu r}{k}\left(\frac{dv_z}{dr}\right)^2$$

(b) Show that the boundary conditions to be imposed are

$$\text{At } r = R \qquad v_z = 0 = v_\theta \qquad \text{at } r = 0 \qquad v_z \text{ finite}$$
$$v_r = 0 \qquad\qquad\qquad\qquad T \text{ finite}$$
$$T = T_0$$

(c) On the basis of parts a and b, show that the z-velocity distribution is

$$v = \frac{R^2}{4\mu}\left(-\frac{dp}{dz}\right)\left(1 - \frac{r^2}{R^2}\right) = 2v_{\text{av}}\left(1 - \frac{r^2}{R^2}\right)$$

with a maximum velocity v_m and average velocity v_{av} given by

$$v_m = \frac{R^2}{4\mu}\left(-\frac{dp}{dz}\right) = 2v_{\text{av}}$$

Also, show that for a pipe of length ΔL the volumetric flow rate Q is given by

$$Q = \frac{\pi R^4}{8\mu}\frac{\Delta P}{\Delta L}$$

(d) On the basis of parts a, b, and c, show that the temperature distribution is given by

$$T = T_0 + \frac{\mu}{k}v_{\text{av}}^2\left(1 - \frac{r^4}{R^4}\right)$$

and the maximum temperature (at the center line) is

$$T_m = T_0 + \frac{\mu}{k}v_{\text{av}}^2$$

3-4 A fluid with a yield stress τ_c and a wall–slip coefficient S (characteristic of a fluid with deformable particles such as a slurper [15] or wood pulp–paper slurry [16] for which a thin lubricating film near the wall permits an apparent slip) flows laminarly down a tube of radius R in response to an imposed pressure gradient $dp/dz = p'$ as in Problem 3-3.

(a) Show that the axial motion equation is

$$\frac{d(r\tau_{rz})}{dr} = rp', \qquad 0 \leqslant r \leqslant R$$

subject to the conditions $v_z(R) = -S\tau_w$ and $\tau_{rz}(0)$ finite in which τ_w is the shear stress at the wall.

(b) For the velocity–shear stress relationship

$$\mu \frac{dv_z}{dr} = \begin{cases} 0, & |\tau_{rz}| \leqslant |\tau_c| \\ \tau_{rz} - \tau_c, & |\tau_{rz}| > |\tau_c| \end{cases}$$

show that for $|\tau_w/\tau_c| > 1$ the velocity distribution is

$$v_z = \frac{2x}{\pi R^2} \begin{cases} \left[(1 - \tau_c/\tau_w)^2 + c/2 \right], & r' \leqslant |\tau_c/\tau_w| \\ \left[\left(1 + r' - 2\frac{\tau_c}{\tau_w} \right)(1 - r') + c/2 \right], & |\tau_c/\tau_w| < r' \leqslant 1 \end{cases}$$

while for $|\tau_w/\tau_c| \leqslant 1$

$$v_z = c \frac{x}{\pi R^2}$$

where

$$x = \pi R^4(-p')/8\mu, \qquad r' = r/R, \qquad c = 4S\mu/R, \qquad \tau_w = Rp'/2$$

(c) Show that the volumetric flow rate Q is

$$Q = \begin{cases} cx, & x/x_c \leqslant 1 \\ x \left[1 - \frac{4}{3}\frac{x_c}{x} + \frac{1}{3}\left(\frac{x_c}{x}\right)^4 + c \right], & x/x_c > 1 \end{cases}$$

where $x_c = -\pi R^3 \tau_c/4\mu = \pi R^4(-p'_c)/8\mu$ is a measure of the pressure gradient p'_c required for the wall shear stress τ_w to exceed the yield stress τ_c.

(d) For $x_c = 1$ m^3/s and $c = \frac{1}{10}$, plot Q versus x on log–log paper. Determine the value of the power-law exponent n that enables a power-law fluid model without yield stress or velocity slip to fit the plotted curve. Comment on the ease and certainty with which non-Newtonian flow parameters can be evaluated from flow measurements such as this. *Answer*: $n = 0.36$, seemingly shear-thinning behavior.

3-5 For oil at 60°F ($\mu = 5.82 \times 10^{-2}$ lb$_m$/ft sec, $k = 0.077$ Btu/hr ft °F, $\rho = 57$ lb$_m$/ft^3) flowing down a pipe of 12 in. diameter, determine the maximum oil temperature. Assume that the oil is flowing at a Reynolds number of Re $= V_{av}D\rho/\mu = 1000$. Determine the pressure gradient down the pipe and the volumetric flow rate as well. *Answer:* $-p' = 10^{-5}$ psi/ft.

3-6 For the Couette flow in Problem 3-1, evaluate the entropy flow rate at each plate and evaluate the rate at which entropy is generated per unit volume in the fluid. *Answer:* $S(0) = U^2\mu/2LT_0$.

3-7 Show that if bulk density ρ is constant the rate of mass transfer of species 1 in Stefan's diffusion problem from Eq. (3-29) is

$$\dot{m}_1 = \frac{\rho D_{12}}{L}\frac{\omega_{10} - \omega_{1L}}{1 - \omega_{10}}$$

3-8 Show that in Stefan's diffusion problem the "blowing" represented by the upward bulk velocity v becomes large at large rates of species 1 mass transfer. Comment on the importance of this effect in view of the predicted rate of mass transfer when the liquid is at its nominal boiling temperature. Show that $v = \dot{m}_1/\rho$ and is always directed upward, always aiding the diffusive transport.

3-9 The solutions to Stefan's diffusion problem have been used in experimental determination of gas diffusivities [17]. In one such experiment the diffusivity of gaseous carbon tetrachloride through oxygen was measured. The distance between the CCl$_4$ liquid and the tube top was 17.1 cm, the tube cross-sectional area was 0.82 cm^2, the total pressure was 775 mm Hg and the saturation pressure of the CCl$_4$ at the ambient temperature of 0°C was 33 mm Hg. If 0.0208 cm^3 of liquid CCl$_4$ evaporated 10 h after steady state was reached (liquid CCl$_4$ density at 0° C is 1.59 g/cm^3), show that the diffusivity is $D_{12} = 0.0636$ cm^2/s.

3-10 Perform a simple experiment [18] to test the accuracy of the analytical solution to Stefan's diffusion problem.

  (a) Partially fill a tall cylinder with water and maintain the water at a constant temperature of about 150°F so that evaporation proceeds at a convenient rate.

  (b) Record the time required for about 1 cm of water to evaporate, and compare this result with that predicted by Eq. (3-36) by use of a reported air–water vapor mass diffusivity.

  (c) Comment on the importance of such sources of discrepancy between measurement and prediction as the cooling effect of evaporation, free convection effects arising from the fact that the evaporating water is less dense than air, and two-dimensional flow patterns.

3-11 Consider the long tube shown in Fig. 3P-11 through which two gases steadily diffuse. At the left end species 1 exists at partial pressure p_{10} and corresponding mass fraction ω_{10}. At the right end species 1 exists at partial pressure p_{1L} and corresponding mass fraction ω_{2L}. The total pressure p and temperature T are uniform and constant. The rate at which moles of species 1 are transported to the right is equal in magnitude and opposite in direction to the rate at which moles of species 2 are transported to the left, perhaps because of some chemical reactions occurring outside the tube. This is an equimolar counter diffusion problem in which $\dot{m}_1(x)/M_1 = -\dot{m}_2(x)/M_2$, where $M_{1,2}$ are the molecular weights of species 1, 2.

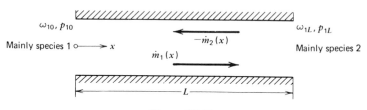

Figure 3P-11

(a) Show that the pertinent describing differential equations are

$$\text{Continuity} \quad \frac{d(\rho u)}{dx} = 0$$

$$\text{Diffusion} \quad \rho u \frac{d\omega_1}{dx} = D_{12}\frac{d(\rho\, d\omega_1/dx)}{dx}$$

and that the boundary conditions are

$$\omega_1(x = 0) = \omega_{10} \quad \text{or} \quad p_1(x = 0) = p_{10}$$
$$\omega_1(x = L) = \omega_{1L} \quad \text{or} \quad p_1(x = L) = p_{1L}$$

(b) Integrate the continuity and diffusion equations to find

$$\rho u = C_1$$

and

$$C_1\omega_1 - C_2 = \rho D_{12}\frac{d\omega_1}{dx}$$

(c) Show that the constraint of equimolar counterdiffusion leads to the result that there is bulk motion by using Eq. (3-24) in

$$\frac{\dot{m}_1(x)}{M_1} = -\frac{\dot{m}_2(x)}{M_2}$$

to obtain

$$\rho u = \frac{\rho D_{12}(1 - M_1/M_2)}{\omega_1(1 - M_1/M_2) + M_1/M_2} \frac{d\omega_1}{dx} = C_1$$

(d) Taking into account the result of part c, show that

$$\dot{m}_1(x) = -\frac{\rho D_{12}(M_1/M_2)}{\omega_1(1 - M_1/M_2) + M_1/M_2} \frac{d\omega_1}{dx} = C_2$$

(e) From part d, show that the partial pressure distribution is linear, as expressed by

$$C_2 = -\frac{D_{12}M_1}{RT} \frac{dp_1}{dx}$$

and that the mass transport rate of species 1 is given by either

$$\dot{m}_1 = \frac{D_{12}M_1}{RT} \frac{(p_{10} - p_{1L})}{L}$$

or the more complex relationship

$$\dot{m}_1 = \frac{D_{12}M_2 p}{RTL} \frac{\omega_{10} - \omega_{1L}}{[1 + \omega_{10}(M_2/M_1 - 1)][1 + \omega_{1L}(M_2/M_1 - 1)]}$$

the correction for "blowing" in the denominator of the expression involving mass fractions being different from the case in which one surface is impermeable to one species.

(f) Show that the "blowing" velocity u is related to the rate of species 1 mass transport by

$$u = \left(1 - \frac{M_2}{M_1}\right) \frac{\dot{m}_1}{\rho}$$

Depending on the ratio of molecular weights, the "blowing" can either help or hinder diffusive transport.

Show that the fraction $\rho_1 u / \dot{m}_1$ of the rate of species 1 mass transport caused by blowing is the usually small quantity

$$\frac{\rho_1 u}{\dot{m}_1} = \omega_1 \left(1 - \frac{M_2}{M_1} \right)$$

(g) Comment on the assumption that total pressure is constant. Would the x-motion equation provide information concerning total pressure variation? Comment on the importance of natural convection.

(h) With a mass-transfer coefficient h_D defined as

$$\dot{m}_1 = \frac{\rho h_D (\omega_{10} - \omega_{1L})}{\left[1 + \omega_{10} \left(\dfrac{M_2}{M_1} - 1 \right) \right]\left[1 + \omega_{1L} \left(\dfrac{M_2}{M_1} - 1 \right) \right]}$$

show that the result of part e leads to the Sherwood number Sh being

$$\text{Sh} = \frac{h_D L}{D_{12}} = 1$$

3-12 Consider a sphere of radius R immersed in a large fluid body of uniform temperature T_f. The sphere is slightly warmer than the fluid, so the fluid is stagnant and is uniformly at temperature T_s.

(a) Show that the energy equation in spherical coordinates describing the temperature distribution in the fluid is

$$\frac{1}{r^2} \frac{d(r^2 \, dT/dr)}{dr} = 0$$

with boundary conditions of

$$T(r = R) = T_s$$

and

$$T(r = \infty) = T_f$$

(b) Show from the result of part a that the temperature distribution is

$$\frac{T - T_f}{T_s - T_f} = \frac{R}{r}$$

and that the heat-transfer rate q_w from the sphere to the fluid is

$$q_w = 4\pi kR(T_s - T_f)$$

(c) Show from the result of part b that the Nusselt number based on sphere diameter is

$$\text{Nu} = \frac{hD}{k} = 2$$

Internal circulation caused by an electric field can increase this result by an order of magnitude [19].

3-13 Consider a sphere of radius R immersed in a large body of fluid. The sphere has an excess of mass of species 1 on its surface, but it is sufficiently slight that there is no buoyancy-induced motion in the fluid at a partial pressure of p_{10} and corresponding mass fraction of ω_{10}. Far away from the sphere, species 1 partial pressure is p_{1f} and its corresponding mass fraction is ω_{1f}. The rate at which moles of species 1 are transported outward is equal and opposite to the rate at which moles of species 2 are transported inward. This is equimolar counterdiffusion from a sphere.

(a) Show that the pertinent describing differential equations are

$$\text{Continuity} \quad \frac{d(r^2 \rho v_r)}{dr} = 0$$

$$\text{Diffusion} \quad \rho v_r \frac{d\omega_1}{dr} = \frac{D_{12}}{r^2} \frac{d(\rho r^2 \, d\omega_1/dr)}{dr}$$

with boundary conditions of

$$\omega_1(r = R) = \omega_{10} \quad \text{or} \quad p_1(r = R) = p_{10}$$
$$\omega_1(r = \infty) = \omega_{1f} \quad \text{or} \quad p_1(r = \infty) = p_{1f}$$

(b) Show that the equimolar counterdiffusion requirement

$$\frac{\dot{m}_1(r)}{M_1} = -\frac{\dot{m}_2(r)}{M_2}$$

leads to the relationship that

$$\dot{m}_{1,\text{total}} = \frac{4\pi D_{12} p M_2 R}{RT}$$

$$\times \frac{(\omega_{10} - \omega_{1f})}{[1 + \omega_{10}(M_2/M_1 - 1)][1 + \omega_{1f}(M_2/M_1 - 1)]}$$

(c) Show from the result of part b that defining a mass-transfer coefficient h_D as

$$\dot{m}_{1,\text{total}} = \frac{A\rho h_D(\omega_{10} - \omega_{1f})}{\left[1 + \omega_{10}(M_2/M_1 - 1)\right]\left[1 + \omega_{1f}(M_2/M_1 - 1)\right]}$$

leads to the Sherwood number Sh being

$$\text{Sh} = \frac{h_D D}{D_{12}} = 2$$

3-14 A carbon particle of diameter $D = 0.1$ in. burns in an oxygen atmosphere. Temperature is uniform at $1800°R$, and the total pressure is 1 atm. At the carbon surface the reaction $C + O_2 \rightarrow CO_2$ occurs, so this is equimolar counterdiffusion about a sphere. Assume that at the carbon surface $p_{CO_2} = 1$ atm while far from the carbon surface $p_{CO_2} = 0$ atm.

(a) Verify that $D_{CO_2-O_2} \approx 4$ ft^2/hr.

(b) Verify that the rate at which carbon burns, assuming the combustion to be limited by the rate at which oxygen can be transported to the carbon surface, is about 0.9×10^{-6} lb$_m$/sec.

3-15 A droplet of water initially of 0.1 in. diameter is maintained at 70°F in 70°F dry stagnant air.

(a) Estimate the water evaporation rate per unit area from this sphere. Is "blowing" significant? What fraction of the maximum possible evaporation is this estimate?

(b) Derive the equation

$$\frac{d\left[(r/r_0)^2\right]}{dt} = \frac{\rho_{\text{air}} D_{12} \text{ Sh}}{\rho_{\text{liqH}_2\text{O}} r_0^2} \frac{\omega_{10} - \omega_{1f}}{1 - \omega_{10}}$$

where $\text{Sh} = 2r_0 h_D/D_{12}$ to describe the rate at which the droplet radius decreases. Estimate the time required for the droplet to completely evaporate. *Answer:* $t = 2000$ sec.

3-16 A droplet of water is placed in dry stagnant air of constant temperature T_f. The droplet evaporates and cools to an equilibrium temperature T_w.

(a) To estimate T_w, derive the equation

$$-k_{\text{air}} \text{ Nu}(T_w - T_f) = \Lambda D_{12}\rho_{\text{air}} \text{ Sh} \frac{\omega_{10} - \omega_{1f}}{1 - \omega_{10}}$$

where Λ is the water latent heat of vaporization, $Nu = hD/k$, and $Sh = Dh_D/D_{12}$, neglecting both thermal radiation (it is likely to be considerable since convection is so low in a stagnant situation) and blowing effects on heat transfer.

(b) Show that the result of part a can be reduced to

$$-\frac{C_{p_{air}}(T_w - T_f)}{\Lambda} = Le\frac{\omega_{10} - \omega_{1f}}{1 - \omega_{10}}$$

where the Lewis number Le is defined as $Le = D_{12}/\alpha$ and $\alpha = k/\rho C_p$.

(c) For the case of $T_f = 70°F$, determine the value of T_w (an iterative solution procedure is required since ω_{10} depends on the unknown T_w). How long will a 0.1-in.-diameter droplet last if evaporating at this equilibrium rate? *Answer:* $t = 5600$ sec.

3-17 Consider steady Couette flow of a constant-property fluid as illustrated in Fig. 3-1. The two plates are porous and fluid enters the interplate space through the bottom plate and leaves at the same steady and uniform rate ρv through the top plate.

(a) Show that the x-motion equation is

$$v\frac{du}{dy} = \nu\frac{d^2u}{dy^2}$$

subject to the boundary conditions that

$$u(y = 0) = 0 \quad \text{and} \quad u(y = L) = U$$

from which it follows that the velocity distribution is

$$\frac{u}{U} = \frac{e^{vy/\nu} - 1}{e^{vL/\nu} - 1}$$

(b) Plot u/U against y/L for several values of the dimensionless Reynolds number vL/ν, and comment on the effect the imposed "blowing" has on the velocity profile. Show that the viscous drag at the bottom plate τ_w is

$$\tau_w = \frac{\mu U}{L}\frac{vL/\nu}{e^{vL/\nu} - 1}$$

When v is positive, the upward "blowing" is representative of that which occurs in evaporation or transpiration cooling of a turbine blade. When v is negative, the downward "suction" is representative of that which occurs in condensation or boundary-layer suction to control flow over an airfoil.

(c) Show that if the bottom and top plates are maintained at temperatures T_0 and T_1, respectively, the energy equation is

$$v \frac{dT}{dy} = \alpha \frac{d^2 T}{dy^2}$$

with the boundary conditions of

$$T(y = 0) = T_0 \quad \text{and} \quad T(y = L) = T_1$$

from which it follows that the temperature distribution is

$$\frac{T - T_0}{T_1 - T_0} = \frac{e^{\Pr v y / v} - 1}{e^{\Pr v L / v} - 1}$$

(d) Plot $(T - T_0)/(T_1 - T_0)$ against y/L and comment as in part b. Show also that the diffusive heat flux q_w at the bottom plate is

$$q_w = \frac{(T_0 - T_1)k}{L} \frac{\Pr v L / v}{e^{\Pr v L / v} - 1}$$

As "blowing" increases the wall heat flux decreases, an effect called *transpiration cooling* that can be used to protect surfaces from hot atmospheres. Comment on the ratio of diffusive to convective energy transport rates.

3-18 An application of transpiration cooling to a spherical geometry is illustrated in Fig. 3P-18, in which two concentric porous spheres of radii r_1 and r_2 are maintained at temperatures T_1 and T_2, respectively. A gas moves radially outward at mass flow rate w. Assuming steady state, laminar flow, constant properties, and negligible viscous dissipation, determine the rate of heat gain by the inner sphere as a function of the mass flow rate of the gas.

(a) Show that the describing equations are

$$\text{Continuity} \qquad \frac{d(r^2 v_r)}{dr} = 0$$

$$\text{Energy} \qquad r^2 v_r \frac{dT}{dr} = \alpha \frac{d(r^2 \, dT/dr)}{dr}$$

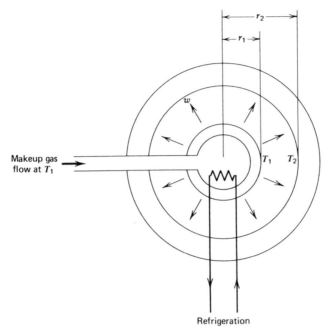

Figure 3P-18

with boundary conditions of

$$r^2 v_r = \frac{w}{4\pi\rho} = \text{const}, \qquad T(r_1) = T_1, \qquad T(r_2) = T_2$$

(b) Show that the rate of heat flow into the inner sphere is

$$Q = 4\pi r_1^2 k \frac{dT(r_1)}{dr} = 4\pi r_2 k (T_2 - T_1) \frac{\text{Re Pr}}{e^{\text{Re Pr}(r_2/r_1 - 1)} - 1}$$

(c) Demonstrate that in the absence of a radial gas flow, heat flows into the inner sphere at a rate given by

$$Q_0 = 4\pi k r_1^2 \frac{dT(r_1)}{dr} = 4\pi r_2 k \frac{T_2 - T_1}{r_2/r_1 - 1}$$

(d) Show from the results of parts b and c that the transpiration reduces the heat flow into the inner sphere as

$$\lim_{w \to 0} \left[\frac{Q}{Q_0} \right] = \frac{2}{2 + \text{Re Pr}(r_2/r_1 - 1)}$$

3-19 It is proposed to reduce the rate of evaporation of liquified oxygen in small containers by taking advantage of the transpiration the evaporation can produce. The geometry is that of Fig. 3P-18, but the makeup tube is absent as is the refrigeration coil. Estimate the rate of heat gain and the rate of evaporation for an inner sphere radius of 6 in. and an outer sphere radius of 12 in. with and without transpiration. The conditions to be used are $T_1 = -297°F$ and $T_2 = 30°F$ with oxygen properties $k = 0.02$ Btu/hr ft °F, $C_p = 0.22$ Btu/lb$_m$ °F, Pr = 0.7, and heat of vaporization = 92 Btu/lb$_m$. *Answer:* Heat gain equals 61 Btu/hr and 82 Btu/hr with and without transpiration, respectively.

3-20 Consider the adiabatic flow of an ideal gas through a one-dimensional standing shock wave in a duct illustrated in Fig. 3P-20. The velocity, temperature, and pressure distributions in and near the standing shock wave are to be determined. Assume all properties save density are constant.

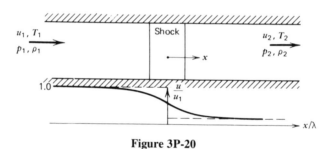

Figure 3P-20

(a) Show that the describing equations are

Continuity $$\frac{d(\rho u)}{dx} = 0$$

x Motion $$\rho u \frac{du}{dx} = -\frac{dp}{dx} + \frac{4}{3}\frac{d(\mu\, du/dx)}{dx}$$

Energy $$C_p \rho u \frac{dT}{dx} = \frac{d(k\, dT/dx)}{dx} + u\frac{dp}{dx} + \frac{4}{3}\mu\left(\frac{du}{dx}\right)^2$$

(b) Integrate the continuity equation to find that

$$\rho u = \text{const} = \rho_1 u_1$$

(c) Eliminate dp/dx in the energy equation by making use of the x-motion equation and incorporate the result of part b to put the energy equation in the form

$$\rho_1 u_1 \left(C_p \frac{dT}{dx} + u \frac{du}{dx} \right) = \frac{4}{3} \frac{d}{dx} \left(\mu u \frac{du}{dx} \right) + \frac{d(k \, dT/dx)}{dx}$$

Integrate twice with $\mathrm{Pr} = \frac{3}{4}$ to find

$$C_p T + \frac{u^2}{2} = C_1 + C_2 e^{\rho_1 u_1 C_p x / k}$$

(d) Evaluate the constants of integration by the following considerations. Since $C_p T + u^2/2$ cannot increase indefinitely in the downstream direction, $C_2 = 0$. Far upstream it must be that $C_p T_1 + u_1^2/2 = C_1$. So

$$C_p T + \frac{u^2}{2} = C_p T_1 + \frac{u_1^2}{2}$$

(e) Substitute the result of part b into the x-motion equation and integrate once to obtain

$$\rho_1 u_1 u = -p + \frac{4}{3} \mu \frac{du}{dx} + C_3$$

Note that evaluation of C_3 from upstream conditions gives

$$C_3 = p_1 + \rho_1 u_1^2 = \rho_1 \left(u_1^2 + \frac{RT_1}{M} \right)$$

(f) Eliminate p from the result of part e by making use of the ideal gas equation of state $(\rho = pM/RT)$ and the energy equation of part d to obtain

$$\frac{4}{3} \frac{\mu}{\rho_1 u_1} u \frac{du}{dx} - \frac{\gamma + 1}{2\gamma} u^2 + \frac{C_3}{\rho_1 u_1} u = \frac{\gamma - 1}{\gamma} C_1$$

where $\gamma = C_p/C_v$. Rearrange this result into the form

$$\phi \frac{d\phi}{d\xi} = \beta \, \mathrm{Ma}_1(\phi - 1)(\phi - \alpha)$$

where

$$\phi = \frac{u}{u_1}, \qquad \xi = \frac{x}{\lambda_1}, \qquad \text{Ma} = u_1 \left(\frac{m}{\gamma R T_1} \right)^{1/2} = \text{Mach number}$$

$$\alpha = \frac{\gamma - 1}{\gamma + 1} + \frac{2}{\gamma + 1} \frac{1}{\text{Ma}_1^2}, \qquad \beta = \frac{9}{8}(\gamma + 1) \left(\frac{\pi}{8\gamma} \right)^{1/2}$$

$$\lambda_1 = 3 \frac{\mu_1}{\rho_1} \left(\frac{\pi M}{8 R T_1} \right)^{1/2} = \text{mean free path}$$

(g) Integrate the result of part f to obtain

$$\frac{1 - \phi}{(\phi - \alpha)^\alpha} = e^{\beta(1 - \alpha)\text{Ma}_1(\xi - \xi_0)}$$

where ξ_0 is an integration constant. With this result, temperature and pressure distributions can be found from the results of parts d and e.

(h) Note from the result of part g that ϕ must approach 1 as $\xi \to -\infty$, which requires $\alpha < 1$. This requirement can be met only if $\text{Ma}_1 > 1$, the upstream flow is supersonic. Note also that as $\xi \to \infty$, $\phi \to \alpha$. For $\text{Ma}_1 = 2$ and $T_1 = 530°\text{R}$ and air, plot ϕ versus $(x - x_0)/\lambda_1$ as illustrated in Fig. 3P-20 to demonstrate that the standing shock wave is only a few mean free paths thick [20–22].

3-21 For the vaporization problem treated in Section 3.4 and illustrated in Fig. 3-8, assumption of a linear temperature distribution in the film gives the requirement that heat conducted into the interface vaporize liquid as

$$k \frac{T_w - T_s}{\delta} = \Lambda \rho_{\text{vapor}} \frac{d\delta}{dt}$$

(a) Show that the solution to the preceding equation is, for $\delta(0) = 0$,

$$\delta^2 = 2 \frac{k \, \Delta T}{\Lambda \rho}$$

(b) Show that the result of part a is that predicted by Eq. (3-31).

3-22 A stagnant pond of water uniformly at its freezing temperature of 32°F is suddenly exposed to 22°F air. Estimate the time required for 4 in. of ice to form. A linear temperature distribution in the ice may be assumed if desired. How much time is required to double the ice thickness? *Answer*: $t_{\text{double}} = 40$ hr.

3-23   Plot the numerical value of the heat flux from a piece of metal at 812°F exposed to liquid water at 212°F against time. The heat flux at the metal surface can be obtained from the derivative of Eq. (3-49).

3-24   For the transient vaporization problem of Section 3.4, show that the velocity in the liquid u_L is given by

$$\rho_L \left(\frac{d\delta}{dt} - u_L \right) = \frac{dM}{dt}$$

from which, for vapor density constant, the velocity in the liquid varies with time as

$$u_L = B \left(1 - \frac{\rho_{vapor}}{\rho_L} \right) \left(\frac{\alpha_{vapor}}{t} \right)^{1/2}$$

Plot u_L against t for the conditions of Problem 3-23, comment on the accelerating force required to achieve this result with an initially stagnant liquid.

3-25   The heat transfer in a piston–cylinder system such as is important to a Stirling-cycle heat engine has been analyzed and the results have been compared to measurements [23]. Prepare a brief report outlining the salient features of the analysis and the results.

3-26   Consider the slow, laminar flow of air from a plastic injection molding machine die as it is evacuated by connection to a vacuum line prior to injection of plastic. In this way, plastic can be injected into the die with greater rapidity than if air filled the die. The die of pressure p and the vacuum line of pressure p_0 are connected by capillary tubes of diameter D and length L. Because resistance to flow is large, pressure gradient in the tube is balanced by drag along the wall; temperature is constant everywhere due to intimate contact between gas and walls. At the same time, the pressure drop in the tube is great enough that air density varies appreciably.

(a) Show that the mass flow rate is given by d'Arcy's formula

$$\dot{m} = \frac{\pi}{256} \frac{D^4}{\mu RT} \frac{p^2 - p_0^2}{L}$$

(b) Show that the pressure p in the die of volume V and initial pressure p_i varies as

$$\frac{p}{p_0} = \frac{1 + a}{1 - a}$$

where N is the number of capillary tubes while

$$a = \frac{p_i - p_0}{p_i + p_0} \exp(-t/r) \quad \text{and} \quad \tau = \frac{128}{\pi} \frac{\mu L V}{N D^4 p_0}$$

(c) For the case in which the die volume is 3 in.^{3} and there are 20 evacuation tubes of 0.1 in. length and 0.005 in. diameter, justify the assumption that the air temperature is unchanging during the decompression. *Answer:* $\tau = 1.7$ sec; heat flow would reduce a temperature difference by a factor of about 300.

3-27 Consider a fluid flowing steadily through a porous medium of thickness L. The excess of the inlet temperature above the initially uniform temperature of the porous medium is T_i.

(a) Sketch this physical situation.

(b) Show that the energy equation [Eq. (2-84)] reduces to

$$\sigma \frac{\partial T}{\partial t} + u \frac{\partial T}{\partial x} = \frac{k_e}{\varrho C_p} \frac{\partial^2 T}{\partial x^2}$$

(c) The relative importance of convection relative to diffusion as a transport mechanism can be estimated from the Peclet number Pe defined as

$$\text{Pe} = \frac{\varrho C_p u \, \Delta T}{-k_e \, \partial T / \partial x}$$

Use the approximation that $\partial T / \partial x \sim -\Delta T / L$ where L is a length characteristic of the problem, probably the distance over which the diffusive gradient occurs, to obtain

$$\text{Pe} = \frac{u L}{\alpha_e}$$

Use the conditions of Problem 2-19 with L equal to 50 particle diameters to show that then Pe $\approx$ 100.

(d) Use the Laplace transform method to solve the energy equation without diffusion

$$\sigma \frac{\partial T}{\partial t} + u \frac{\partial T}{\partial x} = 0$$

subject to the initial condition $T(x, t = 0) = 0$ and the step boundary condition $T(x = 0, t) = T_i$. Sketch the behavior of the solution to demonstrate that when convection predominates a thermal disturbance propagates as a wave downstream through the porous medium at speed u/σ. *Answer*:

$$T/T_i = U(t - \sigma x/u) = \begin{cases} 0, & t < \sigma/u \\ 1, & t > \sigma/u \end{cases}.$$

(e) Use the Laplace transform method to solve the energy equation with diffusion

$$\sigma \frac{\partial T}{\partial t} + u \frac{\partial T}{\partial x} = \alpha_e \frac{\partial^2 T}{\partial x^2}$$

subject to the initial condition, representing an initial step change in temperature across the wave front in part d that can occur when convection predominates,

$$T(x, t = 0) = \begin{cases} 0, & x > 0 \\ T_i, & x < 0 \end{cases}$$

In this, make use of the coordinate transformation illustrated in Problem 2-15 with $u' = u$. Utilize the conditions that the solutions to the left and to the right of the wave front must have equal values and slopes at the wave front. From this result devise a criterion for the effects of diffusion to be negligible by letting the penetration depth δ of the disturbance from the wave front be less than one-tenth of L. Discussion of the related Burgers equation $\partial u/\partial t + u\, \partial u/\partial x = \mu \partial^2 u/\partial x^2$ used by J. M. Burgers to study turbulence is available [24, 25]. *Answer*:

$$\frac{2T}{T_i} = \mathrm{erfc}\left[\left(\frac{\sigma z^2}{4\alpha_e t}\right)^{1/2}\right], \qquad z > 0$$

$$\frac{2T}{T_i} = 2 - \mathrm{erfc}\left[\left(\frac{\sigma z^2}{4\alpha_e t}\right)^{1/2}\right], \qquad z < 0$$

$Pe \gtrsim 500/\sigma$.

REFERENCES

1. R. B. Bird, W. E. Steward, and E. N. Lightfoot, *Transport Phenomena*, Wiley, 1960, pp. 272, 306.
2. E. R. G. Eckert and M. Faghri, *Int. J. Heat Mass Transfer* **29**, 1177–1183 (1986).

3. E. R. G. Eckert, *Mech. Eng.* **106**, 58–65 (1984).

4. H. Schlichting, *Boundary-Layer Theory* (translated by J. Kestin), McGraw-Hill, 1968, pp. 277–278.

5. J. P. Meyer and M. D. Kostin, *Int. J. Heat Mass Transfer* **18**, 1293–1297 (1975).

6. J. H. Arnold, *Trans. AIChE* **40**, 361–378 (1944).

7. *International Critical Tables*, McGraw-Hill, 1933.

8. A. N. Nesmeyanov, *Vapor Pressure of the Elements*, Infosearch Ltd., 1963.

9. T. E. Jordan, *Vapor Pressure of Organic Compounds*, Interscience, 1954.

10. H. S. Carslaw and J. C. Jaeger, *Conduction of Heat in Solids*, 2nd ed., Clarendon, 1959, pp. 405–406.

11. C. Limpiyakorn and L. Burmeister, *ASME J. Heat Transfer* **94**, 415–418 (1972).

12. M. Rooney and L. Burmeister, *Int. J. Heat Mass Transfer* **18**, 671–675 (1975).

13. S. Kibbee, J. A. Orozco, and L. C. Burmeister, *Int. J. Heat Mass Transfer* **20**, 1069–1075 (1977).

14. L. C. Burmeister, *Int. J. Heat Mass Transfer* **21**, 1411–1420 (1978).

15. R. Hardin and L. Burmeister, *Fundamentals of Heat Transfer in Non-Newtonian Fluids* (*Proc. 28th National Heat Transfer Conference, Minneapolis, Minnesota, July 28–31, 1991*), ASME HTD-Vol. 174, pp. 29–39.

16. R. B. Bird, R. C. Armstrong, and O. Hassager, *Dynamics of Polymeric Liquids*, 2nd ed., Vol. 1, Wiley-Interscience, 1987, p. 247.

17. C. Y. Lee and C. R. Wilke, *Ind. Eng. Chem.* **46**, 2381–2387 (1954).

18. L. C. Burmeister, in R. A. Granger, Ed., *Experiments in Heat Transfer and Thermodynamics*, Cambridge, 1994.

19. S. K. Griffiths and F. A. Morrison, *ASME J. Heat Transfer* **101**, 484–488 (1979).

20. M. Morduchow and P. A. Libby, *J. Aeronaut. Sci.* **16**, 674–684 (1949).

21. R. von Mises, *J. Aeronaut. Sci.* **17**, 551–554 (1950).

22. S. Sherman, A Low-Density Wind-Tunnel Study of Shock Wave Structure and Relaxation Phenomena in Gases, NACA TN 3298, 1955; P. Thompson, *Compressible-Fluid Dynamics*, McGraw-Hill, 1972, pp. 361–368; O. Baysal, *AIAA J.* **24**, 800–806, Eq. (21)(1986).

23. J. Polman, *Int. J. Heat Mass Transfer* **24**, 184–187 (1981); M. Nikanjam and R. Greif, *ASME J. Heat Transfer* **100**, 527–535 (1978).

24. D. A. Anderson, J. C. Tannehill, and R. H. Pletcher, *Computational Fluid Mechanics and Heat Transfer*, Hemisphere, 1984, pp. 154–155.

25. E. R. Benton and G. W. Platzmann, *Quart. App. Mech.* **30**, 195–212 (1972).

4

LAMINAR HEAT
TRANSFER IN DUCTS

Convective transport of heat or mass can be first classified as either forced or natural. In forced convection fluid motion is caused by some mechanism not connected with the heat or mass transfer at the location of interest; usually, a pump or fan at a distant location "forces" the fluid into motion. In natural convection fluid motion is caused by a force acting on either density or surface tension differences induced by local heat or mass transfer; the fluid flows "naturally."

A second classification of convective heat or mass transfer is as either *internal* or *external*. In internal flow the fluid is constrained on all sides by solid boundaries, as in flow through a pipe. In external flow the fluid has at least one side extending to infinity without encountering a solid surface. An airplane flying through the air presents an external flow example insofar as the atmospheric air is concerned.

These two classification schemes are independent. An internal flow can be either forced or natural, and an external flow can be either forced or natural.

4.1 MIXING-CUP TEMPERATURE

As seen in Sections 3.1 and 3.2 when accounting for the effects of viscous dissipation, a careful choice of the driving temperature difference simplifies heat-transfer calculations. Although mass transfer does not admit an analog to viscous dissipation, the similarity of the describing equations enables the benefits of a careful choice of the driving potential difference to apply to mass transfer as well. The most useful driving potential difference depends on the physical situation under consideration.

107

In an external flow the temperature of the fluid far removed from the heat- or mass-transfer surface is usually known (and may be constant). However, in an internal flow there is usually no well-defined temperature except, perhaps, for a wall or an inlet temperature. In the case of heat transfer, there will be a temperature variation perpendicular to the wall, and there will probably be a temperature variation in the direction of fluid flow. Thus there are several choices available for the driving temperature difference.

To facilitate discussion, attention is focused on the common situation of flow down a duct as illustrated in Fig. 4-1. In the choice of a temperature for the fluid, the fact that a designer is most interested in the rate of flow of mass and energy down the duct should be recognized. The mass flow rate $\dot{m}$ down the duct is obtained from the velocity distribution as

$$\dot{m} = \int_A \rho v \, dA = \rho_{av} v_{av} A$$

from which it follows that the average velocity is given by

$$v_{av} = \frac{1}{A} \int_A \frac{\rho}{\rho_{av}} v \, dA \tag{4-1}$$

The energy flow rate $\dot{E}$ down the duct is obtained from the velocity and temperature distribution by

$$\dot{E} = \int_A \rho C_p T v \, dA = \rho_{av} C_{p_{av}} v_{av} A T_m$$

from which it is found that the mixing-cup temperature T_m is given by

$$T_m = \frac{1}{A} \int_A \frac{\rho}{\rho_{av}} \frac{C_p}{C_{p_{av}}} T \, dA \tag{4-2}$$

For a circular duct of radius R when velocity and temperature depend only

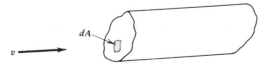

Figure 4-1   Flow down a duct.

on radius, $dA = 2\pi r\,dr$, Eqs. (4-1) and (4-2) become, with $r' = r/R$,

$$v_{av} = 2\int_0^1 \frac{\rho}{\rho_{av}} vr'\,dr'$$

and

$$T_m = 2\int_0^1 \frac{\rho}{\rho_{av}} \frac{C_p}{C_{av}} \frac{v}{v_{av}} Tr'\,dr'$$

The average, mixed temperature defined by Eq. (4.2) is called the *mixing-cup temperature*, the temperature that a cup of fluid scooped out of the stream would attain. The mixing-cup temperature can vary along the length of the duct. Because the mixing-cup temperature is likely to be known, it is anticipated that heat flux at a local wall position can be calculated most simply from

$$q_w = h(T_{wall} - T_m) \tag{4-3}$$

4.2 ENTRANCE LENGTH

Before attempting detailed solutions of particular cases, a review of salient features is in order. First, consider the expected behavior of the velocity profile in a straight round duct of constant cross section. The change in axial velocity profile at successive stations down the duct is shown in Fig. 4-2. As flow proceeds down the duct, the effect of the wall diffuses into the main stream. A boundary layer builds up and eventually occupies the entire tube, with v_{av} remaining constant. After this entrance length L_e the flow is fully developed and no longer changes. If the flow is laminar (Re < 2300) a parabolic velocity profile exists; if the flow is turbulent (Re > 2300), a flatter velocity profile results. To judge whether or not flow is fully developed the rules of thumb

$$\frac{L_e}{D} \geq 0.05\,\text{Re} \qquad \text{laminar flow}$$

$$\tag{4-4}$$

$$\frac{L_e}{D} \geq 40 - 100 \qquad \text{turbulent flow}$$

are commonly used.

A similar behavior is observed for thermal cases. A complication is that there may be an unheated starting section in which velocity profiles develop before a temperature profile does. The general trend of events is shown in Fig. 4-3. Note the thermal boundary-layer growth. Evidently a thermal entrance region is to be expected, and constant heat-transfer coefficients will

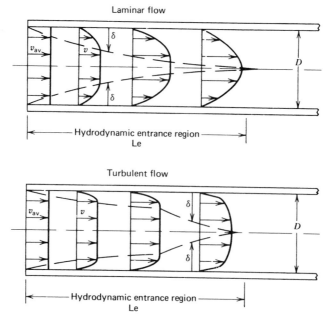

Figure 4-2 Laminar and turbulent velocity profiles in the entrance region of a circular duct.

be achieved only after some distance down the duct. When the temperature distribution develops much more rapidly than does the velocity distribution, a possibility when the unheated length is shorter than L_e and the thermal diffusivity α greatly exceeds the momentum diffusivity ν, the velocity may be assumed constant (slug flow). At the other extreme, if the unheated section is longer than L_e, the velocity profile is fully developed and velocity depends only on the radial position (assuming that properties are not temperature

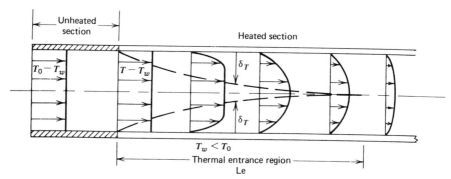

Figure 4-3 Temperature profiles in the entrance region of a circular duct.

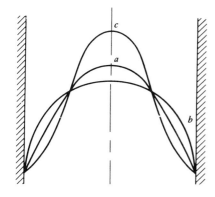

Figure 4-4 Temperature-dependent liquid viscosity effects on fully developed laminar velocity profiles: (*a*) constant viscosity; (*b*) heat flow into liquid; (*c*) heat flow out of liquid.

dependent). In either of these two extreme cases there is no appreciable radial component of velocity. For intermediate lengths of the unheated starting section, the velocity profile is still developing and there may be appreciable radial velocity. As shown in Fig. 4-4 [1] for laminar flow, a temperature-dependent viscosity can result in departure from the usual parabolic velocity profile (curve *a*). If heat flows into a liquid, the resultant velocity profile is flattened (curve *b*) because the viscosity near the wall is less than in the core of the tube. If heat flows out of the liquid, the velocity profile is peaked (curve *c*) because the viscosity in the core of the tube is less than near the wall. For a gas the effects are opposite from those for a liquid since gas viscosity increases with increasing temperature.

The mass diffusion case has similar behavior. An additional complication is a possible "blowing" at the wall.

In the inlet region of a duct where velocity and temperature profiles are developing simultaneously, the core of the duct near the inlet is filled with fluid that is as yet unaffected by the conditions at the wall. In this core region the fluid can be regarded as an external flow and the fluid very near the duct wall, as a boundary layer where the boundary-layer concepts of Chapter 5 can be applied to estimate wall friction, heat transfer, and so forth.

A survey of laminar heat transfer in ducts by Shah and London [2] can be consulted for greater detail and variety of method and result.

4.3   CIRCULAR DUCT

The manner in which heat-transfer coefficients vary along the length of a duct can be determined from knowledge of the temperature distribution in the fluid. To illustrate the procedure employed, consider flow of a Newtonian fluid down a circular duct of radius R.

Assumptions imposed are (1) steady laminar flow, (2) constant properties, (3) velocity distribution known and dependent only on radial position, (4) fluid temperature dependent only on axial and radial position in duct, and (5) inlet temperature known. As a consequence of assuming the axial velocity radial distribution to be known and unchanging, only the energy equation is of concern and has the form

$$V \frac{\partial T}{\partial z} = \alpha \left[\frac{1}{r} \frac{\partial (r \, \partial T / \partial r)}{\partial r} + \frac{\partial^2 T}{\partial z^2} \right] + \frac{\mu}{\rho C_p} \left(\frac{\partial V}{\partial r} \right)^2 \qquad (4\text{-}5)$$

with boundary conditions of

$$T(r,0) = T_0 \qquad\qquad (4\text{-}6a)$$

$$T(0, z) \text{ finite} \qquad\qquad (4\text{-}6b)$$

$$T(r, z_{\max}) = ? \qquad\qquad (4\text{-}6c)$$

$$T(R, z) = T_w, \qquad \text{specified wall temperature} \qquad (4\text{-}6d)$$

or

$$k \frac{\partial T(R, z)}{\partial r} = q_w, \qquad \text{specified heat flux into tube}$$

where V is the known axial velocity that can vary with radius.

Clues as to the importance of various terms can be obtained by examining the dimensionless forms of Eqs. (4-5) and (4-6), which are

$$v \frac{\partial \theta}{\partial z'} = \frac{1}{r'} \partial \frac{(r' \, \partial \theta / \partial r')}{\partial r'} + \frac{4}{\text{Pe}^2} \frac{\partial^2 \theta}{\partial z'^2} + \text{E Pr} \left(\frac{\partial v}{\partial r'} \right)^2 \qquad (4\text{-}7)$$

$$\theta(r', 0) = \frac{T_0 - T_r}{\Delta T} \qquad\qquad (4\text{-}8a)$$

$$\theta(0, z') \text{ finite} \qquad\qquad (4\text{-}8b)$$

$$\theta(r', z'_{\max}) = ? \qquad\qquad (4\text{-}8c)$$

$$\theta(1, z') = \frac{T_w - T_r}{\Delta T}, \qquad \text{specified wall temperature} \qquad (4\text{-}8d)$$

or

$$\frac{\partial \theta(1, z')}{\partial r'} = \frac{q_w R}{k \, \Delta T}, \qquad \text{specified heat flux into tube}$$

The dimensionless parameters are defined in conventional form as

z' = dimensionless axial distance = $4(z/D)/\text{Re Pr} = 4\alpha z/D^2 V_{av}$

Gz = Graetz number = $\text{Pr Re } D/z = D^2 V_{av}/\alpha z = 4/z'$

Pr = Prandtl number = $\nu/\alpha = \mu C_p/k$

Re = Reynolds number = $D V_{av}/\nu$

Pe = Peclet number = $D V_{av}/\alpha = \text{Re Pr}$

E = Eckert number = $V_{av}^2/C_p \Delta T$

$r' = r/R$

$\theta = (T - T_r)/\Delta T$

T_r = reference temperature

ΔT = reference temperature difference

$\upsilon = V/V_{av}$

Inspection of Eq. (4-7) reveals several possible simplifications. If Pe is large (Pe $\geq$ 100 [3]), axial conduction (the next-to-last term) is negligible. This can occur if Re Pr is large. If Pe is small, axial conduction is important, but in such a case the major source of axial conduction can be down the tube walls [4]. Basically, the Peclet number is a measure of the relative importance of axial convection to axial conduction. As a practical matter, Pe is small only if Pr is small as is the case for liquid metals. If E Pr is small, viscous dissipation (the last term) can be discarded. For the low velocities associated with the assumed laminar flow, E is small unless the reference temperature difference is even smaller than V_{av}^2/C_p [5, 6].

A solution to Eq. (4-7) can be used to evaluate a local heat-transfer coefficient. Presume that viscous dissipation is neglected so that $T_w - T_m$ drives the heat transfer. Then the local heat flux into the tube is given in terms of the local heat-transfer coefficient h by

$$q_w = h(T_w - T_m) = k \frac{\partial T(R, z)}{\partial r}$$

which gives, in terms of the previously defined dimensionless parameters,

$$\frac{hR}{k}(\theta_w - \theta_m) = \frac{\partial \theta(1, z')}{\partial r'}$$

From this it is found that the local Nusselt number Nu is

$$\text{Nu} = \frac{hD}{k} = \frac{2}{\theta_w - \theta_m} \frac{\partial \theta(1, z')}{\partial r'} \tag{4-9a}$$

where

$$\theta_w(z') = \theta(1, z')$$

and, from Eq. (4-2),

$$\theta_m(z') = 2\int_0^1 v(r')\theta(r', z')r' \, dr'$$

The most useful procedure for using Eq. (4-7) to evaluate an average heat-transfer coefficient in ducts of specified wall temperature is to be convenient for calculation of the mixing-cup temperature of the fluid at the outlet. This leads to an effective temperature difference between the fluid and the wall, which is referred to as the *log–mean-temperature difference* (LMTD). To see this, consider the energy balance for the cross-hatched control volume shown in Fig. 4-5, which is

$$V_{av}\rho C_p \pi R^2 \, dT_m = h2\pi R(T_w - T_m) \, dz$$
$$T_m(z = 0) = T_0$$

In dimensionless terms this equation is

$$d\theta_m = \text{Nu}(\theta_w - \theta_m)dz'$$
$$\theta_m(z' = 0) = \theta_{m0}$$

If $T_r = T_w$ and $\Delta T = T_0 - T_w$, $\theta_w = 0$, and one then has

$$-\frac{d\theta_m}{\theta_m} = \text{Nu} \, dz'$$
$$\theta_m(z' = 0) = 1$$

whose solution is

$$\theta_m = \frac{T_m - T_w}{T_0 - T_w} = \exp\left[-\int_0^{z'} \text{Nu} \, dz'\right] \tag{4-9b}$$

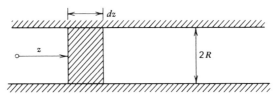

Figure 4-5   Control volume for derivation of average heat-transfer coefficient.

The difference between the fluid and wall temperatures decays exponentially. Naturally, Eq. (4-9b) would be more easily used to compute the outlet mixing-cup temperature T_m if the exponential were given in terms of an average Nusselt number $\overline{\mathrm{Nu}} = \bar{h}D/k$ defined in terms of the local Nusselt number as

$$\overline{\mathrm{Nu}} = \frac{1}{z'}\int_0^{z'} \mathrm{Nu}\ dz' \tag{4-9c}$$

It is then possible to employ Eqs. (4-9b) and (4-9c) to obtain

$$\overline{\mathrm{Nu}} = -\frac{\ln(\theta_m)}{z'} \tag{4-9d}$$

The total heat transferred is obtained as

$$\rho C_p V_{\mathrm{av}} \pi R^2 (T_{m,\,\mathrm{out}} - T_{m,\,\mathrm{in}}) = \bar{h}2\pi Rz\,\Delta T_{\mathrm{av}} \tag{4-9e}$$

In dimensionless terms, this relationship is

$$\Delta T(\theta_m - 1) = \overline{\mathrm{Nu}}\,z'\,\Delta T_{\mathrm{av}}$$

Introduction of Eq. (4-9d) into this result leads to

$$\Delta T_{\mathrm{av}} = \frac{(T_w - T_0) - (T_w - T_m)}{\ln[(T_w - T_0)/(T_w - T_m)]}$$

a result that is often written as

$$\mathrm{LMTD} = \frac{\Delta T_{\mathrm{max}} - \Delta T_{\mathrm{min}}}{\ln(\Delta T_{\mathrm{max}}/\Delta T_{\mathrm{min}})} \tag{4-9f}$$

The average heat flux $\bar{q}_w$ is directly obtained from Eq. (4-9e) as

$$\bar{q}_w = \bar{h}\,\Delta T_{\mathrm{av}} = \bar{h}\,\mathrm{LMTD}$$

with LMTD from Eq. (4-9f).

4.4   LAMINAR SLUG FLOW IN A CIRCULAR DUCT

When the unheated entrance section is short and the diffusivity of momentum is less than the diffusivity of heat ($\mathrm{Pr} = \nu/\alpha \ll 1$), temperature profiles develop more quickly than do velocity profiles. This case corresponds to flow entering a tube from a reservoir. It is then accurate to assume the axial velocity to be constant ($v = V/V_{av} = 1$). Naturally, the results of such an assumption cannot be extended to great distances down the duct since nonuniform velocity profiles will eventually evolve.

Constant Wall Temperature

Equation (4-7) is to be solved for a constant tube wall temperature T_w and a uniform inlet temperature T_0. The equation, neglecting viscous dissipation, and boundary conditions of interest are

$$\frac{\partial \theta}{\partial z'} = \frac{1}{r'} \frac{\partial (r' \partial \theta / \partial r')}{\partial r'} \tag{4-10}$$

$$\theta(r', 0) = 1 \tag{4-11a}$$

$$\theta(0, z') \text{ finite} \tag{4-11b}$$

$$\theta(1, z') = 0 \tag{4-11c}$$

with $T_r = T_w$ and $\Delta T = T_0 - T_w$. Equations (4-10) and (4-11) also describe the transient temperature in a cylinder of initially uniform temperature whose periphery is suddenly reduced to zero temperature. The z' parameter plays the role of elapsed time—it can be seen that $z' = \alpha t / R^2$, where $t = z / V_{av}$. In a separation of variables technique, a product solution is assumed of the form

$$\theta(r', z') = R_1(r') Z_1(z')$$

Insertion of this assumed form into Eq. (4-10) and division by $R_1 Z_1$ gives

$$\frac{1}{r' R_1} \frac{d(r' \, dR_1 / dr')}{dr'} = \frac{1}{Z_1} \frac{dZ_1}{dz'} = -\lambda^2$$

or

$$\text{Function of } r' = \text{function of } z'$$

Here the variables have been separated, inasmuch as the left-hand side is solely a function of r' whereas the right-hand side is solely a function of z'. These two functions must be equal for all values of r' and z', a requirement that can be met only if they both equal a constant $-\lambda^2$ (the constant must be negative to avoid predicting a temperature which increases exponentially down the duct). Two separate ordinary differential equations must now be solved:

$$\frac{dZ_1}{dz'} = -\lambda^2 Z_1, \qquad r'^2\frac{d^2R_1}{dr'^2} + r'\frac{dR_1}{dr'} + \lambda^2 r'^2 R_1 = 0$$

whose solutions are

$$R_1 = C_1 J_0(\lambda r') + C_2 Y_0(\lambda r')$$

$$Z_1 = C_3 e^{-\lambda^2 z'}$$

The zero-order Bessel function of the first kind J_0 and the zero-order Bessel function of the second kind Y_0 behave as illustrated in Fig. 4-6 [7]. Equation (4-11b) requires $C_2 = 0$, since $Y_0(0) = -\infty$. Hence

$$\theta = C \exp(-\lambda^2 z') J_0(\lambda r')$$

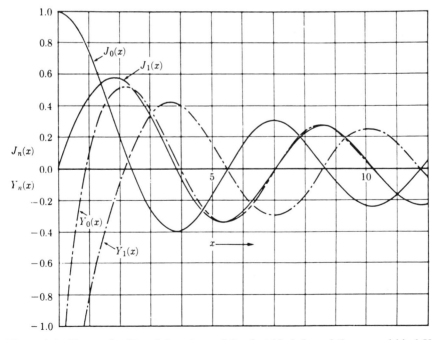

Figure 4-6   Zero-order Bessel functions of the first kind J_0 and the second kind Y_0 and first-order Bessel functions of the first kind J_1 and the second kind Y_1.

Equation (4-11c) requires that $J_0(\lambda) = 0$, from which it is found that

$$\lambda = 2.4048, 5.5201, 8.6537, 11.7915, 14.93309, \ldots \qquad (4\text{-}12)$$

Each of the λ values satisfies the problem conditions thus far, and each must be considered. Therefore

$$\theta = \sum_{n=1}^{\infty} C_n \exp(\lambda_n^2 z') J_0(\lambda_n r')$$

with $J_0(\lambda_n) = 0$. The boundary condition of Eq. (4-11a) demands

$$\sum_{n=1}^{\infty} C_n J_0(\lambda_n r') = 1$$

which is a requirement to express a constant in a Fourier series of Bessel functions over the range $0 \leqslant r' \leqslant 1$. It is found [8] that the individual constants C_n are

$$C_n = \frac{2}{\lambda_n J_1(\lambda_n)}$$

The temperature distribution is then finally found to be

$$\theta = 2 \sum_{n=1}^{\infty} \exp(-\lambda_n^2 z') \frac{J_0(\lambda_n r')}{\lambda_n J_1(\lambda_n)} \qquad (4\text{-}13)$$

The mixing-cup temperature is obtained by insertion of Eq. (4-13) into Eq. (4-2) as

$$\theta_m = 2 \int_0^1 \theta r' \, dr'$$

$$= 4 \int_0^1 \exp(-\lambda_n^2 z') \frac{r' J_0(\lambda_n r') \, dr'}{\lambda_n J_1(\lambda_n)} \qquad (4\text{-}14)$$

$$\theta_m = 4 \sum_{n=1}^{\infty} \frac{\exp(-\lambda_n^2 z')}{\lambda_n^2}$$

Next, realization that $dJ_0(\lambda_n r')/dr' = -\lambda_n J_1(\lambda_n r')$ gives

$$\frac{\partial \theta(1, z')}{\partial r'} = -2 \sum_{n=1}^{\infty} \exp(-\lambda_n^2 z') \qquad (4\text{-}15)$$

Introduction of Eqs. (4-14) and (4-15) into Eq. (4-9a) finally yields the local

Nusselt number as

$$\text{Nu} = \frac{hD}{k} = \frac{\sum_{n=1}^{\infty} \exp(-\lambda_n^2 z')}{\sum_{n=1}^{\infty} \exp(-\lambda_n^2 z')/\lambda_n^2} \tag{4-16}$$

Although the limit as $z' \to \infty$ has no physical importance for slug flow as discussed earlier, Eq. (4-16) shows that the local Nusselt number is then

$$\lim_{z' \to \infty} \text{Nu} = \lambda_1^2 = 5.7831 \tag{4-17}$$

The average coefficient based on the log–mean-temperature difference is found from substitution of Eqs. (4-14) and (4-15) into Eq. (4-9d). Realization that $\theta_w = 0$ and $\theta_{m_{\text{inlet}}} = 1$ for this case then gives

$$\overline{\text{Nu}} = \frac{\bar{h}D}{k} = -\frac{\ln(\theta_m)}{z'} \tag{4-18}$$

with θ_m given by Eq. (4.14). Taking the limit as $z' \to \infty$, one obtains

$$\lim_{z' \to \infty} \overline{\text{Nu}} = 5.7831$$

The manner in which Nu, $\overline{\text{Nu}}$, and θ_m asymptotically approach their limiting values far down the duct is shown in Table 4-1.

Problem 4-9 involves parallel developments in rectangular coordinates.

TABLE 4-1   Solutions for Slug Flow in a Circular Duct; Constant Wall Temperature

$1/\text{Gz}$	Nu	$\overline{\text{Nu}}$	θ_m
0	∞	∞	1
0.001	19.5	37.322	0.861
0.0025	13.1	24.3	0.784
0.005	9.88	17.7	0.701
0.01	7.74	13.2	0.59
0.025	6.18	9.31	0.394
0.05	5.82	7.62	0.218
0.1	5.79	6.71	0.0684
0.25	5.783	6.15	0.0021
0.5	5.783	5.97	6.5×10^{-6}
1.0	5.783	5.88	6.1×10^{-11}
∞		5.783	5.783	0

Constant Wall Heat Flux into Fluid

The case of a specified constant heat flux q_w into the fluid leads to slightly different results for the heat-transfer coefficients. The only changes in Eqs. (4-10) and (4-11) are redefinition of $T_r = T_0$ and $\Delta T = q_w R/k$ and restatement of the boundary condition at $r' = 1$. Then the equations to be solved for the temperature distribution in the fluid are

$$\frac{\partial \theta}{\partial z'} = \frac{1}{r'} \frac{\partial(r' \, \partial\theta/\partial r')}{\partial r'} \tag{4.19a}$$

$$\theta(r', 0) = 0 \tag{4.19b}$$

$$\theta(0, z') \text{ finite} \tag{4.19c}$$

$$\frac{\partial\theta(1, z')}{\partial r'} = 1 \tag{4.19d}$$

Equations (4.19) also describe the temperature in a cylinder of initially zero temperature to whose periphery a constant heat flux is suddenly applied. A separation of variables procedure is employed, assuming

$$\theta(r', z') = \underbrace{R_1(r')Z_1(z')}_{\substack{\text{decaying} \\ \text{initial} \\ \text{transient}}} + \underbrace{Z_2(z')}_{\substack{\text{axial temperature} \\ \text{rise due to} \\ \text{accumulated} \\ \text{wall flux}}} + \underbrace{R_2(r')}_{\substack{\text{radial temperature} \\ \text{variation to let} \\ \text{wall flux into} \\ \text{fluid}}}$$

Insertion of this assumed form into Eq. (4-19a) gives

$$R_1 \frac{dZ_1}{dz'} + \frac{dZ_2}{dz'} = Z_1 \frac{1}{r'} \frac{d(r' \, dR_1/dr')}{dr'} + \frac{1}{r'} \frac{d(r' \, dR_2/dr')}{dr'}$$

The next step is to set

$$\frac{dZ_2}{dz'} = \frac{1}{r'} \frac{d(r' \, dR_2/dr')}{dr'} \tag{4-20}$$

and then to separate variables to obtain, after dividing by $R_1 Z_1$,

$$\underbrace{\frac{1}{r'R_1} \frac{d(r' \, dR_1/dr')}{dr'}}_{\text{function of } r'} = \underbrace{\frac{1}{Z_1} \frac{dZ_1}{dz'}}_{\text{function of } z'} = -\lambda^2$$

As before, if a function of r' always equals a function of z' when the variables are independent, both functions equal a separation constant $-\lambda^2$.

Or

$$r'^2 \frac{d^2 R_1}{dr'^2} + r' \frac{dR_1}{dr'} + \lambda^2 r'^2 R_1 = 0 \quad \text{and} \quad \frac{dZ_1}{dz'} = -\lambda^2 Z_1$$

The solutions for R_1 and Z_1 are, as before,

$$R_1 = C_1 J_0(\lambda r') + C_2 Y_0(\lambda r')$$

$$Z_1 = C_3 e^{-\lambda^2 z'}$$

The boundary condition of Eq. (4-19c) requires that $C_2 = 0$. The same arguments for separation constant apply to Eq. (4-20), so that

$$\frac{dZ_2}{dz'} = C_4 = \frac{1}{r'} \frac{d(r'\, dR_2/dr')}{dr'}$$

from which it is found that the solutions for Z_2 and R_2 are

$$Z_2 = C_4 z' + C_5 \quad \text{and} \quad R_2 = \frac{C_4 r'^2}{4} + C_5 \ln(r') + C_6$$

Equation (4-19c) requires that $C_5 = 0$ and Eqs. (4.19b) and (4-19d), respectively, then require the solution for θ to satisfy

$$C_1 C_3 J_0(\lambda r') + C_5 + C_6 + \frac{C_4 r'^2}{4} = 0$$

and

$$-\lambda C_1 C_3 e^{-\lambda^2 z'} J_1(\lambda) + \frac{C_4}{2} = 1$$

Because there are no additional conditions to satisfy, arbitrarily set $C_5 + C_6 = 0$ and $C_4 = 2$. Then

$$C_1 C_3 J_0(\lambda r') = -\frac{r'^2}{2}$$

$$J_1(\lambda) = 0$$

The latter requirement gives an infinite number of roots of $J_1(\lambda) = 0$ as

$$\lambda = 0, 3.8317, 7.0156, 10.1735, 13.3237, 16.4706, \ldots \quad (4\text{-}21)$$

Each value of λ is admissible, and so the solution is a linear combination of all of the possibilities; the first of these two requirements is then

$$\sum_{n=0}^{\infty} C_n J_0(\lambda_n r') = -\frac{r'^2}{2}$$

which requires that $-r'^2/2$ be expressed in a Fourier series of Bessel functions over the interval $0 \leqslant r' \leqslant 1$. It is found [5] that

$$C_n = \frac{-2}{\lambda_n^2 J_0(\lambda_n)}, \qquad n \neq 0$$

and

$$C_0 = -\tfrac{1}{4}$$

The final result for dimensionless temperature is

$$\theta = 2z' + \frac{r'^2}{2} - \frac{1}{4} - 2 \sum_{n=1}^{\infty} \exp(-\lambda_n^2 z') \frac{J_0(\lambda_n r')}{\lambda_n^2 J_0(\lambda_n)} \qquad (4\text{-}22)$$

From Eq. (4-22) the mixing-cup temperature is, by insertion into Eq. (4-2),

$$\theta_m = 2z' \qquad (4\text{-}23)$$

This shows, as expected on physical grounds, that the average fluid temperature increases linearly down the duct since the heat absorbed is proportional to the distance traveled. This is like the temporal variation of a cylinder's average temperature with a constant peripheral heat flux—it must increase linearly with time. The wall temperature is evaluated from Eq. (4-22) as

$$\theta_w = \theta(1, z') = 2z' + \frac{1}{4} - 2 \sum_{n=1}^{\infty} \frac{\exp(-\lambda_n^2 z')}{\lambda_n^2} \qquad (4\text{-}24)$$

Use of Eqs. (4-23) and (4-24) in Eq. (4-9a) yields the local Nusselt number as

$$Nu = \frac{hD}{k} = \frac{8}{1 - 8\sum_{n=1}^{\infty} \exp(-\lambda_n^2 z')/\lambda_n^2} \qquad (4\text{-}25)$$

The difference between wall and mixing-cup temperature is

$$\theta_w - \theta_m = \frac{1}{4} - 2 \sum_{n=1}^{\infty} \frac{\exp(-\lambda_n^2 z')}{\lambda_n^2} \qquad (4\text{-}26)$$

Although the limit as $z' \to \infty$ has no physical importance, Eq. (4-25) shows that the local Nusselt number is then

$$\lim_{z' \to \infty} Nu = 8 \qquad (4\text{-}27)$$

Also, Eq. (4-26) gives the limiting value

$$\lim_{z' \to \infty} (\theta_w - \theta_m) = \frac{1}{4} \qquad (4\text{-}28)$$

The average fluid temperature at any position along the duct is known in a specified wall heat flux case. Thus the heat-transfer coefficient is of most utility in determining the wall temperature, an important quantity when fluid properties degrade at high temperature.

The average Nusselt number $\overline{Nu}$ is usually based on the average difference between the wall and mixing-cup temperatures

$$\overline{T}_w - \overline{T}_m = \frac{1}{z'} \int_0^{z'} (T_w - T_m) \, dz''$$

so that the average heat-transfer coefficient $\overline{h}$ is defined as

$$q_w = \overline{h}\left(\overline{T}_w - \overline{T}_m\right)$$

From this definition of $\overline{h}$, the average Nusselt number is

$$\overline{Nu} = \frac{\overline{h}D}{k} = \frac{2}{\overline{\theta}_w - \overline{\theta}_m}$$

$$\overline{Nu} = \frac{8}{1 - 8\sum_{n=1}^{\infty}\left\{\left[1 - \exp(-\lambda^2 z')\right]/(\lambda_n^4 z')\right\}} \qquad (4\text{-}28a)$$

The manner in which Nu, $\overline{Nu}$, and θ_m asymptotically approach their limiting values far down the duct is shown in Table 4-2.

TABLE 4-2 Solutions for Slug Flow in a Circular Duct; Constant Wall Heat Flux

1/Gz	Nu	$\overline{Nu}$	$\theta_w - \theta_m$
0	∞	∞	0
0.001	30.6	41.5	0.0655
0.0025	20.4	28.6	0.0981
0.005	15.3	21.5	0.131
0.01	11.9	16.3	0.168
0.025	9.16	11.9	0.218
0.05	8.24	9.95	0.243
0.1	8.01	8.92	0.2496
0.25	8	8.34	0.25
0.5	8	8.17	0.25
1.0	8	8.08	0.25
∞	8	8	0.25

The results for constant wall heat flux are, far from the entrance, the same as for linearly increasing wall temperature and constant-temperature difference between the wall and the fluid as in a counterflow heat exchanger in which the heat capacities of the two streams are equal.

4.5 FULLY DEVELOPED LAMINAR FLOW IN A CIRCULAR DUCT

When the unheated entrance region is long, velocity profiles develop fully before a fluid particle enters the heated region. This case corresponds to flow down a long pipe, only the latter portion of which is heated.

Fully developed velocity profiles for power-law fluids (a Newtonian fluid is a special case) are derived from a momentum balance on the control volume sketched in Fig. 4-7. Either by derivation from first principles or from Appendix C, the z-motion equation is

$$\frac{d(r\tau_{rz})}{dr} = r\frac{dp}{dz}$$

One integration gives

$$\tau_{rz} = \frac{r}{2}\frac{dp}{dz} + \frac{c_1}{r} \tag{4-29}$$

where a constant axial pressure gradient has been assumed. For finite shear stress at the center line, $c_1 = 0$; it is then seen that for any fluid, shear stress varies linearly in the tube. In terms of the shear stress τ_w at the wall,

$$\tau = \frac{\tau_w r}{R}$$

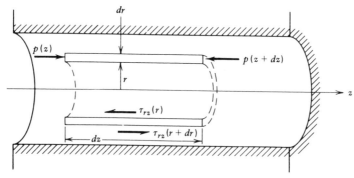

Figure 4-7 Control volume for deriving velocity profile in fully developed laminar flow in a circular duct.

For a power-law fluid, $\tau = \mu |dv/dr|^{n-1} dv/dr$. Expecting that velocity decreases with increasing radius so that $dv/dr < 0$, set

$$\tau = -\mu \left| -\frac{dv}{dr} \right|^{n-1} \left(-\frac{dv}{dr} \right) = -\mu \left(-\frac{dv}{dr} \right)^{n} \tag{4-30}$$

to avoid exponentiating a negative number. Introduction of Eq. (4-30) into Eq. (4-29) results in

$$-\mu \left(-\frac{dv}{dr} \right)^{n} = \frac{r}{2} \frac{dp}{dz}$$

from which it is found [incorporating the no-slip condition at the wall, $V(r = R) = 0$] that the velocity distribution is

$$v = \left(-\frac{R^{n+1}}{2\mu} \frac{dp}{dz} \right)^{1/n} \frac{n}{n+1} \left[1 - \left(\frac{r}{R} \right)^{(n+1)/n} \right]$$

$$= \frac{3n+1}{n+1} \left[1 - \left(\frac{r}{R} \right)^{(n+1)/n} \right] v_{\text{av}} \tag{4-30a}$$

with the average velocity v_{av} from Eq. (6-1) being

$$v_{\text{av}} = \left(-\frac{R^{n+1}}{2\mu} \frac{dp}{dz} \right)^{1/n} \frac{n}{3n+1} \tag{4-31}$$

With a fully developed velocity profile in hand, consideration of heat-transfer aspects can proceed. The energy equation should not be taken directly from Eq. (4-5), which has dissipation for a Newtonian fluid; instead, from either first principles or Appendix C, for a general fluid the energy equation is

$$v \frac{\partial T}{\partial z} = \frac{\alpha}{r} \frac{\partial (r \, \partial T/\partial r)}{\partial r} + \frac{\tau_{rz}}{\rho C_p} \frac{dv}{dr}$$

with the last term representing viscous dissipation. Introduction of the stress relation for a power-law fluid [Eq. (4-30)] gives the energy equation as

$$v \frac{\partial T}{\partial z} = \frac{\alpha}{r} \frac{\partial (r \, \partial T/\partial r)}{\partial r} + \frac{\mu}{\rho C_p} \left(-\frac{dv}{dr} \right)^{n+1}$$

With inclusion of viscous dissipation, the dimensionless energy equation is

$$\frac{3n+1}{n+1}[1 - r'^{(n+1)/n}]\frac{\partial\theta}{\partial z'}$$

$$= \frac{1}{r'}\frac{\partial(r\,\partial\theta/\partial r')}{\partial r'} + \Pr E\left(\frac{3n+1}{n}\right)^{n+1}r'^{(n+1)/n} \qquad (4\text{-}32)$$

where all quantities are as defined following Eq. (4-7) except for the Eckert number, which is $E = v_{av}^{n+1}R^{1-n}/C_p\,\Delta T$. The boundary conditions are as written in Eq. (4-8). The viscous dissipation term can be neglected if $E\,\Pr$ is small as it would be for the low velocities of laminar flow unless $\Delta T \to 0$.

Constant Wall Heat Flux into Fluid—Limit

The case of a constant wall heat flux q_w into the fluid with uniform inlet temperature T_0 is considered first.

If viscous dissipation is neglected, the describing equations are

$$\frac{3n+1}{n+1}[1 - r'^{(n+1)/n}]\frac{\partial\theta}{\partial z'} = \frac{1}{r'}\frac{\partial(r'\,\partial\theta/\partial r')}{\partial r'} \qquad (4\text{-}33a)$$

$$\theta(r',0) = 0 \qquad (4\text{-}33b)$$

$$\theta(0, z') \text{ finite} \qquad (4\text{-}33c)$$

$$\frac{\partial\theta(1, z')}{\partial r'} = 1 \qquad (4\text{-}33d)$$

where $T_r = T_0$ and $\Delta T = q_w R/k$.

A separation of variables solution is assumed of the form

$$\theta(r', z') = \underbrace{R_1(r')Z_1(z')}_{\substack{\text{decaying initial}\\\text{transient}}} + \underbrace{Z_2(z')}_{\substack{\text{axial}\\\text{temperature rise}\\\text{due to}\\\text{accumulated}\\\text{wall flux}}} + \underbrace{R_2(r')}_{\substack{\text{radial temperature}\\\text{variation to let wall}\\\text{flux into fluid}}}$$

Difficulties are imminent in the determination of the decaying initial transient, however, since the functions that are determined are not common, although they are tabulated, as is discussed later. Therefore, focus on the idea that the initial transients decay, leaving only $\theta = R_2 + Z_2$ and allowing only a determination of the asymptotic Nusselt number. Inclusion of $\theta =$

$R_2 + Z_2$ into the differential equation and separation of variables gives

$$\frac{dZ_2}{dz'} = \frac{n+1}{3n+1}\frac{1}{r'[1-r'^{(n+1)/n}]}\frac{d(r'\,dR_2/dr')}{dr'} = \lambda$$

in which λ is the separation constant. From the two resulting ordinary differential equations, it is found that

$$Z_2 = \lambda z' + C_1$$

$$R_2 = \frac{3n+1}{n+1}\lambda\left[\frac{r'^2}{4} - \left(\frac{n}{3n+1}\right)^2 r'^{(3n+1)/n}\right] + C_2\ln r' + C_3$$

Equation (4-33c) requires that $C_2 = 0$, and Eq. (4-33d) leads to $\lambda = 2$. These results give the local temperature far down the duct as

$$\theta = C + 2z' + 2\frac{3n+1}{n+1}\left[\frac{r'^2}{4} - \left(\frac{n}{3n+1}\right)^2 r'^{(3n+1)/n}\right] \tag{4-34}$$

To evaluate C, the mixing-cup temperature is examined. Insertion of Eq. (4-34) into Eq. (4-2) yields

$$\theta_m = C + 2z' + \frac{(3n+1)^3 - 8n^3}{4(n+1)(3n+1)(5n+1)}$$

Since Eq. (4-33b) requires $\theta_m(z'=0) = 0$,

$$C = -\frac{(3n+1)^3 - 8n^3}{4(n+1)(3n+1)(5n+1)}$$

Then the asymptotic mixing-cup temperature is

$$\theta_m = 2z' \tag{4-35}$$

Insertion of C into Eq. (4-34) results in an asymptotic difference between wall and mixing-cup temperatures of

$$\theta_w - \theta_m = \frac{1 + 12n + 31n^2}{4(3n+1)(5n+1)} \tag{4-36}$$

Use of Eq. (4-36) in Eq. (4-9a) gives the asymptotic Nusselt number Nu_∞ as

$$Nu_\infty = \frac{h_\infty D}{k} = \frac{8(3n+1)(5n+1)}{1 + 12n + 31n^2} \tag{4-37}$$

TABLE 4-3   Limiting Nusselt Numbers for a Circular Duct;
Constant Wall Heat Flux

	n	Nu_∞
Slug flow, fully developed	0	8
	1/10	6.22
	1/2	4.75
Newtonian fluid, fully developed	1	4.36
	2	4.14
	5	3.98
Linear profile, fully developed	∞	3.87

The manner in which the asymptotic Nusselt number is affected by the velocity distribution of Eq. (4-37) is shown in Table 4-3; Nusselt number varies by a factor of only 2 as n varies from 0 (slug flow) to ∞ (linear velocity distribution). The slug flow result shown in Table 4-3 agrees perfectly with the result of Eq. (4-27). For a Newtonian fluid ($n = 1$) in fully developed flow with a constant wall heat flux $Nu_\infty = \frac{48}{11} = 4.36$.

An alternative relationship is often used to determine the limiting Nusselt numbers and temperature distributions far downstream. A fully developed temperature profile exists when $(T_w - T)/(T_w - T_m)$ is solely dependent on radial position as expressed by

$$\frac{T_w - T}{T_w - T_m} = f(r)$$

This means that

$$\frac{\partial[(T_w - T)/(T_w - T_m)]}{\partial z} = 0 \tag{4-37a}$$

Expansion of the derivative in Eq. (4.37a) gives

$$\frac{\partial T_w}{\partial z} - \frac{\partial T}{\partial z} - \left(\frac{T_w - T}{T_w - T_m}\right)\left(\frac{\partial T_w}{\partial z} - \frac{\partial T_m}{\partial z}\right) = 0 \tag{4-37b}$$

When $q_w = h(T_w - T_m) = $ const, $T_w - T_m = $ const and Eq. (4-37b) yields

$$\frac{\partial T}{\partial z} = \frac{\partial T_w}{\partial z} = \frac{\partial T_m}{\partial z} = \text{const} \tag{4-37c}$$

Here the value of $\partial T_w/\partial z$ need not be specified to obtain the radial distribution of temperature and the resulting Nusselt number. For the case of

T_w = const, one has $\partial T_w/\partial z = 0$; thus Eq. (4-37b) yields

$$\frac{\partial T}{\partial z} = \frac{T_w - T}{T_w - T_m}\frac{\partial T_m}{\partial z} = f(r)\frac{\partial T_m}{\partial z} \tag{4-37d}$$

A trial-and-error method is utilized to obtain the radial distribution of temperature and the resulting Nusselt number. Usually two or three iterations of an initial assumed radial temperature distribution are sufficient.

Constant Wall Temperature—Limit

The case of constant wall temperature is considered next. Neglecting viscous dissipation, the describing equations are

$$\frac{3n + 1}{n + 1}(1 - r'^{(n+1)/n})\frac{\partial \theta}{\partial z'} = \frac{1}{r'}\frac{\partial(r'\,\partial\theta/\partial r')}{\partial r'} \tag{4-38a}$$

$$\theta(r',0) = 1 \tag{4-38b}$$

$$\theta(0, z') \text{ finite} \tag{4-38c}$$

$$\theta(1, z') = 0 \tag{4-38d}$$

where $T_r = T_w$ and $\Delta T = T_0 - T_w$.

A separation of variables solution of the form

$$\theta(r', z') = R(r')Z(z')$$

is substituted into Eq. (4-38a), giving

$$\frac{dZ}{dz'} = -\lambda^2 Z$$

and

$$\frac{d(r'\,dR/dr')}{dr'} + \frac{3n + 1}{n + 1}\lambda^2(r' - r'^{(2n+1)/n})R = 0 \tag{4-38e}$$

The equation for Z can be easily solved to find

$$Z = Ce^{-\lambda^2 z'}$$

but that for R offers substantial difficulties. There are an infinite number of λ and C values, and they are unknown at this point. Far down the duct, however, only the first term in the equation for Z persists, representing the fact that the fluid temperature has closely approached the wall temperature.

Then

$$\theta \approx C e^{-\lambda_0^2 z'} R(r')$$

Substitution of this approximation for θ into Eq. (4-38a) gives

$$\frac{d(r'\, dR/dr')}{dr'} = -\lambda_0^2 \frac{3n+1}{n+1}(r' - r'^{(2n+1)/n}) R \qquad (4\text{-}39\text{a})$$

$$R(0) \text{ finite} \qquad (4\text{-}39\text{b})$$

$$R(1) = 0 \qquad (4\text{-}39\text{c})$$

An iterative solution to Eq. (4-39) relies on the smoothing process of integration. An assumed temperature profile R_0 is set into the right-hand side of Eq. (4-39), and an improved temperature profile R_1 is found by integration and satisfaction of the boundary conditions. This profile is used in place of R_0 to generate a second improvement R_2. The Nusselt number for each temperature profile (requiring evaluation of the mixing-cup temperature and gradient at the wall) can be calculated; the iterative procedure is terminated when the resulting Nusselt number converges to a final value.

For $n = 1$ (a Newtonian fluid with parabolic velocity distribution), the assumption of a polynomial for R_{i-1} as

$$R_{i-1} = \sum_{m=0}^{\infty} C_m r'^m$$

when substituted into Eq. (4-39) leads to

$$R_i = -4 \sum_{m=0}^{\infty} C_m \frac{m+3}{(m+2)^2(m+4)^2} + \sum_{m=0}^{\infty} C_m \left[\frac{r'^{m+2}}{(m+2)^2} - \frac{r'^{m+4}}{(m+4)^2} \right]$$

After the mixing-cup temperature and gradient at the wall are determined, the Nusselt number that corresponds to the ith iteration is

$$\mathrm{Nu}_{\infty i} = \frac{\sum_{m=0}^{\infty} [C_m/(m+2)(m+4)]}{\sum_{m=0}^{\infty} C_m[(m+11)/(m+2)(m+4)(m+6)(m+8)]}$$

This iterative procedure converges rapidly. Use of the temperature profile for a constant wall flux from Eq. (4-34), $R_0 = \frac{3}{4} - r'^2 + r'^4/4$, gives rise to the Nusselt number sequence: $\mathrm{Nu}_{\infty 1} = 3.729$; $\mathrm{Nu}_{\infty 2} = 3.667$; $\mathrm{Nu}_{\infty 3} = 3.6585$. The limiting value is

$$\mathrm{Nu}_\infty = 3.658 \qquad (4\text{-}40)$$

TABLE 4-4   Limiting Nusselt Numbers for a Circular Duct; Constant Wall Temperature

	n	Nu_∞
Slug flow, fully developed	0	5.783
Newtonian fluid, fully developed	1	3.658
Linear profile, fully developed	∞	3.264

For $n = \infty$ (a fluid with linear velocity profile), use of the temperature profile for a constant wall flux from Eq. (4-34), $R_0 = \frac{5}{9} - r'^2 + 4r'^3/9$, gives rise to the sequence: $Nu_{\infty 1} = 3.324$; $Nu_{\infty 2} = 3.272$; $Nu_{\infty 3} = 3.265$; $Nu_{\infty 4} = 3.2641$. The limiting value is

$$Nu_\infty = 3.264 \tag{4-41}$$

The effect of the fully developed velocity profile on the asymptotic Nusselt number is shown in Table 4-4. Roughly a factor of 2 variation occurs in Nusselt number as n varies from 0 (slug flow) to ∞ (linear velocity distribution). Schenk and van Laar [9] and others as reported in the survey by Skelland [10] found similar results.

The two boundary conditions in Tables 4-3 and 4-4 represent extreme conditions; for a Newtonian fluid ($n = 1$), it is seen that Nu_∞ for constant wall temperature is only 16% lower than for constant wall heat flux. For slug flow ($n = 0$), the constant wall temperature Nu_∞ is 25% lower; for a linear velocity profile ($n = \infty$), it is only 2% lower.

Constant Wall Temperature—Entrance Region

Equations (4-38) have been solved for the entrance region of a constant wall temperature round duct by Whiteman and Drake [11], Lyche and Bird [12] for fully developed flow of a power-law fluid, and Blackwell [13] for a Bingham fluid in fully developed flow. The study of Sellars et al. [14] for a Newtonian fluid is summarized below for fully developed flow.

As discussed following Eq. (4-38), the temperature distribution in the entrance region can be obtained as

$$\theta(r', z') = \sum_{n=0}^{\infty} C_n R_n(r') \exp\left(-\lambda_n^2 z'\right) \tag{4-42}$$

Here the C_n are constants of integration, the λ_n are separation constants, and the R_n are solutions to Eq. (4-38e). From Eq. (4-42) the local wall heat flux into the fluid is

$$q_w = \frac{2k(T_w - T_0)}{R} \sum_{n=0}^{\infty} G_n \exp(-\lambda_n^2 z')$$

where $G_n = -0.5C_n \, dR_n(1)/dr'$. Also, the mixing-cup temperature is

$$\theta_m = 4 \sum_{n=0}^{\infty} G_n \left(\frac{\exp(-\lambda_n^2 z')}{\lambda_n^2} \right)$$

The local and average (based on LMTD) Nusselt numbers are then

$$\text{Nu} = \frac{hD}{k} = \frac{\sum_{n=0}^{\infty} G_n \exp(-\lambda_n^2 z')}{\sum_{n=0}^{\infty} G_n \exp(-\lambda_n^2 z')/\lambda_n^2}$$

and

$$\overline{\text{Nu}} = \frac{\overline{h}D}{k} = -\frac{1}{z'} \ln\left[4 \sum_{n=0}^{\infty} \frac{G_n \exp(-\lambda_n^2 z')}{\lambda_n^2} \right]$$

The first five values appropriate to a determination of heat-transfer coefficients are displayed in Table 4-5. From them it is found that the Nusselt numbers and mixing-cup temperature vary with distance down the duct as displayed in Table 4-6.

The local Nusselt number achieves its limiting value of 3.66 (a value predicted in the preceding section) at $1/\text{Gz} = z/D \, \text{Re} \, \text{Pr} \approx 0.05$. Although at this Graetz number value the local Nusselt number is within 2% of its limiting value, the average Nusselt number is about 25% greater. Nevertheless, it is reasonable to take the entrance region as being the interval

$$0 \leqslant \frac{L_e}{D \, \text{Re} \, \text{Pr}} \leqslant 0.05 \tag{4-43}$$

TABLE 4-5   Eigenvalues and Eigenfunctions for Entrance Region of a Circular Duct; Fully Developed Laminar Flow; Constant Wall Temperature [13]

n	λ_n^2	G_n
0	3.656	0.749
1	22.31	0.544
2	56.9	0.463
3	107.6	0.414
4	174.25	0.382

TABLE 4-6   Solutions for Entrance Region of a Circular Duct; Fully Developed Laminar Flow; Constant Wall Temperature [13]

$1/\mathrm{Gz}$	Nu	$\overline{\mathrm{Nu}}$	θ_m
0	∞	∞	1
0.0005	10.1	15.4	0.940
0.002	8.06	12.2	0.907
0.005	6.00	8.94	0.836
0.02	4.17	5.82	0.628
0.04	3.79	4.89	0.457
0.05	3.71	4.64	0.395
0.1	3.658	4.16	0.190
∞	3.657	3.657	0

Nusselt numbers in the entrance region substantially exceed the limiting value.

These results for a constant wall temperature can be employed to ascertain the consequences of an axially varying wall temperature. The linearity of the describing equations is employed in a superposition scheme. First consider a situation in which $T_w = T_0$ until a position z^* is attained, after which point $T_w - T_0 = \Delta(T_w - T_0)$. The temperature distribution is then found from Eq. (4-42) as

$$T(r', z) - T_0 = \Delta(T_w - T_0) - \Delta(T_w - T_0)$$
$$\times \left\{ \sum_{n=0}^{\infty} C_n R_n(r') \exp\left[\frac{-4\lambda_n^2(z - z^*)}{D\,\mathrm{Re}\,\mathrm{Pr}} \right] \right\} \quad (4\text{-}43a)$$

for $z \geqslant z^*$. As shown in Fig. 4-8, an arbitrary wall temperature variation can be considered to be a sequence of such differential steps. Summation of their

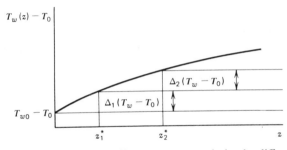

Figure 4-8   Representation of wall temperature variation by differential steps.

effects gives

$$T(r', z) - T_0 = (T_{w0} - T_0)[1 - f(z)] + \Delta_1(T_w - T_0)[1 - f(z - z_1^*)]$$
$$+ \Delta_2(T_w - T_0)[1 - f(z - z_2^*)] + \cdots$$

where f is the term in braces in Eq. (4-43a). In the limit as the incremental increase in wall temperature approaches zero, this sum can be represented by

$$T(r', z) - T_0 = (T_{w0} - T_0)[1 - f(z)]$$
$$+ \int_{T_{w0} - T_0}^{T_w - T_0}[1 - f(z - z^*)] \, d(T_w - T_0) \quad (4\text{-}43b)$$

Integration by parts with $u = [1 - f(z - z^*)]$ and $dv = dT_w$ gives

$$T(r', z) - T_0 = (T_{w0} - T_0)[1 - f(z)] + (T_w - T_0)[1 - f(z - z^*)]\big|_{T_{w0} - T_0}^{T_w - T_0}$$
$$- \int_0^z \frac{d[1 - f(z - z^*)]}{dz^*}(T_w - T_0) \, dz^*$$

or

$$T(r', z) - T_0 = \int_0^z \frac{df(z - z^*)}{dz^*}(T_w - T_0) \, dz^*$$

From Eq. (4-42) it is then finally found that

$$T(r', z) - T_0 = \int_0^z \left\{ \sum_{n=0}^{\infty} C_n R_n(r') \frac{4\lambda_n^2}{D \, \text{Re} \, \text{Pr}} \exp\left[\frac{-4\lambda_n^2(z - z^*)}{D \, \text{Re} \, \text{Pr}} \right] \right\}$$
$$\times (T_w - T_0) \, dz^* \quad (4\text{-}43c)$$

Equation (4-43c) is one form of Duhamel's theorem and may be expressed in other forms by various manipulations. For example, the form of Eq. (4-43b) can be used directly. In any case, system response to a step change in forcing function is utilized.

The local wall heat flux and Nusselt number can be determined from Eq. (4-43c). The local wall heat flux is obtained from

$$q_w(z) = \frac{k}{R} \frac{\partial T(1, z)}{\partial r'}$$

Taking the indicated derivative of Eq. (4-42c), one obtains

$$\frac{q_w(z)R}{k} = 2 \sum_{n=0}^{\infty} G_n \left\{ T_w - T_0 - \frac{4\lambda_n^2}{D \, \text{Re} \, \text{Pr}} \right.$$

$$\left. \times \int_0^z \exp\left[-\frac{4\lambda_n^2(z - z^*)}{D \, \text{Re} \, \text{Pr}} \right] (T_w - T_0) \, dz^* \right\} \quad (4\text{-}43\text{d})$$

where $G_n = -0.5 C_n \, dR_n(1)/dr'$; the G_n and the λ_n are given in Table 4-5. The local mixing-cup temperature is evaluated from

$$2\pi R \int_0^z q_w(z) \, dz = \pi R^2 v_{\text{av}} \rho C_p \left[T_m(z) - T_0 \right]$$

which recognizes that all the heat entering the fluid must be convected downstream. Rearrangement of this relation yields

$$T_m(z) - T_0 = 8 \int_0^z \frac{q_w R}{k} \frac{dz}{D \, \text{Re} \, \text{Pr}} \quad (4\text{-}43\text{e})$$

The local Nusselt number can now be obtained since

$$q_w = h(T_w - T_m)$$

giving

$$\text{Nu} = \frac{hD}{k} = \frac{2(q_w R/k)}{(T_w - T_0) - (T_m - T_0)} \quad (4\text{-}43\text{f})$$

In Eq. (4-43f) the denominator can approach zero from above or below to give a local Nusselt number that approaches $\pm\infty$. Such a case can arise when T_w starts out above T_0 and decreases axially to a value below T_0. This strange Nusselt behavior means only that the local wall heat flux is based on an inconvenient driving temperature difference $T_w - T_m$, which is not directly proportional to the temperature gradient at the wall, the real motivator of the heat flux.

Constant Wall Heat Flux into Fluid—Entrance Region

Equations (4-33) have been solved for the entrance region of a round duct for a Newtonian fluid ($n = 1$) with constant wall heat flux into the fluid by Siegel et al. [15]. Their results are summarized in the paragraphs that follow.

The dimensionless temperature distribution is assumed to be of the form

$$\theta = \theta^+(r', z') + 2z' + r'^2 - \frac{r'^4}{4} - \frac{7}{24}$$

where $\theta^+ = R_1(r')Z(z')$ as discussed following Eqs. (4-33). Introduction of this form into Eqs. (4-33) and separation of variables gives

$$\theta^+(r', z') = \sum_{n=1}^{\infty} C_n R_{1n} \exp\left(-\lambda_n^2 z'/2\right)$$

Here the λ_n^2 are the separation constants and the R_{1n} are the solutions to

$$\frac{d(r' \, dR_{1n}/dr')}{dr'} + \lambda_n^2 r'(1 - r'^2) R_{1n} = 0$$

with the boundary conditions $dR_{1n}/dr = 0$ at $r' = 0$ and $r' = 1$. After the λ_n and R_{1n} were determined by numerical methods, the C_n were obtained from

$$\theta^+(r', 0) = \sum_{n=1}^{\infty} C_n R_{1n} = -\left(r'^2 - \frac{r'^4}{4} - \frac{7}{24}\right)$$

to be given by

$$C_n = -\frac{\int_0^1 r'^3 (1 - r'^2)(1 - r'^2/4) R_{1n} \, dr'}{\int_0^1 r'(1 - r'^2) R_{1n}^2 \, dr'}$$

The results for the first seven values of n are shown in Table 4-7. From these

TABLE 4-7   Eigenvalues and Eigenfunctions for Entrance Region of a Circular Duct; Fully Developed Laminar Flow; Constant Wall Heat Flux [15]

n	λ_n^2	$R_{1n}(1)$	C_n
1	25.6796	-0.492517	0.403483
2	83.8618	0.395508	-0.175111
3	174.167	-0.345872	0.105594
4	296.536	0.314047	-0.0732804
5	450.947	-0.291252	0.0550357
6	637.387	0.273808	-0.043483
7	855.850	-0.259852	0.035597

**TABLE 4-8   Solutions for Entrance Region of a Circular
Duct; Constant Wall Heat Flux [15]**

$1/Gz$	Nu
0	∞
0.0013	11.5
0.0025	9.0
0.005	7.5
0.01	6.1
0.025	5.0
0.05	4.5
0.1	4.364
∞	4.364

results the local Nusselt number is

$$\text{Nu} = \frac{48/11}{1 + (24/11)\sum_{n=1}^{\infty} C_n \exp\left(-\lambda_n^2 z'/2\right)R_{1n}(1)} \tag{4-44}$$

which varies as shown in Table 4-8. The entrance region, based on the closeness of the local Nusselt number to its limiting value, is taken to be

$$0 < \frac{L_e}{D\,\text{Re Pr}} \leqslant 0.05$$

just as for the constant wall temperature case.

The longitudinal variation of wall temperature which accompanies these results is

$$\theta_w = \frac{T_w - T_0}{q_w R/k} = 2z' + \frac{11}{24} + \sum_{n=1}^{\infty} C_n \exp\left(-\lambda_n^2 z'/2\right)R_{1n}(1) \tag{4-45}$$

This result can be employed to ascertain the wall temperature variation that accompanies axially varying wall heat flux into the fluid. Such a circumstance occurs in a nuclear reactor and in electronic equipment cooling [16], for example. The linearity of the describing equations is employed in a superposition technique. First consider a process in which q_w is zero up to a position z^* after which point $q_w = \Delta q$. The wall temperature is found from Eq. (4-45) to be

$$\frac{T_w - T_0}{R/k} = \left\{\frac{8(z - z^*)}{D\,\text{Re Pr}} + \frac{11}{24} + \sum_{n=1}^{\infty} C_n R_{1n}(1) \exp\left[-\frac{2\lambda_n^2(z - z^*)}{D\,\text{Re Pr}}\right]\right\}\Delta q \tag{4-46}$$

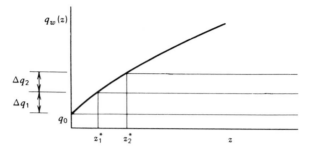

Figure 4-9   Representation of wall heat-flux variation by differential steps.

for $z \geqslant z^*$. As shown in Fig. 4-9, an arbitrary wall heat-flux variation can be considered as a sequence of such differential steps. Summation of their effects gives

$$\frac{T_w(z) - T_0}{R/k} = q_0 f(z) + \Delta q_1 f(z - z_1^*) + \Delta q_2 f(z - z_2^*) + \cdots$$

where f is the term in braces of Eq. (4-46). In the limit as the incremental increase in wall heat flux approaches zero, the sum can be replaced by

$$\frac{T_w(z) - T_0}{R/k} = q_0 f(z) + \int_{q_0}^{q} f(z - z^*) \, dq$$

Integration by parts with $u = f(z - z^*)$ and $dv = dq$ gives

$$\frac{T_w(z) - T_0}{R/k} = q_0 f(z) + q f(z - z^*) \Big|_{q_0}^{q} - \int_{z'=0}^{z} q \frac{df(z - z^*)}{dz^*} \, dz^*$$

or

$$[T_w(z) - T_0] k = \int_{z^*=0}^{z} \left\{ \frac{4}{\mathrm{Re\,Pr}} - \sum_{n=1}^{\infty} C_n R_{1n}(1) \frac{\lambda_n^2}{\mathrm{Re\,Pr}} \right.$$
$$\left. \times \exp\left[\frac{-2\lambda_n^2(z - z^*)}{D\,\mathrm{Re\,Pr}} \right] \right\} q_w(z^*) \, dz^* \qquad (4\text{-}47)$$

Since all needed numerical values are in Table 4-7, the wall temperature variation accompanying a specified wall heat flux can be ascertained. The heat-transfer coefficient is not constant in this situation [17].

4.6   FULLY DEVELOPED FLOW IN NONCIRCULAR DUCTS

Although the most common cross-sectional geometry is the circle, other cross sections are encountered often enough to generate interest in heat-transfer coefficients for them.

It is assumed that the duct flows full. In other words, unlike the case of a storm sewer that may be only partially filled with flowing water that wets only a portion of the sewer perimeter, the only cases considered are those in which the duct perimeter is completely wetted. The characteristic length dimension is the hydraulic diameter D_h defined as

$$D_h = 4\frac{\text{flow area}}{\text{wetted perimeter}}$$

which for a duct flowing full is

$$D_h = 4\frac{\text{duct area}}{\text{duct perimeter}} \tag{4-48}$$

The hydraulic diameter is used in place of a circular duct diameter in the definitions of the Nusselt. Graetz, and Reynolds numbers.

Nusselt Number—Limit

The limiting Nusselt number results for specified wall heat flux and specified wall temperature are displayed in Table 4-9 as reported by Shah and London [18] and are not greatly different from the circular duct results, varying by a factor of about 2. These results demonstrate that the adjustments provided by use of the hydraulic diameter as the characteristic length are insufficient to allow circular duct results to be applied to noncircular ducts unless a substantial error can be tolerated. In Table 4-9 the characteristic length is the hydraulic diameter, f is the shear-stress coefficient ($\tau_w = \rho f v_{av}^2 / 2 g_c$), H_1 refers to a constant axial wall heat flux with constant peripheral wall temperature at a given cross section, H_2 refers to constant axial wall heat flux with uniform peripheral wall heat flux at a given cross section, and T refers to constant wall temperature axially and peripherally. Axial conduction in both the fluid and the duct wall has been studied (see reference 18 for citations).

Results for the concentric-circular annulus geometry discussed by Kays and by Lundberg et al. [19] are of interest not only in their own right, but also because the circular tube and the gap between parallel-plane geometries are limiting cases.

For the circular sector duct, results obtained by Trupp and Lau [20] are shown in Fig. 4-10. The Nusselt number Nu_T for isothermal wall conditions and the Nusselt number Nu_{H1} for uniform heat input axially and uniform wall temperature at any cross section are shown. The friction factor f ($-dp/dz = f \rho V_{av}^2 / 2 g_c D_h$) is given in terms of the Reynolds number $\text{Re} = V_{av} D_h / \nu$ as, ϕ is in radians,

$$f \, \text{Re} / f^* \, \text{Re}^* = \phi^3 \Big/ \big[(\phi + \beta)(1 + \phi)^2 \big]$$

TABLE 4-9 Solutions for Heat Transfer and Friction for Fully Developed Laminar Flow Through Specified Ducts [18]

GEOMETRY ($L/D_h > 100$)	Nu_{H1}	Nu_{H2}	Nu_T	fRe
Triangle, $2b$ high, $2a$ base, $\frac{2b}{2a} = \frac{\sqrt{3}}{2}$	3.014	1.474	2.39*	12.630
Triangle $60°$, $2a$ high, $2b$ base, $\frac{2b}{2a} = \frac{\sqrt{3}}{2}$	3.111	1.892	2.47	13.333
Square, $2b$ by $2a$, $\frac{2b}{2a} = 1$	3.608	3.091	2.976	14.227
Hexagon	4.002	3.862	3.34*	15.054
Rectangle $2b$ by $2a$, $\frac{2b}{2a} = \frac{1}{2}$	4.123	3.017	3.391	15.548
Circle	4.364	4.364	3.657	16.000
Ellipse $2b$ by $2a$, $\frac{2b}{2a} = .9$	5.099	4.35*	3.66	18.700
Rectangle $2b$ by $2a$, $\frac{2b}{2a} = \frac{1}{4}$	5.331	2.930	4.439	18.233
Rectangle $2b$ by $2a$, $\frac{2b}{2a} = \frac{1}{8}$	6.490	2.904	5.597	20.585
Parallel plates, $\frac{2b}{2a} = 0$	8.235	8.235	7.541	24.000
One plate insulated, $\frac{b}{a} = 0$	5.385	-	4.861	24.000

*Interpolated values.

and the Nusselt number $Nu_T = \bar{h}D_h/k$ is given as

$$Nu_T/Nu_T^* = \phi(\phi + \beta)/(1 + \phi)^2$$

where asterisked quantities are for a circular duct as

$$Nu_T^* = 2h^* r_0/k, \qquad f^* = \left[\pi^2 \rho r_0^5/m^2\right][-dp/dx], \qquad Re^* = 2m/\pi r_0 \mu$$

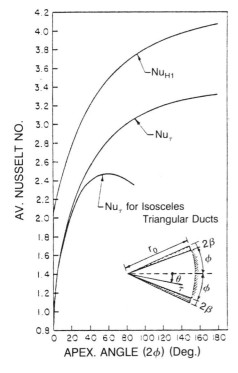

Figure 4-10 Circular sector ducts—fully developed laminar heat transfer. [From A. C. Trupp and A. C. Y. Lau, *ASME J. Heat Transfer* **106**, 467–469 (1984) [20].]

Nusselt Number—Entrance Region

The Nusselt number results for noncircular cross sections are needed for instances in which the duct is sufficiently short that temperature profiles are not fully developed.

For the rectangular duct with constant wall temperature, eigenvalues and associated constants obtained by Dennis et al. [21] needed for a Nusselt number calculation are given in Table 4-10. The Nusselt number $\overline{Nu}$ based on LMTD is achievable from these results by substitution into

$$\overline{Nu} = -\frac{1}{z'} \ln\left[8 \sum_{n=0}^{\infty} \frac{G_n \exp(-\lambda_n^2 z'/2)}{\lambda_n^2}\right]$$

When only the first eigenvalue λ_0 is known, $\overline{Nu}$ can only be evaluated from this equation for $1/Gz \geq 0.05$ (the thermal entrance region lies in the range $0 \leq 1/Gz \leq 0.05$). Information can still be extracted from the preceding equation since only the first term in the summation need be retained when $1/Gz$ becomes large. Then

$$\overline{Nu} \approx \frac{\lambda_0^2}{2} + \frac{\ln(\lambda_0^2/8G_0)}{z'}, \qquad \frac{1}{Gz} \geq 0.05 \qquad (4\text{-}48a)$$

which is a general limiting form for all geometries.

TABLE 4-10   Eigenfunctions for Entrance Region of a Rectangular Duct; Fully Developed Laminar Flow; Constant Wall Temperature [21]

Eigenvalue	Short Side/Long Side					
	1	$\frac{2}{3}$	$\frac{1}{2}$	$\frac{1}{4}$	$\frac{1}{8}$	0
λ_0^2	5.96	6.25	6.78	8.88	11.19	15.09
λ_1^2	35.54					171.3
λ_2^2	78.9					498
G_0	0.598	0.627	0.669	0.839	1.030	1.717
G_1	0.462					1.139
G_2	0.138					0.952
Nu_∞	2.98	3.125	3.39	4.44	5.595	7.545

TABLE 4-11   Limiting Nusselt Numbers for Laminar Flow in a Concentric-Circular Annulus; Constant Wall Heat Flux [19]

$r^* = (r_1/r_0)$	$2/Gz$	Nu_{ii}	Nu_{oo}
0.05	0.001	33.2	13.4
	0.005	24.2	7.99
	0.01	21.5	6.58
	0.05	18.1	4.92
	0.10	17.8	4.80
	∞	17.8	4.79
0.10	0.001	25.1	13.5
	0.005	17.1	8.08
	0.01	14.9	6.65
	0.05	12.1	4.96
	0.10	11.9	4.84
	∞	11.9	4.83
0.25	0.001	18.9	13.8
	0.005	12.1	8.28
	0.01	10.2	6.80
	0.05	7.94	5.04
	0.10	7.76	4.91
	∞	7.75	4.90
0.50	0.001	16.4	14.2
	0.005	10.1	8.55
	0.01	8.43	7.03
	0.05	6.35	5.19
	0.10	6.19	5.05
	∞	6.18	5.04
1.00 (parallel planes)	0.0005	23.5	23.5
	0.005	11.2	11.2
	0.01	7.49	7.49
	0.05	5.55	5.55
	0.125	5.39	5.39
	∞	5.38	5.38

For the concentric-circular annulus with constant wall heat flux, the local Nusselt numbers are presented in Table 4-11 (see reference 19 for a full discussion). In Table 4-11 Nu_{ii} and Nu_{oo} refer to heat applied only at the inner and outer surface, respectively; these results can be used with influence coefficients [19] for any combination of inner and outer heat fluxes to find the resultant Nusselt number.

Fully developed laminar flow between parallel planes with specified heat flux has been studied by Cess and Shaffer [22]. The limiting value is $Nu = 140/17 = 8.235$, which agrees exactly with the limiting value in Table 4-9.

4.7 SIMULTANEOUS VELOCITY AND TEMPERATURE DEVELOPMENT

In many applications the velocity and temperature distributions develop simultaneously in the heated section. This situation can occur when the fluid enters a heated duct from a reservoir. Although the start of the heating section does not always coincide with the duct entrance, it is the most common situation.

The criterion for the thermal entrance region for laminar flow is

$$ 0 \leqslant \frac{L_e}{D_h \, Re \, Pr} \lesssim 0.05 $$

The criterion for the velocity entrance region in laminar flow is, similarly,

$$ 0 \leqslant \frac{L_e}{D_h \, Re} \leqslant 0.05 $$

Comparison of these two criteria reveals that when $Pr \gg 1$, as is the case for oils, the temperature profile takes the longer distance to develop—a fully developed velocity profile may be appropriate. When $Pr \approx 1$, as is the case for gases, the temperature and velocity profiles develop at about the same rate. When $Pr \ll 1$, as is the case for liquid metals, the temperature profile takes the shorter distance to develop—a slug flow velocity profile may be appropriate.

Insight into the nature of the results that can be expected can be obtained by a comparison of Nusselt number results for a circular duct with fully developed and slug flow. For a Newtonian fluid and a constant wall temperature, $\overline{Nu}(Gz^{-1} = 0.005) = 8.94$ is shown in Table 4-6 for a fully developed, parabolic velocity profile, and $\overline{Nu}(Gz^{-1} = 0.005) = 17.7$ is shown in Table 4-1 for slug flow. It is evident that the Nusselt number is greater with a uniform velocity profile. Then the uniform velocity at the duct inlet during

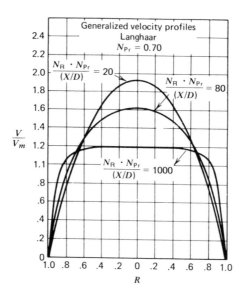

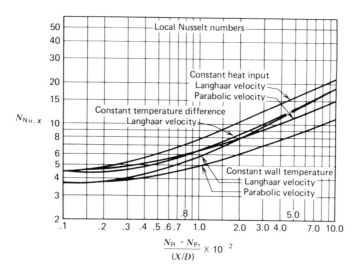

Figure 4-11 Laminar velocity and temperature profiles and local and average Nusselt numbers in the entrance region of a circular duct for constant wall temperature, constant wall heat flux, and constant fluid–wall temperature difference. [From W. M. Kays, *Trans. ASME* **77**, 1265–1274 (1955) [23].]

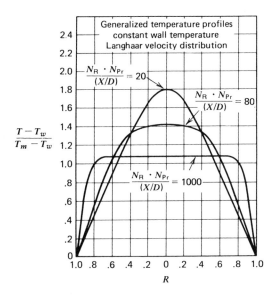

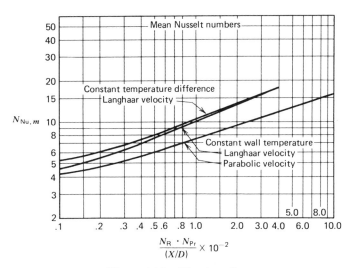

Figure 4-11 *(Continued)*

simultaneous development of velocity and temperature profiles should result in higher Nusselt numbers than in the fully developed velocity case.

This expected trend was quantitatively determined by Kays [23] for constant wall temperature, constant wall heat flux, and constant fluid–wall temperature difference. His results are summarized in Fig. 4-11 for Pr = 0.7 (the calculations were extended to other Prandtl numbers by Goldberg [24]).

TABLE 4-12   Constants K_1 and K_2 for Eq. (4-49)

Wall Condition	Inlet Velocity	Pr	$\overline{Nu}_\infty$	K_1	K_2	n
$T_w = $ const	Developing	0.7	3.66	0.104	0.016	0.8
$T_w = $ const	Parabolic	Any	3.66	0.0668	0.04	$\frac{2}{3}$
$T_w - T_m = $ const	Developing	0.7	4.36	0.1	0.016	0.8
$q_w = $ const	Developing	0.7	4.36	0.036	0.0011	1
$q_w = $ const	Parabolic	Any	4.36	0.023	0.0012	1

These results for the average Nusselt number are encapsulated in the correlating equation due to Hausen [25]

$$\overline{Nu} = \overline{Nu}_\infty + \frac{K_1(\text{Re Pr } D/z)}{1 + K_2(\text{Re Pr } D/z)^n} \tag{4-49}$$

for $T_w = $ const. For $q_w = $ const, the left-hand side must be replaced by the local Nusselt number Nu. The constants to use in Eq. (4-49) are shown in Table 4-12. In developing these results, the velocity profiles in the entrance region found by Langhaar [26]—who solved a linearized form of the radial-motion equation—were used.

The constant wall heat-flux condition was studied by Heaton et al. [27] for simultaneously developing velocity and temperature profiles for the concentric-circular annulus family. Churchill and Ozoe [28] correlated numerial solutions as

$$\frac{Nu + 1.7}{5.337\left[1 + (\pi Gz/388)^{8/9}\right]^{3/8}}$$

$$= \left\{1 + \left[\frac{\pi Gz/284}{\{1 + (Pr/0.0468)^{2/3}\}^{1/2}\{1 + (\pi Gz/388)^{8/9}\}^{3/4}}\right]^{8/6}\right\}^{3/8}$$

4.8   TEMPERATURE-DEPENDENT PROPERTIES

Because substantial temperature variations are often experienced in both the flow and transverse-flow directions and because the accompanying property variations are often substantial, the heat-transfer coefficient often differs appreciably from its constant-property value. This can occur not only because thermal conductivity is temperature dependent, but also because a temperature-dependent viscosity can affect the velocity profile, even in fully developed flow. Also, a temperature-dependent density can give rise to a larger and longer-lasting radial-velocity component that lengthens the entrance region and to natural convection.

For a gas, specific heat and Prandtl number vary only slightly with temperature, but viscosity and thermal conductivity vary roughly with the 0.8 power of absolute temperature, whereas density varies inversely with absolute temperature. For a liquid, the specific heat, density, and thermal conductivity vary only slightly with temperature, but viscosity varies strongly (often exponentially) with temperature.

Two schemes for temperature correction of constant-property results are the reference-temperature and the property-ratio schemes. In the reference-temperature scheme a characteristic temperature is selected at which *all* properties are evaluated. The characteristic temperature can be the average mixing-cup temperature, the surface temperature, or a combination of these. In the property-ratio scheme, properties are evaluated at the characteristic temperature described previously, and all effects of property variation transverse to the flow are then expressed as ratios of properties evaluated at the characteristic and surface temperatures.

For liquids, where the viscosity variation can be substantial, the corrections for temperature-dependent properties take the form, according to Kays and London [29],

$$\frac{\text{Nu}}{\text{Nu}(T_m)} = \left(\frac{\mu_w}{\mu_m}\right)^n \tag{4-50}$$

$$\frac{f}{f(T_m)} = \left(\frac{\mu_w}{\mu_m}\right)^m \tag{4-51}$$

where $\text{Nu}(T_m)$, $f(T_m)$, and μ_m are the Nusselt number, friction factor, and viscosity evaluated at the mixing-cup temperature, which is the arithmetic

TABLE 4-13   Fully Developed Exponents for (T_w / T_m), Gases [19, 29]

	n	m
Laminar Boundary Layer on Flat Plate		
Gas heating	-0.08	-0.08
Gas cooling	-0.045	-0.045
Flow Normal to Circular Tube or Bank of Circular Tubes		
Gas heating	0.0	0.0
Gas cooling	0.0	0.0
Fully Developed Laminar Flow in Circular Tube		
Gas heating	0.0	1.0
Gas cooling	0.0	1.0
Fully Developed Turbulent Flow in Circular Tube		
Gas heating	-0.5	-0.1
Gas cooling	0.0	-0.1

TABLE 4-14   Fully Developed Exponents for (μ_w/μ_m), Liquids [29]

Pr	n		m	
	Heating	Cooling	Heating	Cooling
Laminar	-0.14	-0.14	0.58	0.50
1	-0.20	-0.19	0.09	0.12
3	-0.27	-0.21	0.06	0.09
10	-0.36	-0.22	0.03	0.05
30	-0.39	-0.21	0.00	0.03
100	-0.42	-0.20	-0.04	0.01
1000	-0.46	-0.20	-0.12	-0.02

average (satisfactory for gases if the absolute temperature varies by a factor of less than 2 from inlet to outlet) of its inlet and outlet values. For gases, all property variations can be represented by the absolute temperature; thus the correction for temperature-dependent properties is

$$\frac{\mathrm{Nu}}{\mathrm{Nu}(T_m)} = \left(\frac{T_w}{T_m}\right)^n \qquad (4\text{-}52)$$

$$\frac{f}{f(T_m)} = \left(\frac{T_w}{T_m}\right)^m \qquad (4\text{-}53)$$

For gases, the exponents to be used in conjunction with Eqs. (4-52) and (4-53) are displayed in Table 4-13. These values are for circular tubes but may also be used for noncircular geometries in the absence of other alternatives.

For liquids, the exponents to be used in conjunction with Eqs. (4-50) and (4-51) are displayed in Table 4-14 and can also be assumed to apply to liquid metals. Results for power-law non-Newtonian liquids have also been reported [5, 30].

4.9 TWISTED-TAPE INSERTS

Devices for establishment of fluid swirl often increase heat transfer in duct flows more than the increase in pumping power, according to Bergles [31].

PROBLEMS

4-1   On the basis of the solution for the difference between wall and fluid temperatures provided by Eq. (4-26) for laminar slug flow in a round duct with specified wall heat flux, estimate the Graetz number value required for entrance effects to have diminished by a factor of 100. Sketch the variation of $\theta_w - \theta_m$ versus Gz. *Answer*: Gz $\approx$ 14.

4-2 On the basis of the mixing-cup temperature provided by Eq. (4-14) for laminar slug flow in a round duct with specified wall temperature, estimate the Graetz number value required for mixing-cup temperature to diminish by a factor of 100. Sketch the variation of θ_m versus Gz. *Answer*: Gz $\approx$ 4.3.

4-3 Plot V/V_{av} against r/R for a power-law fluid for $n = 0, \frac{1}{10}, 1, 5,$ and ∞ in fully developed flow through a round duct.

4-4 A Newtonian fluid enters a circular duct in fully developed laminar flow. The heat flux into the fluid at the wall is specified and varies sinusoidally along the tube length L as $q_{w_{max}} \sin(\pi z/L)$—as in a nuclear reactor where heat flux is actually more likely to vary as $A + B \sin(\pi z/L)$.

(a) Show, on the basis of Eq. (4-47), that the wall temperature along the tube length varies as, $a = 2\lambda^2/\text{Re Pr } D$ and $b = \pi/L$,

$$\frac{T_w(z) - T_0}{(q_{w_{max}} R/k)} = \frac{8L(1 - \cos bz)}{\pi D \text{ Re Pr}}$$

$$- \sum_{n=1}^{\infty} C_n R_{1n}(1)\left(\frac{a}{b}\right) \frac{e^{-az} + [(a/b)\sin bz - \cos bz]}{(a/b)^2 + 1}$$

(b) Compare the result for part a with that obtained by using a constant wall heat flux that equals the average heat flux—$q_w = 2q_{w_{max}}/\pi$. For the purpose of this comparison, set L/D Re Pr $= 0.05$ and plot dimensionless wall temperature against z/L.

(c) Comment on the agreement of the prediction of part b with the more exact prediction of part a.

4-5 Repeat Problem 4-4 for the case of slug flow.

4-6 A gas-cooled nuclear reactor (perhaps intended for a coal gasification or a nuclear rocket application) is a core with $\frac{1}{8}$-in.-diameter holes through it. Each flow passage is 4 ft long. The wall heat flux varies along the length of each flow channel as

$$q_w = 300 \text{ Btu/hr ft}^2 + 800 \text{ Btu/hr ft}^2 \sin \frac{\pi z}{400 \text{ ft}}$$

where z is distance from the inlet. Air enters at 200°F and 100 psig at a mass velocity of 5500 $\text{lb}_m/\text{hr ft}^2$. Plot wall heat flux, mixing-cup temperature, and wall temperature against distance along the flow passage. *Answers*: $q_w \approx 300 \text{ Btu/hr ft}^2$; $T_{w,0} \approx 30°F$; $T_{m,0} \approx 40°F$.

4-7 Consider a circular duct through which a Newtonian fluid flows steadily with a fully developed laminar velocity profile. The entrance temperature is uniform at T_0; the wall temperature starts out above T_0 and then decreases linearly along the duct, finally reaching T_0 at $Gz^{-1} = (z/D)/Re\,Pr = 0.1$. In other words

$$T_w - T_0 = (1 - 10\,Gz^{-1})(T_{w0} - T_0)$$

(a) Plot against Gz^{-1} the exact local values of $(q_w R/k)/(T_{w0} - T_0)$, $(T_m - T_0)/(T_{w0} - T_0)$, and Nu for $0 \leqslant Gz^{-1} \leqslant 0.1$.
(b) Show in the plot of part a the local values of the same quantities based on an average value of $T_w - T_0 = (T_{w0} - T_0)/2$. *Answer*: $Nu(Gz^{-1} \to 0.63^\pm) \to \pm\infty$.

4-8 The hydraulic diameter for various cross-section geometries is defined by Eq. (4-48). Show that for (a) a rectangle of sides a and b, $D_h = 2a/(1 + a/b)$, (b) an annulus of inner diameter D_i and outer diameter D_o, $D_h = D_o - D_i$, (c) an equilateral triangle of side a, $D_h = a/4(3)^{1/2}$, (d) a circle of diameter D, $D_h = D$, (e) a sector of a circle of radius $D/2$ and included angle θ, $D_h = D\theta/(2 + \theta)$, and (f) a gap of infinite width and spacing a, $D_h = 2a$.

4-9 A fluid enters a gap bounded by two infinite parallel planes of wall temperature T_w as shown in Fig. 4P-9. The fluid experiences slug flow and has a uniform entrance temperature T_0. The temperature distribution in the fluid is described by

$$\rho C_p V \frac{\partial T}{\partial z} = k \frac{\partial^2 T}{\partial x^2}, \qquad -a < x < a, \qquad z > 0$$

$$T(a, z) = T_w = T(-a, z)$$

$$T(x, 0) = T_0$$

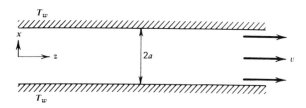

Figure 4P-9

(a) Show that the dimensionless form of the describing equations is

$$\frac{\partial \theta}{\partial z''} = \frac{\partial^2 \theta}{\partial x'^2}$$

$$\theta(1, z'') = 0 = \theta(-1, z'')$$

$$\theta(x', 0) = 1$$

where $\theta = (T - T_w)/(T_0 - T_w)$, $x' = x/a$. $\mathrm{Gz} = D_h \, \mathrm{Re} \, \mathrm{Pr}/z$, $D_h = 4a$, $\mathrm{Re} = VD_h/\nu$, and $z'' = 16/\mathrm{Gz}$.

(b) Show that the solution for the temperature of part a is

$$\theta = \frac{4}{\pi} \sum_{n=0}^{\infty} \frac{(-1)^n}{(2n+1)} \cos\left[(2n+1)\frac{\pi}{2}x'\right] \exp\left[-\left(n+\frac{1}{2}\right)^2 \pi^2 z''\right]$$

from which the mixing-cup temperature is found to be

$$\theta_m = \frac{8}{\pi^2} \sum_{n=0}^{\infty} \frac{1}{(2n+1)^2} \exp\left[-\left(n+\frac{1}{2}\right)^2 \pi^2 z''\right]$$

and the heat flux into the fluid at the upper surface is found to be

$$\frac{q_w a}{(T_w - T_0)k} = 2 \sum_{n=0}^{\infty} \exp\left[-\left(n+\frac{1}{2}\right)^2 \pi^2 z''\right]$$

Hint: Consult [8, p. 97].

(c) From the results of part b show that if $q_w = h(T_w - T_m)$, the local Nusselt number is given by

$$\mathrm{Nu} = \frac{hD_h}{k} = \frac{\pi^2 \sum_{n=0}^{\infty} \exp\left[-\left(n+\frac{1}{2}\right)^2 \pi^2 z''\right]}{\sum_{n=0}^{\infty}\left[1/(2n+1)^2\right] \exp\left[-\left(n+\frac{1}{2}\right)^2 \pi^2 z''\right]}$$

(d) From part c, show that the limiting local Nusselt number is

$$\mathrm{Nu}_\infty = \pi^2$$

(e) Show that the average Nusselt number based on LMTD is obtained from substitution of the mixing-cup temperature into

$$\overline{\mathrm{Nu}} = \frac{\overline{h}D_h}{k} = -\frac{4}{z''} \ln(\theta_m)$$

(f) Plot $\overline{\mathrm{Nu}}$, Nu, and θ_m against Gz^{-1} and determine a thermal entrance length criterion from that plot.

4-10 Repeat Problem 4-9 for the case in which a constant heat flux q_w flows into the fluid at each surface. *Hint*: Consult Carslaw and Jaeger [8, p. 12].

(a) Show that the temperature distribution is

$$\frac{T - T_0}{q_w a/k} = z'' + \frac{3x'^2 - 1}{6}$$

$$- \frac{2}{\pi^2} \sum_{n=1}^{\infty} \frac{(-1)^n}{n^2} \exp(-n^2\pi^2 z'') \cos(n\pi x')$$

from which the mixing-cup and wall temperatures are found to be

$$\frac{T_m - T_0}{q_w a/k} = z''$$

$$\frac{T_w - T_0}{q_w a/k} = z'' + \frac{1}{3} - \frac{2}{\pi^2} \sum_{n=1}^{\infty} \frac{1}{n^2} \exp(-n^2\pi^2 z'')$$

and the limiting temperature distributions are

$$\lim_{z'' \to \infty} \frac{T - T_0}{q_w a/k} = z'' + \frac{3x'^2 - 1}{6}$$

$$\lim_{z'' \to \infty} \frac{T_w - T_m}{q_w a/k} = \frac{1}{3}$$

(b) From the results of part a show that the local Nusselt number based on the heat flux from one surface is

$$\mathrm{Nu} = \frac{hD_h}{k} = \frac{12}{1 - (6/\pi^2)\sum_{n=1}^{\infty}(1/n^2)\exp(-n^2\pi^2 z'')}$$

from which it is found that the limiting Nusselt number is

$$\mathrm{Nu}_\infty = 12$$

(c) Compare Nu_∞ for the constant q_w and constant T_w cases.

4-11 A power-law fluid enters a gap bounded by two infinite parallel planes, both of whose wall heat fluxes are constant and equal to q_w. The fluid experiences fully developed flow and has a uniform entrance temperature T_0. The geometry is illustrated in Fig. 4P-9.

(a) Show that the velocity distribution is described by

$$\frac{d\left(\mu \left| dv/dx \right|^{n-1} dv/dx\right)}{dx} = \frac{dp}{dz} = \text{const}$$

$$v(x = a) = 0 = v(x = -a)$$

from which it is found that, with $x' = x/a$,

$$\frac{v}{v_{av}} = \frac{2n + 1}{n + 1}\left(1 - x'^{(n+1)/n}\right)$$

(b) Show that the temperature distribution far downstream from the entrance is described by

$$\frac{(T - T_0)}{q_w a/k} = -\frac{(2n + 1)(24n^2 + 13n + 2)}{6(3n + 1)(4n + 1)(5n + 2)} + z''$$

$$+ \frac{2n + 1}{n + 1}\left[\frac{x'^2}{2} - \frac{n^2}{(2n + 1)(3n + 1)}x'^{(3n+1)/n}\right]$$

(c) From part b show that the limiting value of the Nusselt number is

$$\text{Nu}_\infty = \frac{hD_h}{k} = 12\frac{(4n + 1)(5n + 2)}{32n^2 + 17n + 2}$$

where $D_k = 4a$ is the hydraulic diameter, $\text{Gz} = D_h \, \text{Re} \, \text{Pr}/z$, $z'' = 16/\text{Gz}$, and $\text{Re} = v_{av} D_h/v$.

(d) Compare the result of part c against the value reported for a Newtonian fluid and develop a table to show how Nu_∞ varies with n (identify the slug and linear velocity profiles).

4-12 Reconsider Problem 4-11 with both surfaces at constant temperature T_w. Far downstream from the entrance the temperature distribution can be determined by iteratively solving the energy equation, as was done to find Nu_∞ for fully developed flow in a circular duct. Use this technique to determine $\text{Nu}_\infty = hD_h/k$ for fully developed slug, parabolic, and linear velocity profiles. Compare the result for a Newtonian fluid with the value reported in the text. The temperature distribution of $1 - x'^2$ can be used to begin the iterations. *Answer*: $n = 0$; $\text{Nu}_\infty = 9.87$; $n = 1$, $\text{Nu}_\infty = 7.54$; $n = \infty$, $\text{Nu}_\infty = 6.96$.

4-13 Repeat Problem 4-11 for the case in which one surface is adiabatic while the other surface delivers a constant heat flux q_w into the fluid. This approximates the situation in a wide and long solar collector with the adiabatic surface representing the glazing through which the insolation comes, the constant-heat-flux surface representing the absorber plate, and the heat flux representing the absorbed insolation.

(a) Determine Nu_∞ for slug, parabolic, and linear velocity profiles.

(b) Determine Nu in the entrance region, using the assumption that the dimensionless temperature has the form

$$\theta = \underbrace{X_1(x)Z_1(z)}_{\substack{\text{decaying} \\ \text{transient}}} + \underbrace{Z_2(z)}_{\substack{\text{axial temperature} \\ \text{rise due to} \\ \text{accumulated} \\ \text{wall flux}}} + \underbrace{X_2(x)}_{\substack{\text{transverse temperature} \\ \text{variation to let} \\ \text{wall flux into fluid}}}$$

Answer: (a) $Nu_\infty = 6(4n + 1)(5n + 2)/(23n^2 + 14n + 2)$.

4-14 Consider a $\frac{1}{4}$-in.-i.d. (inner diameter), 4-ft-long circular duct that receives a constant wall heat flux from an electric resistance heating element wrapped around the duct. An organic liquid flows through the duct at 10 lb_m/hr and is to be heated from its inlet temperature of 50°F to an outlet temperature of 150°F. The fluid properties are $Pr = 10$, $\rho = 47$ lb_m/ft^3, $C_p = 0.5$ Btu/lb_m °F, $k = 0.079$ Btu/hr ft °F, and μ (Newtonian) $= 1.6$ lb_m/hr ft.

(a) Can the duct be considered to be infinitely long?

(b) Determine the maximum fluid temperature, the maximum wall temperature, and the wall heat flux. *Answers*: (a) $L_{e,v} = 0.4$ ft, $L_{e,T} = 4$ ft; (b) $T_{\text{fluid, max}} = T_w = 255$°F, $q_w = 1900$ Btu/hr ft^2.

4-15 Consider fully developed laminar flow of a constant-property Newtonian fluid through a circular duct. The wall temperature is constant. There is uniformly distributed heat generated in the fluid (due, perhaps to chemical or nuclear reaction) at a rate S. Determine the limiting value of the Nusselt number $Nu_\infty = h_\infty D/k$.

4-16 Consider fully developed laminar flow of a Newtonian fluid in a circular duct. The wall temperature T_w is constant, and the fluid entrance temperature T_0 is uniform. Fluid properties are constant, the duct is of radius R, and the fluid liberates thermal energy uniformly at the rate q'''.

(a) Show that the describing energy equation is

$$2(1 - r'^2)\frac{\partial \theta}{\partial z''} = \frac{1}{r'}\frac{\partial(r' \, \partial\theta/\partial r')}{\partial r'} + E\,Pr\,16r'^2 + Q$$

$$\theta(r' = 1, z') = 0, \qquad \theta(r',0) = 1, \qquad \frac{\partial\theta(r' = 0, z')}{\partial r'} \quad \text{finite}$$

where $r' = r/R$, $\theta = (T - T_w)/(T_0 - T_w)$, $E = v_{av}^2/C_p(T_0 - T_w)$, $Gz = D\,Re\,Pr/z$, $z'' = 4/Gz$, and $Q = q'''R^2/k(T_0 - T_w)$.

(b) Show that far downstream from the beginning of the heated section where $\partial\theta/\partial z'' = 0$,

$$\theta(r', z'' \rightarrow \infty) = \mathrm{E}\,\mathrm{Pr}(1 - r'^4) + \frac{Q(1 - r'^2)}{4}$$

(c) From the result of part b show that the mixing-cup temperature is

$$\theta_m(r, z'' \rightarrow \infty) = \frac{5}{6}\mathrm{E}\,\mathrm{Pr} + \frac{Q}{6}$$

(d) From the result of part c show that the limiting Nusselt number, based on the driving temperature difference of $T_w - T_m$, is

$$\mathrm{Nu}_\infty = \frac{hD}{k} = \frac{48\mathrm{E}\,\mathrm{Pr} + 6Q}{5\mathrm{E}\,\mathrm{Pr} + Q}$$

(e) Discuss the result of part d with particular attention to the Nu_∞ that is found when $\mathrm{E}\,\mathrm{Pr} = 0 = Q$ and to the appropriateness of $T_w - T_m$ as a driving temperature difference. *Note*: Consult Singh [35] for entrance region and Ebadian et al. [36] for dissipation details.

4-17 Compare the constant heat flux and constant wall temperature limiting Nusselt number values for an equilateral triangle from Table 4-9 with those for a sector of a circle with a 60° included angle from Fig. 4-10. For which condition is there the closer agreement?

A fluid can flow laminarly through either a circular duct or a rectangular duct whose cross section has a length/width ratio of 8. Both ducts have equal cross-sectional area, so the average velocity is the same. Determine the ratio of the heat-transfer coefficient–surface area product for these two geometries.

4-18 Show, for a 2:1 rectangle with constant wall temperature and fully developed laminar flow of a Newtonian fluid, that the Nusselt number based on LMTD is

$$\overline{\mathrm{Nu}} = 3.39 + 0.06\,\frac{D_h\,\mathrm{Re}\,\mathrm{Pr}}{z}, \qquad \frac{z}{D_h\,\mathrm{Re}\,\mathrm{Pr}} \gtrsim 0.05$$

and determine the value of z/D_h at which Nu exceeds its limiting value by 10%. *Answer*: $z/D_h \approx 0.174\,\mathrm{Re}\,\mathrm{Pr}$.

4-19 On the basis of the criterion for thermal entrance region for laminar flow of $0 \leqslant z/D\,\mathrm{Re}\,\mathrm{Pr} \leq 0.05$, estimate the length of the thermal entrance region when $\mathrm{Re} = 500$ for (a) an oil ($\mathrm{Pr} = 500$), (b) air ($\mathrm{Pr} = 0.7$), and (c) liquid metal ($\mathrm{Pr} = 0.01$). If the duct length is

$100 D_h$, determine whether or not either the velocity or the temperature profiles are fully developed over a major portion of the tube length. *Answers*: $L_{e,v}/D \approx 25$; $L_{e,T}/D = 12,500$ (oil).

4-20 The Nusselt number for a Newtonian fluid based on LMTD very near the inlet of a circular duct of constant wall temperature can be obtained from Eq. (4-49). Show that for $\mathrm{Re\,Pr}\,D/z$ large

(a)

$$\overline{\mathrm{Nu}} \approx 3.66 + 6.5\left(\frac{\mathrm{Re\,Pr}\,D}{z}\right)^{1/5}$$

for simultaneously developing velocity and temperature profiles.

(b)

$$\overline{\mathrm{Nu}} = 3.66 + 1.67\left(\frac{\mathrm{Re\,Pr}\,D}{z}\right)^{1/4}$$

for a fully developed parabolic velocity profile.

(c) Compare by plotting a curve of $\overline{\mathrm{Nu}}$ against $\mathrm{Re\,Pr}\,D/z$ the accuracy of the result of part b with Eq. (4-49) and with

$$\overline{\mathrm{Nu}} = 1.86\left(\frac{\mathrm{Re\,Pr}\,D}{z}\right)^{1/3}\left(\frac{\mu_m}{\mu_w}\right)^{0.14}$$

due to Seider and Tate [37].

(d) On the basis of the results of parts a and b, show that near the inlet the Nusselt number with a developing velocity profile is four times greater than the Nusselt number with a fully developed velocity.

4-21 Show that for a circular duct, $\mathrm{Re\,Pr}\,D/z = (4/\pi)C_p\dot{M}/z$, where $\dot{M}$ is the mass flow rate.

4-22 Show that for a duct of any cross section, $\mathrm{Re\,Pr}\,D_h/z = (16A/C^2)C_p\dot{M}/kz$, where A is the cross-sectional area and C is the cross-sectional circumference.

4-23 From the results of Fig. 4-11, show that $0 \leqslant z/D\,\mathrm{Re\,Pr} \leqslant 0.05$ is a reasonable criterion for the thermal entrance region for $\mathrm{Pr} \approx 0.7$ with simultaneous development of velocity and temperature profiles.

4-24 Calculate the heat-transfer coefficient for laminar flow of a Newtonian fluid ($k = 0.1$ Btu/hr ft °F) inside a $\frac{1}{4}$-in.-i.d. tube in the hydrodynamically and thermally developed region with a uniform wall temperature. Also, determine the heat-transfer rate between the tube wall and the fluid if the inlet temperature is 300°F and the wall temperature is

100°F. *Hint*: $h_\infty = 17.6$ Btu/hr ft^2 °F, $Q = \rho C_p v_{av} \pi D^2 / 4(200°F) =$ indeterminant without flow rate.

4-25 Compare the limiting heat-transfer coefficients for a Newtonian fluid flowing laminarly through (a) a circular duct of diameter D, (b) a square duct of side D, and (c) and equilateral-triangle duct of side D.

4-26 Oil at 70°F with a mean inlet velocity of 2 ft/sec enters a $\frac{1}{2}$-in.-i.d., 5-ft-long tube with a constant wall temperature of 150°F. Determine the pressure drop and outlet temperature, accounting for viscosity variation and entrance effects. Repeat the outlet temperature and pressure drop calculations, first neglecting the viscosity variation and then neglecting all entrance effects. Compare the importance of these two factors for air and oil. Fluid properties are: at 70°F, $\mu = 0.015$ lb$_m$/ft sec; at 150°F, $\mu = 0.0055$ lb$_m$/ft sec, $\rho = 55$ lb$_m$/ft^3, $C_p = 0.45$ Btu/lb$_m$ °F, $k = 0.1$ Btu/hr ft °F. *Answers*: $T_{m,0} = 77$°F; $\Delta p = 0.36$ psi.

4-27 Air at 70°F with a mean inlet velocity of 2 ft/sec enters a $\frac{1}{2}$-in.-i.d., 1-ft-long tube with a constant wall temperature of 150°F. Determine the air outlet temperature, pressure drop, and the length of the entrance region, accounting for property variation and entrance effects. Repeat the outlet temperature and pressure drop calculations, first neglecting the property variations and then neglecting all entrance effects. Discuss the importance of these two factors for air and oil (do not repeat the calculation for oil). *Answers*: $T_{m,0} = 150$°F; $\Delta p = 6.3 \times 10^{-4}$ psi.

4-28 Water at 100°F from a large header enters a 1-cm-i.d. straight circular tube of 2 m length (whose wall temperature is constant at 150°F) that is part of a flat-plate solar collector. The water flow is forced at the rate of 0.003 kg/s, which is a typical value near the optimum for flat-plate solar collectors. Estimate the outlet temperature, pressure drop, the thermal and hydraulic entrance lengths, and the value of the heat-transfer coefficient. *Answers*: $T_{m,0} = 57$°C; $\Delta p = 8$ N/m^2; $L_{e,v}/D = 28$; $L_{e,T}/D = 127$; $\bar{h} = 244$ W/m^2 K.

4-29 Air at 60°F from a large header enters a flat-plate solar collector that is approximated by two parallel planes, one side insulated to represent the collector's glazing and the other side at 200°F constant temperature (or constant absorbed solar heat flux, depending on one's point of view).

   The flow channel is 4 m long, 1 m $\times$ 0.01 m cross section, and flow rate is 40 kg/h. Estimate the outlet temperature, the pressure drop, the thermal and hydraulic entrance lengths, and the value of the heat-transfer coefficient. *Answers*: $T_{m,0} = 67$°C; $\Delta p = 12$ N/m^2; $L_{e,v}/D = 60$; $L_{e,T}/D = 40$; $\bar{h} = 5.6$ W/m^2 K.

4-30   Determine the delivered solar power: pumping power ratio for water (using the conditions of Problem 4-28 and for air (using the conditions of Problem 4-29). On the basis of equal delivered solar power, judge which of these two methods of cooling a solar collector consumes the lesser pumping power. References 53 and 54 of Chapter 9 might be helpful. *Answer*: $P_{solar}/P_{pump} = 10^7$ water, 10^3 air.

4-31   Hydrogen flows at the rate of 1.26×10^{-3} kg/s through a tube of 4 cm diameter and 1 m length packed with 1-mm glass spheres. Because of a chemical reaction between components of the gas, catalyzed by a coating on the spheres, heat is uniformly released. The gas enters at $20°C$ and leaves at $80°C$ with the tube wall maintained at $80°C$. Consult Bauer [32] for information. Determine the numerical values of the:

(a) effective thermal conductivity k_e for the packed-tube–gas combination,

(b) overall heat-transfer coefficient U, and

(c) the volumetric heat release rate q'''.

Briefly discuss the procedure for finding the tube diameter that would give a maximum gas outlet temperature of $75°C$.

REFERENCES

1. C. S. Keevil and M. M. McAdams, *Chem. Metallurg. Eng.* **36**, 464–467 (1929).
2. R. K. Shah and A. L. London, in T. F. Irvine and J. P. Harnett, Eds., *Advances in Heat Transfer*, Academic, 1978.
3. E. M. Sparrow, T. S. Chen, and V. K. Jonsson, *Int. J. Heat Mass Transfer* **7**, 583–585 (1964); E. M. Sparrow and A. Haji–Sheikh, *ASME J. Heat Transfer* **88**, 351–358 (1966).
4. M. Faghri and E. M. Sparrow, *ASME J. Heat Transfer* **102**, 382–384 (1980).
5. S. H. Lin and W. K. Ksu, *ASME J. Heat Transfer* **102**, 382–384 (1980).
6. A. B. Metzner, *Adv. Heat Transfer* **2**, 376 (1965).
7. M. Abramowitz and I. A. Stegun, *Handbook of Mathematical Functions*, National Bureau of Standards, Applied Mathematics Series 55, 1965.
8. H. S. Carslaw and J. C. Jaeger, *Conduction of Heat in Solids*, 2nd ed., Clarendon, 1959, pp. 199, 203.
9. J. Schenk and J. van Laar, *Appl. Sci. Res.* **7**, 449–462 (1958).
10. A. H. P. Skelland, *Non-Newtonian Flow and Heat Transfer*, Wiley, 1967, pp. 353–383.
11. I. R. Whiteman and W. B. Drake, *Trans. ASME* **80**, 728–732 (1980).
12. B. C. Lyche and R. B. Bird, *Chem. Eng. Sci.* **6**, 35–41 (1956).
13. B. F. Blackwell, *ASME J. Heat Transfer* **107**, 466–468 (1985).
14. J. A. Sellars, M. Tribus, and J. S. Klein, *Trans. ASME* **78**, 441–448 (1959).

15. R. Siegel, E. M. Sparrow, and T. M. Hallman, *Appl. Sci. Res.* **7A**, 386–392 (1958).

16. R. J. Moffat and A. M. Anderson, *ASME J. Heat Transfer* **112**, 882–890 (1990).

17. S. Golos, *Int. J. Heat Transfer* **18**, 1467–1471 (1975).

18. R. K. Shah and A. L. London, *ASME J. Heat Transfer* **96**, 159–165 (1974).

19. W. M. Kays, *Convective Heat and Mass Transfer*, McGraw-Hill, 1966, pp. 112–116, 128–133; R. E. Lundberg, W. C. Reynolds, and W. M. Kays, Heat transfer with laminar flow in concentric annuli with constant and variable wall temperature and heat flux, NASA TN D-1972, August 1963.

20. A. C. Trupp and A. C. Y. Lau, *ASME J. Heat Transfer* **106**, 467–469 (1984).

21. S. C. R. Dennis, A. McD. Mercer, and G. Poots, *Quart. Appl. Math.* **17**, 285–297 (1959).

22. R. D. Cess and E. C. Shaffer, *Appl. Sci. Res.* **8A**, 339–344 (1958).

23. W. M. Kays, *Tran. ASME* **77**, 1265–1274 (1955).

24. P. Goldberg, M.S. Thesis, Mechanical Engineering Department, Massachusetts Institute of Technology, Cambridge, MA, January 1958.

25. H. Hausen, *Z. Ver. Deutsch. Ing., Beih. Verfahrenstech.* **4**, 91–98 (1943).

26. H. L. Langhaar, *ASME J. Appl. Mech.*, **64**, A55–A58 (1942).

27. H. S. Heaton, W. C. Reynolds, and W. M. Kays, *Int. J. Heat Mass Transfer* **7**, 763–781 (1964).

28. S. W. Churchill and H. Ozoe, *ASME J. Heat Transfer* **95**, 416–519 (1973).

29. W. M. Kays and A. L. London, *Compact Heat Exchangers*, McGraw-Hill, 1964, pp. 86–91.

30. S. D. Joshi and A. E. Bergles, *ASME J. Heat Transfer* **102**, 397–401 (1980).

31. A. Bergles, *Heat Exchanger Design Handbook*, Hemisphere, 1986, pp. 2.5.11-1–2.5.11-12.

32. R. Bauer, *Heat Exchanger Design Handbook*, Vol. 2, Hemisphere, 1986, pp. 2.8.2-1–2.8.2-8.

33. E. U. Schlunder, *Chem. Ing. Tech.* **38**, 967–978 (1966).

34. J. S. M. Botterill, *Heat Exchanger Design Handbook*, Vol. 2, Hemisphere, 1986, pp. 2.8.4-1–2.8.4-8.

35. S. N. Singh, *Appl. Sci. Res.* **7A**, 325–340 (1958).

36. M. A. Ebadian, H. C. Topakoglu, and O. A. Arnas, *ASME J. Heat Transfer* **110**, 1001–1004 (1988).

37. E. N. Seider and G. E. Tate, *Ind. Eng. Chem.* **28**, 1429–1435 (1936).

5

LAMINAR BOUNDARY LAYERS

As can be appreciated, the complete equations of continuity, motion, energy, and diffusion are difficult to solve in important cases. Although there is little difficulty for such simple cases as Couette flow, they can generally be solved only with numerical methods. Accordingly, an order of magnitude analysis is necessary to reveal insignificant terms. Discarding these terms will permit more easily solved equations to be taken as the describing equations without an appreciable sacrifice of accuracy.

5.1 LAMINAR BOUNDARY-LAYER EQUATIONS

The basic ideas that yield the boundary-layer equations were developed by Prandtl about 1904. The essential idea is to divide a flow into two parts. The larger part concerns a free stream of fluid, far from the bounding surface, which is accurately considered to be inviscid. The smaller part is a thin layer next to the bounding surface in which the effects of molecular transport (viscosity, thermal conductivity, and mass diffusivity) are considered at the expense of some approximations.

To gain insight, note the experimentally observed distributions sketched in Fig. 5-1. The major parts of all variations occur in a thin layer adjacent to the boundary. As indicated, the free stream steadily flows parallel to a plate. The plate velocity, temperature, and mass fraction are constant and the fluid properties are also constant. Because of the stabilizing effect of viscosity, tending to make the fluid next to the plate acquire the plate velocity, the flow in the boundary layer is laminar. Actually, it might only be laminar in the vicinity of the leading edge; farther downstream, turbulent flow might occur.

160

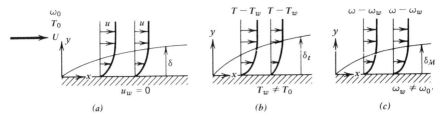

Figure 5-1 Profiles in a laminar boundary layer: (*a*) horizontal velocity (δ, velocity boundary-layer thickness); (*b*) temperature, (δ_T, thermal boundary-layer thickness); (*c*) mass fraction (δ_M, diffusion boundary-layer thickness).

As discussed by Schlichting [1], neglect of viscous effects in the free stream is accurate and advantageously simplifies the Navier–Stokes equations; which describe the velocity distribution, since inviscid flow is irrotational. A disadvantageous result is the vanishing of the second-order terms for which viscosity is a coefficient, rendering impossible the simultaneous satisfaction of the conditions of no slip and impermeability at a solid boundary. If the impermeability condition is satisfied, as it usually is since it most characterizes a solid surface, the potential flow solution shows a velocity slip at the solid boundary. Analytically and experimentally it is found that even in the limit as viscosity effects approach zero, the no-slip condition applies at a solid boundary (provided the fluid can be considered to be continuous); the smaller the viscosity is, the closer to the solid boundary one must be before the potential flow solution loses accuracy. In this thin fluid layer near the boundary, viscous effects must not be ignored. In mathematical terms, inclusion of viscous effects in the Navier–Stokes equations provides two orders in the direction normal to the solid surface and enables both the no-slip and the impermeability conditions to be satisfied there.

Historically, the velocity boundary layer was of first concern and is here, too. Referring to Fig. 5-1 for the coordinate system, the describing equations are

$$\text{Continuity} \qquad \frac{\partial u}{\partial x} + \frac{\partial v}{\partial y} = 0$$

$$x \text{ Motion} \qquad \rho\left(u\frac{\partial u}{\partial x} + v\frac{\partial u}{\partial y}\right) = B_x - \frac{\partial p}{\partial x} + \mu\left(\frac{\partial^2 u}{\partial x^2} + \frac{\partial^2 u}{\partial y^2}\right)$$

$$y \text{ Motion} \qquad \rho\left(u\frac{\partial u}{\partial x} + v\frac{\partial v}{\partial y}\right) = B_y - \frac{\partial p}{\partial y} + \mu\left(\frac{\partial^2 v}{dx^2} + \frac{\partial^2 v}{\partial y^2}\right)$$

with $u = 0 = v$ at $y = 0$ and $u \to U$ at $y \to \infty$. These equations are made dimensionless by dividing through by the free-stream velocity U, and a characteristic length L (perhaps the length of the plate). Let $u' = u/U$, $x' = x/L$, and so forth. To ascertain whether or not the continuity equation contains negligible terms, its dimensionless form

$$\frac{\partial u'}{\partial x'} + \frac{\partial v'}{\partial y'} = 0$$

is subjected to an order of magnitude analysis. Its reasoning is that one is principally interested in locations removed from the leading edge, so that $x' = x/L \sim O(1)$. The symbol $O(\ )$ denotes "order of." In most of the boundary layer $u' = u/U \sim O(1)$. Furthermore, the values of y of interest are only in the boundary layer. So $y' = y/L \sim O(\delta' = \delta/L)$. Substitution of these orders of magnitude into the dimensionless continuity equation gives

$$\frac{O(1)}{O(1)} + \frac{v'}{O(\delta')} \sim 0 \tag{5-1}$$

which can be true only if

$$v' \sim O(\delta') \tag{5-2}$$

which predicts that the velocity normal to the solid surface is small. An essential assumption here is that δ/L is small—the boundary layer is thin—and so the layer is sometimes called a thin shear layer. On physical grounds Eq. (5-2) is plausible since the plate is impermeable and the nearby free stream flows parallel to the plate. Turning next to the dimensionless x-motion equation

$$u'\frac{\partial u'}{\partial x'} + v'\frac{\partial u'}{\mu y'} = \left(B'_x - \frac{\partial p'}{\partial x'}\right) + \frac{v}{UL}\left(\frac{\partial^2 u'}{\partial x'^2} + \frac{\partial^2 u'}{\partial y'^2}\right)$$

an order of magnitude analysis, with $\mathrm{Re} = UL/v$, gives

$$O(1)\frac{O(1)}{O(1)} + O(\delta')\frac{O(1)}{O(\delta')} \sim \frac{1}{\mathrm{Re}}\left[\frac{O(1)}{\cancel{O(1^2)}}^{\text{moderate}} + \frac{O(1)}{\cancel{O(\delta'^2)}}^{\text{large}}\right] \tag{5-3}$$

Consolidation of the terms of the left-hand side and retention of only the large term on the right-hand side leads to

$$O(1) \sim \frac{1}{\mathrm{Re}}\frac{1}{O(\delta'^2)}$$

This relationship can be true only if

$$\text{Re} \sim \frac{1}{O(\delta'^2)} \tag{5-4}$$

Note that none of the convective terms on the left-hand side of the x-motion equation can be neglected. Most importantly, it is seen that $\partial^2 u/\partial x^2$ can be dropped from the viscous term, changing the equation from an elliptic to a parabolic form, but that viscous effects are retained since $\partial^2 u/\partial y^2$ must be kept. Also note that it is necessary for $\text{Re} \sim O(1/\delta'^2)$—the Reynolds number must be large for the boundary layer approximation to be accurate. For the parallel flow over a flat plate of this discussion, $B'_x - dp'/dx' = (B_x - dp/dx)L/\rho U^2 = 0$ since outside the boundary layer in the free stream

$$u'\underbrace{\frac{\partial u'}{\partial x'}}_{0} + v'\underbrace{\frac{\partial u'}{\partial y'}}_{0} = \left(B'_x - \frac{dp'}{dx'} \right) + \frac{1}{\text{Re}} \left(\underbrace{\frac{\partial^2 u}{\partial x'^2}}_{0} + \underbrace{\frac{\partial^2 u}{\partial y'^2}}_{0} \right)$$

inasmuch as u is constant there. Similarly, subjection of the dimensionless y-motion equation

$$u'\frac{\partial v'}{\partial x'} + v'\frac{\partial v'}{\partial y'} = B'_y - \frac{\partial p'}{\partial y'} + \frac{1}{\text{Re}} \left(\frac{\partial^2 v'}{\partial x'^2} + \frac{\partial^2 v'}{\partial y'^2} \right)$$

to an order of magnitude analysis

$$O(1)\frac{O(\delta')}{O(1)} + O(\delta')\frac{O(\delta')}{O(\delta')} \sim B'_y - \frac{\partial p'}{\partial y'} + O(\delta'^2)\left[\frac{O(\delta')}{O(1)} + \frac{O(\delta')}{O(\delta'^2)} \right] \tag{5-5}$$

gives

$$O(\delta') \sim B'_y - \frac{\partial p'}{\partial y'} \tag{5-6}$$

Pressure does not vary much across the boundary layer according to Eq. 5-6; thus the boundary-layer pressure at any x position is imposed by the free stream.

The dimensionless energy equation

$$u'\frac{\partial \theta}{\partial x'} + v\frac{\partial \theta}{\partial y'} = \frac{1}{\text{Re Pr}} \left(\frac{\partial^2 \theta}{\partial x'^2} + \frac{\partial^2 \theta}{\partial y'^2} \right) + \frac{E}{\text{Re}}\Phi' + \beta TE\left(u'\frac{\partial p'}{\partial x'} + v'\frac{\partial p'}{\partial y'} \right)$$

where the dimensionless temperature $\theta = (T - T_{ref})/\Delta T_{ref}$ and the Eckert number $E = U^2/C_p \Delta T_{ref}$ is subjected to the same order of magnitude analysis. With the information provided by Eqs. (5-2) and (5-4), it is found that the dimensionless energy equation terms have the orders of magnitude of

$$O(1)\frac{O(1)}{O(1)} + O(\delta')\frac{O(1)}{O(\delta'_T)} \sim \frac{O(\delta'^2)}{\text{Pr}}\left[\overset{\text{moderate}}{\frac{O(1)}{O(1^2)}} + \overset{\text{large}}{\frac{O(1)}{O(\delta'^2_T)}}\right]$$

$$+ E\frac{O(\delta'^2)}{O(\delta'^2)}$$

$$+ \beta TE[O(1)O(0) + O(\delta')O(\delta')] \quad (5\text{-}7)$$

Consolidation, assuming $O(\delta')/O(\delta'_T) \sim O(1)$ on the left-hand side, gives

$$O(1) \sim \frac{1}{\text{Pr}}O\left(\frac{\delta'^2}{\delta'^2_T}\right) + E + \beta TE\, O(\delta'^2) \quad (5\text{-}8)$$

From this it can be deduced that $\text{Pr} \sim O(\delta'^2/\delta'^2_T)$ from which a preliminary inference of $\delta'_T/\delta' = 1/\text{Pr}^{1/2}$ can be made, that $E \sim O(1)$ if viscous dissipation is to be important, and that $\beta TE \sim O(1/\delta'^2)$ if compressive work is to be important. Dimensionless viscous dissipation is

$$\Phi' = 2\left[\left(\frac{\partial u'}{\partial x'}\right)^2 + \left(\frac{\partial v'}{\partial y'}\right)^2\right] + \left[\frac{\partial v'}{\partial x'} + \frac{\partial u'}{\partial y'}\right]^2$$

$$\sim \left[\frac{O(1)}{O(1)} + \frac{O(\delta')}{O(\delta')}\right]^2 + \left[\frac{O(\delta')}{O(1)} + \frac{O(1)}{O(\delta')}\right]^2$$

$$\sim \frac{1}{O(\delta'^2)}$$

Note that $\partial^2 T/\partial x^2$ is negligible compared to $\partial^2 T/\partial y^2$ in the energy equation. This is plausible on physical grounds since gradients are greater in the y direction.

The mass diffusion equation can be similarly treated with the result that the second derivative with respect to x in the diffusion term is negligible.

The laminar boundary-layer equations, finally, are

Continuity
$$\frac{\partial \rho}{\partial t} + \frac{\partial \rho u}{\partial x} + \frac{\partial \rho v}{\partial y} = 0$$

x Motion
$$\rho \left(\frac{\partial u}{\partial t} + u \frac{\partial u}{\partial x} + v \frac{\partial u}{\partial y} \right) = B_x - \frac{\partial p}{\partial x} + \frac{\partial}{\partial y} \left(\mu \frac{\partial u}{\partial y} \right)$$

y Motion
$$B_y - \frac{\partial p}{\partial y} = 0$$

Energy
$$\rho C_p \left(\frac{\partial T}{\partial t} + u \frac{\partial T}{\partial x} + v \frac{\partial T}{\partial y} \right) = \frac{\partial}{\partial y} \left(k \frac{\partial T}{\partial y} \right) + \mu \left(\frac{\partial u}{\partial y} \right)^2$$
$$+ \beta T \left(\frac{\partial p}{\partial t} + u \frac{\partial p}{\partial x} \right)$$

Diffusion
$$\rho \left(\frac{\partial \omega_a}{\partial t} + u \frac{\partial \omega_a}{\partial x} + v \frac{\partial \omega_a}{\partial y} \right) = \frac{\partial}{\partial y} \left(D_{ab} \rho \frac{\partial \omega_a}{\partial y} \right) + r_a''' \quad (5\text{-}9)$$

Incorporation of time derivatives is justified on the grounds that they must be present if unsteady problems are to be solved. The possibility of variable properties has been included since the property variation is seldom large enough to invalidate the essential approximation of a thin shear layer.

As mentioned earlier, the free stream communicates with the boundary layer through the fact that

$$\rho \left(\frac{\partial U}{\partial t} + U \frac{\partial U}{\partial x} \right) = B_x - \frac{\partial p}{\partial x} \qquad (5\text{-}10)$$

where all terms are just outside the boundary layer. Because the free-stream velocity U is either known or is determined from the potential flow (also referred to as inviscid or irritational flow) solution, $\partial p / \partial x$ can be regarded as a known forcing function in the boundary-layer equations.

5.2 SIMILARITY SOLUTION FOR PARALLEL FLOW OVER A FLAT PLATE—VELOCITY DISTRIBUTION

The sketches in Fig. 5-1, representing measurements, lead one to suspect that all profiles are similar if the coordinates are properly stretched. If such similarity exists, a mathematical transformation of coordinates can be made to reflect this fact. Of course, not all boundary layers have "similar" profiles, but many do. The boundary-layer equations are nonlinear; a similarity transformation results in nonlinear equations, still, but they are ordinary—a big advantage.

To introduce similarity solutions, consider a flat plate over which a constant-property fluid steadily flows parallel to the plate. Note that $\partial p / \partial x = 0$ since U is constant. The applicable boundary-layer equations are

Continuity
$$\frac{\partial u}{\partial x} + \frac{\partial v}{\partial y} = 0$$

x Motion
$$u \frac{\partial u}{\partial x} + v \frac{\partial u}{\partial y} = \nu \frac{\partial^2 u}{\partial y^2}$$

with the boundary conditions

$$u(x,0) = 0 = v(x,0) \qquad \text{and} \qquad u(x,\infty) = U$$

A clue as to the form of the coordinate transformation that will give a similarity solution can be obtained from the results of the transient vaporization study of Section 3.4. There it was found that the vapor film thickness always increased as $t^{1/2}$. If one views y as the vertical distance from the plate the effect of the plate has penetrated, one has some reason to suspect that $y \sim t^{1/2}$; the time t elapsed for an affected fluid particle can be viewed as x/U. Therefore, a variable $y/t^{1/2} \sim y/(x/U)^{1/2}$ may be the one sought. A rigorous analysis (see Appendix D) shows that the variable sought is

$$\eta = y \left(\frac{U}{\nu x} \right)^{1/2} \tag{5-11}$$

The salient point is that a rigorous analysis (see Hansen [2] or Schlichting [1]) cannot start without a general idea as to the form of the coordinate transformation.

The continuity equation is automatically satisfied by a stream function Ψ such that $u = \partial \Psi / \partial y$ and $v = -\partial \Psi / \partial x$, as shown by direct substitution to obtain

$$\frac{\partial^2 \Psi}{\partial x \, \partial y} - \frac{\partial^2 \Psi}{\partial y \, \partial x} = 0$$

requiring only that $\partial \Psi / \partial x$ and $\partial \Psi / \partial y$ exist as well as $\partial^2 \Psi / \partial x \, \partial y$ be continuous for the order of differentiation to be interchanged.

The stream function is (see Appendix D for details) as

$$\Psi(x, \eta) = \sqrt{\nu U x} \, F(\eta) \tag{5-12}$$

Equation (5-12) displays a separation-of-variables form inasmuch as $\Psi = H(x)F(\eta)$ is the product of two functions that separately depend on independent variables. The velocity components are found from

$$u = \frac{\partial \Psi}{\partial y} = \frac{\partial \Psi}{\partial x}\frac{\partial x^0}{\partial y} + \frac{\partial \Psi}{\partial \eta}\frac{\partial \eta}{\partial y} = \sqrt{\nu U x}\,\frac{dF}{d\eta}\sqrt{\frac{U}{\nu x}}$$

to be

$$\frac{u}{U} = \frac{dF}{d\eta} = F' \tag{5-13}$$

and from

$$-\upsilon = \frac{\partial \Psi}{\partial x} = \frac{\partial \Psi}{\partial x}\frac{\partial x^1}{\partial x} + \frac{\partial \Psi}{\partial \eta}\overset{-\eta/2x}{\frac{\partial \eta}{\partial x}}$$

$$= \frac{1}{2}\sqrt{\frac{\nu U}{x}}\,F + \left(\sqrt{\nu U x}\,\frac{dF}{d\eta}\right)\left(-\frac{1}{2}\frac{y}{x}\sqrt{\frac{U}{\nu x}}\right)$$

to be

$$\frac{\upsilon}{U}\sqrt{\frac{Ux}{\nu}} = \frac{\eta F' - F}{2} \tag{5-14}$$

with $F' = dF/d\eta$. Equations (5-13) and (5-14) are substituted into the x-motion boundary-layer equation to achieve (see Problem 5-2 for details)

$$F''' + \tfrac{1}{2}FF'' = 0 \tag{5-15a}$$

which, although nonlinear, is an ordinary differential equation. The requirement that the plate be impermeable, the no-slip condition be observed at the plate surface, and far from the plate the free-stream velocity be approached gives

$$F(0) = 0 = F'(0) \tag{5-15b}$$

$$F'(\infty) = 1 \tag{5-15c}$$

Equation (5-15), referred to as the *Blasius equation* in honor of $H.$ Blasius, who first solved it, can be numerically solved. It is a boundary value problem —some conditions are given at the plate surface and others at infinity, and the difficulty really is in finding $F''(0)$. The results are shown in Fig. 5-2 and Table 5-1 as numerically determined by Howarth [3].

A numerical procedure to solve Eq. (5-15) is a so-called shooting method, in which a value of the missing $F''(0)$ is guessed and the profile (trajectory) is

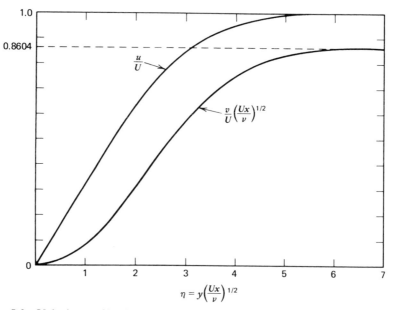

Figure 5-2   Velocity profiles in a laminar boundary layer on a flat plate.

generated out to a large η, say, $\eta = 20$; if $F'(20) \neq 1$, an adjustment is made to $F''(0)$ and a new profile is generated. This procedure is repeated until $F'(20)$ is as close to unity as desired; at the termination the last $F''(0)$ value used is the one sought (see Bejan [4] for more discussion).

In Fig. 5-1 and Eq. (5-14), the vertical component of velocity v is not zero at the edge of the boundary layer ($\eta \approx 5$), even though the free stream flows parallel to the plate. Since v/U must be small for the boundary-layer approximation to be accurate, accuracy would not be expected near the plate leading edge where x is small and v/U is large.

The *boundary-layer thickness* δ is now available since $u/U = 0.99$ at $\eta = 5$. Realizing that $y/x = \eta/(UX/v)^{1/2}$, one has

$$\frac{\delta}{x} = \frac{5}{\sqrt{\mathrm{Re}_x}} \tag{5-16}$$

following conventional practice that the edge of the boundary layer occurs where $u/U = 0.99$—the 1% deviation from free-stream conditions is not magical. For this case the boundary layer is slightly more a point of view than a tangible thing. In order to better verify the present analysis, which agrees well with pitot tube velocity measurements as shown in Fig. 5-3, something

TABLE 5-1   Function $F(\eta)$ for Boundary Layer Along a Flat Plate at Zero Incidence [3]

$\eta = y\sqrt{U_\infty/\nu x}$	F	$F' = u/U_\infty$	F''
0	0	0	0.33206
0.2	0.00664	0.06641	0.33199
0.4	0.02656	0.13277	0.33147
0.6	0.05974	0.19894	0.33008
0.8	0.10611	0.26471	0.32739
1.0	0.16557	0.32979	0.32301
1.2	0.23795	0.39378	0.31659
1.4	0.32298	0.45627	0.30787
1.6	0.42032	0.51676	0.29667
1.8	0.52952	0.57477	0.28293
2.0	0.65003	0.62977	0.26675
2.2	0.78120	0.68132	0.24835
2.4	0.92230	0.72899	0.22809
2.6	1.07252	0.77246	0.20646
2.8	1.23099	0.81152	0.18401
3.0	1.39682	0.84605	0.16136
3.2	1.56911	0.87609	0.13913
3.4	1.74696	0.90177	0.11788
3.6	1.92954	0.92333	0.09809
3.8	2.11605	0.94112	0.08013
4.0	2.30576	0.95552	0.06424
4.2	2.49806	0.96696	0.05052
4.4	2.69238	0.97587	0.03897
4.6	2.88826	0.98269	0.02948
4.8	3.08534	0.98779	0.02187
5.0	3.28329	0.99155	0.01591
5.2	3.48189	0.99425	0.01134
5.4	3.68094	0.99616	0.00793
5.6	3.88031	0.99748	0.00543
5.8	4.07990	0.99838	0.00365
6.0	4.27964	0.99898	0.00240
6.2	4.47948	0.99937	0.00155
6.4	4.67938	0.99961	0.00098
6.6	4.87931	0.99977	0.00061
6.8	5.07928	0.99987	0.00037
7.0	5.27926	0.99992	0.00022
7.2	5.47925	0.99996	0.00013
7.4	5.67924	0.99998	0.00007
7.6	5.87924	0.99999	0.00004
7.8	6.07923	1.00000	0.00002
8.0	6.27923	1.00000	0.00001
8.2	6.47923	1.00000	0.00001
8.4	6.67923	1.00000	0.00000
8.6	6.87923	1.00000	0.00000
8.8	7.07923	1.00000	0.00000

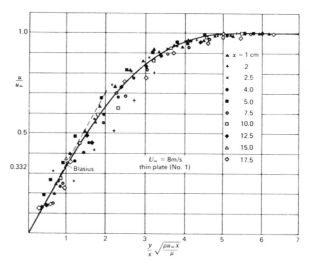

Figure 5-3   Comparison of the velocity profile in a laminar boundary layer predicted by Blasius with measurements of Hansen. [From M. Hansen, NACA TN 585, 1930.]

more tangible, the local viscous wall shear τ_w, will be predicted:

$$\tau_w = \mu \frac{\partial u(y=0)}{\partial y} = \frac{\rho U^2 C_f}{2}$$

$$\frac{C_f}{2} = \sqrt{\frac{\nu}{Ux}} F''(\eta = 0) \qquad (5\text{-}17)$$

$$\frac{C_f}{2} = \frac{0.33206}{\sqrt{\text{Re}_x}}$$

where $\text{Re}_x = Ux/\nu$.

The inaccuracy of the boundary-layer equations at the leading edge of the plate is shown by the tendency of the friction coefficient C_f to approach infinity as x approaches zero. Recall that $v(x \to 0)$ behaves similarly.

For comparison with measurement, it is convenient to have the average value of C_f, denoted by $\overline{C}_f$. This is obtained from the drag on one side of the plate as

$$\text{Drag} = x\overline{\tau}_w = \int_0^x \tau_w \, dx = \tfrac{1}{2}\rho U^2 \overline{C}_f x$$

$$\frac{1}{x} \int_0^x \sqrt{\frac{\nu}{Ux}} F''(0) \, dx = \frac{\overline{C}_f}{2}$$

$$2\sqrt{\frac{\nu}{Ux}} F''(0) = \frac{\overline{C}_f}{2} \qquad (5\text{-}18)$$

$$2\frac{C_f}{2} = \frac{\overline{C}_f}{2}$$

In other words, the average friction coefficient is twice the local friction coefficient. Thus $\bar{C}_f/2 = 0.66412/\mathrm{Re}_x^{1/2}$. The secondary flow perturbations produced by these edge boundary layers increase total drag. For references dealing with these considerations, consult Schlichting [1, p. 247].

The decrease in fluid flow through the boundary-layer region due to the viscous influence exerted by the plate is

$$\int_0^\infty \rho U \, dy - \int_0^\infty \rho u \, dy$$

Because of this decrease, the streamlines of the external flow field are displaced a distance δ_1 away from the plate since the flow decrease in the boundary layer must be compensated elsewhere. Thus

$$U\delta_1 = \int_0^\infty (U - u) \, dy$$

From Eqs. (5-11) and (5-13) it follows that the *displacement thickness* δ_1 is

$$\delta_1 \left(\frac{\nu x}{U} \right)^{-1/2} = \int_0^\infty (1 - F') \, d\eta = (\eta - F)|_{\eta = \infty} = 1.7208 \qquad (5\text{-}19)$$

The displacement thickness is related to the boundary-layer thickness according to Eq. (5-16) as

$$\frac{\delta_1}{o} = 0.3442 \approx \frac{1}{3} \qquad (5\text{-}20)$$

The decrease in x momentum flowing through the boundary-layer region due to the viscous influence of the plate is

$$\int_0^\infty \frac{\rho UU}{g_c} \, dy - \int_0^\infty \frac{\rho uu}{g_c} \, dy - \frac{\rho v(\eta = \infty)}{g_c} U$$

where, by conservation of mass, $\rho v(\eta = \infty) = \int_0^\infty (\rho U - \rho u) \, dy$. On physical grounds, this decrement in x-momentum flow must be equal to the drag force on the plate as discussed in Problem 5-9. The thickness, possessing the free stream velocity, that would accommodate this decrement is called the *momentum thickness* δ_2 and is seen to be

$$UU\delta_2 = \int_0^\infty u(U - u) \, dy$$

This relationship can be put into the form

$$\delta_2 \left(\frac{\nu x}{U} \right)^{-1/2} = \int_0^\infty F'(1 - F') \, d\eta = F' \Big|_0^\infty - \int_0^\infty (F')^2 \, d\eta$$

An integration by parts gives

$$\delta_2 \left(\frac{\nu x}{U} \right)^{-1/2} = F(1 - F') \Big|_0^\infty + \int_0^\infty FF'' \, d\eta$$

Information from Eq. (5-15) permits this to be expressed as

$$\delta_2 \left(\frac{\nu x}{U} \right)^{-1/2} = -2 \int_0^\infty F''' \, d\eta = -2F'' \Big|_0^\infty = 0.66412 \qquad (5\text{-}21)$$

The momentum thickness is related to the boundary-layer thickness according to Eq. (5-16) as

$$\frac{\delta_2}{\delta} = 0.1328 \approx \frac{1}{7} \qquad (5\text{-}22)$$

5.3   SIMILARITY SOLUTION FOR PARALLEL FLOW OVER A FLAT PLATE—TEMPERATURE DISTRIBUTION

The similarity solutions introduced to determine the velocity distribution can be extended to determine the temperature distribution as well. The assumption of constant properties uncouples the equations of motion and energy, allowing the velocity profiles to be determined first and the temperature profiles second. The same problem as before is solved, but with the additional stipulation that both the free-stream and the wall temperature are constant.

The energy equation, from Eq. (5-9) with viscous dissipation ignored, is

$$u \frac{\partial T}{\partial x} + v \frac{\partial T}{\partial y} = \alpha \frac{\partial^2 T}{\partial y^2} \qquad (5\text{-}23a)$$

with boundary conditions of

$$T(x,0) = T_w \quad \text{and} \quad T(x,\infty) = T_\infty \qquad (5\text{-}23b)$$

With definition of a dimensionless temperature as $\theta = (T - T_w)/(T_\infty - T_w)$,

Eq. (5-23) becomes

$$u\frac{\partial\theta}{\partial x} + v\frac{\partial\theta}{\partial y} = \frac{\nu}{\text{Pr}}\frac{\partial^2\theta}{\partial y^2} \tag{5-24a}$$

with

$$\theta(x,0) = 0 \quad \text{and} \quad \theta(x,\infty) = 1 \tag{5-24b}$$

Note that θ occupies the same position as did u/U in the x-motion boundary-layer equation. The boundary conditions are the same as for u/U, and the differential equation is also the same if $\text{Pr} = 1$. For $\text{Pr} = 1$, the solution for θ is identical to that for u/U. The plate drag and heat transfer are closely related as can be seen by the parallel developments of

$$\tau_w = \mu\frac{\partial u}{\partial y}\bigg|_{y=0} \quad \text{and} \quad q_w = -k\frac{\partial T}{\partial y}\bigg|_{y=0}$$

$$\frac{\tau_w}{\mu U} = \frac{\partial u/U}{\partial y}\bigg|_{y=0} \quad \text{and} \quad \frac{q_w}{k(T_w - T_\infty)} = \frac{\partial\theta}{\partial y}\bigg|_{y=0}$$

If $\text{Pr} = 1$, the two derivatives at the wall are equal and

$$\frac{\tau_w}{\mu U} = \frac{q_w}{k(T_w - T_\infty)}$$

or

$$\frac{\rho U^2 C_f/2}{\mu U} = \frac{h(T_w - T_\infty)}{k(T_w - T_\infty)}$$

which can be rearranged into, with L a characteristic length,

$$\frac{C_f}{2} = \frac{hL}{k}\frac{1}{UL/\nu} = \frac{\text{Nu}}{\text{Re}} \tag{5-25}$$

which is Reynolds' analogy.

If $\text{Pr} \neq 1$, the simple Reynolds analogy of Eq. (5-25) needs correction. For this purpose, the boundary-layer energy equation is considered in detail. With $u/U = F'$, $(v/U)(Ux/\nu)^{1/2} = (\eta F' - F)/2$, $\eta = y(U/\nu x)^{1/2}$, and a prime denoting $d/d\eta$, Eq. (5-24) acquires the similarity form

$$\theta'' + \frac{\text{Pr}}{2}F\theta' = 0 \tag{5-26a}$$

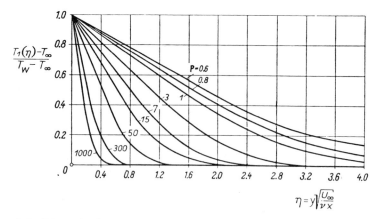

Figure 5-4   Dimensionless temperature variation in a laminar boundary layer on a flat plate.

with

$$\theta(0) = 0 \qquad\qquad\qquad (5\text{-}26b)$$

$$\theta(\infty) = 1 \qquad\qquad\qquad (5\text{-}26c)$$

Numerical solutions of Eq. (5-26) show θ to vary with η as depicted in Fig. 5-4. As expected from the order of magnitude analysis whose result is Eq. (5-8), the ratio of the thermal boundary-layer thickness δ_T to the velocity boundary-layer thickness δ varies inversely with the Prandtl number as

$$\frac{\delta_T}{\delta} = \mathrm{Pr}^{-1/3} \qquad 0.6 \leqslant \mathrm{Pr} \leqslant 10 \qquad (5\text{-}27)$$

The important quantity $d\theta(0)/d\eta$ depends on Pr as illustrated in Fig. 5-5. Approximate relations are

$$\frac{d\theta(0)}{d\eta} = 0.564\,\mathrm{Pr}^{1/2} \qquad \mathrm{Pr} < 0.05 \qquad (5\text{-}28a)$$

$$= 0.33206\,\mathrm{Pr}^{1/3} \qquad 0.6 \leqslant \mathrm{Pr} \leqslant 10 \qquad (5\text{-}28b)$$

$$= 0.339\,\mathrm{Pr}^{1/3} \qquad \mathrm{Pr} > 10 \qquad (5\text{-}28c)$$

From this information the local heat-transfer coefficient h is obtainable since

$$q_w = h(T_w - T_\infty) = -k\frac{\partial T(x,0)}{\partial y}$$

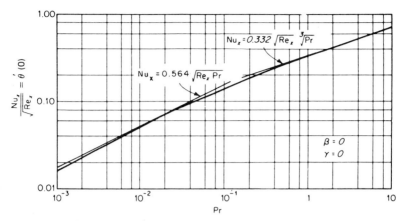

Figure 5-5 Local Nusselt number for a laminar boundary layer on a flat plate. [By permission from E. R. G. Eckert and R. M. Drake, *Analysis of Heat and Mass Transfer*, Taylor and Francis, 1972 [5].]

This gives

$$h = k\frac{\partial\theta(x,0)}{\partial y} = k\left(\frac{U}{\nu x}\right)^{1/2}\frac{d\theta(0)}{d\eta}$$

Note that large heat flux is predicted near the plate leading edge for the same reasons that the drag is predicted to be large there—the boundary-layer assumptions lose accuracy.

When the information of Eq. (5-28) is used in the foregoing development of Reynolds' analogy, the needed correction for $Pr \neq 1$ is found to result in

$$\frac{Nu}{Re\,Pr^{1/2}} = 1.7\frac{C_f}{2} \qquad\qquad Pr < 0.05 \qquad\qquad (5\text{-}29a)$$

$$\frac{Nu}{Re\,Pr^{1/3}} = \frac{C_f}{2} \qquad\qquad 0.6 \leqslant Pr \leqslant 10 \qquad\qquad (5\text{-}29b)$$

$$\frac{Nu}{Re\,Pr^{1/3}} = 1.021\frac{C_f}{2} \qquad\qquad Pr > 10 \qquad\qquad (5\text{-}29c)$$

Now that the salient features of the temperature distribution are known some common methods for solving Eq. (5-26), a boundary-value problem, are described. In the first method, it is recognized that the equation is linear in θ. Therefore, any two solutions can be added together in such a way as to satisfy the given boundary conditions. So, let

$$\theta = \theta_1 + C\theta_2$$

where

$$\theta_1'' + \frac{\text{Pr}}{2} F\theta_1' = 0 \qquad \theta_2'' + \frac{\text{Pr}}{2} F\theta_2' = 0$$

$$\theta_1(0) = 0 \quad \text{and} \quad \theta'(0) = 1 \qquad \theta_2(0) = 0 \quad \text{and} \quad \theta_2'(0) = -1$$

Determination of θ_1 and θ_2 requires only the solution of two initial-condition problems. To satisfy the condition that $\theta(\infty) = 1$, C must be given by

$$1 = \theta_1(\infty) + C\theta_2(\infty)$$

The equations for θ_1 and θ_2 are numerically solved so that C can be numerically evaluated. The missing value of $d\theta(0)/d\eta$ is then found from

$$\frac{d\theta(0)}{d\eta} = \frac{d\theta_1(0)}{d\eta} + C\frac{d\theta_2(0)}{d\eta}$$

to be

$$\frac{d\theta(0)}{d\eta} = 1 - C = \frac{\theta_1(\infty) + \theta_2(\infty) - 1}{\theta_2(\infty)}$$

A second method, first used by Pohlhausen [6], makes use of the fact that $F(\eta)$ is known from a prior solution for the velocity distribution. Equation 5-26 can be written as

$$\frac{1}{\theta'}\frac{d\theta'}{d\eta} = -\frac{\text{Pr}}{2} F$$

and integrated once with respect to η from 0 to η to find

$$\theta'(\eta) = \theta'(0) \exp\left(-\frac{\text{Pr}}{2}\int_0^{\eta} F\,d\eta\right)$$

A second integration with respect to η from 0 to η gives

$$\theta(\eta) - \theta(\overset{0}{\cancel{0}}) = \theta'(0)\int_0^{\eta} \exp\left(-\frac{\text{Pr}}{2}\int_0^{\eta} F\,d\eta\right) d\eta$$

From Eq. (5-15) it is found that $-F/2 = F'''/F''$, allowing the simplification

$$-\int_0^{\eta}\frac{F}{2}\,d\eta = \int_0^{\eta}\frac{F'''}{F''}\,d\eta = \ln\left[\frac{F''}{F''(0)}\right]$$

Use of this relationship in that for $\theta(\eta)$ then gives

$$\theta(\eta) = \theta'(0) \int_0^\eta \left(\frac{F''}{0.33206} \right)^{Pr} d\eta \qquad (5\text{-}30)$$

It remains only to evaluate the missing $\theta'(0)$. From the condition that $\theta(\infty) = 1$ it follows that

$$\theta'(0) = 1 \Big/ \int_0^\infty \left(\frac{F''}{0.33206} \right)^{Pr} d\eta = a_1(Pr) \qquad (5\text{-}31)$$

Churchill and Ozoe [7] suggest the fit to numerical results

$$a_1(Pr) = 0.5642\, Pr^{1/2} \Big/ \left[1 + (Pr/0.0468)^{4/6} \right]^{1/4}$$

The average heat-transfer coefficient $\bar{h}$ is obtained from the total heat transfer from one side of the plate as

$$q_w = \bar{h}x(T_w - T_\infty) = -k \int_0^x \frac{\partial T(x',0)}{\partial y} dx'$$

This is rearranged into the form

$$\bar{h} = \frac{k}{x} \int_{x'=0}^x \left(\frac{U}{\nu x'} \right)^{1/2} \frac{d\theta(0)}{d\eta} dx'$$

$$= 2k \left(\frac{U}{\nu x} \right)^{1/2} \frac{d\theta(0)}{d\eta} \qquad (5\text{-}32)$$

$$\bar{h} = 2h$$

Thus the average heat-transfer coefficient is twice the local one. Equation (5-29) also applies to the average heat-transfer coefficient.

5.4 SIMILARITY SOLUTION FOR PARALLEL FLOW OVER A FLAT PLATE—VISCOUS DISSIPATION EFFECTS ON TEMPERATURE DISTRIBUTION

Introductory discussion in Sections 5.1 and 5.2 for Couette and Poiseuille flows suggested that the influence of viscous dissipation on heat transfer is embodied in $q_w = h(T_w - T_{aw})$, where T_{aw} is the temperature achieved by an adiabatic wall. This relationship must be verified and the relationship between T_{aw} and T_∞ ascertained for steady parallel flow of a constant-property fluid over an isothermal flat plate.

The describing equations, including viscous dissipation, are

$$\frac{\partial u}{\partial x} + \frac{\partial v}{\partial y} = 0$$

$$u\frac{\partial u}{\partial x} + v\frac{\partial u}{\partial y} = \nu\frac{\partial^2 u}{\partial y^2}$$

$$u\frac{\partial T}{\partial x} + v\frac{\partial T}{\partial y} = \alpha\frac{\partial^2 T}{\partial y^2} + \frac{\nu}{C_p}\left(\frac{\partial u}{\partial y}\right)^2$$

with the boundary conditions that

$$u(y = 0) = 0 = v(y = 0)$$

$$T(y = 0) = T_w \qquad \text{for specified wall temperature}$$

$$\frac{\partial T(y = 0)}{\partial y} = 0 \qquad \text{for adiabatic wall}$$

$$u(y \to \infty) = U = \text{const}$$

$$T(y \to \infty) = T_\infty = \text{const}$$

Restricting the problem to one which admits a similarity solution, set $\eta = y\sqrt{U/\nu x}$ and $\psi = \sqrt{\nu x U}\,F(\eta)$ with u and v from Eqs. (7-13) and (7-14). The energy equation then reduces to

$$\frac{d^2 T}{d\eta^2} + \frac{\text{Pr}}{2}F\frac{dT}{d\eta} = -\underbrace{\text{Pr}\frac{U^2}{C_p}(F'')^2}_{\substack{\text{forcing function due} \\ \text{to viscous dissipation}}} \qquad (5\text{-}33)$$

$$T(\eta = 0) = T_w \qquad \text{for specified wall temperature}$$

$$\frac{dT(\eta = 0)}{d\eta} = 0 \qquad \text{for adiabatic wall}$$

$$T(\eta \to \infty) = T_\infty$$

Equation (5-33) is linear in T; thus a solution for particular boundary conditions can be formed by adding individual solutions. Therefore, let

$$T - T_\infty = C\underbrace{\theta_1(\eta)}_{\substack{\text{complementary} \\ \text{solution}}} + \frac{U^2}{2C_p}\underbrace{\theta_2(\eta)}_{\substack{\text{particular} \\ \text{solution}}}$$

where

$$\theta_1(\eta = 0) = 1 \qquad \theta_2'(\eta = 0) = 0$$
$$\theta_1(\eta \to \infty) = 0 \qquad \theta_2(\eta \to \infty) = 0$$

and

$$\theta_1'' + \frac{\Pr}{2}F\theta_1' = 0 \qquad \theta_2'' + \frac{\Pr}{2}F\theta_2' = -2\Pr(F'')^2 \qquad (5\text{-}34)$$

Note that θ_1 is the solution for $(T - T_\infty)/(T_w - T_\infty) = 1 - (T - T_w)/(T_\infty - T_w)$ without viscous dissipation and θ_2 is the contribution of viscous dissipation. From Eq. (5-30)

$$\theta_1(\eta) = 1 - \theta = \frac{\int_\eta^\infty (F'')^{\Pr} d\eta}{\int_0^\infty (F'')^{\Pr} d\eta}$$

To similarly express the solution for θ_2, introduce the Blasius equation $-F'''/F'' = F/2$ into the θ_2 equation to get

$$\frac{d\theta_2'}{d\eta} - \Pr\left(\frac{F'''}{F''}\right)\theta_2' = -2\Pr(F'')^2$$

An integrating factor is $(F'')^{-\Pr}$; hence a first integration yields

$$\theta_2' = -2\Pr(F'')^{\Pr}\int_0^\eta (F'')^{2-\Pr} d\eta$$

A second integration then gives

$$\theta_2(\eta) = \theta_2(0) - 2\Pr\int_0^\eta [F''(\xi)]^{\Pr}\left\{\int_0^\xi [F''(\tau)]^{2-\Pr} d\tau\right\} d\xi$$

The value of $\theta_2(0)$ is obtained from the condition that $\theta_2(\infty) = 0$. Hence

$$\theta_2(\eta) = 2\Pr\int_\eta^\infty [F''(\xi)]^{\Pr}\left\{\int_0^\xi [F''(\tau)]^{2-\Pr} d\tau\right\} d\xi \qquad (5\text{-}35)$$

The result is $\theta_2(0)$, displayed in Fig. 5-6 and approximated by

$$\theta_2(0) = \Pr^{1/2} \qquad 0.5 \leqslant \Pr < 47$$
$$= 1.9\Pr^{1/3} \qquad \Pr \geqslant 47 \qquad (5\text{-}36)$$

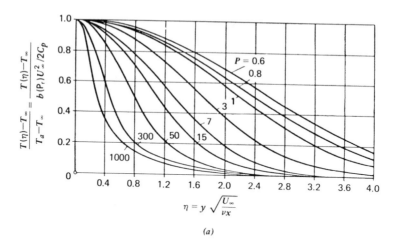

(a)

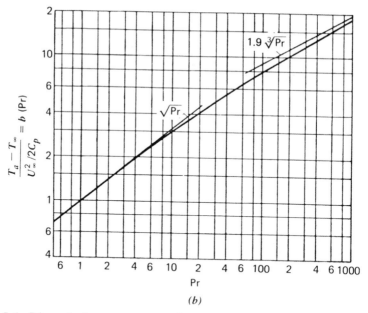

(b)

Figure 5-6 Dimensionless temperatures for a laminar boundary layer on a flat plate: (a) temperature excess θ_2 variation on an adiabatic plate; (b) adiabatic wall temperature T_{aw} dependence on Prandtl number. [By permission from H. Schlichting, *Boundary-Layer Theory*, 6th ed., McGraw-Hill, 1968 [1].]

Focus on the use of the solutions to solve two cases. First, consider the plate to be adiabatic so that at $y = 0$, $dT/dy = 0$. Now

$$T - T_\infty = \underbrace{C\theta_1(\eta)}_{\text{disregard}} + \frac{U^2}{2C_p}\theta_2(\eta)$$

Here θ_1 is disregarded since it does not allow the imposed boundary condition to be satisfied. Then

$$\overset{T_{aw}}{\cancel{T}(y = 0)} - T_\infty = \frac{U^2}{2C_p}\theta_2(0)$$

or

$$\frac{T_{aw} - T_\infty}{U^2/2C_p} = \theta_2(0) = b(\text{Pr}) = r \tag{5-37}$$

Here r is the *recovery factor* and represents the fraction of the kinetic energy that is "recovered" at the wall. As indicated in Eq. (5-36), $r = \text{Pr}^{1/2}$ for moderate Prandtl numbers. Note that r depends on the velocity distribution since $r = \text{Pr}$ for Couette flow from Section 3.1. Second, consider the plate temperature to be specified as T_w. Then the energy equation and boundary conditions are satisfied by selecting C to give

$$T - T_\infty = [(T_w - T_\infty) - (T_{aw} - T_\infty)]\theta_1(\eta) + \frac{U^2}{2C_p}\theta_2(\eta)$$

since at $\eta = 0$

$$\overset{T_w}{\cancel{T}} - T_\infty = [T_w - T_\infty - (T_{aw} - T_\infty)][1] + \overset{T_{aw} - T_\infty}{\cancel{\frac{U^2}{2C_p}}}\theta_2(0)$$

Now the heat-transfer coefficient is ascertained by

$$q_w = -k\frac{\partial T}{\partial y}\bigg|_{y=0} = \underbrace{k\left(\frac{U}{\nu x}\right)^{1/2}[-\theta_1'(0)]}_{h}(T_w - T_{aw}) \tag{5-37a}$$

This is exactly the same h as when viscous dissipation was neglected. The moral remains as suggested earlier; use low-speed results to evaluate h [e.g., Eq. (5-29)], but use a driving temperature difference of $T_w - T_{aw}$, where $T_{aw} - T_\infty = (U^2/2C_p)r$.

The variation of properties with temperature for most gases can be taken into account, according to Eckert [8], by evaluating all properties at the reference temperature T^* where

$$T^* = 0.22(T_{aw} - T_\infty) + \frac{T_w + T_\infty}{2} \tag{5-38}$$

Equation (5-38) also applies for turbulent boundary layers but with $r = \mathrm{Pr}^{1/3}$.

5.5 VARIABLE-PROPERTY EFFECT ON BOUNDARY LAYER FOR PARALLEL FLOW OVER A FLAT PLATE

The manner in which temperature and velocity distributions are affected by variable properties is explored for steady parallel flow over a flat plate. Although these effects are of interest for general fluids, they have been most intensively studied for gases for high-temperature-difference applications.

The describing boundary-layer equations are, from Eq. (5-9),

$$\text{Continuity} \qquad \frac{\partial(\rho u)}{\partial x} + \frac{\partial(\rho v)}{\partial y} = 0 \tag{5-39a}$$

$$x \text{ Motion} \qquad \rho\left(u\frac{\partial u}{\partial x} + v\frac{\partial u}{\partial y}\right) = \frac{\partial(\mu\,\partial u/\partial y)}{\partial y} \tag{5-39b}$$

The energy equation is most generally handled by a form that involves stagnation enthalpy $H_0 = H + u^2/2$ instead of temperature. Reference to Eqs. (2-35) and (2-37) gives the energy equation as

$$\rho\left(u\frac{\partial H_0}{\partial x} + v\frac{\partial H_0}{\partial y}\right) = \frac{\partial(k\,\partial T/\partial y)}{\partial y} + \frac{\partial(\mu u\,\partial u/\partial y)}{\partial y} + uB_x$$

Equation (2-50) gives $DH = C_p\,DT$ if pressure is constant and allows the diffusive heat flux to be related to the enthalpy gradient. Introduction of this relationship into the energy equation for negligible body forces results in

$$\rho\left(u\frac{\partial H_0}{\partial x} + v\frac{\partial H_0}{\partial y}\right) = \frac{\partial(\mu/\mathrm{Pr})\,\partial H_0/\partial y}{\partial y} + \frac{\partial[(1-1/\mathrm{Pr})\mu u\,\partial u/\partial y]}{\partial y}$$

$$\tag{5-39c}$$

For an impermeable plate with no slip and no thermal jump, the boundary conditions imposed on Eq. (5-39) are

$$\text{At } y = 0 \quad u = 0 = v \qquad \text{and} \qquad H_0 = H_w \tag{5-40a}$$

$$\text{At } y \to \infty \quad u \to U \qquad \text{and} \qquad H_0 = H_{0\infty} \tag{5-40b}$$

A similarity solution is begun by defining a stream function ψ as

$$\frac{\rho u}{\rho_\infty} = \frac{\partial \psi}{\partial y} \quad \text{and} \quad \frac{\rho v}{\rho_\infty} = -\frac{\partial \psi}{\partial x}$$

which automatically satisfies the continuity equation. Then, parallel to the idea utilized in Eq. (3-40) for transient change of phase with variable properties, a coordinate transformation is effected by defining

$$\eta = \left(\frac{U}{\nu_\infty x}\right)^{1/2} \int_0^y \frac{\rho}{\rho_\infty} \, dy \tag{5-41}$$

which is a *Dorodnitzyn–Stewartson* [9, 10] transformation. Additionally, the stream function is

$$\psi = (\nu_\infty U x)^{1/2} F(\eta) \tag{5-42}$$

With the definitions of Eqs. (5-41) and (5-42), it is found that

$$u = UF'$$

$$\frac{\rho}{\rho_\infty} \frac{v}{U} \left(\frac{Ux}{\nu_\infty}\right)^{1/2} = \frac{1}{2} \left\{ \left[\eta - 2 \left(\frac{Ux}{\nu_\infty}\right)^{1/2} \frac{\partial}{\partial x} \int_0^y \frac{\rho}{\rho_\infty} \, dy \right] F' - F \right\}$$

with $F' = dF/d\eta$. Then the x-motion equation becomes with $C = \mu\rho/\mu_\infty\rho_\infty$,

$$\frac{d(CF'')}{d\eta} + \frac{1}{2} FF'' = 0 \tag{5-43}$$

Similarly, the energy equation [Eq. (5-39c)] becomes

$$\frac{d[(C/\mathrm{Pr}) \, dH_0/d\eta]}{d\eta} + \frac{F}{2} \frac{dH_0}{d\eta} = -\frac{U^2}{2} \frac{d[(1 - 1/\mathrm{Pr})C\,d(F')^2/d\eta]}{d\eta} \tag{5-44}$$

The boundary conditions imposed on Eqs. (5-43) and (5-44) are

$$F(0) = 0 = F'(0) \quad \text{and} \quad H_0(0) = H_w \tag{5-45a}$$

$$F'(\infty) = 1 \quad \text{and} \quad H_0(\infty) = H_{0\infty} \tag{5-45b}$$

Since enthalpy is related to temperature in a known way, it is possible to relate such other temperature-dependent properties as μ and Pr to enthalpy. Use of the total enthalpy H_0 renders the effects of shear work negligible if

$Pr \approx 1$. As Eqs. (5-43) and (5-44) stand, they are coupled and must be solved simultaneously.

For gases, numerical solutions show that less than 5% error results if $C = 1$ and Pr constant is assumed in Eqs. (5-43) and (5-44), provided all properties are evaluated at the reference enthalpy H^*, corresponding to the reference temperature T^* of Eq. (5-38),

$$H^* = 0.22(H_{aw} - H_\infty) + \frac{H_w + H_\infty}{2} \tag{5-46}$$

This has been demonstrated for air by Eckert [8], for nitrogen and carbon dioxide by Simon et al. [11], for ionized and dissociated gases as well as plasmas as surveyed by Eckert and Pfender [12], and by the numerical solutions of Reshotko and Cohen [13, 14] and Levy [15]. Poots and Raggett [16] suggest the following reference temperature for water

$$T^* = \begin{cases} T_w + 0.6(T_\infty - T_w) & \text{heated wall} \\ T_w + 0.69(T_\infty - T_w) & \text{cooled wall} \end{cases}$$

The energy equation [Eq. (5-44)] solution with Pr constant is achieved by a linear combination of the homogeneous and the particular solutions, in both of which F is known. The homogeneous problem is taken as

$$\theta_3'' + \frac{Pr}{2} F\theta_3' = 0$$

$$\theta_3(0) = 1 \quad \text{and} \quad \theta_3(\infty) = 0$$

whose solution is given by Eq. (5-30) subtracted from unity as

$$\theta_3 = \frac{\int_\eta^\infty (F'')^{Pr} \, d\eta}{\int_0^\infty (F'')^{Pr} \, d\eta} = \theta_1(\eta) \tag{5-47}$$

The particular solution is that derived from the inhomogeneous problem

$$\theta_4'' + \frac{Pr}{2} F\theta_4' = -(Pr - 1)\frac{d^2(F')^2}{d\eta^2}$$

$$\theta_4'(0) = 0 \quad \text{and} \quad \theta_4(\infty) = 0$$

This can be rearranged into

$$\frac{d^2[\theta_4 - (F')^2]}{d\eta^2} + \frac{Pr}{2}F\frac{d[\theta_4 - (F')^2]}{d\eta} = -2\,Pr(F'')^2$$

whose solution, from Eq. (5-35), is

$$\theta_4(\eta) = [F'(\eta)]^2 - 1 + 2\Pr\int_\eta^\infty [F(\xi)]^{\Pr}\left\{\int_0^\xi [F(\tau)]^{2-\Pr}\,d\tau\right\}d\xi$$

$$= \left(\frac{u}{U}\right)^2 - 1 + \theta_2(\eta) \tag{5-48}$$

For specified wall enthalpy H_w the solution can be constructed as the sum of Eqs. (5-47) and (5-48)

$$H_0 - H_\infty = [(H_{0w} - H_{0\infty}) - (H_{0aw} - H_{0\infty})]\theta_3(\eta) + \frac{U^2}{2}\theta_4(\eta) \tag{5-49}$$

Here the stagnation enthalpy H_{0aw} at an insulated surface is

$$H_{0aw} - H_{0\infty} = \frac{U^2}{2}\theta_4(0) \tag{5-50}$$

Reference to Eqs. (5-35)–(5-37) shows that $\theta_4(0) = -1 + r_H$, where $r_H \approx \Pr^{1/2}$ is the recovery factory. Hence Eq. (5-50) gives the enthalpy at an insulated wall as

$$\frac{H_{aw} - H_\infty}{U^2/2} = r_H \tag{5-51}$$

The heat flux q_w at the flat-plate surface is found as

$$q_w = -k_w\frac{\partial T(x,0)}{\partial y} = \underbrace{\frac{C_w}{\Pr_w}\left(\frac{U\rho_\infty\mu_\infty}{x}\right)^{1/2}[-\theta_3'(0)]}_{h_H}(H_w - H_{aw})$$

For $C = 1$, constant-property results are taken over directly with enthalpy used in place of temperature as the primary variable. Then one has

$$q_w = h_H(H_w - H_{aw}) \tag{5-52}$$

with the local heat-transfer coefficient based on enthalpy given by

$$\mathrm{Nu} = \frac{h_H x}{k} = \frac{0.33206\,\mathrm{Re}^{1/2}\,\Pr^{1/3}}{C_p} \tag{5-53}$$

Note that when specific heat is constant, $h_H = h/C_p$ and $r_H = r$.

5.6 SIMILARITY SOLUTION FOR FLOW OVER A WEDGE

The boundary-layer equations describing steady parallel flow of a constant-property fluid over a flat plate were shown by Prandtl [17] in 1904 to be transformable into a single ordinary differential equation. Blasius [18] then obtained the first similarity solution, following this lead, in 1908. Because of the importance of exact solutions, extensive effort has been devoted to finding the conditions of surface geometry and free-stream velocity that admit a similarity solution.

As remarked by Hansen [19], the boundary-layer equations seem to have similarity solutions only when at least one dimension extends to infinity. The wedge geometry shown in Fig. 5-7 has this characteristic and occurs more commonly than does a flat plate. Potential (inviscid) flow theory [1, 20–23] shows that the free-stream velocity at the wedge surface varies with distance from the tip as

$$U = Cx^m \tag{5-54}$$

where the exponent m is related to the wedge angle $\beta\pi$ by

$$m = \frac{\beta}{2 - \beta} \quad \text{or} \quad \beta = \frac{2m}{1 + m}$$

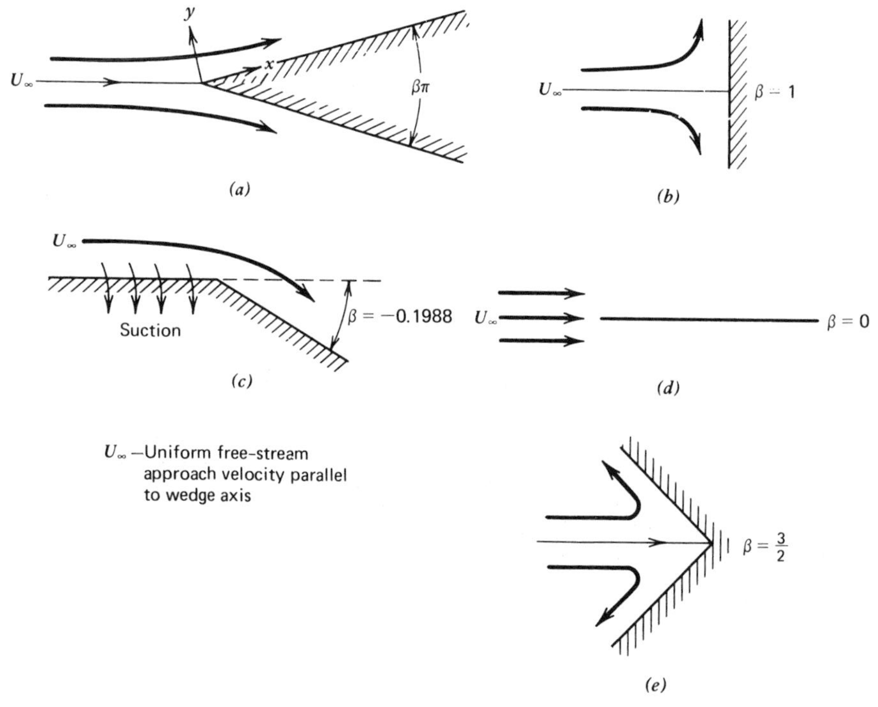

Figure 5-7 Wedge flow with various wedge angles.

When β is positive, the free-stream velocity increases along the wedge surface; for negative β, it decreases. A wedge of negative opening angle is physically impossible, of course, but this situation could be nearly realized when the boundary layer experiences suction and then makes a "turn," as shown in Fig. 5-7c. Near the leading edge of a blunt object, the free-stream velocity also varies as x^m. In particular, the free-stream velocity near the leading stagnation point of a cylinder and a sphere of radius R is predicted by potential flow theory to vary as

$$U = 2U_\infty \sin \frac{x}{R} \approx 2U_\infty \frac{x}{R}$$

and

$$U = \frac{3}{2} U_\infty \sin \frac{x}{R} \approx \frac{3}{2} U_\infty \frac{x}{R}$$

respectively. Thus wedge flow is applicable near a stagnation point.

Velocity Distribution

Falkner and Skan [24] discovered the similarity transformation appropriate to wedge flow and presented numerical results in 1931. Hartree [25] performed a more detailed study in 1937. Then in 1939 Goldstein [26] found that free-stream velocity variations of the forms $U = Cx^m$ and $U = Ce^{\alpha x}$ ($\alpha \geqslant 0$) are the only allowable ones for $u/U = F(\eta)$ with $\eta = y/\phi(x)$.

In parallel with the procedure for the case of a flat plate at zero incidence, the similarity variable η for wedge flow

$$\eta = y \left(\frac{U}{\nu x} \right)^{1/2} = y \left(\frac{C}{\nu} \right)^{1/2} x^{(m-1)/2} \tag{5-55}$$

A stream function ψ that automatically satisfies the continuity equation is

$$\psi = (\nu Ux)^{1/2} F(\eta) = (\nu Cx^{m+1})^{1/2} F(\eta) \tag{5-56}$$

Considerations that lead to these definitions of η and ψ are discussed in Appendix D in detail.

The boundary-layer equations of continuity and x motion are

$$\frac{\partial u}{\partial x} + \frac{\partial v}{\partial y} = 0 \tag{5-57a}$$

$$u\frac{\partial u}{\partial x} + v\frac{\partial u}{\partial y} = U\frac{dU}{dx} + \nu\frac{\partial^2 u}{\partial y^2} \tag{5-57b}$$

The conditions of an impermeable wedge surface with no-slip conditions and achievement of the free-stream velocity far from the wedge surface require

$$u(y = 0) = v(y = 0) \quad \text{and} \quad u(y \to \infty) \to U = Cx^m \quad (5\text{-}57c)$$

Application of Eqs. (5-55) and (5-56) yields, with $F' = dF/d\eta$,

$$u = \frac{\partial \psi}{\partial y} = \frac{\partial \psi}{\partial x} \frac{\partial x}{\partial y} + \frac{\partial \psi}{\partial \eta} \frac{\partial \eta}{\partial y} = UF'$$

$$\frac{\partial u}{\partial y} = \frac{\partial(UF')}{\partial x} \frac{\partial x}{\partial y} + \frac{\partial(UF')}{d\eta} \frac{\partial \eta}{\partial y} = U\left(\frac{U}{\nu x}\right)^{1/2} F''$$

$$\frac{\partial^2 u}{\partial y^2} = \frac{\partial(\partial u/\partial y)}{\partial x} \frac{\partial x}{\partial y} + \frac{\partial(\partial u/\partial y)}{\partial \eta} \frac{\partial \eta}{\partial y} = \frac{U^2}{\nu x} F'''$$

$$-v = \frac{\partial \psi}{\partial x} = \frac{\partial \psi}{\partial x} \frac{\partial x}{\partial x} + \frac{\partial \psi}{\partial \eta} \frac{\partial \eta}{\partial x} = U\left(\frac{Ux}{\nu}\right)^{-1/2}\left(\frac{m+1}{2}\right)\left[F - \frac{1-m}{1+m}\eta F'\right]$$

Substitution into Eq. (5-57b) gives the nonlinear ordinary differential equation

$$F''' + \frac{m+1}{2}FF'' + m\left[1 - (F')^2\right] = 0$$

$$F(0) = 0 = F'(0) \quad \text{and} \quad F'(\infty) = 1 \quad (5\text{-}58)$$

called the *Falkner–Skan equation*. The trend of the results is displayed in Fig. 5-8, where it is seen that for accelerating flows $(m, \beta > 0)$, the boundary layer is thinner than for a flat plate at zero incidence. For decelerating flows $(m, \beta < 0)$, the boundary layer is thicker than for a flat plate at zero incidence. At $\beta = -0.1988$, the velocity gradient at the wedge surface is zero, indicating that back flow and flow separation are imminent. Values of useful quantities are presented in Table 5-2; $F''(0)$, missing from Eq. (5-58), is approximated by Forbrich [27] (see Chuang [28] regarding small inaccuracies) as

$$F''(0) = \frac{1}{\sqrt{2-\beta}}\left[\frac{\beta + 0.1988377}{0.8160218}\right]^{0.53394399}$$

Note that the boundary-layer thickness δ (really a mental thing) occurs when $u/U = 0.99$, and this depends on m as

$$\delta = \eta_\delta\sqrt{\frac{\nu x}{U}} = \eta_\delta\sqrt{\frac{\nu}{C}}x^{(1-m)/2}$$

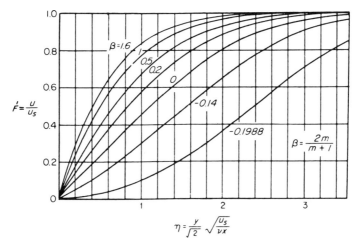

Figure 5-8 Velocity profiles for laminar wedge flow based on the calculations of Hartree [25]. [By permission from E. R. G. Eckert and R. M. Drake, *Analysis of Heat and Mass Transfer*, Taylor and Francis, 1972 [5].]

For stagnation flow, $m = 1$ and $\delta = $ const. A tangible item is the friction coefficient obtained from

$$\tau_w = \mu \frac{\partial u}{\partial y}\Big|_{y=0} = \frac{C_f}{2}\rho U^2$$

$$\frac{C_f}{2} = \sqrt{\frac{\nu}{C}}\, x^{-(m+1)/2}F''(0) = \frac{F''(0)}{\sqrt{\mathrm{Re}_x}} \qquad (5\text{-}59)$$

Here again $\mathrm{Re}_x = Ux/\nu$. The total drag on one wedge surface is given by

$$\mathrm{Drag} = \int_0^x \tau_w\, dx = \mu F''(0)\int_0^x \left(\frac{U^3 x^{-1}}{\nu}\right)^{1/2} dx$$

TABLE 5-2   Laminar Wedge Flow Result [25]

β	m	$\eta_\delta = (Ux/\nu)^{1/2}\delta/x$	$F''(0) = \mathrm{Re}^{1/2}\,C_f/2$	
2	∞		∞	
1.6	5	1.3	2.6344	
1.0	1	2.4	1.2326	Two-dimensional stagnation
0.5	1/3	3.4	0.75746	
0.2	1/9	4.2	0.51199	
0	0	5.0	0.33206	Flat plate
−0.14	−0.06542	5.5	0.16372	
−0.18	−0.08257	5.8	0.08228	
−0.1988	−0.09041	6.5	0	Separation

thus allowing the average friction coefficient $\bar{C}_f$ to be evaluated as

$$\frac{\bar{C}_f}{2} = \left(\frac{2}{3m + 1}\right)\frac{C_f}{2} \qquad (5\text{-}60)$$

For $-0.1988 \leqslant \beta < 0$, similarity solutions also exist (as a so-called lower branch) with negative wall shear, corresponding to the reverse flow that characterizes flows beyond the separation point as first found by Stewartson [29] and refined by Cebeci and Keller [30]. For $-0.5 < \beta < 0$, similarity solutions exist that correspond to a boundary layer bounded on one side by a free-stream line of zero shear rather than a wall—this can occur downstream from the flow separation point. See Yu and Yili [31] for recent work and references.

Temperature Distribution

Heat transfer from a wedge surface can be determined from the laminar boundary-layer energy equation. Constant properties are assumed, rendering the energy equation linear in temperature since velocity is known.

The energy equation and boundary conditions are

$$u\frac{\partial T}{\partial x} + v\frac{\partial T}{\partial y} = \alpha\frac{\partial^2 T}{\partial y} + \frac{v}{C_p}\left(\frac{\partial u}{\partial y}\right)^2 \qquad (5\text{-}61a)$$

$$T(y = 0) = T_w \qquad \text{and} \qquad T(y \to \infty) \to T_\infty \qquad (5\text{-}61b)$$

Inasmuch as similarity solutions are sought, a dimensionless temperature is defined as $\theta = (T - T_w)/(T_\infty - T_w)$, and the preceding energy equation is

$$u\frac{\partial\theta}{\partial x} + v\frac{\partial\theta}{\partial y} + u\theta\frac{d(T_\infty - T_w)/dx}{T_\infty - T_w} + \frac{u\,dT_w/dx}{T_\infty - T_w} = \frac{v}{\text{Pr}}\frac{\partial^2\theta}{\partial y^2} + \frac{v}{C_p}\frac{(\partial u/\partial y)^2}{T_\infty - T_w}$$

with $\theta(y = 0) = 0$ and $\theta(y \to \infty) \to 1$. The coordinate transformation $x, y \to \xi = x$, $\eta = y\sqrt{U/vx}$ is effected, and the allowable form of $d(T_\infty - T_w)/dx$, which is consistent with a similarity solution, is determined. Then

$$\frac{\partial\theta}{\partial x} = \frac{\partial\theta}{\partial\xi}\frac{\partial\xi}{\partial x}^{1} + \frac{\partial\theta}{\partial\eta}\frac{\partial\eta}{\partial x} = \frac{\partial\theta}{\partial x} + \frac{\partial\theta}{\partial\eta}\left(y\sqrt{\frac{C}{v}}\frac{m-1}{2}x^{(m-3)/2}\right)$$

and since similarity is required, $\partial\theta/\partial x = 0$. Thus

$$\frac{\partial\theta}{\partial x} = \theta'\frac{m-1}{2}\frac{\eta}{x}$$

$$\frac{\partial\theta}{\partial y} = \frac{\partial\theta}{\partial\xi}\frac{\partial\overset{0}{\xi}}{\partial y} + \frac{\partial\theta}{\partial\eta}\frac{\partial\eta}{\partial y} = \theta'\sqrt{\frac{C}{\nu}}x^{(m-1)/2}$$

$$\frac{\partial^2\theta}{\partial y^2} = \frac{\partial}{\partial y}\left(\frac{\partial\theta}{\partial y}\right) = \frac{\partial}{\partial\xi}(\ )\frac{\partial\overset{0}{\xi}}{\partial y} + \frac{\partial}{\partial\eta}(\ )\frac{\partial\eta}{\partial y} = \theta''\frac{C}{\nu}x^{m-1}$$

Also, $u = Cx^m F'$ and $v = [(m+1)/2]\sqrt{\nu C}\,x^{(m-1)/2}[\{(1-m)/(1+m)\}\eta F' - F]$; thus the energy equation [Eq. (5-61)] becomes

$$\underbrace{\theta'' + \frac{m+1}{2}\mathrm{Pr}\,F\theta}_{\text{func}(\eta)}$$

$$= \underbrace{\frac{x}{T_\infty - T_w}\left[(\theta-1)\frac{d(T_\infty - T_w)}{dx} + \frac{dT_\infty}{dx}\right]\mathrm{Pr}\,F' - \frac{\mathrm{Pr}}{C_p}C^2(F'')^2\frac{x^{2m}}{T_\infty - T_w}}_{\text{possible func}(x)} \tag{5-62}$$

It is seen that Eq. (5-62) possesses similarity in which $\theta = \theta(\eta)$ only if special conditions are satisfied by the right-hand side.

Neglecting viscous dissipation [the last term on the right-hand side of Eq. (5-62)] for the moment, it is seen that similarity is achievable if

$$\underbrace{\frac{\theta' + [(m+1)/2]\mathrm{Pr}\,F\theta'}{\mathrm{Pr}\,F'}}_{\text{func}(\eta)} = \underbrace{\frac{x}{T_\infty - T_w}\left[(\theta-1)\frac{d(T_\infty - T_w)}{dx} + \frac{dT_\infty}{dx}\right]}_{\text{possible func}(x)}$$

This leads to three cases. The first case is of constant-temperature difference $T_\infty - T_w$, which requires

$$\frac{x}{T_\infty - T_w}\frac{dT_\infty}{dx} = \gamma$$

where γ is a separation constant. Or

$$T_\infty = K_{1\infty} + (T_\infty - T_w)\gamma \ln x$$

$$T_w = K_{1w} + (T_\infty - T_w)\gamma \ln x$$

The energy equation without viscous dissipation is then

$$\theta'' + \frac{m+1}{2} \Pr F\theta' - \gamma \Pr F' = 0$$

$$\theta(0) = 0 \quad \text{and} \quad \theta(\infty) = 1 \tag{5-63}$$

Similarity is not preserved if viscous dissipation is included unless $m = 0$.

The second case is of constant wedge temperature T_w, which requires that, neglecting viscous dissipation again and with a separation constant,

$$\frac{\theta'' + [(m+1)/2] \Pr F\theta'}{\Pr F'\theta} = \frac{x}{T_\infty - T_w} \frac{dT_\infty}{dx} = \gamma$$

which has similarity if

$$T_\infty = T_w + K_2 x^\gamma$$

The energy equation with viscous dissipation is then

$$\theta'' + \frac{m+1}{2} \Pr F\theta' - \gamma \Pr F'\theta = -\Pr E(F'')^2 x^{2m-\gamma}$$

$$\theta(0) = 0 \quad \text{and} \quad \theta(\infty) = 1 \tag{5-64}$$

with $E = C^2/C_p K_2$. It is seen that a similarity solution exists when viscous dissipation is included only if $2m = \gamma$.

The third case is of constant free-stream temperature T_∞, which requires, neglecting viscous dissipation, that

$$\frac{\theta'' + [(m+1)/2] \Pr F\theta'}{\Pr F'(\theta - 1)} = -\frac{x}{T_\infty - T_w} \frac{dT_w}{dx} = \gamma$$

or

$$T_w = T_\infty - K_3 x^\gamma$$

The energy equation with viscous dissipation is then

$$\theta'' + \frac{m+1}{2} \Pr F\theta' - \gamma \Pr F'(\theta - 1) = -\Pr E(F'')^2 x^{2m-\gamma} \tag{5-65}$$

$$\theta(0) = 0 \text{ and } \theta(\infty) = 1$$

with $E = C^2/C_p K_3$. As for the case of $T_w = $ const, no similarity solution exists when viscous dissipation is included unless $2m = \gamma$. Stojanovic [32] investigated the conditions for the existence of similarity solutions to the energy equation for wedge flows, rotating bodies of revolution, and bodies of revolution in a rotating fluid.

Heat flow from a wedge surface is obtained according to

$$q_w = -k\frac{\partial T(y=0)}{\partial y} = k(T_w - T_\infty)\left(\frac{Cx^{m-1}}{\nu}\right)^{1/2}[\theta'(0)] \qquad (5\text{-}66)$$

For the second and third cases in which $T_w - T_\infty = Kx^\gamma$, $q_w \sim x^{(2\gamma+m-1)/2}$. Then if wall heat flux is constant, $\gamma = (1-m)/2$ and $T_w - T_\infty = Kx^{(1-m)/2}$. For parallel flow over a flat plate where $m = 0$, constant wall heat flux requires that $T_w - T_\infty = Kx^{1/2}$.

A local heat-transfer coefficient is found by setting q_w from Eq. (5-66) equal to $h(T_w - T_\infty)$. This yields

$$h = k\left(\frac{Cx^{m-1}}{\nu}\right)^{1/2}\theta'(0) \qquad (5\text{-}67)$$

In two-dimensional stagnation flow ($m = 1$, $\beta = 1$) the heat-transfer coefficient is constant, indicating that the thermal boundary layer is of constant thickness as was previously remarked for the velocity boundary layer

$$\text{Nu} = \theta'(0)\,\text{Re}^{1/2}$$

The solution to Eq. (5-65) for $\gamma = 0$ is shown in Fig. 5-9. The thermal boundary layer thins as the wedge angle increases in a manner similar to that depicted in Fig. 5-8 for velocities. Evans [34] gives the local Nusselt number over $\text{Re}^{1/2}$, again for constant free-stream and wall temperatures and neglecting viscous dissipation, in Table 5-3. Levy [35] solved Eq. (5-65) for $\gamma = 0$, finding the temperature distributions in Fig. 5-10 for parallel flow over a flat plate ($\beta = 0$). These trends, also observed for $\beta \neq 0$, show that the thickness of the thermal boundary layer is reduced as γ increases just as was found for Pr. For $\gamma > -1/(2 - \beta)$, heat always flows into the wall despite the wall temperature's continual excess over the free-stream temperature. This is a consequence of a fluid particle heated to nearly the wall temperature being convected downstream to a place at which wall temperature is lower. Then, since the fluid particle is warmer than the wall, heat flows into the wall. Such an occurrence results in negative heat-transfer coefficients and means only that the temperature gradient at the wall is no longer proportional to $T_w - T_\infty$. Levy suggests that the local Nusselt number results be correlated by

$$\frac{\text{Nu}}{\text{Re}^{1/2}} = B(m,\gamma)\,\text{Pr}^\lambda \qquad (5\text{-}68)$$

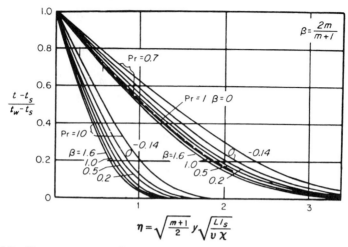

Figure 5-9 Temperature profiles for laminar wedge flow with constant surface temperatures. [By permission from E. R. G. Eckert and R. M. Drake, *Analysis of Heat and Mass Transfer*, Taylor and Francis, 1972 (after E. Eckert, *VDI-For-schungsheft*, No. 416, Berlin, 1942) [5, 33].]

where, for $-0.0904 \leqslant m \leqslant 4$,

$$B(m,\gamma) = 0.57\left(\frac{2m}{m+1} + 0.205\right)^{0.104}\left(1 + \frac{2\gamma}{m+1}\right)^{0.37+0.12m/(m+1)}\left(\frac{m+1}{2}\right)^{1/2}$$

and $\lambda \approx \frac{1}{3}$ for $\mathrm{Pr} \approx 1$, but with variations as reported in Table 5-4 that are based on results for $0.7 < \mathrm{Pr} \leqslant 4$: accuracy is $\pm 5\%$, generally, but deteriorates at large negative γ values.

The work of Brun [36] can be consulted to obtain results that include the effects of viscous dissipation.

TABLE 5-3   Nu / Re$^{1/2}$ for Laminar Wedge Flow [34]

	Pr				
m	0.7	0.8	1.0	5.0	10.0
-0.0753	0.242	0.253	0.272	0.457	0.570
0	0.292	0.307	0.332	0.585	0.730
0.111	0.331	0.348	0.378	0.669	0.851
0.333	0.384	0.403	0.440	0.792	1.013
1.0	0.496	0.523	0.570	1.043	1.344
4.0	0.813	0.858	0.938	1.736	2.236

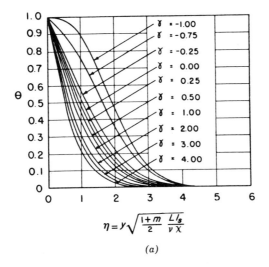

$$\eta = y\sqrt{\frac{1+m}{2}\frac{L l_s}{\nu \chi}}$$

(a)

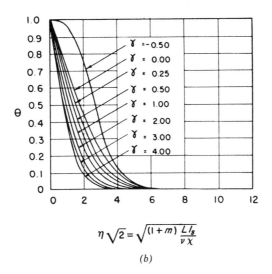

$$\eta\sqrt{2} = \sqrt{(1+m)\frac{L l_s}{\nu \chi}}$$

(b)

Figure 5-10 Dimensionless temperature distribution with variable surface temperature at $Pr = 0.7$ for (a) stagnation flow and (b) flat-plate flow. [By permission from S. Levy, *J. Aeronaut. Sci.* **19**, 341–348 (1952) [35].]

TABLE 5-4   Values of λ for Eq. (5-68) [35]

β	1.6	1.0	0	-0.199
λ	0.367	0.355	0.327	0.254

The results for wedge flow with varying wall temperature can be applied to the polynomial wall-temperature variation

$$T_w - T_\infty = a_0 + a_1 x + a_2 x^2 + \cdots = \sum_{\gamma=0}^{\infty} a_\gamma x^\gamma$$

as expounded by Chapman and Rubesin [37].

Unfortunately, the surface temperature may not be well approximated over the required range of x by a finite number of terms, each of the form $a_\gamma x^\gamma$; an arbitrary variation of surface temperature is more easily analytically treated by the approximate integral method of chapter 6. Surveys of methods for calculating thermal boundary layers are available in Schlichting [1, p. 295] and Eckert [5, p. 321]; often numerical solutions are required.

5.7   ROTATIONALLY SYMMETRIC STAGNATION FLOW AND MANGLER'S TRANSFORMATION

The flow of a fluid past a body of revolution, as during entry of a space vehicle into a planetary atmosphere, is important. For simplicity, only the case in which the fluid flows parallel to the body axis is considered, as illustrated in Fig. 5-11, and steady flow of a constant property fluid is assumed.

Application of the boundary-layer assumptions to the continuity, x motion, and energy equations for cylindrical coordinates from Appendix C gives the rotationally symmetric boundary-layer equations as

$$\frac{\partial(ur)}{\partial x} + \frac{\partial(vr)}{\partial y} = 0 \tag{5-69a}$$

$$u\frac{\partial u}{\partial x} + v\frac{\partial u}{\partial y} = U\frac{dU}{dx} + v\frac{\partial^2 u}{\partial y^2} \tag{5-69b}$$

$$u\frac{\partial T}{\partial x} + v\frac{\partial T}{\partial y} = \alpha\frac{\partial^2 T}{\partial y^2} + \frac{\mu}{\rho C_p}\left(\frac{\partial u}{\partial y}\right)^2 \tag{5-69c}$$

Here r is the radial distance from the body centerline to the surface location.

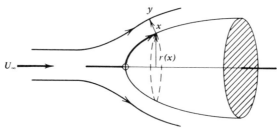

Figure 5-11   Coordinate system for flow parallel to the axis of a body of revolution.

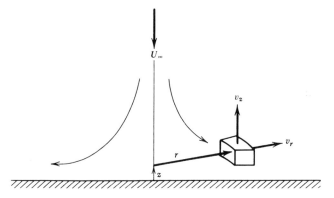

Figure 5-12   Coordinate system for stagnation flow perpendicular to a plane.

Only the continuity equation is changed from the two-dimensional form. To see that this is so, consider the rotationally symmetric stagnation case depicted in Fig. 5-12. There, the continuity equation in cylindrical coordinates is

$$\frac{\partial(rv_r)}{\partial r} + r\frac{\partial(\rho v_z)}{\partial z} = 0$$

Taking $r = x$, $z = y$, $v_r = u$, and $v_z = v$ gives Eq. (5-69a) since r is independent of z and can be taken inside the z derivative. Similarly, the r equation of motion in cylindrical coordinates under the boundary-layer assumption is

$$v_r\frac{\partial v_r}{\partial r} + v_z\frac{\partial v_z}{\partial z} = -\frac{dp}{dr} + \mu\frac{\partial^2 v_r}{\partial z^2}$$

which becomes, with the preceding substitutions, as written in Eq. (5-69b). In like manner, the energy equation is given by Eq. (5-69c). A mass balance on a pillbox-shaped control volume centered at the origin and extending just above the boundary layer shows that the velocity just outside the boundary layer varies as $U = Cr = Cx$.

The solution to Eq. (5-69) is complex because, since r depends on x, the body contour is involved. Fortunately, Mangler's coordinate transformation [38] provides a relationship between the two-dimensional boundary-layer equations and those for rotationally symmetric flow. In Mangler's transformation the equivalent two-dimensional coordinates ($\bar{x}$ and $\bar{y}$) are related to those for rotational symmetry by

$$\bar{x} = L^{-2}\int_0^x r^2(x)\,dx \quad\text{and}\quad \bar{y} = L^{-1}r(x)y \qquad (5\text{-}70)$$

where L is an arbitrary constant with a dimension of length. The basis for these transformations can be glimpsed by referring to Eq. (5-69b) in the form

$$u\frac{\partial u}{\partial x} + \cdots = \cdots + \nu\frac{\partial^2 u}{\partial y^2}$$

Application of the chain rule for $\bar{x} = \bar{x}(x, y)$ and $\bar{y} = \bar{y}(x, y)$ puts this into the form $u(\partial u/\partial \bar{x})(\partial \bar{x}/\partial x) + \cdots = \cdots + (\partial \bar{y}/\partial y)^2 \partial^2 u/\partial \bar{y}^2$. If this is to have the same form as for a two-dimensional case, $\partial \bar{x}/\partial x = (\partial \bar{y}/\partial y)^2$. Furthermore, if the boundary condition $u(y = \infty) = (L/r)\partial \psi/\partial y = (L/r)(\partial \bar{y}/\partial y)\partial \psi/\partial \bar{y}$ is to have the same form as the two-dimensional case, it is necessary that $\partial \bar{y}/\partial y = r/L$. Hence $d\bar{y} = (r/L)\,dy$ and $d\bar{x} = (r^2/L^2)\,dx$ as stated previously.

To show that Eq. (5.70) removes the influence of $r(x)$ from Eq. (5-69), a stream function to satisfy Eq. (5-69a) is defined as

$$\frac{\partial \psi}{\partial y} = \frac{ru}{L} \quad \text{and} \quad -\frac{\partial \psi}{\partial x} = \frac{rv}{L}$$

The coordinate transformation of Eq. (5-70) is effected with

$$\frac{\partial}{\partial y} = \frac{\partial}{\partial \bar{x}}\frac{\partial \bar{x}}{\partial y}^{\,0} + \frac{\partial}{\partial \bar{y}}\frac{\partial \bar{y}}{\partial y}^{\,r/L} = \frac{r}{L}\frac{\partial}{\partial \bar{y}}$$

$$\frac{\partial}{\partial x} = \frac{\partial}{\partial \bar{x}}\frac{\partial \bar{x}}{\partial x}^{\,r^2/L^2} + \frac{\partial}{\partial \bar{y}}\frac{\partial \bar{y}}{\partial x}^{\,(1/L)\,dr/dx} = \left(\frac{r}{L}\right)^2\frac{\partial}{\partial \bar{x}} + \frac{r'}{L}\frac{\partial}{\partial \bar{y}}$$

giving, with $r' = dr/dx$, Eq. (5-69b) in the form identical with that for two-dimensional boundary layers (when $u = \partial \psi/\partial \bar{y}$ and $v = -\partial \psi/\partial \bar{x}$)

$$\frac{\partial \psi}{\partial \bar{y}}\frac{\partial^2 \psi}{\partial \bar{x}\,\partial \bar{y}} - \frac{\partial \psi}{\partial \bar{x}}\frac{\partial^2 \psi}{\partial \bar{y}^2} = U\frac{dU}{dx} + \nu\frac{\partial^3 \psi}{\partial \bar{y}^3}$$

and verifying the transformation's claimed property. The velocities at x, y in the rotationally symmetric case equal the velocities at $\bar{x}, \bar{y}$ in the two-dimensional case. This follows from

$$u(x, y) = \frac{\partial \psi}{\partial y}\frac{L}{r} = \left(\frac{r}{L}\frac{\partial \psi}{\partial \bar{y}}\right)\frac{L}{r} = \frac{\partial \psi}{\partial \bar{y}} = u(\bar{x}, \bar{y})$$

For the rotationally symmetric stagnation flow of Fig. 5-12, representing an infinite stream impinging perpendicularly on a flat plate, $r = x$ and

$U = Cx$. Then Eq. (5-70) gives $\bar{x} = x^3/3L^2$ and $\bar{y} = xy/L$. Furthermore,

$$U = C(3L^2\bar{x})^{1/3}$$

Thus the equivalent two-dimensional flow is one for which $U \sim x^m$ with $m = \frac{1}{3}$, or $\beta = \frac{1}{2}$, a wedge of included angle $\pi/2$. For convenience, set $\bar{x} = x$ at one, x_0 so that $L = x_0/3^{1/2}$. At that point the corresponding value of $\bar{y}$ is related to y as $3^{-1/2}\bar{y} = y$, showing that the rotationally symmetric boundary layer is thinner than the equivalent two-dimensional boundary layer by the factor $3^{1/2}$. The heat-transfer coefficient is obtained from the two-dimensional problem solution as

$$q_w = -k\frac{\partial T(x, y = 0)}{\partial y}$$

$$h(T_w - T_\infty) = \underbrace{\left[-k\frac{\partial T(\bar{x}, \bar{y} = 0)}{\partial \bar{y}} \right]}_{h_{2-D}\left(\bar{x}, m = \dfrac{1}{3}\right)(T_w - T_\infty)}\frac{r}{L}$$

Reference to Table 5-2 at $Pr = 1$ and $m = \frac{1}{3}$ gives

$$h = 0.44k(3^{1/2})\left(\frac{C}{\nu}\right)^{1/2} \tag{5-71}$$

This result for rotationally symmetric stagnation flow can be expressed as

$$\frac{Nu}{Re^{1/2}\,Pr^{0.4}} = 0.76 \tag{5-72}$$

where $Nu = hx/k$ and $Re = Ux/\nu$ and the Prandtl number influence is approximately accounted for by a 0.4 exponent [13]. The constant value of h in Eq. (5-71) is a result of a boundary layer of constant thickness as seen before.

Use of the Mangler transformation and local similarity with variable properties has been studied by Lees [39] and Eckert and Tewfik [40]. The survey by Dewey and Gross [41] provides references, discussion of local-similarity schemes, and tables of numerical solutions to the boundary-layer equations.

The simplification provided by a well-chosen transformation is evident. Sun [42] describes the most prominent ones. The Meksyn–Görtler transformation [1, p. 164] allows accounting for nonsimilarity effects in the equation of motion, setting $\xi = \int_0^x (U/\nu)\,dx$ and $\eta = yU/[2\nu\int_0^x U\,dx]$ together with

$\psi = \nu(2\xi)^{1/2}F(\xi, \eta)$ so that the equation of motion is $F_{\eta\eta\eta} + FF_{\eta\eta} + \beta(\xi)(1 - F_\eta^2) = 2\xi(F_\eta F_{\xi\eta} - F_\xi F_{\eta\eta})$ and the boundary conditions are $F(\eta = 0) = 0 = F_\eta(\eta = 0)$ and $F_\eta(\eta = \infty) = 1$; since $\beta = 2(dU/dx)(\int_0^x U\,dx)/U^2$, it is possible to express $F(\xi, \eta)$ and $\beta(\xi)$ as a series of powers of ξ whose coefficients are given by similarity equations. The Hantzche–Wendt transformation [1, p. 319] treats compressible fluids with zero pressure gradient. The von Mises transformation [1, p. 143] uses the streamline ψ as a vertical coordinate for two-dimensional boundary layers so that the convective terms vanish from the equation of motion. The Crocco transformation [1, p. 324] simplifies the equations for variable viscosity. The Mangler transformation discussed earlier was extended by Probstein and Elliot [1, p. 229] for boundary-layer flow on slender bodies of revolution. The Howarth [1, p. 324], Illingworth–Stewartson [1, p. 324], and Cope–Hartree [1, p. 358] transformations put the equations for compressible flow into the same form as for incompressible flow. Moore's transformation [1, p. 397] is one of the few available for applying similarity concepts to unsteady boundary-layer problems; it has been applied to unsteady flight velocity for the flat-plate geometry and to unsteady film boiling by Burmeister and Schoenhals [43].

The heat-transfer coefficient between impinging gas jets and solid surfaces resembles that for the rotationally symmetric stagnation case of an infinite stream impinging on a flat surface as surveyed by Martin [44]. For a single round nozzle, the average heat-transfer coefficient for perpendicular impingement is given by

$$\frac{\overline{Nu}}{Pr^{0.42}} = \frac{D}{r}\,\frac{1 - 1.1D/r}{1 + 0.1(H/D - 6)D/r}\left[2\,Re^{1/2}\left(1 + \frac{Re^{0.55}}{200}\right)^{0.5}\right] \quad (5\text{-}73)$$

where $\overline{Nu} = \bar{h}D/k$, $Re = V_{nozzle\ exit}D/\nu$, H = nozzle height above surface, D = nozzle diameter, r = radius from nozzle centerline, $2 \times 10^3 \leqslant Re \leqslant 4 \times 10^5$, $2.5 \leqslant r/D \leqslant 7.5$, and $2 \leqslant H/D \leqslant 12$. Close to the stagnation point, $r/D < 2.5$, higher heat-transfer coefficients are encountered.

5.8 TRANSPIRATION ON A FLAT PLATE—LAMINAR BOUNDARY-LAYER SIMILARITY SOLUTIONS

The behavior of a laminar boundary layer can be influenced by either suction or injection of fluid at the solid surface. Suction removes decelerated fluid particles from the boundary layer before they have a chance to cause flow separation. A major benefit of suction on airfoils is to reduce drag. Injection can also reduce drag since the injected fluid may have sufficient momentum to prevent flow separation. The primary heat-transfer application of injection is to reduce the net wall heat flux as a result of the injected fluid motion in a direction opposite to the heat flux from the external fluid. Another term for injection cooling is *transpiration cooling*, since the injected coolant fluid transpires through a porous wall.

Velocity Distribution

The laminar boundary-layer equation of motion has a similarity solution for wedge flow with $U = Cx^m$ if $v_w \sim x^{(m-1)/2}$. Although the injection, or "blowing," velocity v_w at the wall is unlikely to vary in exactly this way in an application, such a variation is sufficiently realistic to enable use of solutions based on that assumption. The x-motion equation for wedge flow is unchanged except that v_w does not equal zero. Thus

$$F''' + \frac{m+1}{2}FF'' + m\left[1 - (F')^2\right] = 0$$

$$F(0) = -\frac{v_w}{U}\left(\frac{Ux}{v}\right)^{1/2}\left(\frac{2}{m+1}\right), \qquad F'(0) = 0, \qquad F'(\infty) = 1 \quad (5\text{-}74)$$

The local friction coefficient displayed in Fig. 5-13 demonstrates the influence of injection. Note that the boundary-layer assumptions do not permit a solution for $(v_w/U)\sqrt{Ux/v}$, the "blowing" parameter, in excess of 0.619 since the boundary layer is then "blown" off the plate.

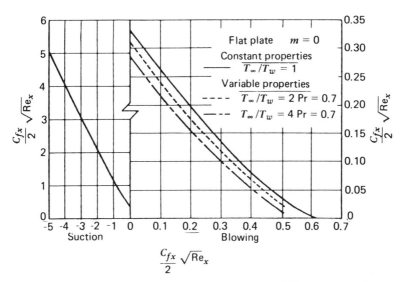

Figure 5-13 Dimensionless local friction coefficient $C_f \, \text{Re}_x^{1/2}$ for laminar flow over a flat plate with $(v_w/U)\text{Re}_x^{1/2}$ as an injection parameter. [From J. P. Hartnett and E. R. G. Eckert, *Trans. ASME* **79**, 247–254 (1957) [45].]

Velocity distributions for wedge flow [46] show that the trend of events with transpiration is as indicated for the flat plate.

Temperature Distribution

It is assumed that the injected fluid properties are constant and equal to those of the fluid of the free stream, and there is no difference between the wall temperature and the fluid temperature near the wall. The last assumption is similar in spirit to the no-slip assumption at a porous wall during transpiration.

The energy equation and boundary conditions are identical with Eq. (5-64) or (5-65), but F is now given by Eq. (5-74). Specialization to the case of constant wall and free-stream temperature requires that $\gamma = 0$ and prohibits a similarity solution if viscous dissipation is included (unless $m = 0$). With $\theta = (T - T_w)/(T_\infty - T_w)$ the equation to be solved is, with $\gamma = 0$,

$$\theta'' + \frac{m+1}{2} \Pr F\theta' = -\Pr E(F'')^2 x^{2m}$$

$$\theta(0) = 0 \quad \text{and} \quad \theta(\infty) = 1 \tag{5-75}$$

For a flat plate ($m = 0$), the results due to Hartnett and Eckert [45] show the "blowing parameter" $(v_w/U)(Ux/v)^{1/2}$ to have similar effects on the

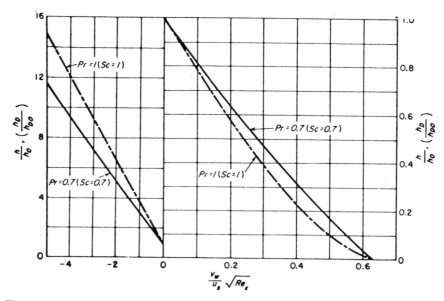

Figure 5-14 Heat- and mass-transfer coefficients for laminar flow over a flat plate. The subscript 0 indicates heat- and mass-transfer coefficients with vanishing values of the injection parameter $(v_w/U)\mathrm{Re}_x^{1/2}$; Pr belongs to h and Sc belongs to h_D. [From J. P. Hartnett and E. R. G. Eckert, *Trans. ASME* **79**, 247–254 (1957) [45].]

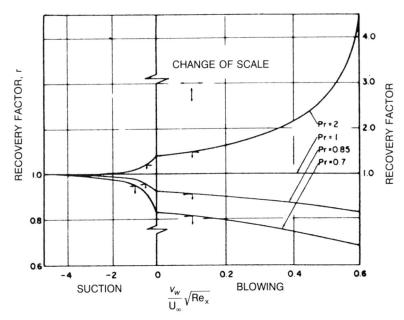

Figure 5-15 Recovery factor r for laminar flow over a flat plate for several Prandtl numbers with $(v_w/U)\text{Re}_x^{1/2}$ as an injection parameter. [From J. P. Hartnett and E. R. G. Eckert, *Trans. ASME* **79**, 247–254 (1957) [45].]

TABLE 5-5 $\text{Nu}_x\,\text{Re}_x^{-1/2}$ **for Various Rates of Blowing or Suction and Various Values of** m; **Laminar Constant-Property Boundary Layer** $(t_\infty, t_0 = \text{const}; \text{Pr} = 0.7)$

$\dfrac{v_0}{u_\infty}\sqrt{\dfrac{\rho u_\infty x}{\mu}}$	m							
	-0.04175	-0.0036	0	0.0257	0.0811	0.333	0.500	1.000
0			0.292			0.384		0.496
0.239	0.103							
0.250			0.166					
0.333						0.242		
0.375			0.107				0.259	
0.500		0.0251	0.0517					0.293
0.518				0.087				
0.558					0.109			
0.667						0.131		
1.000								0.146

Source: By permission from W. M. Kays and M. E. Crawford, *Convective Heat and Mass Transfer*, McGraw-Hill, New York, 1980.

TABLE 5-6 Summary of Heat-Transfer and Friction Parameters and Boundary-Layer Thicknesses

$\dfrac{2}{m+1}\dfrac{u_w}{U}\sqrt{Re}$	m	γ	$\dfrac{Nu}{\sqrt{Re}}$	$\dfrac{C_f}{2}\sqrt{Re}$ [a]
0	0	−0.5000	0	0.3320
		0	0.2927	
		0.5000	0.4059	
		1.000	0.4803	
	0.5	−0.7500	0	0.89975
		0	0.4162	
		0.5000	0.5426	
		1.000	0.6350	
	1.0	−1.000	0	1.2326
		−0.5000	0.3228	
		0	0.4958	
		0.5000	0.6159	
		1.000	0.7090	
−0.5	0	−0.3702	0	0.1645
		0	0.1661	
		0.5000	0.2611	
		1.000	0.3211	
	0.5	−0.5356	0	0.6974
		−0.5000	0.0272	
		0	0.2594	
		0.5000	0.3834	
		1.0000	0.4711	
	1.0	−0.6789	0	0.9692
		0	0.2934	
		0.5000	0.4132	
		1.0000	0.5030	
−1.0	0	−0.2384	0	0.0355
		0	0.0516	
		0.5000	0.1052	
		1.0000	0.1383	
	0.5	−0.3585	0	0.5345
		0	0.1392	
		0.5000	0.2528	
		1.0000	0.3314	
	1.0	−0.4235	0	0.7565
		0	0.1457	
		0.5000	0.2553	
		1.0000	0.3360	

[a] $Re = Ux/\nu$.

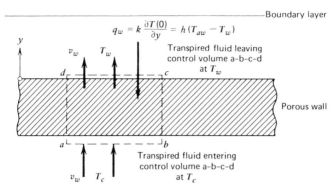

Figure 5-16 Control volume illustrating the use of transfer coefficients at a transpiring wall.

velocity and temperature profiles. Suction thins the thermal boundary layer, whereas injection thickens it. The local heat-transfer coefficient varies with the "blowing parameter" as depicted in Fig. 5-14; a linear variation is a reasonable approximation. The recovery factor r for a flat plate is also affected by transpiration (see Fig. 5-15).

For flow over a surface other than a flat plate, a single adiabatic wall temperature does not exist. Therefore, similarity solutions ignore the effect of viscous dissipation. For two-dimensional stagnation flow ($m = 1$), the local heat-transfer coefficient depends on the "blowing parameter" as shown by Hartnett and Eckert [45] and in Table 5-5. The results of Donoughe and Livingood [46] in Table 5-6 show the effect of variable wall temperature along with transpiration on wedge flow. The value of γ (in $T_w - T_\infty = Kx^\gamma$) that produces a constant wall heat flux is easily determined [as explained in conjunction with Eq. (5-66), $2\gamma = 1 - m$ for such a condition] as is the corresponding local Nusselt number.

The local heat-transfer and friction coefficients for transpiration represent diffusive flux relative to the transpiration. With reference to Fig. 5-16, an energy balance on control volume a–b–c–d shown by dashed lines gives, if radiative heat fluxes and horizontal conduction in the wall are ignored.

$$q_{\text{net from wall}} = -k\frac{\partial T(0)}{\partial y} + \rho C_p v_w T_w - \rho C_p v_w T_c$$

On setting $h(T_w - T_{aw}) = -k\,\partial T(0)/\partial y$, one has

$$q_{\text{net from wall}} = \rho C_p v_w(T_w - T_c) + h(T_w - T_{aw}) \tag{5-76}$$

5.9 MASS TRANSFER ON A FLAT PLATE–LAMINAR BOUNDARY-LAYER SIMILARITY SOLUTION

Not all applications are boundary-layer problems. However, a general analogy between heat and mass transfer is likely to be common everywhere. Looking ahead, it is possible that the transpiration caused by the vertical velocity components at the surface from which mass leaves will distort temperature, velocity, and concentration profiles in such a way as to render high-mass-transfer-rate results substantially different from low-mass-transfer-rate results.

Examination of Stefan's diffusion problem in Section 3.3 showed that induced velocity effects could be combined with the diffusive mass transfer by using

$$\dot{m}_1 = \underbrace{\left(\frac{1}{1 - \omega_1} \right)}_{\substack{\text{induced} \\ \text{flow} \\ \text{correction}}} \underbrace{\left(-\rho D_{12} \frac{d\omega_1}{dy} \right)}_{\text{diffusion}} \qquad (5\text{-}77)$$

where species 1 is diffusing through species 2. Realization that the diffusive flow is given by $\dot{m}_1 = -\rho D_{12}\, d\omega_1/dy$ leads to

$$\qquad \overset{\substack{\text{transfer coefficient} \\ \text{neglecting} \\ \nearrow\,\text{induced flow}}}{} \qquad\qquad (5\text{-}78)$$

$$\dot{m}_1 = \rho h_D (\omega_{1w} - \omega_{1\infty}) / \underbrace{(1 - \omega_{1w})}_{\substack{\text{correction for} \\ \text{induced flow}}}$$

The form of Eq. (5-78) requires verification for a specific case. For this purpose, consider the boundary-layer situation illustrated in Fig. 5-17.

A flat plate is exposed to a parallel-flowing free stream of constant velocity U, temperature T_∞, and a fixed proportion of species 1 and 2 so that $\omega_{1\infty}$ is also constant. Species 1 is the diffusing substance. The plate temperature is constant at T_w, and the mass fraction of 1 on the plate is also constant at

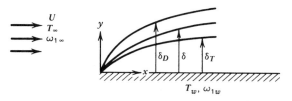

Figure 5-17   Simultaneous heat and mass transfer in laminar boundary-layer flows.

ω_{1w}. The boundary-layer equations are, with $\mathrm{Pr} = \nu/\alpha$, $\mathrm{Sc} = \nu/D_{12}$, $\theta = (T - T_w)/(T_\infty - T_w)$, and $\phi = (\omega_1 - \omega_{1w})/(\omega_{1\infty} - \omega_{1w})$,

Continuity
$$\frac{\partial u}{\partial x} + \frac{\partial v}{\partial y} = 0$$

x Motion
$$u\frac{\partial u}{\partial x} + v\frac{\partial u}{\partial y} = \nu\frac{\partial^2 u}{\partial y^2}$$

Energy
$$u\frac{\partial \theta}{\partial x} + v\frac{\partial \theta}{\partial y} = \frac{\nu}{\mathrm{Pr}}\frac{\partial^2 \theta}{\partial y^2}$$

Diffusion
$$u\frac{\partial \phi}{\partial x} + v\frac{\partial \phi}{\partial y} = \frac{\nu}{\mathrm{Sc}}\frac{\partial^2 \phi}{\partial y^2} \qquad (5\text{-}79)$$

with boundary conditions

$$\text{At } y = 0; \qquad u = 0, \qquad v = v_w, \qquad \theta = 0 = \phi$$
$$\text{At } y \to \infty; \qquad u \to U, \qquad \theta \to 1 \leftarrow \phi$$

The "blowing" condition at $y = 0$ makes this situation similar to the transpiration studies described in Section 5.8. The energy and the diffusion equations are coupled to the x-motion equation in the sense that the heat and mass transfer at the wall may cause v_w to differ from zero. If the mass-transfer rate $\dot{m}_1$ is specified

$$\dot{m}_1 = \rho_1|_w v_w - \rho D_{12}\frac{\partial \omega_1}{\partial y}\bigg|_w = \rho\left(\omega_1 v_w - D_{12}\frac{\partial \omega_1}{\partial y}\bigg|_w\right)$$

the mass fraction ω_{1w} at the wall may not be known. On the other hand, if a particular value of ω_{1w} is forced to exist, $\dot{m}_1$ and v_w may not be known. There is a possibility of $v_w = 0$ if the plate surface is permeable to both species, but the mass flow rates of the two species must be balanced in such a case.

Since the differential equations and boundary conditions for θ and ϕ are the same in Eq. (5-79), the solutions for ϕ and θ will be the same with Sc and Pr interchanged. Consider that (with $v_w \approx 0$) for mass and heat transfer, respectively, the diffusive fluxes are given by

$$\dot{m}_1 = -\rho D_{12}\frac{\partial \omega_1(y = 0)}{\partial y} = \rho h_D(\omega_{1w} - \omega_{1\infty})$$

and

$$q_w = -k\frac{\partial T(y=0)}{\partial y} = h(T_w - T_\infty)$$

In terms of dimensionless ϕ and θ, these relationships become

$$\underbrace{\frac{\partial \phi(y=0)}{\partial y}}_{\approx\, 0.33206\, \text{Sc}^{1/3}} = \frac{h_D}{D_{12}} \quad \text{and} \quad \underbrace{\frac{\partial \theta(y=0)}{\partial y}}_{\approx\, 0.33206\, \text{Pr}^{1/3}} = \frac{h}{k}$$

Equation of the preceding mass- and heat-transfer results then leads to

$$\frac{h_D}{D_{12}}\text{Sc}^{-1/3} = \frac{h}{k}\text{Pr}^{-1/3}$$

$$\frac{\text{Sh}}{\text{Sc}^{1/3}} = \frac{\text{Nu}}{\text{Pr}^{1/3}} \tag{5-80}$$

Two points are important here: (1) the coefficients h_D and h in $\text{Sh} = h_D x/D_{12}$ and $\text{Nu} = hx/k$ give the diffusive transport; any convective transport is an additional contribution; and (2) the extended Reynolds analogy of Eq. (5-80) has been verified only for laminar boundary-layer flow over a flat plate; its applicability to more general cases such as turbulent flow and other geometries is not assured, but it is likely (as experiment confirms [47]). For solution of the boundary-layer equations, see Hartnett and Eckert [45].

5.10 FINITE-DIFFERENCE SOLUTIONS

In many applications the free-stream velocity, wall blowing velocity, or wall temperature vary in such a manner that accurate analytical solutions do not exist for the boundary-layer equations. In such cases accuracy is obtained by numerical solution of the boundary-layer equations. For a flat plate the steady free stream $U(x)$ and wall blowing velocities $v_w(x)$ which might be encountered are illustrated in Fig. 5-18.

Velocities in the boundary layer could be directly determined as outlined by Pletcher [48]. But it is more efficient to first subject the boundary-layer equations to a similarity-like coordinate transformation since the departure from similarity is rarely more than a factor of 3. The similarity variable and stream function are [see Eqs. (5-55) and (5-56)]

$$\eta = y\left(\frac{U}{\nu x}\right)^{1/2} \quad \text{and} \quad \psi = (\nu x U)^{1/2} F(\eta, x)$$

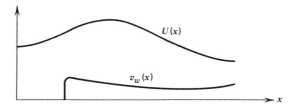

Figure 5-18 Free-stream U and wall blowing v_w velocity variations that can cause a laminar boundary layer to be nonsimilar.

The transformation $x, y \rightarrow x' = x, \eta$ reduces variations in the x direction as illustrated in Fig. 5-19 and is effected by applying the chain rule to obtain

$$u = (\nu x U)^{1/2} \left(\frac{\partial F}{\partial x'} \underset{\bcancel{}}{\frac{\partial x'}{\partial y}}^{0} + \left(\frac{\partial F}{\partial \eta} \frac{\partial \eta}{\partial y} \right)^{(U/\nu x)^{1/2}} \right) = UF'$$

$$-v = \left(\frac{\nu U}{4x} \right)^{1/2} \left(1 + \frac{x}{U} \frac{dU}{dx} \right) F + (\nu x U)^{1/2} \left(\frac{\partial F}{\partial x'} \underset{\bcancel{}}{\frac{\partial x'}{\partial x}}^{1} + \frac{\partial F}{\partial \eta} \underset{\bcancel{}}{\frac{\partial \eta}{\partial x}}^{-\frac{\eta}{2x} + \frac{\eta}{2U} \frac{dU}{dx}} \right)$$

where $F' = \partial F(\eta, x)/\partial \eta$. Repeated applications of the chain rule result in

$$\frac{\partial u}{\partial x} = F' \frac{dU}{dx} + U \frac{\partial F'}{\partial x} - \frac{U\eta}{2x} F'' \left(1 - \frac{x}{U} \frac{dU}{dx} \right)$$

$$\frac{\partial u}{\partial y} = U \left(\frac{U}{\nu x} \right)^{1/2} F''$$

$$\frac{\partial^2 u}{\partial y^2} = \frac{U^2}{\nu x} F'''$$

Substitution into the x-motion equation gives, with $P = (x/U)\, dU/dx$ and

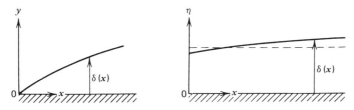

Figure 5-19 Boundary-layer thickness variation in the downstream direction in untransformed and transformed coordinates.

the realization that the expression for v can be rewritten as

$$-v = \frac{\partial\left[(vxU)^{1/2}F\right]}{\partial x} + \eta\left(\frac{vU}{4x}\right)^{1/2}F'$$

and a rearrangement for the purpose of controlling error in a finite-difference solution by setting $f = F - \eta$,

$$f''' - P\left[2f' + (f')^2\right] + \left(\frac{1+P}{2}\right)(f+\eta)f'' = x\left[(f'+1)\frac{\partial f'}{\partial x} - f''\frac{\partial f}{\partial x}\right]$$

$$f'(x, \eta = 0) = -1$$

$$f(x, \eta = 0) = -(vxU)^{-1/2}\int_0^x v_w\,dx$$

$$f'(x, \eta \to \infty) \to 0 \qquad\qquad (5\text{-}81)$$

The transformed x-motion equation [Eq. (5-81)] is approximated in the x-direction by finite differences. For this purpose, the Taylor series for $f(\eta, x_{n-1}) = f_{n-1}$ and $f(\eta, x_{n-2}) = f_{n-2}$ can be written in terms of $f(\eta, x_n) = f_n$ as

$$f_{n-1} = f_n + \frac{\partial f_n}{\partial x}(x_{n-1} - x_n) + \frac{\partial^2 f_n}{\partial x^2}\frac{(x_{n-1} - x_n)^2}{2!} + \cdots$$

$$f_{n-2} = f_n + \frac{\partial f_n}{\partial x}(x_{n-2} - x_n) + \frac{\partial^2 f_n}{\partial x^2}\frac{(x_{n-2} - x_n)^2}{2!} + \cdots$$

A three-point backward difference formula, with an error $O(\Delta x^2)$, is

$$\frac{\partial f(\eta, x)}{\partial x} = \frac{2x_n - x_{n-1} - x_{x-2}}{(x_n - x_{n-1})(x_{n-2} - x_n)}f_n$$

$$- \frac{x_n - x_{n-2}}{(x_n - x_{n-1})(x_{n-1} - x_{n-2})}f_{n-1}$$

$$+ \frac{x_n - x_{n-1}}{(x_n - x_{n-2})(x_{n-1} - x_{n-2})}f_{n-2} \qquad (5\text{-}82)$$

A two-point backward difference formula, with an error $O(\Delta x)$, is

$$\frac{\partial f(\eta, x)}{\partial x} = \frac{f_n - f_{n-1}}{x_n - x_{n-1}} \qquad\qquad (5\text{-}83)$$

The equation to be solved [Eq. (5-81)] is discretized in the x direction. The discrete locations along the axis need not be equally spaced. To illustrate the solution procedure, consider, however, equal spacing Δx in the x direction. The values of f at $x = 0$ can be obtained from similarity solutions by fitting $U(x)$ with Cx^m. At the next downstream location the two-point Eq. (5-83) approximation is employed in a manner similar to the following procedure. For further downstream locations, the three-point Eq. (5-82) gives Eq. (5-81) as

$$f_n''' - P\left[2f_n' + (f_n')^2\right] + \frac{(1 + P)}{2}(f_n + \eta)f_n''$$

$$= \frac{x}{\Delta x}\left[(f_n' + 1)(3f_n' - 4f_{n-1}' + f_{n-2}') - f_n''(3f_n - 4f_{n-1} + f_{n-2})\right]$$

$$(5\text{-}84)$$

with $f_n' = df(\eta, x_n)/d\eta$. Since f_{n-1} and f_{n-2} are now known, f_n is found by numerically solving (perhaps by a fourth-order Runge–Kutta procedure as described by Chow [49]) this ordinary differential equation in the η direction. Equally spaced increments in the η direction of $\Delta\eta = 0.05$ for $0 \leqslant \eta \leqslant \eta_{max} \approx 12$ have been found to give accurate results. It is suggested that $\Delta x \geqslant x/50$ for the x-direction spacing.

The solution proceeds by marching downstream to successive x locations, with the results being dependent mostly on local conditions at x_n. Conditions at x_{n-1} and x_{n-2} are of successively weaker influence. This explains the success of methods that assume local similarity. The equations are parabolic (as a result of the boundary-layer assumptions), so that downstream events do not affect upstream events; if the boundary-layer assumptions are not made, the full equations are elliptic and downstream events affect upstream ones (usually weakly), and numerical computation is more complicated.

Finite-difference solution of the energy equation is accomplished by similar means. To begin, let $\theta = (H - H_\infty)/H_\infty$ and again effect the coordinate transformation $x, y \to x' = x, \eta = y(U/\nu x)^{1/2}$. Application of the chain rule gives

$$\frac{\partial H_0}{\partial x} = \frac{\partial H_0}{\partial x'}\frac{\partial x'}{\partial x} + \frac{\partial H_0}{\partial \eta}\frac{\partial \eta}{\partial x} = H_\infty\frac{\partial \theta}{\partial x} + (\theta + 1)\frac{dH_\infty}{dx} - (\eta/2x)H_\infty\frac{\partial \theta}{\partial \eta}$$

$$\frac{\partial H_0}{\partial y} = \frac{\partial H_0}{\partial x'}\frac{\partial x'}{\partial y} + \frac{\partial H_0}{\partial \eta}\frac{\partial \eta}{\partial y} = H_\infty\left(\frac{U}{\nu x}\right)^{1/2}\frac{\partial \theta}{\partial \eta}$$

$$\frac{\partial^2 H_0}{\partial y^2} = H_\infty\frac{U}{\nu x}\frac{\partial^2 \theta}{\partial \eta^2}$$

Substitution into the energy equation in terms of enthalpy [Eq. (5-39c)] finally results in

$$\frac{\partial\left[\theta'/\text{Pr} + (1 - 1/\text{Pr})(U^2/H_\infty)(f' + 1)f''\right]}{\partial\eta}$$

$$= -\left(\frac{1 + P}{2}\right)(f + \eta)\theta' + x\left[(f' + 1)\frac{\partial\theta}{\partial x} - \theta'\frac{\partial f}{\partial x}\right] \qquad (5\text{-}85)$$

with either

$$\theta(x, \eta = 0) = \frac{H_w(x) - H_\infty}{H_\infty} \qquad \text{or} \qquad \theta'(x, \eta = 0) = -\frac{C_p q_w(x)}{k H_\infty}$$

$$\theta(x, \eta \to \infty) \to 0$$

H_∞ is constant in the free stream. The finite-difference solution procedure for Eq. (5-85) is similar to that discussed from Eq. (5-81).

Complexities such as temperature-dependent properties can be numerically included in this way. The energy and x-motion equations then have additional terms and are coupled. For details, including axisymmetric boundary-layer applications, consult Smith and Clutter [50]. Additional useful information is given by Cebeci and Bradshaw [51]; their FORTRAN programs for solving the laminar boundary-layer equations in finite-difference form are based on the Keller–Cebeci [52, 53] box method and employ a nonuniform spacing in the η direction. Computer programs are given by Schetz [53a] for both laminar and turbulent boundary layers, as well.

Boundary-fitted coordinate systems transform the problem doamin into one (often rectangular) in which numerical calculations are more easily done and enable solution points to be clustered in regions where gradients are large.

When a conjugate problem is encountered, involving a coupling of heat conduction in a solid and convection in the adjacent fluid, a finite-difference method devised by Patankar [54, 55] allows formulating the problem in terms of a variable viscosity and thermal conductivity fluid over a composite domain that includes both the solid and the fluid. That part of the composite domain occupied by the solid has the thermal conductivity of the solid and an infinite viscosity, whereas the part occupied by the fluid has the properties of the fluid. To illustrate the method, consider a grid laid out so that the fluid–solid interface is between control volumes on either side with a grid point at the center of each control volume. Let L_1 and L_2 be the perpendicular distances from each grid point to the interface. Then the effective viscosity μ^* at the interface, in terms of the viscosities μ_1 and μ_2, at the grid points, is

$$\frac{1}{\mu^*} = \frac{L_1/\mu_1 + L_2/\mu_2}{L_1 + L_2}$$

In similar fashion, the effective thermal conductivity k^* at the interface is

$$\frac{1}{k^*} = \frac{L_1/k_1 + L_2/k_2}{L_1 + L_2}$$

The basis for these representations of effective interface properties is grasped by considering one-dimensional heat conduction through the composite slab previously described in which the temperatures of the two grid points are T_1 and T_2, respectively. An effective thermal conductivity satisfies the relationship

$$\frac{k^*(T_1 - T_2)}{L_1 + L_2} = \frac{T_1 - T_2}{k_1/L_1 + k_2/L_2}$$

The previous expression for the effective thermal conductivity at the interface follows. This method has been applied to laminar flow in a square duct of finite wall thickness [54] and to ducts with internal fins [54, 56].

PROBLEMS

5-1 (a) Show that the dimensionless unsteady continuity and x-motion boundary-layer equations are

$$\text{Continuity} \qquad \frac{L}{t_r U}\frac{\partial \rho'}{\partial t'} + \frac{\partial(\rho' u')}{\partial x'} + \cdots = 0$$

$$x \text{ Motion} \quad \rho'\left(\frac{L}{t_r U}\frac{\partial u'}{\partial t'} + u'\frac{\partial u'}{\partial x'} + \cdots\right) = \cdots$$

where $t' = t/t_r$, $\rho' = \rho/\rho_r$, $u' = u/U$, and $x' = x/L$ with t_r a time characteristic of that required for the contemplated change to occur (e.g., in free-stream velocity U). Thus, if $L/t_r U \ll 1$, unsteady effects can be neglected and the viscous drag can be determined at any instant from the corresponding instantaneous free-stream velocity.

(b) Show that for $L = 3$ ft and $U = 300$ ft/sec, if $t_r > 10^{-2}$ sec, unsteady effects have negligible effect on velocity distributions.

5-2 **(a)** Show by means of an order of magnitude analysis that the boundary-layer equations for a general fluid can be expressed as

$$x \text{ Motion} \qquad \rho \frac{Du}{Dt} = B_x - \frac{\partial p}{\partial x} + \frac{\partial \tau_{yx}}{\partial y}$$

$$\text{Energy} \quad \rho C_p \frac{DT}{Dt} = \frac{\partial(k \, \partial T/\partial y)}{\partial y} + \tau_{yx} \frac{\partial u}{\partial y} + q''' + \beta T \left(\frac{\partial p}{\partial t} + u \frac{\partial p}{\partial x} \right)$$

(b) Show that, for steady parallel flow of a constant-property Newtonian fluid over a flat plate where $U = \text{const}$, $u = \partial\psi/\partial y$, and $v = -\partial\psi/\partial x$ with the stream function $\psi = (\nu U x)^{1/2} F(\eta)$ and the similarity variable $\eta = y(U/\nu x)^{1/2}$, the chain rule gives

$$(1) \quad u = \frac{\partial\psi}{\partial y} = \frac{\partial\psi}{\partial x}\overset{0}{\frac{\partial x}{\partial y}} + \frac{\partial\psi}{\partial\eta}\overset{(U/\nu x)^{1/2}}{\frac{\partial\eta}{\partial y}} = UF'$$

$$(2) \quad -v = \frac{\partial\psi}{\partial x} = \frac{\partial\psi}{\partial x}\overset{1}{\frac{\partial x}{\partial x}} + \frac{\partial\psi}{\partial\eta}\overset{-\eta/2x}{\frac{\partial\eta}{\partial x}} = \left(\frac{U\nu}{x}\right)^{1/2}\frac{F - \eta F'}{2}$$

$$(3) \quad \frac{\partial u}{\partial y} = \frac{\partial UF'}{\partial y} = \frac{\partial UF'}{\partial x}\overset{0}{\frac{\partial x}{\partial y}} + \frac{\partial UF'}{\partial\eta}\frac{\partial\eta}{\partial y} = \left(\frac{U^3}{\nu x}\right)^{1/2} F''$$

$$(4) \quad \frac{\partial^2 u}{\partial y^2} = \frac{\partial\left[(U^3/\nu x)^{1/2} F''\right]}{\partial x}\overset{0}{\frac{\partial x}{\partial y}} + \frac{\partial\left[(U^3/\nu x)^{1/2} F''\right]}{\partial\eta}\frac{\partial\eta}{\partial y}$$
$$= \frac{U^2}{\nu x} F'''$$

$$(5) \quad \frac{\partial u}{\partial x} = \frac{\partial UF'}{\partial x}\overset{1}{\frac{\partial x}{\partial x}} + \frac{\partial UF'}{\partial\eta}\frac{\partial\eta}{\partial x} = -\frac{U\eta F''}{2x}$$

(c) Substitute the results of parts 1–5 into the x-motion laminar boundary-layer equation to verify Eq. (5-15).

5-3 Construct an additional column for Table 5-1 to show the value of $v(Ux/\nu)^{1/2}/U$ for steady parallel flow of a constant-property fluid over a flat plate.

5-4 An iterative procedure [57] for approximating the velocity distribution is as follows.

(a) Starting with Eq. (5-15) show that a first integration gives

$$F'' = F''(0)\exp\left(-\int_0^\eta \frac{F}{2}\, d\eta\right)$$

and a second integration gives

$$F'(\eta)^{\overbrace{}^{u/U}} - F'(0)^{\overbrace{}^{0}} = F''(0)\int_0^\eta \exp\left(-\int_0^\eta \frac{F}{2}\, d\eta\right) d\eta$$

(b) Show that, since $F'(\infty) = 1$,

$$F''(0) = \frac{1}{\int_0^\infty \exp\left(-\int_0^\eta (F/2)\, d\eta\right) d\eta}$$

and so

$$F(\eta) - F(0)^{\overbrace{}^{0}} = F''(0)\int_0^\eta \left[\int_0^\eta \exp\left(-\int_0^\eta \frac{F}{2}\, d\eta\right) d\eta\right] d\eta$$

(c) Starting with $F(\eta) = \eta$, perform at least two iterations with a new and updated function for $F(\eta)$ obtained from the last iteration.

(d) Compare the value of $F''(0)$ from part c after each iteration with the exact answer. Plot u/U against η from part c and compare it with the exact profile.

5-5 The average drag coefficient for steady parallel flow of a constant-property fluid over a flat plate given by Eq. (5-18) includes the effect of an integrable singularity at the leading edge, $x = 0$. As pointed out by Schlichting [1, p. 131], a refined analysis that accounts for the singularity results in

$$\frac{\overline{C}_f}{2} = \frac{0.664}{Re_x^{1/2}} + \frac{1.163}{Re_x}$$

For Re_x of 10^3, 10^4, and 10^5, evaluate the error involved in using only Eq. (5-18) to compute the friction coefficient.

5-6 A method for converting a boundary-value problem into an initial-value problem eases the numerical solution of the laminar boundary-layer equations 58–60. For the Blasius equation of Eq. (5-15), consider the transformation

$$\eta = A^a n \quad \text{and} \quad F = A^b f$$

where a and b are constants to be determined and A is a parameter.
(a) Show that, under these transformations, Eq. (5-15) becomes

$$A^{b-3a} \frac{d^3 f}{dn^3} + \frac{1}{2} A^{2b-2a} f \frac{d^2 f}{dn^2} = 0$$

$$f(n = 0) = 0 = \frac{df(n = 0)}{dn}$$

$$A^{b-a} \frac{df(n = \infty)}{dn} = 1$$

(b) Now, realizing that the unknown value of $d^2 F(0)/d\eta^2$ is the difficulty in the first place, set it equal to a constant so that, after transformation,

$$A^{b-2a} \frac{d^2 f(0)}{dn^2} = A \qquad\qquad (\text{P6-1})$$

(c) Show that for the results of parts (a) and (b) to be invariant under the transformation

$$b - 3a = 2b - 2a \quad \text{and} \quad b - 2a = 1$$

from which it is found that $-a = \frac{1}{3} = b$.

(d) Show that from the boundary condition

$$\frac{dF(\infty)}{d\eta} = 1$$

the constant A is given in terms of f as

$$A = \left[\frac{df(\infty)}{dn} \right]^{-3/2} \qquad\qquad (\text{P6-2})$$

where $f(\infty)$ is found from the solution to

$$\frac{d^3 f}{dn^3} + \frac{1}{2} f \frac{d^2 f}{dn^2} = 0$$

$$\qquad\qquad (\text{P6-3})$$

$$f(0) = 0 = \frac{df(0)}{dn} \quad \text{and} \quad \frac{d^2 f(0)}{dn^2} = 1$$

(e) Solve Eq. (P6-3) numerically with a computer and determine $d^2 F(0)/d\eta^2$ from the result. Also, in view of the relation that $u/U = dF/d\eta = A^{2/3} df/dn$, determine the velocity profile. Compare both results with the exact answers.

(f) Solve the Blasius equation [Eq. (5-15)] numerically with a computer, using a shooting method.

(g) Cast Eq. (P6-3) into the form suited for an iterative solution (refer to Problem 5-4) of

$$f(n) = \int_0^n \left[\int_0^n \exp\left(-\int_0^n \frac{f}{2} \, dn \right) dn \right] dn$$

(h) Starting with $f = 2n$ and using the results of part g, perform at least two iterations with a new and updated f obtained from the last equation in part g after each iteration. Compare the value of $d^2F(0)/dn^2$ after each iteration with the exact answer.

5-7 (a) For steady parallel flow of air over a flat plate at atmospheric pressure at 100°F and a free-stream velocity of 100 ft/sec, determine the magnitude of the vertical component of velocity at the outer edge of the boundary layer at distances of 0.3, 3, and 6 cm from the plate leading edge. At each point determine the boundary-layer thickness and the local wall shear stress. *Answer*: At $x = 6$ cm, $v = 0.26$ ft/sec.

(b) Assess the probable accuracy of the numerical results of part a. How well are the boundary-layer assumptions satisfied?

(c) Determine the total drag force acting on one side of the plate up to each point of computation. *Answer*: At $x = 6$ cm, drag = 0.0087 lb$_f$/ft.

(d) If a second plate were to be brought opposite and parallel to the plate of part a, estimate the separation between them that would be required to keep the boundary layers from meeting 6 cm from the leading edge. *Answer*: 0.11 cm.

5-8 To assess the importance of fluid properties, work Problem 5-7, but take the fluid to be water at 22°C.

5-9 Consider application of the conservation principles of Chapter 2 to control volume $A-B-C-D$ for the steady laminar boundary-layer problem sketched in Fig. 5P-9.

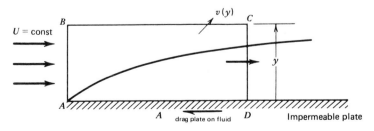

Figure 5P-9

(a) Show that the mass flow rate across B–C equals

$$\int_0^y (\rho U - \rho u)\, dy = \int_{x_A}^{x_D} \rho v(y)\, dx$$

(b) Show that the drag force exerted on the fluid by the plate is

$$\int_0^y \frac{\rho UU}{g_c}\, dy - \left[\int_0^y \frac{\rho uu}{g_c}\, dy + \int_{x_A}^{x_B} \frac{\rho v(y)U}{g_c}\, dx \right] - F_{\substack{\text{drag} \\ \text{plate on fluid}}} = 0$$

(c) Use the results of parts a and b to show that, when $y \to \infty$ to avoid shear stresses that exist near the plate,

$$F_{\text{drag}} = \frac{\rho}{g_c} \int_0^\infty u(U - u)\, dy$$

and

$$\delta_2 = x \frac{\overline{C}_f}{2} = 0.66412 \left(\frac{\nu x}{U} \right)^{1/2}$$

5-10 Prepare a brief report, explaining the manner in which a small movable element mounted flush with a plate surface can be used to measure local drag [61–65].

5-11 Show that the similarity solution for steady parallel flow of a constant-property fluid over a flat plate results in the local Nusselt number being related to $\theta = (T - T_w)/(T_\infty - T_w)$ by

$$Nu = \frac{hx}{k} = \left(\frac{Ux}{\nu} \right)^{1/2} \theta'(0)$$

5-12 Show that Eq. (5-31) reduces to the simple relation $\theta(\eta) = u/U$ if $Pr = 1$.

5-13 Consider steady parallel flow of a constant-property fluid over a flat plate. When $Pr \to 0$, the thermal boundary layer is much thicker than the velocity boundary layer, so that throughout the thermal boundary layer $F \approx \eta$ since then $u/U = F' \approx 1$. Such is the case for liquid metals.

(a) Show that in such a case Eq. (5-26) gives

$$\frac{d\theta(0)}{d\eta} = \left(\frac{Pr}{\pi} \right)^{1/2}$$

(b) Show that the result of part a gives, with $\text{Nu} = hx/k$ and $\text{Re} = Ux/\nu$,

$$\lim_{\text{Pr}\to 0} \frac{\text{Nu}}{\text{Re Pr}^{1/2}} = \frac{1}{0.33206\pi^{1/2}} \frac{C_f}{2} = 1.7\frac{C_f}{2}$$

5-14   Consider steady parallel flow of a constant-property fluid over a flat plate. When $\text{Pr} \to \infty$, the thermal boundary layer is much thinner than the velocity boundary layer, so that throughout the thermal boundary layer u/U is nearly linear. This case was solved by M. A. Leveque in 1928 according to Schlichting [1, p. 272], who notes that this situation can also occur if there is an unheated starting length x_0 at the leading edge. A velocity distribution that retains accuracy while giving the necessary linearity is $u = \tau_w y/\mu$. The continuity equation then gives

$$v = -\left(\frac{y^2}{2\mu}\right)\frac{d\tau_w}{dx}$$

(a) Show that the substitution

$$n = \frac{y(\tau_w/\mu)^{1/2}}{\left[9\alpha\int_{x_0}^{x}(\tau_w/\mu)^{1/2}\,dx\right]^{1/3}}$$

transforms the energy equation into the similarity form

$$\frac{d^2\theta}{dn^2} + 3n^2\frac{d\theta}{dn} = 0$$

$$\theta(0) = 0 \quad \text{and} \quad \theta(\infty) = 1$$

(b) Show that the solution to the differential equation of part a is

$$\theta = \frac{d\theta(0)}{dn}\int_0^n e^{-Z^3}\,dZ$$

and that

$$\frac{d\theta(0)}{dn} = \frac{1}{\int_0^\infty e^{-Z^3}\,dZ} = \frac{1}{\Gamma(4/3)} = \frac{1}{0.893}$$

Here $\Gamma(Z)$ is the gamma function that is related to the integral in question as

$$\Gamma(Z+1) = \int_0^\infty \exp(-x^{1/Z})\,dx$$

(c) Show that the result of part b leads to

$$\lim_{\text{Pr} \to \infty} \text{Nu} = \frac{hx}{k} = x \left(\frac{\tau_w}{\mu} \right)^{1/2} \frac{d\theta(0)/dn}{\left[9\alpha \int_0^x (\tau_w/\mu)^{1/2} \, dx \right]^{1/3}}$$

From Eq. (5-17) obtain the information relating τ_w to x and evaluate the integral to yield

$$\lim_{\text{Pr} \to \infty} \frac{\text{Nu}}{\text{Re Pr}^{1/3}} = 1.021 \frac{C_f}{2}$$

(d) Show that the results of parts a–c can also be achieved by realizing that near the plate, $d^2F/d\eta^2 = 0.33206$ so that $dF/d\eta \approx 0.33206\eta$ and $F \approx 0.33206\eta^2/2$.

5-15 Find the values of Pr at which Eqs. (5-28a) and (5-28b) are equal and Eqs. (5-28b) and (5-28c) are equal.

5-16 Show that if Pr $= 1$, Eq. (5-34) gives $\theta_2(\eta) = 1 - [F'(\eta)]^2 = 1 - (u/U)^2$ and the recovery factor as $r = 1$.

5-17 Consider steady parallel flow of a constant-property fluid over a flat plate with viscous dissipation. The value of $\theta_2(0)$ from Eq. (5-34) is desired for Pr $\to \infty$. Since the thermal boundary layer is much thinner than the velocity boundary layer, $F'' \approx 0.33206$, $F' \approx 0.33206\eta$, and $F \approx 0.33206\eta^2/2$. Thus Eq. (5-34) is approximately given by

$$\theta_2'' + \frac{\text{Pr}}{2} \frac{0.33206\eta^2}{2} \theta_2' = -2\,\text{Pr}(0.33206)^2$$

$$\theta_2'(0) = 0 = \theta_2(\infty)$$

(a) Show that the solution, obtained by Meksyn [66], is

$$\theta_2(\eta) - \theta_2(0) = -2.4105\,\text{Pr}^{1/3} \int_0^x e^{-y^3} \left[\int_0^y e^{Z^3} \, dZ \right] dy$$

(b) Show that the result of part a yields

$$\lim_{\text{Pr} \to \infty} \theta_2(0) = 1.9\,\text{Pr}^{1/3}$$

5-18 Consider a constant-property fluid in steady parallel flow over an adiabatic flat plate. Show that the total temperature in the fluid

$T° = T + u^2/2C_p$ differs between the plate and the free stream by

$$T_\infty° - T_{aw}° = \frac{(1 - r)U^2}{2C_p}$$

Note particularly that if $r \equiv 1$, no energy separation occurs. Comment on the magnitude of the energy separation predicted for the case of laminar boundary layers, including energy separation in the Hilsch–Ranque tube by Eckert (see reference [3] in Chapter 3).

5-19 The effect of variable properties described in Section 5.5 made use of the coordinate transformation $\eta = (U/\nu_\infty x)^{1/2}\int_0^y (\rho/\rho_\infty)\, dy$.

   (a) Show that the incremental distance dy from the solid surface is expressed according to this transformation by

$$dy = \left(\frac{\nu_\infty x}{U}\right)^{1/2} \frac{\rho_\infty}{\rho}\, d\eta$$

   (b) On the basis of the result of part a, verify that the velocity distribution in Fig. 5P-19a for the η coordinate would have the velocity distribution in Fig. 5P-19b for the y coordinate for gases.

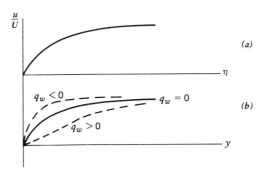

Figure 5P-19

   (c) Repeat part b for a liquid.

5-20 Show for wedge flow that the displacement thickness δ_1 is given by

$$\delta_1\left(\frac{U}{\nu x}\right)^{1/2} = (\eta - F)\big|_\infty$$

and that the momentum thickness δ_2 is given by

$$\delta_2 \left(\frac{U}{\nu x} \right)^{1/2} = \frac{2[F''(0) - m(\eta - F)|_\infty]}{3m + 1}$$

Either consult Hartree [25], noting that his definition of η may cause his F values to differ from those in this text, or solve the Falker–Skan equation [Eq. (5-58)] numerically to construct two additional columns for δ_1 and δ_2 in Table 5-2.

5-21 Compare the magnitudes of the heat-transfer coefficients for laminar boundary-layer flow over a flat plate for the specified heat flux and specified temperature cases for both the local and average values.

5-22 Determine whether or not the Reynolds analogy $Nu/Re\,Pr^{1/3} = C_f/2$ is valid for wedge flow on both local and average bases.

5-23 Determine the variation in local laminar heat-transfer coefficient on the forward part of a cylinder in normal crossflow by use of local similarity as discussed by Burmeister [67].

5-24 For two-dimensional stagnation flow ($m = 1$), $Nu/Re^{1/2}\,Pr^{0.4} = 0.57$ in Table 5-3; and for rotationally symmetric stagnation flow, $Nu/Re^{1/2}\,Pr^{0.4} = 0.76$ in Eq. (5-72). Here $Nu = hx/k$ and $Re = Ux/\nu$.

(a) Show that for crossflow over a cylinder, $U \approx 2U_\infty x/R$. Substitute this result into the two-dimensional correlation to show that the two-dimensional stagnation point heat-transfer coefficient varies with radius of curvature as

$$\frac{Nu_R}{Re_R^{1/2}\,Pr^{0.4}} = 0.81$$

where $Nu_R = hR/k$ and $Re_R = U_\infty R/\nu$.

(b) Repeat part a for a sphere, showing $U \approx 3U_\infty x/2R$, to find

$$\frac{Nu_R}{Re_R^{1/2}\,Pr^{0.4}} = 0.93$$

(c) On the basis of the results of parts a and b, comment on the reduction in heat flux on entry of a space vehicle into a planetary atmosphere offered by a large radius of curvature since $h \sim R^{-1/2}$.

5-25 Compare the heat-transfer coefficient that results from an infinite stream impinging perpendicularly on a flat surface with that resulting from an optimal array of round nozzles under the condition that the average fluid flow per unit plate area is the same in both cases.

5-26 From Table 5-6 determine the ratio of the local heat-transfer coefficients for constant heat flux to constant-temperature wall conditions for several wedge angles and transpiration rates.

REFERENCES

1. H. Schlichting, *Boundary-Layer Theory*, 6th ed., McGraw-Hill, 1968.
2. A. G. Hansen, *Similarity Analyses of Boundary Value Problems in Engineering*, Prentice-Hall, 1964.
3. L. Howarth, *Proc. Roy. Soc. London, Ser. A* **164**, 547–579 (1938).
4. A. Bejan, *Convection Heat Transfer*, Wiley-Interscience, 1984, pp. 432–434.
5. E. R. G. Eckert and R. M. Drake, *Analysis of Heat and Mass Transfer*, Taylor and Francis, 1972.
6. E. Pohlhausen, *Zeitschrift für angewandte Mathematik und Mechanik* **1**, 115–121 (1921).
7. S. Churchill and H. Ozoe, *ASME J. Heat Transfer* **95**, 416–419 (1973).
8. E. R. G. Eckert, *Trans. ASME* **78**, 1273–1284 (1956).
9. A. Dorodnitzyn, *Prikladnaia Matematika i Mekhanika* **6**, 449–485 (1942).
10. K. Stewartson, *Proc. Roy. Soc. London, Ser. A* **200**, 84–100 (1950).
11. H. A. Simon, C. S. Liu, and J. P. Hartnett, *Int. J. Heat Mass Transfer* **10**, 406–409 (1967).
12. E. R. G. Eckert and E. Pfender, *Adv. Heat Transfer* **4**, 229–316 (1967).
13. E. Reshotko and C. B. Cohen, Heat transfer at the forward stagnation point of blunt bodies, NACA TN 3513, July 1955.
14. C. B. Cohen and E. Reshotko, Similar solutions for the compressible laminar boundary layer with heat transfer and pressure gradient, NASA TR 1293, 1956, pp. 919–956.
15. S. Levy, *J. Aeronaut. Sci.* **21**, 459–474 (1954).
16. G. Poots and G. F. Raggett, *Int. J. Heat Mass Transfer* **10**, 597–610 (1967).
17. L. Prandtl, "Uber Flüssigkeits bewegung bei sehr kleiner Reibung," *Proc. Third Int. Math. Kong, Heidelberg*, 1904.
18. H. Blasius, *Zeitschrift für angewandte Mathematik und Physik* **56**, 1 (1908); see also NASA TM 1256.
19. A. G. Hansen, *Similarity Analyses of Boundary Value Problems in Engineering*, Prentice-Hall, 1964.
20. R. L. Panton, *Incompressible Flow*, Wiley-Interscience, 1984, pp. 451–473.
21. J. T. Bertin, *Engineering Fluid Mechanics*, 2nd ed., Prentice-Hall, 1984, pp. 237–239.
22. J. E. A. John and W. L. Haberman, *Introduction to Fluid Mechanics*, 3rd ed., Prentice-Hall, 1988, pp. 225–237.
23. F. M. White, *Fluid Mechanics*, 2nd ed., McGraw-Hill, 1986, pp. 441–494.
24. V. M. Falkner and S. W. Skan, *Philos. Mag.* **12**, 865–896 (1931); see also Report of the Memorial Aeronautical Research Committee, London, No. 1, 314, 1930.

25. D. R. Hartree, *Proc. Cambridge Philos. Soc.* **33**, Part II, 223–239 (1937).

26. S. Goldstein, *Proc. Cambridge Philos. Soc.* **35**, 338–340 (1939).

27. S. A. Forbrich, *AIAA J.* **20**, 1306–1307 (1982).

28. H. Chuang, *AIAA J.* **23**, 2004–2005 (1985).

29. K. Stewartson, *Proc. Cambridge Philos. Soc.* **50**, 454–465 (1954).

30. T. Cebeci and H. B. Keller, *J. Comput. Phys.* **7**, 289–300 (1971).

31. L. Yu and L. Yili, *AIAA J.* **27**, 1453–1455 (1989).

32. D. Stojanovic, *J. Aerospace Sci.* **26**, 571–574 (1959).

33. E. Eckert, *VDI-Forschungsheft* **416**, Berlin, 1942.

34. H. L. Evans, *Int. J. Heat Mass Transfer* **5**, 35–57 (1962).

35. S. Levy, *J. Aeronaut. Sci.* **19**, 341–348 (1952).

36. E. A. Brun, *Selected Combustion Problems*, Vol. 2, AGARD, Pergamon, 1956, pp. 105–198.

37. D. R. Chapman and W. M. Rubesin, *J. Aeronaut. Sci.* **16**, 547–565 (1949).

38. W. Mangler, *Zeitschrift für angewandte Mathematik und Mechanik* **28**, 97–103 (1948).

39. L. Lees, *Jet Propulsion* **26**, 259–269 (1956).

40. E. R. G. Eckert and O. E. Tewfik, *J. Aerospace Sci.* **27**, 464–466 (1960).

41. C. F. Dewey and J. F. Gross, *Adv. Heat Transfer* **4**, 317–446 (1967).

42. E. Y. C. Sun, A compilation of coordinate transformations applied to the boundary-layer equations for laminar flows, Deutsche Versuchanstalt für Luftfahrt Report 121, 1960.

43. L. C. Burmeister and R. G. Schoenhals, *Prog. Heat Mass Transfer* **2**, 371–394 (1969).

44. H. Martin, *Heat Exchanger Design Handbook*, Hemisphere, 1986, pp. 2.5.6-1–2.5.6-10.

45. J. P. Hartnett and E. R. G. Eckert, *Trans. ASME* **79**, 247–254 (1957).

46. P. L. Donoughe and J. N. B. Livingood, Exact solutions of laminar-boundary-layer equations with constant property values for porous wall with variable temperature, NASA TN 3151, 1954.

47. J. N. Shadid and E. R. G. Eckert, *ASME J. Turbomachinery* **11**, 27–33 (1991).

48. R. H. Pletcher, *AIAA J.* **7**, 305–311 (1969).

49. C.-Y. Chow, *An Introduction to Computational Fluid Mechanics*, Seminole Publishing Co., P.O. Box 3315, High-Mar Station, Boulder, CO 80307, 1983.

50. A. M. O. Smith and D. W. Clutter, *AIAA J.* **3**, 639–647 (1965).

51. T. Cebeci and B. Bradshaw, *Momentum Transfer in Boundary Layers*, Hemisphere, 1977.

52. H. B. Keller and T. Cebeci, Accurate numerical methods for boundary-layer flows, Part 1, Two-dimensional laminar flows, *Lecture Notes in Physics, Vol. 8, Proc. Second International Conference on Numerical Methods in Fluid Dynamics*, Springer-Verlag, 1971, p. 92.

53. T. Cebeci, *Handbook of Numerical Heat Transfer*, Wiley, 1988, pp. 117–154.

53a. J. A. Schetz, *Boundary Layer Analysis*, Prentice Hall, 1993.

54. S. V. Patankar, A numerical method for conduction in composite materials, flow in irregular geometries and conjugate heat transfer, *Proc. Sixth International Heat Transfer Conference, Toronto, Canada*, Vol. 3, 1978, Paper CO-14, pp. 297–302.

55. S. V. Patankar, *Numerical Heat Transfer and Fluid Flow*, Hemisphere, 1980.

56. E. M. Sparrow and C. F. Hsu, *ASME J. Heat Transfer* **103**, 18–25 (1981).

57. J. Piercy and N. A. V. Preston, *Philos. Mag.* 995–1005 (1936).

58. S. Goldstein, *Modern Developments in Fluid Dynamics*, Oxford University Press, 1938, p. 135.

59. T. Y. Na, *SIAM Rev.* **10**, 85–87 (1968).

60. T. Y. Na, An initial value method for the solution of a class of nonlinear equations in fluid mechanics, ASME Paper 69-WA/FE-8.

61. M. Acharya, J. Bornstein, M. P. Escudier, and V. Vokurka, *AIAA J.* **23**, 410–415 (1985).

62. H. W. Liepmann and S. Dhawan, Direct measurements of local skin friction in low-speed and high-speed flow, *Proc. First U.S. Nat. Cong. Appl. Mech.*, 1951, p. 869.

63. J. M. Allen, *AIAA J.* **18**, 1342–1345 (1980).

64. G. M. Kelly, J. M. Simmons, and A. Paull, *AIAA J.* **30**, 844–845 (1992).

65. D. J. Monson, *AIAA J.* **22**, 557–559 (1984).

66. D. Meksyn, *Zeitschrift für angewandte Mathematik und Physik* **11**, 63–68 (1960).

67. L. C. Burmeister, *Convective Heat Transfer*, Wiley-Interscience, 1983, pp. 310–312.

6

INTEGRAL METHODS

The problems treated thus far have exact solutions, if the underlying assumptions are accepted. In many instances the solution methods previously used are either not directly applicable or do not give answers in closed form. A solution method is desired that gives accurate answers easily for complex situations, even though the answers might not be exact. An integral method meets these needs.

6.1 LEIBNITZ' FORMULA FOR DIFFERENTIATION OF AN INTEGRAL WITH RESPECT TO A PARAMETER

In formulating an integral description of a phenomenon, it is helpful to be able to differentiate an integral with respect to a parameter. The basic operations to be performed are depicted in Fig. 6-1. One basically seeks, with $I(t) = \int_{\alpha(t)}^{\beta(t)} f(x, t) \, dt$,

$$\lim_{\Delta t \to 0} \frac{I(t + \Delta t) - I(t)}{\Delta t}$$

It can be seen that

$$\lim_{\Delta t \to 0} \frac{I(t + \Delta t) - I(t)}{\Delta t}$$

$$= \lim_{\Delta t \to 0} \left\{ \left[\int_{a(t)}^{\beta(t)} f(x, t + \Delta t) \, dx \right. \right.$$

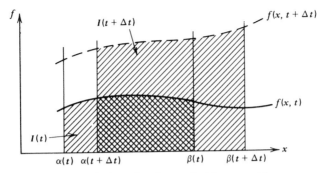

Figure 6-1   Differentiation of an integral with respect to a parameter.

$$+f\left(x = \beta\left(t + \frac{\Delta t}{2}\right)\right)[\beta(t + \Delta t) - \beta(t)]$$

$$-f\left(x = \alpha\left(t + \frac{\Delta t}{2}\right)\right)[\alpha(t + \Delta t) - \alpha(t)]\bigg]$$

$$-\int_{\alpha(t)}^{\beta(t)} f(x,t)\, dx\bigg\} \bigg/ \Delta t$$

or, which is Leibnitz' formula,

$$\frac{\partial}{\partial t}\left[\int_{\alpha(t)}^{\beta(t)} f(x,t)\, dx\right] = \int_{\alpha(t)}^{\beta(t)} \frac{\partial f(x,t)}{\partial x}\, dx$$

$$+ f(x = \beta(t), t)\frac{d\beta(t)}{dt} - f(x = \alpha(t), t)\frac{d\alpha(t)}{dt}$$

$$\tag{6-1}$$

6.2   TRANSIENT ONE-DIMENSIONAL CONDUCTION

Although integral methods appear to have been first used to solve boundary-layer problems by von Karman [1] and Pohlhausen [2], they are introduced by considering transient heat conduction. The simpler mathematics in such an application allows concentration on the method rather than on the problem. Goodman [3–5] treats one-dimensional transient heat conduction extensively, whereas Sfeir [6] treats two-dimensional steady conduction problems; Langford [7] can be consulted for additional discussion.

Consider a semi-infinite solid initially at uniform temperature T_0 whose surface temperature is suddenly raised to a new constant level T_s as shown in

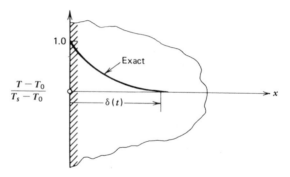

Figure 6-2 Dimensionless transient temperature variation in a semi-infinite solid of initially uniform temperature subjected to a step change in surface temperature.

Fig. 6-2. The describing equation and boundary conditions are

$$\rho C_p \frac{\partial T}{\partial t} = \frac{\partial(k\, \partial T/\partial x)}{\partial x} \tag{6-2a}$$

$$T(x,0) = T_0 \tag{6-2b}$$

$$T(\infty, t) = T_0 \tag{6-2c}$$

$$T(0, t > 0) = T_s \tag{6-2d}$$

This is a classical problem whose solution is available from Carslaw and Jaeger [8]. Here we wish to convert to an integral formulation by integrating the differential equation with respect to x. This gives

$$\int_0^x \rho C_p \frac{\partial T}{\partial t}\, dx = \int_0^x \frac{\partial}{\partial x}\left(k \frac{\partial T}{\partial x}\right) dx$$

The right-hand side is integrated to put the preceding relation into the form

$$\int_0^x \rho C_p \frac{\partial T}{\partial t}\, dx = \left[-k \frac{\partial T(0,t)}{\partial x}\right] - \left[-k \frac{\partial T(x,t)}{\partial x}\right]$$

$$= q_w(t) - q_x(t) \tag{6-3}$$

where q_w is the heat flux into the wall at $x = 0$ and q_x is the heat flux in the x direction at x. Next it is assumed that the thermal disturbance penetrates into the wall only a distance $\delta(t)$—an approximation since the exact solution shows that the disturbance is nonzero, although often small, everywhere for $t > 0$—beyond which nothing of interest occurs. Therefore, the upper limit of the integral in Eq. (6-3) is set equal to $\delta(t)$ and $q_\delta = 0$, giving

$$\int_0^{\delta(t)} \rho C_p \frac{\partial T}{\partial t}\, dx = q_w(t)$$

Assuming that ρ and C_p are constant and applying the Leibnitz formula [Eq. (6-1)] leads to

$$\frac{\partial}{\partial t}\left(\int_0^\delta T\, dx\right) - T_\delta \frac{d\delta}{dt} + T_0 \frac{d0}{dt} = \frac{q_w}{\rho C_p}$$

But according to the earlier assumption, $T_\delta = T_0$, so

$$\rho C_p \frac{\partial}{\partial t}\left[\int_0^{\delta(t)} (T - T_0)\, dx\right] = q_w \qquad (6\text{-}4)$$

The integral formulation of Eq. (6-4) expresses the conservation of energy principle, stating that the difference between the rate of energy flow in and out must equal the rate of energy storage. Also, the steps in achieving the integral formulation from the partial differential formulation have reversed those in deriving the partial differential formulation from an integral one such as were discussed in Chapter 2. The advantage of Eq. (6-4) is that a temperature profile can be assumed for the integral, following which $\delta(t)$ can be evaluated from the resulting first-order differential equation that will be ordinary rather than partial, although nonlinear.

Any temperature profile must satisfy $T(x = 0, t) = T_s$ and $T(x = \delta, t) = T_0$, which corresponds to the two boundary conditions of Eqs. (6-2c) and (6-2d). The simplest profile shown in Fig. 6-3 satisfying these conditions is

$$T - T_0 = (T_s - T_0)\left(1 - \frac{x}{\delta}\right), \qquad x \leqslant \delta$$

$$= 0, \qquad x > \delta$$

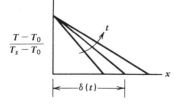

Figure 6-3 Linear approximation to the transient variation of dimensionless temperature in a semi-infinite solid of initially uniform temperature subjected to a step change in surface temperature.

Introduction of this profile into the integral formulation of Eq. (6-4) gives

$$\rho C_p \frac{\partial}{\partial t}\left[\frac{(T_s - T_0)\delta}{2}\right] = k\frac{T_s - T_0}{\delta}$$

This differential equation can be simplified to

$$\frac{d\delta^2}{dt} = 4\alpha$$

where $\alpha = k/\rho C_p$. Assumption that $\delta(t = 0) = 0$ gives

$$\delta = (4\alpha t)^{1/2} \tag{6-5}$$

showing that the penetration depth of a disturbance varies as the square root of elapsed time. Knowing δ, a somewhat fictitious quantity, one can now evaluate the wall heat flux, a real quantity, as

$$q_w = k\frac{T_s - T_0}{\delta} = (T_s - T_0)\left(k\frac{\rho C_p}{4t}\right)^{1/2} \tag{6-6}$$

Comparison of the result of Eq. (6-6) with the exact answer $q_{w,\text{exact}} = (T_s - T_0)(k\rho C_p/\pi t)^{1/2}$ gives

$$\frac{q_w}{q_{w,\text{exact}}} = \left(\frac{\pi}{4}\right)^{1/2}$$

which is only 12% low and in good agreement in view of the coarseness of the assumed temperature profile.

Note that integration over space eliminated the most difficult part of the problem—it is second order in space but only first order in time. The smoothing effect of integration makes use of an assumed profile reasonably accurate since the conservation of energy principle is satisfied on the average. The conservation of energy principle is unlikely to be satisfied at a particular interior position, however; collocation methods [5] have been devised to improve local accuracy.

The assumed profile used before was a two-parameter curve composed of the first two terms of a polynomial—$T - T_0 = a + bx + cx^2 + \cdots$. If a two-parameter profile is to be used, it could be a collection of functions that more nearly fit the true shape of the profile such as $T - T_0 = [1 - \sin(\frac{1}{2}\pi x/\delta)](T_s - T_0)$. The advantage of a polynomial profile is that it is easy to integrate.

Acquisition of a more accurate temperature profile requires consideration of the conditions that can be imposed. Although not exhaustive, these are

$$\text{At } x = 0 \qquad T - T_0 = T_s - T_0 \xleftarrow[\text{\textit{(must be satisfied)}}]{\substack{\text{from} \\ \text{boundary conditions}}} T - T_0 = 0 \qquad \text{at } x = \delta$$

$$\xrightarrow[\text{since } q = 0 \text{ at } x = \delta]{\text{physical insight}} \frac{\partial T}{\partial x} = 0$$

$$\frac{\partial^2 T}{\partial x^2} = 0 \xleftrightarrow[\substack{\frac{\partial T}{\partial t} = 0 \quad \text{at} \quad x = 0, \delta}]{\substack{\text{from D.E. (conservation of energy} \\ \text{at the two boundaries) since}}} \frac{\partial^2 T}{\partial x^2} = 0$$

The last two conditions from the energy equation guarantee that conservation of energy will be satisfied at the two boundaries. Assuming a polynomial profile and keeping in mind that the lowest-order terms of the polynomial are the most important (as are the lowest-order derivatives of the preceding conditions to be imposed) [4], one can evaluate the coefficients from the five equations:

$$T(0) - T_0 = T_s - T_0 = a$$

$$T(\delta) - T_0 = 0 = a + b\delta + c\delta^2 + d\delta^3 + e\delta^4$$

$$\frac{\partial T(\delta)}{\partial x} = 0 = b + 2c\delta + 3d\delta^2 + 4e\delta^3$$

$$\frac{\partial^2 T(0)}{\partial x^2} = 0 = 2c$$

$$\frac{\partial^2 T(\delta)}{\partial x^2} = 0 = 2c + 6d\delta + 12e\delta^2$$

The resulting temperature profile is the quartic

$$\frac{T - T_0}{T_s - T_0} = 1 - 2\left(\frac{x}{\delta}\right) + 2\left(\frac{x}{\delta}\right)^3 - \left(\frac{x}{\delta}\right)^4 \qquad (6\text{-}7)$$

which displays a similarity form inasmuch as x/δ is the only parameter and all coefficients are constant. If the coefficients of the polynomial are not constants, the problem is a *nonsimilar* one; the coefficients must then be evaluated in the course of the problem solution, usually by the solution of a differential equation. Introduction of Eq. (6-7) into the integral formulation

[Eq. (6-4)] gives

$$\rho C_p \frac{d}{dt}\left[(T_s - T_0)\delta\frac{3}{10}\right] = 2k\frac{T_s - T_0}{\delta}$$

The solution for the penetration depth is, assuming $\delta(t = 0) = 0$,

$$\delta = \left(\frac{40\alpha t}{3}\right)^{1/2} \tag{6-8}$$

This is nearly twice the value predicted for a linear profile in Eq. (6-5), but both results agree that $\delta^2/\alpha t = $ const. The wall heat flux is available as

$$q_w = 2k\frac{T_s - T_0}{\delta} = (T_s - T_0)\left(3k\frac{\rho C_p}{10t}\right)^{1/2} \tag{6-9}$$

This is 3% below the exact answer.

This success in wall heat flux prediction attests to the potential accuracy of the integral method. Occasionally, however, mathematically singular behavior is predicted by integral methods even though the problem is physically sound as explained by Potts [9]. See Problems 6-16 and 6-17 for refinements.

6.3 LAMINAR BOUNDARY LAYERS BY THE INTEGRAL METHOD

Enough laminar boundary-layer problems have been solved to provide a feel for proper trends and to enable the accuracy of approximate answers to be judged. Because turbulent boundary layers possess peculiar uncertainties, one wishes to be confident of the integral method before applying it to them. The following discussion is aimed at supplying that confidence.

In the interests of simplicity, only geometries already familiar are treated. Specifically, a flat plate is always assumed as shown in Fig. 6-4.

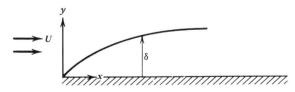

Figure 6-4 Laminar boundary layer on a flat plate.

Motion Equations

The boundary-layer equations and their boundary conditions are, for constant properties, steady state, and laminar flow,

$$\frac{\partial u}{\partial x} + \frac{\partial v}{\partial y} = 0 \qquad (6\text{-}10\text{a})$$

$$u\frac{\partial u}{\partial x} + v\frac{\partial u}{\partial y} = \nu\frac{\partial^2 u}{\partial y^2} + \left\{ \frac{1}{\rho}\left(B_x - \frac{dP}{dx} \right) = \begin{cases} U\dfrac{dU}{dx} & \text{forced convection} \\[2ex] \dfrac{(\rho_\infty - \rho)g}{\rho_\infty} & \text{natural convection} \end{cases} \right.$$

$$(6\text{-}10\text{b})$$

$$u(y = 0) = 0 \qquad (6\text{-}10\text{c})$$

$$v(y = 0) = v_w \qquad (6\text{-}10\text{d})$$

$$u(y = \delta) = \begin{cases} U(x) & \text{forced convection} \\ 0 & \text{natural convection} \end{cases} \qquad (6\text{-}10\text{e})$$

Before specifying natural or forced convection, the integral formulation is taken as far as possible. The x-motion equation is more difficult in the y direction, as it is of second order there. Also, it is desirable to ultimately make assumptions concerning only u, and not v, because one has a better feel for the behavior of u and because wall shear stress depends on u and not v, and it is probably wall shear that is of ultimate interest.

The continuity equation [Eq. (6-10a)] permits v to be found in terms of u. First, integrate with respect to y from 0 to δ (note that integration beyond δ will add nothing since conditions are constant outside the boundary layer) to obtain

$$\int_0^\delta \frac{\partial u}{\partial x}\, dy + \int_0^\delta \frac{\partial v}{\partial y}\, dy = 0$$

which gives

$$\int_0^\delta \frac{\partial u}{\partial x}\, dy + v(y = \delta) - v(y = 0) = 0$$

Application of Leibnitz' formula [Eq. (6-1)] puts this result in the form

$$v_\delta = v_w + U\frac{d\delta}{dx} - \frac{\partial}{\partial x}\left(\int_0^\delta u\, dy \right) \qquad (6\text{-}11)$$

Turning now to the x-motion equation [Eq. (6-10b)], it is convenient to first rearrange it into the conservation-law form

$$\frac{\partial uu}{\partial x} + \frac{\partial vu}{\partial y} - u\left(\frac{\partial u}{\partial x} + \frac{\partial v}{\partial y}\right) = \frac{B_x - dP/dx}{\rho} + \nu\frac{\partial^2 u}{\partial y^2}$$

The last term in parentheses on the left-hand side equals zero by the continuity equation. Integration from $y = 0$ to $y = \delta$ gives

$$\int_0^\delta \frac{\partial u^2}{\partial x}\,dy + \int_0^\delta \frac{\partial vu}{\partial y}\,dy = \int_0^\delta\left(B_x - \frac{dP}{dx}\right)\frac{1}{\rho}\,dy + \nu\int_0^\delta \frac{\partial^2 u}{\partial y^2}\,dy$$

which becomes

$$\int_0^\delta \frac{\partial u^2}{\partial x}\,dy + v(y = \delta)u(y = \delta) - v(y = 0)u(y = 0)$$

$$= \frac{1}{\rho}\left[\int_0^\delta\left(B_x - \frac{dP}{dx}\right)dy + \mu\frac{\partial u(y = \delta)}{\partial y} - \mu\frac{\partial u(y = 0)}{\partial y}\right]$$

The no-slip boundary condition of Eq. (6-10c) and the relationship for v_δ given by Eq. (6-11) then permit this equation to be simplified to

$$\int_0^\delta \frac{\partial u^2}{\partial x}\,dy + Uv_w + U^2\frac{d\delta}{dx} - U\frac{\partial}{\partial x}\left(\int_0^\delta u\,dy\right) = \frac{1}{\rho}\left[\int_0^\delta\left(B_x - \frac{dP}{dx}\right)dy - \tau_w\right]$$

Here $\tau_w = \mu\,\partial u(y = 0)/\partial y$, and it has been assumed (on the physical ground that there is no velocity gradient at the outer edge of the boundary layer) that $\tau_\delta = \mu\,\partial u(y = \delta)/\partial y = 0$. Application of the Leibnitz formula gives

$$\frac{\partial}{\partial x}\left[\int_0^\delta u(u - U)\,dy\right] + \left(\int_0^\delta u\,dy\right)\frac{dU}{dx} - \frac{1}{\rho}\int_0^\delta\left(B_x - \frac{dP}{dx}\right)dy$$

$$= -\frac{\tau_w + \rho Uv_w}{\rho} \tag{6-12}$$

The integral formulation [Eq. (6-12)] contains no v values except for v_w, usually known. Note that if v_w is positive, the net effect is to increase apparent wall shear; if v_w is negative, the net effect is to decrease wall shear. This insight follows on physical grounds since if v_w is positive, fluid with no x momentum is being added to the boundary layer and must be accelerated.

Further progress requires that the type of convection considered (natural or forced) be specified. Specialization to forced convection ($B_x - dP/dx)/$

$\rho = U\,dU/dx$ then puts Eq. (6-12) into the form

$$\frac{\partial}{\partial x}\left[U^2\int_0^\delta \frac{u}{U}\left(1 - \frac{u}{U}\right)dy\right] + \left[\int_0^\delta\left(1 - \frac{u}{U}\right)dy\right]U\frac{dU}{dx} = \frac{\tau_w + \rho U v_w}{\rho}$$

It is more compact to use the displacement thickness δ_1

$$\delta_1 = \int_0^\delta\left(1 - \frac{u}{U}\right)dy$$

and the momentum thickness δ_2

$$\delta_2 = \int_0^\delta\frac{u}{U}\left(1 - \frac{u}{U}\right)dy$$

in the integral formulation for forced convection which is then

$$\frac{\partial\left(U^2\delta_2\right)}{\partial x} + \delta_1 U\frac{dU}{dx} = \frac{\tau_w + \rho U v_w}{\rho} \tag{6-13}$$

A polynomial velocity profile is selected for use in the integrals of Eq. (6-13) as

$$u = a + by + cy^2 + dy^3$$

and subjected to the conditions

At $y = 0$ $u = 0 \xleftarrow[\text{(\textit{must} be satisfied)}]{\substack{\text{from} \\ \text{boundary conditions}}} u = U$ at $y = \delta$

$$\xrightarrow[\substack{\tau=0 \text{ at } y=\delta}]{\text{physical insight since}} \frac{\partial u}{\partial y} = 0$$

$$v_w\frac{\partial u}{\partial y} = U\frac{dU}{dx} + v\frac{\partial^2 u}{\partial y^2} \xrightarrow{\text{from D.E.}} \frac{\partial^2 u}{\partial y^2} = 0$$

Restricting attention to the case of $U = \text{const}$ and $v_w = 0$, for which case exact solutions are given in Section 5.2, the four algebraic equations to be solved for the polynomial coefficients are

$$u(0) = 0 = a$$

$$u(\delta) = U = a + b\delta + c\delta^2 + d\delta^3$$

$$\frac{\partial u(\delta)}{\partial y} = 0 = b + 2c\delta + 3d\delta^2$$

$$\frac{\partial^2 u(0)}{\partial y^2} = 0 = 2c$$

The resulting velocity profile is the cubic

$$
\frac{u}{U} =
\begin{cases}
1, & \dfrac{y}{\delta} > 1 \\[2ex]
\dfrac{3}{2}\dfrac{y}{\delta} - \dfrac{1}{2}\left(\dfrac{y}{\delta}\right)^3, & \dfrac{y}{\delta} \leqslant 1
\end{cases}
\tag{6-14}
$$

which has a similarity form inasmuch as the coefficients are constants. From Eq. (6-14) it follows that

$$
\delta_1 = \delta \int_0^1 \left(1 - \frac{u}{U}\right) d(y/\delta) = \frac{3\delta}{8} \approx \frac{\delta}{3}
$$

$$
\delta_2 = \delta \int_0^1 \frac{u}{U}\left(1 - \frac{u}{U}\right) d(y/\delta) = \frac{39\delta}{280} \approx \frac{\delta}{7}
$$

$$
\tau_w = \mu\,\frac{\partial u(0)}{\partial y} = \frac{\mu}{\delta}\frac{\partial u(0)}{\partial(y/\delta)} = \frac{3\mu U}{2\delta}
$$

Introduction of these results into Eq. (6-13) gives

$$
\frac{d\delta^2}{dx} = \frac{280\nu}{13U}
$$

With the assumption that $\delta(x = 0) = 0$, the solution is

$$
\delta = \left(\frac{280\nu x}{13U}\right)^{1/2} = \frac{4.64}{\mathrm{Re}^{1/2}}
\tag{6-15}
$$

where $\mathrm{Re} = Ux/\nu$. The friction coefficient $C_f/2 = \tau_w/\rho U^2$ is found to be

$$
\frac{C_f}{2} = \frac{0.323}{\mathrm{Re}^{1/2}}
\tag{6-16}
$$

Comparison with the exact answers from Eqs. (5-16) and (5-17), $\delta\,\mathrm{Re}^{1/2}/x = 5$ and $(C_f/2)\,\mathrm{Re}^{1/2} = 0.33206$, reveals that the integral method prediction for the boundary-layer thickness is 7% low whereas its prediction for the friction coefficient is only 3% low. The boundary-layer thickness is best thought of as a parameter determined from the integral equation that allows the assumed profile to satisfy the macroscopic conservation principle to the greatest possible extent.

The integral form of the x-motion equation for a body of revolution is the subject of Problem 6-12.

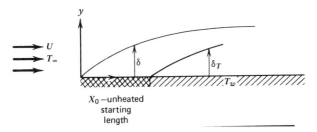

Figure 6-5 Laminar boundary layer on a flat plate with an unheated starting length.

Energy Equation

Solution of the energy equation by an integral method is undertaken next with the same geometry for, as shown in Fig. 6-5, plate and free-stream temperature constant and an unheated starting length X_0. When the unheated starting length disappears, the thermal behavior of the laminar boundary layer is as was exactly determined in Section 5.3.

The laminar boundary-layer energy equation and boundary conditions are

$$u\frac{\partial T}{\partial x} + v\frac{\partial T}{\partial y} = \alpha\frac{\partial^2 T}{\partial y^2} \tag{6-17a}$$

$$T(y = \infty) = T_\infty \tag{6-17b}$$

$$T(y = 0) = \begin{cases} T_\infty, & x < X_0 \\ T_w, & x \geqslant X_0 \end{cases} \tag{6-17c}$$

with viscous dissipation neglected.

The energy equation is first rearranged into the conservation-law form

$$\frac{\partial uT}{\partial x} + \frac{\partial vT}{\partial y} - T\left(\frac{\partial u}{\partial x} + \frac{\partial v}{\partial y}\right) = \alpha\frac{\partial^2 T}{\partial y^2}$$

The last term in parentheses on the left-hand side equals zero according to the continuity equation. Integration from $y = 0$ to $y = \delta_T$ then gives

$$\int_0^{\delta_T}\frac{\partial uT}{\partial x}\,dy + \int_0^{\delta_T}\frac{\partial vT}{\partial y}\,dy = \alpha\int_0^{\delta_T}\frac{\partial^2 T}{\partial y^2}\,dy$$

which becomes

$$\int_0^{\delta_T}\frac{\partial uT}{\partial x}\,dy + v(y = \delta_T)T(\delta_T) - v(y = 0)T(y = 0)$$

$$= \frac{-k\,\partial T(y = 0)/\partial y - [-k\,\partial T(y = \delta_T)/\partial y]}{\rho C_p}$$

There is no temperature gradient at the outer edge of the thermal boundary layer so that $\partial T(y = \delta_T)/\partial y = 0$. Adoption of the nomenclature that $v(y = 0) = v_w$, $T(y = 0) = T_w$, $T(y = \delta_T) = T_\infty$, and $-k\,\partial T(y = 0)/\partial y = q_w$ then puts the integral relation in the form

$$\int_0^{\delta_T} \frac{\partial u T}{\partial x}\,dy + T_\infty\left[v_w + u_{\delta_T}\frac{d\delta_T}{dx} - \frac{\partial}{\partial x}\left(\int_0^{\delta_T} u\,dy\right)\right] = \frac{q_w + \rho C_p v_w T_w}{\rho C_p}$$

where Eq. (6-11) relates v_δ to u_{δ_T}. Application of Leibnitz' formula gives

$$\frac{\partial}{\partial x}\left[\int_0^{\delta_T} u(T - T_\infty)\,dy\right] + \left(\int_0^{\delta_T} u\,dy\right)\frac{dT_\infty}{dx} = \frac{q_w + \rho C_p v_w(T_w - T_\infty)}{\rho C_p} \tag{6-18}$$

The integral formulation of Eq. (6-18) reveals that if v_w is positive, the energy flux from the wall increases (presuming $T_w > T_\infty$). If v_w is negative, on the other hand, the wall energy flux decreases.

For the problem at hand, the assumed polynomial temperature profile

$$T = a + by + cy^2 + dy^3 + \cdots$$

is subjected to the conditions

$$\text{At } y = 0 \qquad T = T_w \xleftarrow[\substack{(\textit{must be satisfied})}]{\substack{\text{from}\\ \text{boundary conditions}}} \to T = T_\infty \qquad \text{at } y = \delta_T$$

$$\xleftarrow[\substack{\text{used in integral formulation}}]{\substack{\text{physical insight, already}}} \to \frac{\partial T}{\partial y} = 0$$

$$v_w\frac{\partial T}{\partial y} = \alpha\frac{\partial^2 T}{\partial y^2} \xleftarrow{\substack{\text{from energy equation}}} \to u_{\delta_T}\frac{dT_\infty}{dx} = \alpha\frac{\partial^2 T}{\partial y^2}$$

Additional, derived, conditions can also be imposed if desired. For example, the derivative of the energy equation with respect to y evaluated at $y = 0$ gives

$$\frac{\partial u(y = 0)}{\partial y}\frac{dT_w}{dx} = \alpha\frac{\partial^3 T(y = 0)}{\partial y^3}$$

It is not usually convenient to satisfy all of these conditions. If only four are satisfied, the lowest-ordered derivatives and the condition at the wall for the second derivative would be used. Assuming that $v_w = 0 = dT_\infty/dx$, these four conditions give the four algebraic equations

$$T(y = 0) = T_w = a$$

$$T(y = \delta_T) = T_\infty = a + b\delta_T + c\delta_T^2 + d\delta_T^3$$

$$\frac{\partial T(\delta_T)}{\partial y} = 0 = b + 2c\delta_T + 3d\delta_T^2$$

$$\frac{\partial^2 T(0)}{\partial y^2} = 0 = 2c$$

Their simultaneous solution gives

$$\frac{T - T_\infty}{T_w - T_\infty} = \begin{cases} 0, & \dfrac{y}{\delta_T} > 1 \\[2mm] 1 - \dfrac{3}{2}\dfrac{y}{\delta_T} + \dfrac{1}{2}\left(\dfrac{y}{\delta_T}\right)^3, & \dfrac{y}{\delta_T} \leqslant 1 \end{cases} \qquad (6\text{-}19)$$

Introduction of this profile into the integral formulation of Eq. (6-18) gives

$$\frac{d}{dx}\left[\frac{\delta_T^2}{\delta}\left(1 - \frac{\delta_T^2}{14\delta^2}\right)\right] = \frac{10\nu}{\Pr U}\frac{1}{\delta_T}$$

Of course, δ is known by Eq. (6-15) so that there is only one unknown. On the basis of prior experience, it is suspected that the two boundary layers approach a constant ratio far downstream from the leading edge. This ratio $\zeta = \delta_T/\delta$ is formed in the differential equation to achieve

$$\frac{d}{dx}\left[\zeta^2\delta\left(1 - \frac{\zeta^2}{14}\right)\right] = \frac{10\nu}{\Pr U}\frac{1}{\zeta\delta}$$

The $\zeta^2/14$ term could be discarded at this point since $\zeta < 1$. Further rearrangement gives

$$\left[\zeta^3\left(1 - \frac{\zeta^2}{14}\right)\right]\frac{d(\ln \delta^{3/2})}{dx} + \frac{d}{dx}\left[\zeta^3\left(1 - \frac{6}{5}\frac{\zeta^2}{14}\right)\right] = \frac{15\nu}{\Pr U\delta^2}$$

Since it has already been assumed that $\zeta < 1$, there is no appreciable loss in accuracy in assuming $\frac{6}{5} \approx 1$ in the last term on the left-hand side. Then a solution for $\zeta^3(1 - \zeta^2/14)$ is achieved, with $\delta^{3/2}$ as an integrating factor, as

$$\zeta^3\left(1 - \frac{\zeta^2}{14}\right) = \frac{15}{\Pr U}\delta^{-3/2}\int\delta^{-1/2}\,dx + c\delta^{-3/2}$$

On the basis that $\delta_T = 0$ when $x = X_0$,

$$\zeta\left(1 - \frac{\zeta^2}{14}\right)^{1/3} = \left(\frac{13}{14}\right)^{1/3}\left[1 - \left(\frac{X_0}{x}\right)^{3/4}\right]^{1/3}\Pr^{-1/3} \qquad (6\text{-}20)$$

As expected, $\zeta = \delta_T/\delta$ approaches a constant far downstream—under the assumptions of this discussion, $\zeta < 1$ always.

The local heat-transfer coefficient is attainable now that the thermal boundary-layer thickness is given by Eq. (6-20) since

$$q_w = h(T_w - T_\infty) = -k\frac{\partial T(0)}{\partial y}$$

$$h = \frac{3}{2}\frac{k}{\zeta\delta}$$

Use of ζ from Eq. (6-20) and δ from Eq. (6-15) then gives, accurately [10],

$$\frac{\mathrm{Nu}}{\Pr^{1/3}\mathrm{Re}^{1/2}} = \frac{0.3317(1 - \zeta^2/14)^{1/3}}{\left[1 - (X_0/x)^{3/4}\right]^{1/3}} \qquad (6\text{-}20a)$$

In the limit as the unheated starting length disappears, this result becomes

$$\frac{\mathrm{Nu}}{\Pr^{1/3}\mathrm{Re}^{1/2}} = 0.3317\left(1 - \frac{\zeta^2}{14}\right)^{1/3}$$

For $\Pr \to \infty$, $\zeta^2 \ll 14$ and

$$\frac{\mathrm{Nu}}{\Pr^{1/3}\mathrm{Re}^{1/2}} = 0.3317 \qquad (6\text{-}21)$$

which is 2% below the exact answer of 0.339 from Eq. (5-29c). For $\Pr \approx 1$, Eq. (6-20) reveals that $\zeta \approx 1$, so that then

$$\frac{\mathrm{Nu}}{\Pr^{1/3}\mathrm{Re}^{1/2}} = 0.323$$

which is also 2% lower than the exact answer of 0.33206 [11].

The integral method has been applied by Goodman [12] to determination of the effect of time-dependent wall temperature on the heat transfer to a constant-property fluid in steady boundary-layer flow over a flat plate.

The unheated starting length problem for wedge flow $U = Cx^m$ was treated by Lighthill [13].

The integral form of the energy equation for a body of revolution is the subject of Problem 6-13.

6.4   HEAT FLUX FROM PLATE OF ARBITRARY WALL TEMPERATURE

The results for a surface with an unheated starting length can be employed to evaluate the heat flux from a surface whose temperature varies. The method is similar to that used for tube flow as sketched in Fig. 4-8. Consider a flat plate subjected to steady parallel flow of a constant-property fluid of constant free-stream temperature.

As shown in Fig. 6-6, the wall temperature can be regarded as equal to T_∞ until a point x^* is reached, where a step change of magnitude $\Delta(T_w - T_\infty)$ occurs that is maintained over the remainder of the plate. Let the temperature distribution following a unit step change be $f(x, x^*, y)$ as given by Eq. (6-19), which is now a known function. The temperature in the laminar boundary layer for a step change of magnitude $\Delta(T_w - T_\infty)$ is then

$$T(x, y) - T_\infty = \Delta(T_w - T_\infty) f(x, x^*, y)$$

For a series of steps, the fluid temperature, relying on the linearity of the energy equation, is just the sum of the effects of the individual steps so that

$$T(x, y) - T_\infty = (T_{w_0} - T_\infty) f(x, y) + \Delta_1(T_w - T_\infty) f(x, x_1^*, y)$$
$$+ \Delta_2(T_w - T_\infty) f(x, x_2^*, y) + \cdots$$

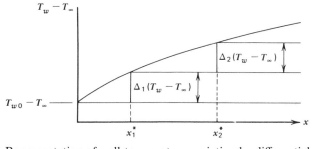

Figure 6-6   Representation of wall temperature variation by differential steps.

In the limit as the spacing between steps become small, the preceding sum can be represented as the integral

$$T(x, y) - T_\infty = (T_{w_0} - T_\infty)f(x, y) + \int_{x^*=0}^{x} f(x, x^*, y)\, d[T_w(x^*) - T_\infty] \tag{6-22}$$

Equation (6-22) allows for only one discontinuity in $T_w - T_\infty$, at the leading edge. Each discontinuity in $T_w - T_\infty$ requires addition of a term, $(T_{w_i} - T_\infty)f(x - x_i^*, y)$. If desired, an integration by parts can be employed to recast the integral in terms of $T_w(x^*) - T_\infty$ and $df(x - x^*, y)/dx^*$, as was done to obtain Eq. (4-43c).

The wall heat flux is obtained from Eq. (6-22) as

$$q_w(x) = -k\frac{\partial T(x, y = 0)}{\partial y}$$

$$= (T_{w_0} - T_\infty)\left[-k\frac{\partial f(x, y = 0)}{\partial y} \right]$$

$$+ \int_{x^*=0}^{x} \left[-k\frac{\partial f(x, x^*, y = 0)}{\partial y} \right] d[T_w(x^*) - T_\infty]$$

If it is recognized that $-k\,\partial f(x, x^*, y = 0)/\partial y = h(x, x^*)$, where $h(x, x^*)$ is the heat-transfer coefficient at x due to a step change in surface temperature at x^* as given by Eq. (6-20a), this relation can be written as

$$q_w(x) = (T_{w_0} - T_\infty)h(x) + \int_{x^*=0}^{x} h(x, x^*)\frac{d[T_w(x^*) - T_\infty]}{dx^*}\, dx^* \tag{6-23}$$

which is essentially an application of Duhammel's theorem dealt with in Problem 6-7.

Eckert et al. [14] present simplified methods to treat Eq. (6-23), pointing out that if $T_w - T_\infty = Ax^\gamma$, immediate integration is possible and extending the method to turbulent boundary layers for which $h(x, x^*)/h(x, 0) = [1 - (x^*/x)^{9/10}]^{-1/9}$ from Eq. (8-24). The integral method expressed by Eq. (6-23) was apparently first applied to this sort of problem by Chapman and Rubesin [15] and was later refined by Klein and Tribus [16]. Additional discussion is provided by Schlichting [17]. The inversion of Eq. (6-23) to give the surface-temperature distribution that corresponds to a specified surface heat flux is the subject of Problem 6-9.

PROBLEMS

6-1 Determine, by use of the integral method, the local heat-transfer coefficient for steady parallel flow of a constant-property fluid over a flat plate in the absence of an unheated starting length for the case in which $\delta_T > \delta$. This requires that

$$\int_0^{\delta_T} u(T - T_\infty)\, dy = \int_0^\delta u(T - T_\infty)\, dy + U \int_\delta^{\delta_T} (T - T_\infty)\, dy$$

6-2 Use the integral method for the conditions of Problem 6-1 to show that, in the limit as $\mathrm{Pr} \to 0$, $\mathrm{Nu}/\mathrm{Re}^{1/2}\,\mathrm{Pr}^{1/2} = \mathrm{const}$, and evaluate the constant. Compare the integral method prediction with the exact answer of Eq. (5-29a).

6-3 An integral formulation of the two-dimensional laminar boundary-layer equations [Eqs. (5-9)] is needed for the unsteady state. For a constant-property fluid:

(a) Show that the continuity equation gives

$$v = v_w - \frac{\partial}{\partial x}\left(\int_0^y u\, dy\right)$$

where $v_w = v(y = 0)$.

(b) Show that the x-motion equation gives

$$\frac{\partial}{\partial t}\left[U \int_0^\delta \left(1 - \frac{u}{U}\right) dy\right] + \frac{\partial}{\partial x}\left[U^2 \int_0^\delta \frac{u}{U}\left(1 - \frac{u}{U}\right) dy\right]$$

$$+ \left[\int_0^\delta \left(1 - \frac{u}{U}\right) dy\right] U \frac{\partial U}{\partial x} = \frac{\tau_w + \rho U v_w}{\rho}$$

(c) Show that the energy equation gives

$$\delta_T \frac{\partial T_\infty}{\partial t} + \frac{\partial}{\partial t}\left[\int_0^{\delta_T} (T - T_\infty)\, dy\right] + \frac{\partial}{\partial x}\left[\int_0^{\delta_T} u(T - T_\infty)\, dy\right]$$

$$+ \left[\int_0^{\delta_T} u\, dy\right] \frac{\partial T_\infty}{\partial x} = \frac{q_w + \rho C_p v_w (T_w - T_\infty)}{\rho C_p}$$

6-4 Consult Goodman [12], and use his temperature and velocity profiles in the integral formulation of the energy equation as given in part c of Problem 6-3 for the case where (a) the velocity boundary-layer thickness is constant and δ_T depends only on time as a result of a step change in wall temperature that is spatially uniform and (b) the velocity boundary layer varies as $x^{1/2}$ and δ_T depends on both space and time as a result of a step change in wall temperature that is spatially uniform. Verify that the final equations to be solved are the same as Goodman's.

6-5 Use the integral method to determine the friction and heat-transfer coefficients for steady parallel flow over a flat plate with injection $v_w > 0$. Plot the ratio of the heat-transfer coefficient for $v_w > 0$ to that for $v_w = 0$ against $(v_w/U)(Ux/v)^{1/2}$ and compare with the exact results of Section 5.8.

6-6 Plot the temperature profile predicted by Eq. (6-20) for the case of no unheated starting length against $y(U/vx)^{1/2}$ and compare it with the exact solution for several Prandtl numbers.

6-7 (a) For the case in which the wall temperature of a flat plate varies linearly as $T_w - T_\infty = a + bx$, use Eq. (6-23) to show that the wall heat flux is given by, with $Re = Ux/v$,

$$q_w = \frac{0.332}{x} k\, Pr^{1/3}\, Re^{1/2}\left\{ a + b\int_{x^*=0}^{x}\left[1 - \left(\frac{x^*}{x}\right)^{3/4}\right]^{-1/3} dx^* \right\}$$

(b) Evaluate the integral of part a through use of the relationship, where $\beta(m, n)$ is the beta function [related to the gamma function by $\beta(m, n) = \Gamma(m)\Gamma(n)/\Gamma(m + n)$],

$$\beta(m, n) = \int_0^1 Z^{m-1}(1 - Z)^{n-1}\, dZ, \qquad m > 0, \qquad n < \infty$$

and the transformation $Z = 1 - (x^*/x)^{3/4}$ to show that

$$q_w = 0.332\frac{k}{x} Pr^{1/3}\, Re^{1/2}(a + 1.612bx)$$

(c) Show that the local Nusselt number for $a = 0$ is 61% larger than for $b = 0$.

6-8 Use the integral method to find the surface temperature and heat-transfer coefficient for a flat plate in steady parallel flow of a constant-property fluid with constant free-stream temperature for the case in which a constant wall heat flux is imposed after an unheated starting length of X_0. Show that if $X_0 = 0$, the local wall temperature varies as

$$T_w - T_\infty = \frac{2.201}{k\, Pr^{1/3}\, Re^{1/2}} x$$

where $Re = Ux/\nu$ and that the local Nusselt number is

$$Nu = \frac{hx}{k} = 0.454\,Pr^{1/3}\,Re^{1/2}$$

which is about 36% higher than for $T_w = const$. *Note*: The insight gained from similarity solutions as expressed by Eq. (5-66) that $T_w - T_\infty = Kx^{1/2}$ if $q_w = const$ can be used in a specified T_w solution by employing Eqs. (6-18) and (6-19).

6-9 It is required that the surface temperature that corresponds to a specified surface heat flux be predicted through inversion of Eq. (6-23).

(a) Show that the local heat-transfer coefficient given by Eq. (6-20a) is of the form

$$h(x, x^*) = f(x)(x^h - x^{*b})^{-a}$$

where $f(x) = 0.3317k\,Pr^{1/3}(U/\nu)^{1/2}x^{-1/4}$, $a = \frac{1}{3}$, and $b = \frac{3}{4}$.

(b) Recognizing that $q_w(x = 0)$ will be unbounded if $T_{w0} - T_\infty \neq 0$, write Eq. (6-23) for $T_{w0} - T_\infty = 0$ as

$$q_w(x) = \int_{x^*=0}^{x} f(x)(x^b - x^{*b})^{-a}\,\frac{dy(x^*)}{dx^*}\,dx^*$$

where $y(x^*) = T_w(x^*) - T_\infty$. Let $\lambda^{1/b} = x^*$ and $t^{1/b} = x$, and cast the integral equation into the form

$$\frac{q_w(t)}{f(t)} = \int_{\lambda=0}^{t}(t - \lambda)^{-a}\,\frac{dy(\lambda)}{d\lambda}\,d\lambda$$

(c) Recognizing that the result of part b is a convolution integral [18], take its Laplace transform with respect to t to obtain

$$\overline{q_w(t)/f(t)} = \overline{t^{-a}}\,\overline{y'}$$

where the bar superscript denotes a Laplace transformed quantity. Now $\overline{t^{-a}} = \Gamma(1 - a)/s^{1-a}$ and $\overline{y'} = s\bar{y} - y(t = 0)$, where s is the Laplace transform parameter. Thus

$$\bar{y} = \frac{1}{\Gamma(1 - a)}\,\frac{\overline{[q_w(t)/(t)]}}{s^a}$$

This result represents the convolution integral

$$y(t) = \frac{1}{\Gamma(1-a)\Gamma(a)} \int_{\lambda=0}^{t} (t-\lambda)^{a-1} \frac{q_w(\lambda)}{f(\lambda)} \, d\lambda$$

(d) From part b recall that $t = x^b$ and $\lambda = x^{*b}$ and put the result of part c into the form

$$T_w(x) - T_\infty = \frac{b}{\Gamma(1-a)\Gamma(a)}$$

$$\times \int_{x^*=0}^{x} (x^b - x^{*b})^{a-1} \frac{q_w(x^*)}{f(x^*)} x^{*b-1} \, dx^*$$

(e) With $\Gamma(\frac{1}{3}) = 2.67894$ and $\Gamma(\frac{2}{3}) = 1.35412$ [19], use the definitions of part a to put the result of part d into the form specific to a laminar boundary layer on a flat plate, with $\mathrm{Re} = Ux/\nu$, of

$$k[T_w(x) - T_\infty] \mathrm{Pr}^{1/3} \mathrm{Re}^{1/2} = 0.623 \int_0^x \left[1 - \left(\frac{x^*}{x} \right)^{3/4} \right]^{-2/3}$$

$$\times q_w(x^*) \, dx^*$$

6-10 Air flows steadily at 25 ft/sec parallel to a flat plate that is 6 in. long in the flow direction. Air temperature is 20°F, and atmospheric pressure is 14.7 lb$_f$/in². The entire surface of the plate is adiabatic, except for a strip of 1 in. width (located between 2 and 3 in. from the plate leading edge) that is electrically heated so that the heat flux from it is uniform. Determine the numerical value of the heat flux from the strip that will maintain the temperature at the plate trailing edge at 32°F. Plot the temperature distribution along the entire plate surface and discuss the relevance of this problem to aircraft wing deicing. *Answer*: $T_{w,\max} = 97°F$.

6-11 Air flows steadily at 50 ft/sec, 2000°F, and 14.7 lb$_f$/in² parallel to a flat plate. The first 6 in. of the plate are maintained at 200°F. The next 18 in. are adiabatic. Plot the plate temperature against the position for the first 24 in. Note that the first 6 in. must be treated as a specified surface temperature problem whereas the last 18 in. must be treated as a specified heat-flux problem. *Answer*: $T_w(x = 24 \text{ in.}) = 1600°F$.

6-12 The integral method can be used to ascertain the velocity distribution on a body of revolution. The physical configuration and the control volume, extending above the boundary layer to a distance Y, to which conservation principles will be applied are illustrated in Fig. 6P-12. This derivation from first principles is an alternative to integration of

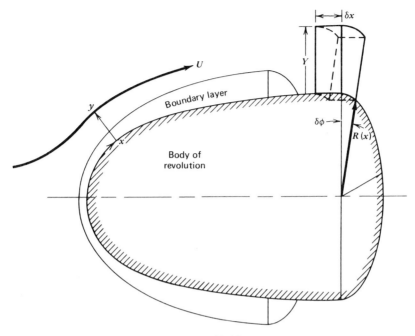

Figure 6P-12

the describing partial differential equations. Steady state is assumed as are constant properties. Most importantly, the boundary layer is assumed to be thin relative to the local radius of revolution $R(x)$. Wall transpiration normal to the surface is also assumed.

(a) Show by a detailed accounting that the conservation of mass principle applied to the control volume gives

$$\left\{ R\,\delta\phi \int_0^Y \rho u\,dy \right\}_x + \left\{ R\,\delta\phi\,\delta x\,\rho v_w \right\}_{x+\delta x/2}$$

$$- \left\{ R\,\delta\phi \int_0^Y \rho u\,dy \right\}_{x+\delta x} - \left\{ R\,\delta\phi\,\delta x\,\rho_Y v_Y \right\}_{x+\delta x/2} = 0$$

from which it follows that, for constant ρ,

$$v_Y = v_w - \frac{1}{R}\frac{d}{dx}\left[R\int_0^Y u\,dy \right]$$

(b) Show by a detailed accounting that the conservation of momentum principle applied to the control volume gives

$$\{R\,\delta\phi\,\delta x\,\tau_w\}_{x+\delta x/2} + \{R\,\delta\phi\,Yp\}_x$$

$$-\{R\,\delta\phi\,Yp\}_{x+\delta x} + \left\{R\,\delta\phi\int_0^Y \rho u^2\,dy\right\}_x$$

$$-\left\{R\,\delta\phi\int_0^Y \rho u^2\,dy\right\}_{x+\delta x} - \{R\,\delta\phi\,\delta x\,\rho_Y v_Y u_Y\}_{x+\delta x/2} = 0$$

Then use the result of part a to eliminate v_Y, recognizing that $u_Y = U$, and incorporate the Bernoulli equation for inviscid flow along a streamline that gives $\rho^{-1}\,dp/dx = U\,dU/dx = dU^2/dx - U\,dU/dx$ to rearrange this equation into

$$\frac{d\delta_2}{dx} + \delta_2\left[\frac{1}{R}\frac{dR}{dx} + \left(2 + \frac{\delta_1}{\delta_2}\right)\frac{1}{U}\frac{dU}{dx}\right] = \frac{\tau_w + \rho v_w U}{\rho U^2}$$

where

$$\delta_2 = \int_0^\delta \frac{u}{U}\left(1 - \frac{u}{U}\right)dy$$

is the momentum thickness and

$$\delta_1 = \int_0^\delta \left(1 - \frac{u}{U}\right)dy$$

is the displacement thickness. For laminar flow, $\tau_w = \mu\,\partial u(y = 0)/\partial y$.

(c) Obtain the result of part b by the alternative method of integrating the partial differential equation form of the boundary-layer equations given by Eq. (5-69) from $y = 0$ to $y = \delta$.

(d) Show that if the Mangler transformation of Eq. (5-70),

$$\bar{x} = L^{-2}\int_0^x R^2(x)\,dx \qquad \text{and} \qquad \bar{y} = L^{-1}R(x)y$$

where L is an arbitrary constant length, is employed the result of part b can be expressed, $\delta' = R\delta/L$, as

$$\frac{d\delta_2'}{d\bar{x}} + \delta_2'\left(2 + \frac{\delta_1'}{\delta_2'}\right)\frac{1}{U}\frac{dU}{d\bar{x}} = \frac{\mu}{\rho U^2}\frac{\partial u(\bar{y} = 0)}{\partial \bar{y}}$$

This result is the same as for flow over a wedge, transforming away the effects of the body of revolution and enabling wedge solutions to be taken over to the axisymmetric case.

6-13 The temperature distribution on a body of revolution is to be ascertained. The discussion in Problem 6-12 applies as does Fig. 6P-12. Properties are constant, steady state exists, viscous dissipation is unimportant, there is no volumetric heat source, and conduction is important only in the y direction. Again, the boundary layer is taken to be thin relative to the radius of revolution $R(x)$.

(a) Show by a detailed accounting that the conservation of energy principle applied to the control volume in Fig. 6P-12 gives

$$\left\{ R\,\delta\phi \int_0^Y \rho u C_p T\,dy \right\}_x + \left\{ R\,\delta\phi\,\delta x\,\rho_w \upsilon_w C_p T_w \right\}_{x+\delta x/2}$$

$$+ \left\{ R\,\delta\phi\,\delta x\,q_w \right\}_{x+\delta x/2} - \left\{ R\,\delta\phi \int_0^Y \rho u C_p T\,dy \right\}_{x+\delta x}$$

$$- \left\{ R\,\delta\phi\,\delta x\,\rho_Y \upsilon_Y C_p T_Y \right\}_{x+\delta x/2} - \left\{ R\,\delta\phi\,\delta x\,q_Y \right\}_{x+\delta x/2} = 0$$

Here the work done by surface and body forces is neglected. Then show in detail that this result becomes

$$\frac{1}{R}\frac{d}{dx}\left(R\int_0^Y \rho u C_p T\,dy \right) + \rho_Y \upsilon_Y C_p T_Y - \rho_w \upsilon_w C_p T_w = q_w - q_Y$$

(b) Recognize that Y extends outside the boundary layer so that $T_Y = T_\infty$ and $q_Y = 0$, and incorporate the result of Problem 6-12 part a to put the result of part a, making use of the identity,

$$\frac{d}{dx}\left\{ T_\infty R\int_0^Y u\,dy \right\} = T_\infty \frac{d}{dx}\left(R\int_0^Y u\,dy \right) + \left(R\int_0^Y u\,dy \right)\frac{dT_\infty}{dx}$$

into the form

$$\frac{d}{dx}\left[\int_0^{\delta_T} u(T - T_\infty)\,dy \right] + \left[\int_0^{\delta_T} u(T - T_\infty)\,dy \right]\frac{1}{R}\frac{dR}{dx}$$

$$+ \left(\int_0^{\delta_T} u\,dy \right)\frac{dT_\infty}{dx} = \frac{q_w + \rho C_p \upsilon_w (T_w - T_\infty)}{\rho C_p}$$

by letting Y approach the outer edge of the boundary layer. Explain how the assumption that the thermal boundary-layer thickness δ_T is less than the velocity boundary-layer thickness δ is incorporated into the preceding equation. Briefly explain the effects of Prandtl number and unheated starting length on the accuracy of this assumption, and also briefly outline the manner in which the result could be used if $\delta_T > \delta$.

(c) Compare the result of part b for a body of revolution with Eq. (6-18) for a flat plate.

(d) Show that the result of part b can be expressed for $\delta_T < \delta$ as

$$\frac{d\Delta}{dx} + \Delta \left[\frac{1}{T_w - T_\infty} \frac{d(T_w - T_\infty)}{dx} + \frac{1}{U} \frac{dU}{dx} + \frac{1}{R} \frac{dR}{dx} \right]$$

$$+ \frac{\Delta_u}{T_w - T_\infty} \frac{dT_\infty}{dx} = \frac{q_w + \rho C_p v_w (T_w - T_\infty)}{\rho C_p U (T_w - T_\infty)}$$

with

$$\Delta = \int_0^{\delta_T} \frac{u}{U} \frac{T - T_\infty}{T_w - T_\infty} \, dy \qquad \text{and} \qquad \Delta_u = \int_0^{\delta_T} \frac{u}{U} \, dy$$

Show that the result of part b can be alternatively expressed as

$$\frac{1}{R} \frac{d}{dx} [RG(T_w - T_\infty)\Delta] + GA_u \frac{dT_\infty}{dx} = \frac{q_w + \rho C_p v_w (T_w - T_\infty)}{C_p}$$

where the mass flux $G = \rho U$ is important rather than either ρ or U separately.

(e) Show that the Mangler transformation of Eq. (5-70),

$$\bar{x} = L^{-2} \int_0^x R^2(x) \, dx \qquad \text{and} \qquad \bar{y} = L^{-1} R(x) y$$

where L is a constant of arbitrary length, transforms the result of part d into the form

$$\frac{d\Delta'}{d\bar{x}} + \Delta' \left[\frac{1}{T_w - T_\infty} \frac{d(T_w - T_\infty)}{d\bar{x}} + \frac{1}{U} \frac{dU}{d\bar{x}} \right] + \frac{\Delta'_u}{T_w - T_\infty} \frac{dT_\infty}{d\bar{x}}$$

$$= -\frac{k}{\rho C_p U (T_w - T_\infty)} \frac{\partial T(\bar{y} = 0)}{\partial \bar{y}}$$

Here $\Delta' = R\Delta/L$ and $\Delta'_u = R\Delta_u/L$. This result is the same as for flow over a wedge, transforming away the effects of the body of revolution and enabling wedge solutions to be taken over to the axisymmetric case.

6-14 The laminar boundary layer on a body of revolution with variable free-stream velocity is to be studied by application of the integral form of the x-motion equation from Problem 6-12. Constant properties, steady state, and no wall transpiration are assumed.

(a) Show that the integral form of the x-motion equation is

$$\frac{d\delta_2}{dx} + \delta_2\left[\frac{1}{R}\frac{dR}{dx} + \left(2 + \frac{\delta_1}{\delta_2}\right)\frac{1}{U}\frac{dU}{dx}\right] = \frac{\tau_w}{\rho U^2}$$

(b) Rearrange the result of part a into the form

$$\frac{dZ}{dx} = \frac{R^2 F}{U}$$

where $Z = \delta_2^2 R^2/\nu$, $F = 2\tau_w\delta_2/\mu U - 2(2 + \delta_1/\delta_2)K$, and $K = (\delta_2^2/\nu)\,dU/dx$.

(c) As discussed by Schlichting [17, pp. 192–200, 229–231], use of a fourth-order polynomial for the velocity profile leads to the realization that F is a function of only K, $F = F(K)$. Walz [20] pointed out that the straight line $F \approx a - bK$, with $a = 0.47$ and $b = 6$, is an accurate approximation. Show that this approximation used in the result of part b gives

$$\frac{dZ}{dx} + \frac{b}{U}\frac{dU}{dx}Z = a\frac{R^2}{U}$$

(d) Show that the solution to the result of part c is

$$U^b Z = a\int_0^x R^2 U^{b-1}\,dx$$

since either U, R, or δ_2 equals zero at the leading edge where $x = 0$. For the given values of a and b, show that this becomes

$$U\frac{\delta_2^2}{\nu} = \frac{0.47}{R^2 U^5}\int_0^x R^2 U^5\,dx$$

(e) If U and R are given, the local value of δ_2 can be calculated. Then K is known and pertinent quantities such as τ_w and $H_{12} = \delta_1/\delta_2$ can be ascertained from the literature as cited in part c. If flow separation occurs at the location for which $\tau_w = 0$, determine the accompanying value of K by consulting Schlichting [17, p. 198].

6-15 The local heat-transfer coefficient for a body of revolution exposed to axisymmetric flow is sometimes needed. A simplified method based on the local similarity assumption that the rate of growth of the thermal boundary layer depends only on local conditions has been developed by Smith and Spalding [21]. The method is based on the suspicion, justified by either dimensional analysis or substitution of appropriate profiles for temperature and velocity into the integral energy and momentum equations for wedge flow, that

$$\frac{U}{\nu}\frac{d\Delta_4^2}{dx} = f\left(\frac{\Delta_4^2}{\nu}\frac{dU}{dx}\right)$$

where

$$\Delta_4 = \frac{k}{h} = \left[\frac{\partial(T - T_w)/(T_\infty - T_w)}{\partial y}\Big|_{y=0}\right]^{-1}$$

and f is an as yet unknown function. From exact solutions, such as in Table 5-2, it is determined that a straight line passing through the exact answers at the stagnation and flat-plate conditions is accurate for accelerating flows $(dU/dx \geqslant 0)$, although it loses accuracy for decelerating flows $(dU/dx < 0)$. Then

$$\frac{U}{\nu}\frac{d\Delta_4^2}{dx} \approx 11.68 - 2.87\left(\frac{\Delta_4^2}{\nu}\frac{dU}{dx}\right)$$

(a) Show that this differential equation has the solution

$$U^{2.87}\Delta_4^2 = 11.68\nu\int_0^x U^{1.87}\,dx$$

if either U or Δ_4 is zero at $x = 0$. *Note:* At the stagnation point where $U = x\,dU(x = 0)/dx$, $\Delta_4^2 = 4.07\nu/(dU/dx)$.

(b) Consult the result of part e of Problem 6-13 to verify that the wedge flow result can be taken over to a body of revolution of local radius $R(x)$ if all quantities involving a y dimension are multiplied by the factor R and all quantities involving $\int_0^x dx$ are replaced by $\int_0^x R^2\,dx$. Then justify by a brief discussion that the wedge result of part a can be taken over for a body of revolution as

$$U^{2.87}(R\Delta_4)^2 = 11.68\nu\int_0^x U^{1.87}R^2\,dx$$

TABLE 6P-15   Values of $C_{1,2,3}$ for Various Prandtl Numbers

Pr	C_1	C_2	C_3
0.7	0.418	0.435	1.87
0.8	0.384	0.450	1.90
1.0	0.332	0.475	1.95
5.0	0.117	0.595	2.19
10.0	0.073	0.685	2.37

(c) Noting that $\Delta_4 = k/h$, show that the result of part b gives (for Pr = 0.7)

$$\frac{\text{Nu}}{\text{Re}_x \, \text{Pr}} = \frac{0.418\nu^{1/2}U^{0.435}R}{\left(\int_0^x U^{1.87}R^2 \, dx\right)^{1/2}}$$

For variable density, the results are given in the form

$$\frac{\text{Nu}}{\text{Re}_x \, \text{Pr}} = \frac{C_1\mu^{1/2}RG^{C_2}}{\left(\int_0^x G^{C_3}R^2 \, dx\right)^{1/2}}$$

with $G = \rho U$ and $C_{1,2,3}$ as shown in Table 6P-15.

6-16 A boundary-layer integral method refinement developed by Volkov [22] and extended by Pitts and Griggs [23] involves replacement of the gradient at the wall by an integral expression.

(a) Carry out two integrations of the boundary-layer x-motion equation to obtain

$$\delta\nu\frac{\partial u(y=0)}{\partial y} = -\int_0^\delta \frac{\partial}{\partial x}\left(\int_0^y u^2 \, dy\right) dy + \int_0^\delta u\frac{\partial}{\partial x}\left(\int_0^y u \, dy\right) dy$$

$$+ \frac{\delta^2}{4}\frac{dU^2}{dx} + \nu U - v_w\int_0^\delta u \, dy$$

(b) Substitute the results of part a into Eq. (6-13) to obtain

$$\frac{A}{2}\frac{d\delta^2 U^2}{dx} + \frac{C}{2}\frac{dU^2}{dx}\delta^2 = (v_w \, \delta D + \nu)U$$

where

$$A = \int_0^1 f(\eta)[1 - f(\eta)]\, d\eta + \int_0^1 \left\{ \int_0^\eta f(\eta')[f(\eta') - f(\eta)]\, d\eta' \right\} d\eta$$

$$C = \frac{1}{2} + \int_0^1 \left[-f^2(\eta) + \int_0^\eta f^2(\eta')\, d\eta' \right] d\eta$$

$$D = \int_0^1 [1 - f(\eta)]\, d\eta$$

with $f = u/U$ and $\eta = y/\delta$.

(c) Show that if $v_w \delta = $ const, the only case for which a similarity solution is given, and with $\varepsilon = 1 + C/A$,

$$\delta^2 = 2 \frac{v_w \,\delta D + v}{AU^\varepsilon} \int_0^x U^{\varepsilon - 1}\, dx$$

(d) Use Eq. (6-13) to show that the friction coefficient is given by

$$\frac{C_f}{2} = B \frac{d\delta}{dx} + E \frac{\delta}{U} \frac{dU}{dx} - \frac{v_w}{U}$$

where

$$B = \int_0^1 f(\eta)[1 - f(\eta)]\, d\eta, \qquad E = \int_0^1 [1 - f(\eta)][1 + 2f(\eta)]\, d\eta$$

(e) With $u/U = \eta$ for $0 \leqslant \eta \leqslant 1$ ($A = \frac{1}{16}$, $B = \frac{1}{6}$, $C = \frac{7}{24}$, $D = \frac{1}{2}$, $E = \frac{5}{6}$) and $v_w = 0$, construct a table of $(C_f/2)(Ux/v)^{1/2}$ from both part d and Table 5-2 for wedge-flow values of m ranging from $-\frac{1}{9}$ to 5.

(f) In parallel fashion, integrate the boundary-layer energy equation twice to obtain ($v_w = 0$, $T_\infty = $ const, and $T_w = $ const)

$$\delta_T \alpha \frac{\partial T(y = 0)}{\partial y} = \alpha(T_\infty - T_w) - \int_0^{\delta_T} \frac{\partial}{\partial x} \left(\int_0^y uT\, dy \right) dy$$

$$- v_w \int_0^{\delta_T} (T - T_w)\, dy + \int_0^{\delta_T} T \frac{\partial}{\partial x} \left(\int_0^y u\, dy \right) dy$$

(g) Substitute the result of part f into Eq. (6-18) to obtain

$$\delta \int_0^\xi \frac{\partial}{\partial x} \left[U\delta \int_0^\eta fg\, d\eta \right] d\eta - \delta \int_0^\xi g \frac{\partial}{\partial x} \left[\delta U \int_0^\eta f\, d\eta \right] d\eta$$

$$+ \xi\delta \frac{d}{dx} \left[U\delta \int_0^\xi f(1 - g)\, d\eta \right] = \alpha$$

in which $g = (T - T_w)/(T_\infty - T_w)$ and $\xi = \delta_T/\delta$. In this integral form of the energy equation, the profile-sensitive wall temperature gradient has been replaced by the integral form of part f.

(h) With $f = \eta$ and $g = \eta/\xi$ for $0 \leqslant \eta \leqslant 1$ for $\xi \leqslant 1$, show that the result of part g leads to

$$\frac{d\xi^3}{dx} + \frac{3}{2U\delta}\frac{dU\delta}{dx}\xi^3 = \frac{12\alpha}{U\delta^2}, \qquad x \geqslant x_0$$

and that

$$\xi^3 = \frac{12\alpha}{(U\delta)^{3/2}}\int_{x_0}^{x}\left(\frac{U}{\delta}\right)^{1/2} dx$$

where x_0 is the extent of an unheated starting length.

(i) From Eq. (6-18), realizing that $q_w = h(T_w - T_\infty)$, show that the local heat-transfer coefficient h is given by

$$h = \varrho C_p \frac{d}{dx}\left[U\delta\int_0^\xi f(1 - g)\, d\eta\right]$$

(j) Show that for $f = \eta$ and $g = \eta/\xi$ the local Nusselt number is

$$\frac{\mathrm{Nu}}{\mathrm{Re}^{1/2}\,\mathrm{Pr}^{1/3}} = \frac{(m + 1)^{1/3}(1 + 17m/3)^{1/6}}{3},$$

$$\xi \leqslant 1 \qquad \text{and} \qquad x_0 = 0$$

(k) Construct a table of $\mathrm{Nu}/\mathrm{Re}^{1/2}$ from both part j and Eq. (5-68) for wedge-flow values of m ranging from $-\frac{1}{9}$ to 4.

6-17 Briefly report on the refined integral method devised by Zien [24] (also see Kou and Fan [25]) that utilizes a moment-like idea to obtain two integral equations from which surface heat flux q_w and temperature profile parameter δ are simultaneously determined after insertion of assumed profiles.

REFERENCES

1. T. von Karman, *Zeitschrift für angewandte Mathematik und Mechanik* **1**, 233–252 (1921); see also NACA TM 1092, 1946.

2. K. Pohlhausen, *Zeitschrift für angewandte Mathematik und Mechanik* **1**, 252–268 (1921).

3. T. R. Goodman, *Heat Transfer and Fluid Mechanics Institute*, California Institute of Technology, Pasadena, CA, June 1957, pp. 383–400.

4. T. R. Goodman, *ASME J. Heat Transfer* **83**, 83–86 (1961).

5. T. R. Goodman, *Advances in Heat Transfer* **1**, 52–122 (1964).

6. A. A. Sfeir, *ASME J. Heat Transfer* **98**, 466–470 (1976).

7. D. Langford, *Int. J. Heat Mass Transfer* **16**, 2424–2428 (1973).

8. H. S. Carslaw and J. C. Jaeger, *Conduction of Heat in Solids*, 2nd ed., Clarendon, 1959, p. 59.

9. R. L. Potts, *AIAA J.* **2**, 630–631 (1983).

10. J. Dey and G. Nath, *Int. J. Heat Mass Transfer* **27**, 2429–2431 (1984).

11. S. C. Lau and E. M. Sparrow, *ASME J. Heat Transfer* **102**, 364–366 (1980).

12. T. R. Goodman, *ASME J. Heat Transfer* **84**, 347–352 (1962).

13. M. J. Lighthill, *Proc. Roy. Soc. London, Ser. A* **202**, 359–377 (1950).

14. E. R. G. Eckert, J. P. Hartnett, and R. Birkebak, *J. Aeronaut. Sci.* **24**, 549–550 (1957).

15. D. R. Chapman and M. W. Rubesin, *J. Aeronaut. Sci.* **16**, 547–565 (1949).

16. J. Klein and M. Tribus, Forced convection from non-isothermal surfaces, *Heat Transfer Symposium*, Engineering Research Institute, University of Michigan, August 1952; see also ASME Paper 53-5A-46, ASME Semi-Annual Meeting, 1953.

17. H. Schlichting, *Boundary-Layer Theory*, McGraw-Hill, 1968, pp. 295–296.

18. R. V. Churchill, *Operational Mathematics*, McGraw-Hill, 1958, pp. 57, 324.

19. M. Abramowitz and I. A. Stegun, Eds., *Handbook of Mathematical Functions*, National Bureau of Standards, Applied Mathematics Series 55, 1965.

20. A. Walz, *Lilienthal-Bericht* **141**, 8–12 (1941).

21. A. G. Smith and D. B. Spalding, *J. Roy. Aeronaut. Soc.* **62**, 60–64 (1958).

22. V. N. Volkov, *Inzhenerno-Fizicheskii Zhurnal* **9** (5), 583–588; see also *J. Eng. Phys.* **9** (5), 371.

23. D. R. Pitts and E. I. Griggs, in F. Payne, C. Corduneanu, A. Haji-Sheikh, and T. Huang, Eds., *Integral Methods in Science and Engineering*, Hemisphere, 1986, pp. 530–540.

24. T. F. Zien, *AIAA J.* **14**, 404–406 (1976); **16**, 1287–1295 (1978); *Int. J. Heat Mass Transfer* **19**, 513–521 (1976).

25. H. S. Kou and N. W. Fan, *J. Thermophys. Heat Transfer* **4**, 417–418 (1990).

7

TURBULENCE FUNDAMENTALS

In turbulent flow, as in the molecular regime, the principle theme is chaos. Of course, a physical phenomenon such as turbulence that has the appearance of chaos is not necessarily random. It might be the deterministic result of several basic physical phenomena with feedback loops (Hofstadter [1] discusses chaos and its application to turbulence). On a small time scale turbulent flow is seemingly irregular and disordered with "clumps," large relative to molecular size, aimlessly moving hither and thither. There is order on a large scale, of course, since a fluid flows steadily down a duct in response to a pressure difference.

The random behavior that was used to advantage in the study of molecular motions has had only limited success when applied to turbulent flows, possibly not only because turbulence might have a "structure" but also because of the orders of magnitude differences in molecular and turbulence quantities. The structure of turbulent flow under typical conditions (Hinze [2]) is one of eddies and vortices of a range of sizes. Within each eddy molecular effects are important, but the interaction of these eddies masks molecular effects.

Air moving at about 100 m/s is typical. For such a flow a typical eddy size is about 1 mm, large relative to a molecular mean free path of 10^{-4} mm, and contains about 10^{17} molecules. Anemometer, either hot-wire or laser-doppler, measurements show that velocity varies roughly 10 m/s (roughly 10% of the average velocity) whereas mean molecular velocities are about 300 m/s. Further, turbulent velocity fluctuations occur about once every 10^{-4}–1 s whereas molecular collisions occur about once every 10^{-10} s. Finally, in contrast to the short average distance, about 10^{-7} m, traveled by a molecule between collisions, an eddy travels a large distance to a location where conditions are very different from those at its origin before losing its identity.

When the molecular mean free path is small relative to a system dimension, fluid properties are characteristic of the molecules alone. When the mean free path is of the size of a system dimension, velocity slip and temperature jump at a solid surface are observed (see Burmeister [3, Chapter 3]), and the system influences the "effective" properties of the fluid. In this light, since the typical turbulent eddy dimension easily is of the size of a system dimension (e.g., a boundary-layer thickness), it is likely that the "effective properties" of turbulent flow will be influenced by the system. Then, not only will the "effective properties" not solely be a property of the fluid but they might differ in detail from one system to another.

Fortunately, the unsteady equations of motion and continuity accurately describe turbulent flow because the eddies are much larger than the molecular spacing or mean free path and the fluctuations are not excessively rapid. Since turbulent flow can be mathematically modeled, there is hope for continued progress in developing practical descriptions of turbulence processes. Recent developments are discussed in Section 8.7. In these discussions, turbulence near a solid boundary is envisioned since that is the usual case for synthetic structures. However, turbulence in jets, oceans, and the atmosphere is also important and has received attention.

7.1 LAMINAR-TO-TURBULENT TRANSITION

Flows often undergo transition from the laminar to the turbulent state and can relaminarize in a transition from the turbulent to the laminar state, the former transition being more common. The operational concern is to predict the position along a flat plate at which the boundary layer will become turbulent, for example, or the Reynolds number at which pipe flow will become turbulent. As intimated earlier, the requisite relations are experimentally determined—theory alone does not suffice.

Historically, the best hope of general description was believed to be to look at things from a time-averaged viewpoint since all quantities have some time-averaged value about which they fluctuate. Velocity, for example, can be represented as sketched in Fig. 7-1. Thus $\mathbf{V} = \bar{\mathbf{V}} + \mathbf{V}'$ (a mass-weighted velocity, $\rho\mathbf{V} = \bar{\rho}\bar{\mathbf{V}} + \overline{\rho'\mathbf{V}'}$, could be used to account for density fluctuations). In terms of components,

$$u = \bar{u} + u'$$
$$v = \bar{v} + v'$$
$$w = \bar{w} + w'$$

where the barred quantities are time averaged and the primed quantities are

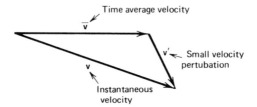

Figure 7-1 Representation of instantaneous velocity in turbulent flow as the vector sum of a time-averaged component and a velocity perturbation.

instantaneous perturbations. To illustrate, $\bar{u}$ is defined as

$$\bar{u} = \frac{1}{\Delta t} \int_{t}^{t + \Delta t} u \, dt \tag{7-1}$$

where Δt is a time interval large relative to turbulent fluctuations. Hence

$$\bar{u} = \frac{1}{\Delta t} \int_{t}^{t + \Delta t} (\bar{u} + u') \, dt = \bar{u} + \underbrace{\frac{1}{\Delta t} \int_{t}^{t + \Delta t} u' \, dt}_{\bar{u}'} \tag{7-2}$$

$$0 = \bar{u}'$$

and a time-averaged perturbation is seen to always be zero. However, $\overline{u'u'}$ is not necessarily of zero value.

The conventional way of expressing the intensity J of flow turbulence is

$$J = \frac{1}{\bar{V}} \left[\frac{1}{3} \left(\overline{u'u'} + \overline{v'v'} + \overline{w'w'} \right) \right]^{1/2} \tag{7-3}$$

Individual terms such as $\overline{u'u'}$ can be measured by a hot-wire or a laser-doppler anemometer with the former needed for turbulence intensities [see Eq. (7-3)] exceeding 30% [4] and in recirculating flows. In addition to these point measurements, whole-field velocity measurements can be made with particle-imaging techniques [5]. If $\overline{u'u'} = \overline{v'v'} = \overline{w'w'}$, the turbulence is isotropic and there is no gradient in the mean velocity $\bar{V}$. Isotropy is observed downstream from screens in wind tunnels, for example. When the mean velocity has a gradient, the turbulence is anisotropic.

The transition from laminar to turbulent flows is illustrated in Fig. 7-2, where a flat plate is shown in steady parallel flow of a constant-property fluid. The flow in the boundary layer is laminar over the first part of the plate. Then transition to turbulent flow begins at location x_i and is completed by location x_c, the critical distance from the leading edge. The transition from laminar to turbulent flow is characterized by a rapid increase in boundary-layer thickness and a flattening of the velocity profile shown in

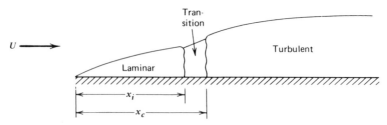

Figure 7-2   Transition from laminar to turbulent flow in a boundary layer.

Fig. 7-3 as reported by Schubauer and Klebanoff [6] for a flat plate. The ratio of the displacement δ_1 to momentum δ_2 boundary-layer thicknesses also undergoes a dramatic change from $\delta_1/\delta_2 = 2.6$ for a laminar boundary layer to 1.4 for a turbulent boundary layer as shown in Fig. 7-4. In the transition region there is an irregular array of laminar and turbulent regions. Kidney-shaped turbulent spots originate at random locations, are swept downstream, and, under proper conditions, amplify in size.

From these turbulent spots, a concept due to Emmons [7], a "bursting" of fluid with high vorticity from the wall region takes place. On the basis of flow visualization and anemometery coupled with numerical simulations, it now seems that the interaction of three-dimensional vortical structures describes

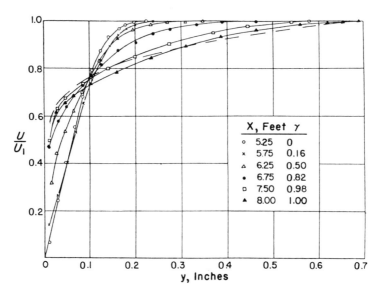

Figure 7-3   Measured velocity profiles in a boundary layer on a flat plate in the transition region: (1) laminar, Blasius profile; (2) turbulent, $\frac{1}{7}$th power law; $\delta = 17$ mm, $U = 27$ m/s, turbulence intensity = 0.03%. [From G. B. Schubauer and P. S. Klebanoff, NACA TN 3489, 1955 and NACA Report 1289, 1956 [6].]

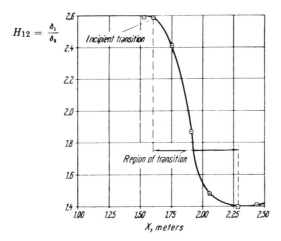

Figure 7-4 Measured change in the shape factor $H_{12} = \delta_1/\delta_2$ for a flat plate in the transition region. [From G. B. Schubauer and P. S. Klebanoff, NACA TN 3489, 1955 and NACA Report 1289, 1956 [6].]

the events in the turbulent spot and nearby regions as surveyed by Robinson [8]. In the hairpin vortex model, vortices in the shape of a hairpin spanwise to the flow and with heads inclined downstream at about 45° form near the wall as illustrated in Fig. 7-5. Fluid lifted in the legs of a hairpin is eventually released into the main flow as a "burst" of high turbulence intensity fluid when the vortex breaks down. Other vortex models have been proposed (see Blackwelder and Swearingen [10], for example) of which that by Robinson [11] is illustrated in Fig. 7-6 where a population of arch-shaped vortical structures through a turbulent boundary layer is shown along with the rollup of fluid to form a vortex in the near-wall region. In all of these models, turbulence is a consequence of flow instabilities that amplify.

Transition in a boundary layer, surveyed by Tani [12], is affected by such parameters as heat transfer, free-stream pressure gradient, surface roughness, and free-stream turbulent intensity. The influence of the latter two is understandable since they are disturbances that can be amplified if the fluid flow has a tendency toward instability. As shown in Fig. 7-7, at low turbulence intensities ($J < 0.0008$) a critical Reynolds number of

$$\mathrm{Re}_{x,c} = \left(\frac{Ux}{\nu}\right)_{\mathrm{critical}} = 2.8 \times 10^6$$

must be exceeded for transition to begin. At the "normal" turbulence intensities ($J \sim 0.01$) commonly encountered, the critical Reynolds number is conventionally taken to be

$$\mathrm{Re}_{x,c} = 3.5 \times 10^5\text{--}5 \times 10^5 \qquad (7\text{-}4a)$$

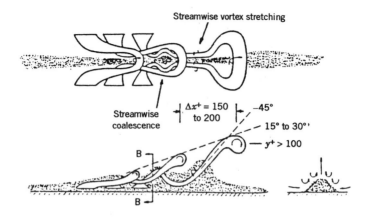

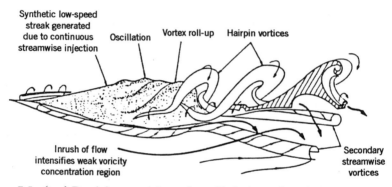

Figure 7-5 (top) Breakdown and formation of hairpin vortices during a streak-bursting process [by permission from C. R. Smith, Proc. Eighth Symp. Turb., Rolla, MO, 1984 [9a]]; (bottom) breakup of synthetic low-speed streak-generating hairpin vortices with secondary streamwise vortices generated by inrushing fluid [by permission from M. S. Acarlar and C. R. Smith, *J. Fluid Mech.* **175**, 43–83 (1987) [9b]].

with the effect of free-stream turbulence represented, according to Sucker and Brauer [14], by

$$\mathrm{Re}_{x,c} = 3.78 \times 10^5 \exp(-6J^{1/2}) \tag{7-4b}$$

Recently, Schmidt and Patankar [15] applied a k–ε turbulence model of the sort discussed in Chapter 8 to verify the laminar-to-turbulent criterion of Abu-Ghannam and Shaw [16]

$$\mathrm{Re}_{\delta 2,\,\mathrm{start}} = 163 + \exp[6.91 - 100J] \tag{7-4c}$$

$$\mathrm{Re}_{\delta 2,\,\mathrm{end}} = 2.667\,\mathrm{Re}_{\delta 2,\,\mathrm{start}} \tag{7-4d}$$

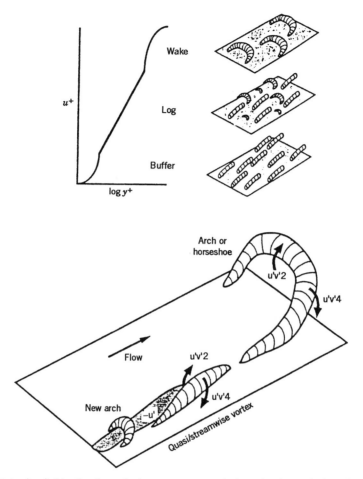

Figure 7-6 (top) Idealized vortical structure populations in the turbulent boundary layer; (bottom) ejection/sweep motions and streamwise vortices in the near-wall region along with ejection/sweep motions and arch-shaped vortical structures in the outer region. [From S. K. Robinson, Kinematics of turbulent boundary layer structure, Ph.D. Dissertation, Stanford University, Stanford, CA, 1990 [11].]

If free-stream turbulence intensity is low, a laminar boundary layer can exist for a large distance on the plate. As the free stream becomes more turbulent, the laminar regime can exist for smaller distances.

Similar results are observed for duct flow, reported early by Osborne Reynolds [17]. In fully developed laminar flow a parabolic velocity profile exists; in fully developed turbulent flow a more uniform velocity profile is found as sketched in Fig. 7-8. Under "usual" conditions the critical Reynolds number based on duct diameter, above which the flow is turbulent and below

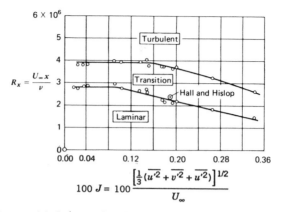

Figure 7-7 Measured influence of turbulence intensity on critical Reynolds number on a flat plate at zero incidence. [By permission from G. B. Schubauer and H. K. Skramstad, *J. Aeronaut. Sci.* **14**, 69–78 (1947) [13].]

which the flow is laminar, is

$$\text{Re}_{D,c} = \left(\frac{UD}{\nu}\right)_{\text{critical}} = 2300 \qquad (7\text{-}5)$$

where U is the velocity averaged over the duct cross-sectional area. The critical Reynolds number is influenced by external disturbances. When the duct is free of vibration and the inlet fluid is free of disturbances, laminar flow has been achieved for $\text{Re} = UD/\nu = 10^5$ according to Hinze [2, p. 707]. A lower bound of $\text{Re}_{D,c} \approx 2000$ is reported by Schlichting [18, p. 433] below which the flow remains laminar even for strong disturbances. For noncircular ducts the critical Reynolds number can be roughly predicted from Eq. (7-5) if the hydraulic diameter $D_H = 4$ (flow area)/(wetted perimeter) is used as the characteristic length (see [19] for rectangular-duct details); in triangular ducts it is possible to have laminar flow in the vertices with turbulent flow in the main portion of the duct [18, pp. 575–578], however.

The laminar-to-turbulent transition criterion stated in Eq. (7-5) can be extended to a power-law fluid as Skelland [20] shows and Cho and Hartnett

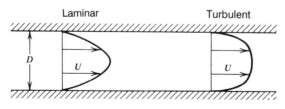

Figure 7-8 Comparison of laminar and turbulent velocity profiles in a circular duct.

[21] discuss. An effective viscosity μ_{eff} is first defined for use in the Reynolds number as that parameter that makes Poiseuille's equation relate pressure drop to average velocity. For a power-law fluid, Eq. (4-31) gives

$$V_{\text{av}} = \frac{n}{3n + 1}\left(-\frac{1}{2\mu}\frac{dP}{dx}\right)^{1/n} R^{(n+1)/n}$$

from which it follows that

$$\frac{\Delta P}{L} = \frac{2\mu\{[(3n + 1)/n]V_{\text{av}}\}^n}{R^{1/(n+1)}}$$

For a Newtonian fluid ($n = 1$),

$$V_{\text{av}} = \frac{1}{8\mu}\frac{\Delta P}{L}R^2 = \frac{1}{32\mu}\frac{\Delta P}{L}D^2$$

or

$$\mu_{\text{eff}}|_{\text{Newtonian}} = \frac{1}{32}\frac{\Delta P}{L}\frac{D^2}{V_{\text{av}}}$$

For a power-law fluid then

$$\mu_{\text{eff}}|_{\text{power law}} = \frac{1}{32}\frac{\Delta P}{L}\frac{D^2}{V_{\text{av}}} = \frac{\mu}{8D^{n-1}}\left(\frac{6n + 2}{n}\right)^n V_{\text{av}}^{n-1}$$

Thus

$$\text{Re} = \frac{V_{\text{av}}D\rho}{\mu_{\text{eff}}} = \frac{D^n V_{\text{av}}^{2-n}}{\mu/\rho}8\left(\frac{n}{6n + 2}\right)^n$$

Computation of pipe flow friction factors and transition between laminar and turbulent flow can now proceed for a power-law fluid simply by taking over all the information developed for Newtonian fluids without change.

As with the boundary layers, laminar-to-turbulent transition in the entrance region of a duct takes place over a distance from the entrance. According to Hinze [2, p. 717], fully developed turbulent flow requires at least 40–100 diameters with a lower bound given by

$$\frac{X_{\text{entry}}}{D} = 0.693\,\text{Re}^{1/4}$$

When the Reynolds number approaches the critical value from below, alternate slugs of laminar and turbulent fluid pass down the duct with the turbulent slugs decaying; when it exceeds the critical value, the turbulent slugs grow until they fill the duct while being swept downstream.

Results of like nature have been found for other geometries. For Couette flow of two parallel walls, the critical Reynolds number is reported to be [2, p. 77; 18, p. 558]

$$\text{Re}_{h,c} = \frac{U_{max}h}{\nu} = 1900$$

where h is the wall separation; if flow is caused by a pressure gradient between parallel walls of separation h, then

$$\text{Re}_c = \frac{U_{av}h}{\nu} = 1150$$

For two concentric cylinders of radial separation d, inner radius R_i, inner tangential velocity U_i, and the outer cylinder at rest, there are three flow regimes [18, p. 503] described by the Taylor number $\text{Ta} = (U_i d/\nu)(d/R_i)^{1/2}$:

$$\text{Ta} < 41.3 \qquad \text{laminar Couette flow}$$
$$41.3 < \text{Ta} < 400 \qquad \text{laminar flow with Taylor vortices}$$
$$\text{Ta} > 400 \qquad \text{turbulent flow}$$

The destabilizing effect of centrifugal forces is illustrated by these results. Vertical stratification caused by temperature or salinity gradients also affects the critical Reynolds number. The effect of temperature on viscosity can cause early transition to turbulent boundary layers when heat flows from the surface into a gas, acting as a pressure increase in the downstream direction, whereas a cooling of a gas by the wall retards transition just as does a pressure decrease in the downstream direction—for a liquid, the effect of heat transfer would be in the opposite direction. Boundary-layer suction is effective in retarding transition, whereas random surface roughnesses can speed transition because of the disturbances they generate.

A single criterion for laminar-to-turbulent transition is sought by bringing together the criteria for round duct flow $\text{Re}_{D,c} = U_{av}D/\nu \approx 2300$ and boundary-layer flow over a flat plate $\text{Re}_{X,c} = UX_c/\nu \approx 3.5 \times 10^5$. To achieve unification, it is necessary to base a Reynolds number for the flat-plate case on a characteristic dimension that can be the distance X_c (the dimension ultimately of design interest) from the plate leading edge at which transition occurs, or a local dimension less influenced by upstream events and more characteristic of the boundary layer such as the boundary-layer thickness at transition δ_c (a somewhat fictitious quantity), the displacement thickness at transition δ_{1c}, or the momentum thickness at transition δ_{2c} (a real quantity related to local wall shear stress).

First, consider use of $\text{Re}_{\delta,c} = U\delta_c/\nu$ for a boundary layer. For a laminar boundary layer, Eq. (5-16) gives $\delta/x = 5/(Ux/\nu)^{1/2}$, which gives the transition criterion

$$\frac{U\delta_c}{\nu} \approx 5(3.5 \times 10^5)^{1/2} = 2960 \qquad (7\text{-}6)$$

In comparison, the round-duct result can be expressed in terms of the center-line velocity for a parabolic profile [$V_m = 2V_{av}$ from Eq. (4-30a)] with the pipe radius taken to be the boundary-layer thickness. This procedure gives the round-duct transition criterion, in fair agreement with Eq. (7-6), as

$$\frac{V_m \delta}{\nu} \approx 2300$$

Second, consider use of $\text{Re}_{\delta_{1_c}} = U\delta_{1_c}/\nu$ for a boundary layer. Equation (5-20) gives $\delta_1/\delta = 0.3442$, which, when used in Eq. (7-6), gives the transition criterion as

$$\frac{U\delta_{1c}}{\nu} \approx 2960(0.3442) = 1020 \qquad (7\text{-}7)$$

In comparison, the round-duct result can be based on the center-line velocity for laminar flow and the pipe displacement thickness (the radial displacement required of the wall to allow a stream of center-line velocity to flow the discrepancy between actual flow and that which would flow if all fluid had the center-line velocity). The pipe displacement thickness is calculated from

$$V_m \int_R^{R+\delta_1} 2\pi r\, dr = \int_0^R (V_m - V)2\pi r\, dr$$

with $V/V_m = 1 - (r/R)^2$ from Eq. (4-30a) to be $\delta_1/R = 0.22$. Introduction of this into the pipe transition criterion gives

$$V_m \delta_1/\nu \approx 2300(0.22) = 506$$

which does not agree well with Eq. (7-7). Nevertheless, stability theory suggests that $U\delta_1/\nu$ is a proper transition criterion, and it is frequently used [18, p. 470] in a modified form. Table 7-1, based on the work of Pretsch [22], shows the influence of a pressure gradient. Furthermore, the $m = 0$ result suggests that transition begins at a value of $UX_c/\nu = 2.3 \times 10^5$ which might be the case, according to Fig. 7-7, for a highly turbulent free stream.

Third, consider the use of $\text{Re}_{\delta_{2c}} = U\delta_{2c}/\nu$ for a boundary layer. Equation (5-22) gives $\delta_1/\delta = 0.1328$, which, when used in Eq. (7-6), gives the transition

TABLE 7-1 Critical Reynolds Number for Laminar Wedge Flow [22][a]

β	-0.1	0	0.2	0.4	0.6	1.0
$m = (x/U)\,du/dx$	-0.048	0	0.111	0.25	0.43	1.0
$(U\delta_1/\nu)_{\text{critical}}$	126	660	$3{,}200$	$5{,}000$	$8{,}300$	$12{,}600$

[a]See Fig. 5-7 for m and β definitions.

criterion as

$$\text{Re}_{\delta_{2c}} = \frac{U\delta_{2c}}{\nu} = 390 \tag{7-8}$$

The momentum thickness for a pipe is the radial displacement of the pipe wall required for a stream at the center-line velocity to flow the momentum discrepancy between that actually transported by the parabolic velocity and what would have flowed had the stream all been at the center-line velocity. Thus

$$V_m^2 \int_R^{R+\delta_2} 2\pi r\, dr = \int_0^R (V_m^2 - V^2)2\pi r\, dr$$

which gives $\delta_2/R = 0.154$. Introduction of this into the pipe transition criterion gives

$$V_m\delta_2/\nu \approx 2300(0.154) = 350$$

which is in good agreement with Eq. (7-8) and is similar to the refinements of Eqs. (7-4b)–(7-4d) that show the effect of free-stream turbulence. Kays and Crawford [23] suggest that Eq. (7-8) be used but with a value of 162, rather than 390. A similar criterion [24] enables prediction of the relaminarization of initially turbulent tube flows that can occur when a gas is strongly heated as in a nuclear reactor.

Prediction of the location at which transition occurs is commonly accomplished by the semiempirical e^n-criterion, as discussed by Herbert [25] and Cebeci [26]. As shown in Fig. 7-9, the Reynolds number based on momentum thickness $\delta_2 = \theta$ at transition was found by Michel [27] to correlate with the Reynolds number based on the distance x from the leading edge for two-dimensional incompressible boundary layers on smooth surfaces. Smith and Gamberoni [28] found that the maximum amplification ratio of initial disturbances computed from linearized stability theory was about equal to $e^9 = 8100$ at the observed location of transition. Later it was found that the best predictions occurred with an amplification ratio of e^n with n usually between 8 and 9. Greater detail of the e^n method is given by Bradshaw et al. [29] and by Cebeci and Egan [30]. Direct numerical simulations [8, 31] are being developed for a universally applicable mathematical description of turbulent flow and, especially, transition to and from laminar flow.

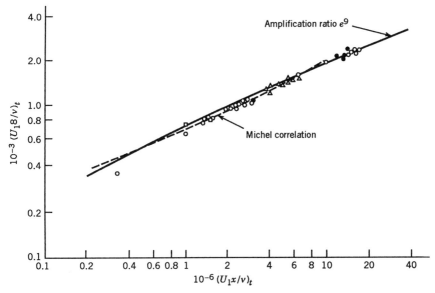

Figure 7-9 Relationship of transitional, momentum-thickness Reynolds and x-Reynolds numbers. [Correlation of R. Michel, Office Nat. Etude Recherche Aeronaut., Report 58, 1952 [27] and analysis of A. M. O. Smith and N. Gamberoni, Douglas Aircraft Co., Report ES-26388, 1956 [28].]

7.2 TIME-AVERAGED DESCRIBING EQUATIONS

Time averaging of the Navier–Stokes and energy equations (exact if their unsteady form is taken) should give the net effect of the turbulent perturbations. In what follows it is assumed for simplicity that properties are constant, in much the spirit that guided the breakthrough to the boundary-layer equations for laminar flow. First, the continuity equation for the unsteady state is

$$\frac{\partial u}{\partial x} + \frac{\partial v}{\partial y} + \frac{\partial w}{\partial z} = 0$$

In terms of a time-averaged and perturbation quantity, this becomes

$$\frac{\partial \bar{u} + u'}{\partial x} + \frac{\partial \bar{v} + v'}{\partial y} + \frac{\partial \bar{w} + w'}{\partial z} = 0$$

Time averaging this equation [a bar denotes $(1/\Delta t)\int_{t}^{t+\Delta t} f(t)\, dt$] gives

$$\overline{\frac{\partial \bar{u} + u'}{\partial x}} + \overline{\frac{\partial \bar{v} + v'}{\partial y}} + \overline{\frac{\partial \bar{w} + w'}{\partial z}} = 0$$

Interchanging the order of operation on space and time gives

$$\frac{\partial \bar{u} + u'}{\partial x} + \frac{\partial \bar{v} + v'}{\partial y} + \frac{\partial \bar{w} + w'}{\partial z} = 0$$

Since the time average of a perturbation quantity is zero, the time-averaged continuity equation is

$$\frac{\partial \bar{u}}{\partial x} + \frac{\partial \bar{v}}{\partial y} + \frac{\partial \bar{w}}{\partial z} = 0$$

or

$$\nabla \cdot \bar{\mathbf{V}} = 0 \qquad (7\text{-}9)$$

Second, the unsteady x-motion equation (τ_{ij}^m is a stress due to molecular level events) is, in conservation form,

$$\rho \left[\frac{\partial u}{\partial t} + \frac{\partial uu}{\partial x} + \frac{\partial vu}{\partial y} + \frac{\partial wu}{\partial z} \right] - \rho \left[\frac{\partial u}{\partial x} + \frac{\partial v}{\partial y} + \frac{\partial w}{\partial z} \right] u$$

$$= B_x - \frac{\partial P}{\partial x} + \frac{\partial \tau_{xx}^m}{\partial x} + \frac{\partial \tau_{yx}^m}{\partial y} + \frac{\partial \tau_{zx}^m}{\partial z}$$

Time averaging and interchanging the order of operation on time and space gives

$$\rho \left[\frac{\partial \bar{u}}{\partial t} + \frac{\partial \bar{u}'}{\partial t} + \frac{\partial \left(\overline{\bar{u}\bar{u}} + 2\overline{\bar{u}u'} + \overline{u'u'} \right)}{\partial x} + \frac{\partial \left(\overline{\bar{v}\bar{u}} + \overline{\bar{v}u'} + \overline{\bar{u}v'} + \overline{v'u'} \right)}{\partial y} \right.$$

$$\left. + \frac{\partial \left(\overline{\bar{w}\bar{u}} + \overline{\bar{w}u'} + \overline{\bar{u}w'} + \overline{w'u'} \right)}{\partial z} \right]$$

$$= \bar{B}_x + \bar{B}_x' - \frac{\partial \left(\bar{P} + \bar{P}' \right)}{\partial x} + \frac{\partial \left(\bar{\tau}_{xx}^m + \bar{\tau}_{xx}'^m \right)}{\partial x} + \cdots$$

$$\rho \left[\frac{\partial \bar{u}}{\partial t} + \frac{\partial \overline{\bar{u}\bar{u}}}{\partial x} + \frac{\partial \overline{\bar{v}\bar{u}}}{\partial y} + \frac{\partial \overline{\bar{w}\bar{u}}}{\partial z} \right] = \bar{B}_x - \frac{\partial \bar{P}}{\partial x} + \frac{\partial}{\partial x} \left[\bar{\tau}_{xx}^m - \rho \overline{u'u'} \right]$$

$$+ \frac{\partial}{\partial y} \left[\bar{\tau}_{yx}^m - \rho \overline{v'u'} \right] + \frac{\partial}{\partial z} \left[\bar{\tau}_{zz}^m - \rho \overline{w'u'} \right]$$

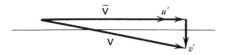

Figure 7-10 Transport across a surface due to turbulent fluctuations of velocity.

Realization that $\nabla \cdot \overline{V} = 0$ then gives

$$\rho\left(\frac{\partial \overline{u}}{\partial t} + \overline{u}\frac{\partial \overline{u}}{\partial x} + \overline{v}\frac{\partial \overline{u}}{\partial y} + \overline{w}\frac{\partial \overline{u}}{\partial z}\right) = \overline{B}_x - \frac{\partial \overline{P}}{\partial x} + \frac{\partial \overline{\tau}_{xx}}{\partial x} + \frac{\partial \overline{\tau}_{yx}}{\partial y} + \frac{\partial \overline{\tau}_{zx}}{\partial z} \quad (7\text{-}10)$$

In Eq. (7-10) it is evident that the total stress is composed of the sum of a molecular and a turbulent part $\overline{\tau} = \overline{\tau}^m + \overline{\tau}^{\text{turb}}$, both of which have the same physical origin—a particle crosses a control surface, carrying with it momentum as sketched in Fig. 7-10—with the latter called a *Reynolds stress*. In brief, time averaging has shown that effective additions to the time-averaged molecular transport are possible as

$$\overline{\tau}_{ij} = \overline{\tau}_{ij}^m + \overline{\tau}_{ij}^{\text{turb}}$$

stress formulation used for laminar flow based on time-averaged velocity

Reynolds stress, $\overline{\tau}_{ij}^t = -\overline{\rho u_i' u_j'}$

The equations of y and z motion have the form of Eq. (7-10) which, except for the addition of the Reynolds stresses, have the same form as the instantaneous equations.

Third, the equations of energy and diffusion have the form of the x-motion equation, with T and ω replacing u (the complication in the energy equation is the viscous dissipation terms, which are omitted for simplicity here). The diffusive fluxes are given by

$$\overline{q}_i = -k\frac{\partial \overline{T}}{\partial x_i} + \rho C_p \overline{u_i' T'} = \overline{q}_i^m + \overline{q}_i^{\text{turb}}$$

$$\overline{m}_i = -\rho D_{12}\frac{\partial \overline{\omega}_1}{\partial x_i} + \rho \overline{u_i' \omega_1'} = \overline{m}_{1_i}^m + \overline{m}_{1_i}^{\text{turb}}$$

The task now is to calculate the turbulent transport (e.g., $\overline{\tau}_{ij}' = -\overline{\rho u_i' u_j'}$) from the time-averaged velocity.

7.3 EDDY DIFFUSIVITIES

As long as one must work without rigorous guides, it is best to put things into familiar and convenient forms. With boundary-layer ideas in mind, plausible relationships are

$$\bar{\tau}_{yx} = \mu\frac{\partial\bar{u}}{\partial y} + \underbrace{\mu'\frac{\partial\bar{u}}{\partial y}}_{-\rho\overline{v'u'}} = \rho(\nu + \varepsilon_m)\frac{\partial\bar{u}}{\partial y}$$

eddy diffusivity
for momentum

$$\bar{q}_y = -k\frac{\partial\bar{T}}{\partial y} + \underbrace{\left(-k'\frac{\partial\bar{T}}{\partial y}\right)}_{\rho C_p\overline{v'T'}} = -\rho C_p\left(\frac{\nu}{\text{Pr}} + \frac{\varepsilon_m}{\text{Pr}'}\right)\frac{\partial\bar{T}}{\partial y}$$

turbulent Pr

$$\bar{m}_y = -\rho D_{12}\frac{\partial\bar{\omega}_1}{\partial y} + \underbrace{\left(-\rho D'_{12}\frac{\partial\bar{\omega}_1}{\partial y}\right)}_{\rho\overline{u'\omega'_1}} = -\rho\left(\frac{\nu}{\text{Sc}} + \frac{\varepsilon_m}{\text{Sc}'}\right)\frac{\partial\bar{\omega}_1}{\partial y}$$

turbulent Sc

The boundary-layer equations for a constant-property fluid in turbulent flow are

$$\frac{\partial\bar{u}}{\partial x} + \frac{\partial\bar{v}}{\partial y} = 0$$

$$\frac{D\bar{u}}{Dt} = \frac{1}{\rho}\left(\bar{B}_x - \frac{\partial\bar{P}}{\partial x}\right) + \frac{\partial}{\partial y}\left[(\nu + \varepsilon_m)\frac{\partial\bar{u}}{\partial y}\right]$$

$$\frac{D\bar{T}}{Dt} = \frac{\partial}{\partial y}\left[\left(\frac{\nu}{\text{Pr}} + \frac{\varepsilon_m}{\text{Pr}'}\right)\frac{\partial\bar{T}}{\partial y}\right] \qquad (7\text{-}11)$$

$$\frac{D\bar{\omega}_1}{Dt} = \frac{\partial}{\partial y}\left[\left(\frac{\nu}{\text{Sc}} + \frac{\varepsilon_m}{\text{Sc}'}\right)\frac{\partial\bar{\omega}_1}{\partial y}\right]$$

The turbulent Prandtl Pr' and Schmidt Sc' numbers are often assumed to be constant near unity since the mechanisms for turbulent transport of momentum, heat, and mass should be the same.

Of course, the question remains as to what a workable expression for ε_m might be. Boussinesq [32] in 1877 postulated that ε_m is constant. This may have been guided by the constancy of molecular viscosity, but it turns out that, particularly near solid surfaces, ε_m is dependent on velocity and geometry and is not a property of the fluid. Nevertheless, a constant ε_m is still sometimes used because of its convenience.

In about 1925 Prandtl [33] devised the Prandtl mixing-length relationship, based on kinetic theory of gases ideas. The reasoning is that the turbulent

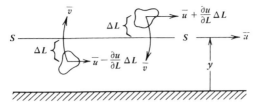

Figure 7-11 Turbulent transport across a surface by random movement of clumps of fixed identity.

transport is much like the molecular one; it occurs because discrete lumps of matter of fixed identity (for at least short times) cross back and forth and carry their properties with them as sketched in Fig. 7-11. For molecular transport,

$$\nu = \mu/\rho \sim V_{av}\lambda$$

It is reasonable to set $\Delta L \sim \lambda$ and $|(d\bar{u}/dL)\Delta L| \sim V_{av}$. Thus the eddy diffusivity for momentum is written by analogy as

$$\varepsilon_m = \frac{\mu^{turb}}{\rho} \sim (\Delta L)^2 \left|\frac{d\bar{u}}{dL}\right|$$

But ΔL must be small near a solid surface for a turbulent eddy. Since the principal flaw of Boussinesq's hypothesis is its failure near a solid surface, it is guessed that $\Delta L = K_1 y$, where y is the distance from the solid surface. Comparison with data shows that $K_1 = 0.36$. So Prandtl's mixing-length relationship is

$$\varepsilon_m = l^2 \left|\frac{d\bar{u}}{dy}\right| \tag{7-12}$$

where $l = K_1 y$ is the mixing length. Equation (7-12) requires that $\varepsilon_m = 0$ at the center of a flow channel where $d\bar{u}/dy = 0$, a difficulty remedied for the case of free turbulent flow (such as occurs in jets) by Prandtl's [34] use of Reichardt's [35] data to devise the simpler relationship

$$\varepsilon_m = K_2 b(\bar{u}_{max} - \bar{u}_{min}) \tag{7-13}$$

Here K_2 is a constant, b is the jet mixing-zone width, and $\bar{u}_{max} - \bar{u}_{min}$ is the maximum difference in velocity across the jet width. Prandtl also proposed a modification, little used because of its complexity, $\varepsilon_m = l^2[(d\bar{u}/dy)^2 + L^2(d^2\bar{u}/dy^2)^2]^{1/2}$, to circumvent the difficulty with Eq. (7-12). The similarity hypothesis devised by von Karman [36] and described by Schlichting [18,

p. 551] gives

$$\varepsilon_m = K_3^2 \left[\frac{d\bar{u}/dy}{d^2\bar{u}/dy^2} \right]^2 \left| \frac{d\bar{u}}{dy} \right|$$

with $K_3 = 0.4$ (see Bird et al. [37] and Problem 7-8, also).

Even with Prandtl's mixing-length relationship of Eq. (7-12), large discrepancies between data and predictions are observed close to the wall. Deissler [38] combined an empirical fit to data in this region with his insight that the damping effect of the wall dies away exponentially as one moves away from the wall. This idea was combined with the molecular idea to the extent that $V_{av} \sim \bar{u}$ and $\lambda \sim y$ to give

$$\varepsilon_m = \underbrace{n^2\bar{u}y}_{\substack{\text{molecular} \\ \text{idea}}} \underbrace{\left[1 - \exp\left(-n^2 \frac{\overset{\displaystyle\downarrow}{\bar{u}y}}{\nu} \right) \right]}_{\substack{\text{exponential decay} \\ \text{of wall damping}}} \qquad (7\text{-}14)$$

where the arrow points to $\dfrac{\bar{u}y}{\nu}$ labeled **dimensionless Reynolds number**.

where $n = 0.124$. No derivatives of velocity appear, circumventing the difficulty encountered with Eq. (7-12). Van Driest [39] similarly proposed at about the same time, with $v^{*2} = \tau_w/\rho$,

$$\varepsilon_m = K^2 y^2 \left[1 - \exp\left(-\frac{y}{A} \right) \right]^2 \left| \frac{\partial\bar{u}}{\partial y} \right|, \qquad K = 0.4 \qquad \text{and} \qquad A = \frac{26\nu}{v^*}$$

$$(7\text{-}14a)$$

which has been applied to boundary-layer problems. The basis for the exponential form is the subject of Problem 7-9.

Liepmann [40] points out that averaging alone only introduces more unknowns and requires closure of the sequence of equations by either additional measurements or physical arguments such as have been adduced in this section. It is likely that turbulent flow will ultimately be explainable on a fundamental basis as the long-range interaction of vortices (see Lugt [41] for a popularized description). Once this understanding is achieved, turbulence control may be possible.

7.4 UNIVERSAL TURBULENT VELOCITY PROFILE

The foregoing eddy diffusivity information allows turbulent velocity profiles to be computed. From these computed profiles comparisons can be made with measurements and constants evaluated. Measurement of wall shear

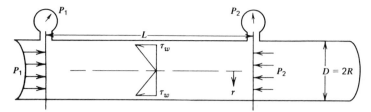

Figure 7-12   Shear-stress variation in a circular duct.

stress τ_w is easy in steady flow in a long and smooth pipe since a momentum balance on the fluid in the pipe as suggested in Fig. 7-12 yields the relationship between pressure drop and wall shear stress of

$$\tau_w = \frac{D(P_1 - P_2)}{4L}$$

and also shows that shear stress varies linearly with radius as

$$\tau = \tau_w \frac{r}{R}$$

Determination of wall shear stress is a major reason for finding velocity distributions, so this easy experimental method is important.

A turbulent velocity profile can be computed by using the relationships given by Deissler and Prandtl in the region "close" to and "far" from the wall, respectively, for fully developed steady turbulent flow in a smooth pipe. Use of Eq. (7-14) in

$$\tau = \rho(\nu + \varepsilon_m)\frac{d\bar{u}}{dy}$$

yields, with $r = R - y$, where y is the distance from the wall,

$$\tau_w\left(1 - \frac{y}{R}\right) = \rho\left\{\nu + n^2\bar{u}y\left[1 - \exp\left(-n^2\frac{\bar{u}y}{\nu}\right)\right]\right\}\frac{d\bar{u}}{dy}$$

With $u^+ = \bar{u}/v^*$, $y^+ = yv^*/\nu$ where $v^* = (\tau_w/\rho)^{1/2}$ is called the *friction velocity*, and $\tau \approx \tau_w$ close to the wall, this becomes

$$\frac{du^+}{dy^+} = \frac{1}{1 + n^2u^+y^+\left[1 - \exp(-n^2u^+y^+)\right]}, \qquad u^+(0) = 0 \quad (7\text{-}15)$$

Extremely close to the wall, molecular effects will predominate so that

$$\frac{du^+}{dy^+} \approx 1, \qquad u^+(0) = 0$$

giving the linear velocity profile

$$u^+ \approx y^+ \qquad\qquad (7\text{-}16)$$

which measurements show to be accurate for $0 \leqslant y^+ \leqslant 5$. Farther away from the wall both molecular and turbulent effects are important, so Eq. (7-15) must be solved numerically; it is suggested here only in integral form as

$$u^+ = 5 + \int_5^{y^+} \left\{ 1 + n^2 u^+ y^+ \left[1 - \exp(-n^2 u^+ y^+) \right] \right\}^{-1} dy^+$$

$$\approx 5 + 5 \ln\left(\frac{y^+}{5}\right) \qquad\qquad (7\text{-}17)$$

which is in agreement with measurements for $5 \leqslant y^+ \leqslant 26$.

Deissler's Eq. (7-14) has given the velocity profile close to the wall where molecular and turbulent contributions must both be accounted for (a *buffer zone*, $5 \leqslant y^+ \leqslant 26$); the region where molecular effects are most important ($y^+ \leqslant 5$) is a *laminar sublayer*—a misnomer since the flow is turbulent everywhere. This is illustrated in Fig. 7-13 where it is seen that the local turbulence remains a substantial fraction of the local flow right up to the wall.

Farther from the wall, but still close, Prandtl's mixing-length relation of Eq. (7-12) can be used. The contribution of molecular effects is ignored. Then

$$\tau = \rho \varepsilon_m \frac{d\bar{u}}{dy}$$

Introduction of Eq. (7-12) into this relation gives

$$\tau_w\left(1 - \frac{y}{R}\right) \approx \rho K_1^2 y^2 \left(\frac{d\bar{u}}{dy}\right)^2$$

Because attention is still directed to regions near the wall, $\tau \approx \tau_w$. Then with $y^+ = yv^*/\nu$, $u^+ = \bar{u}/v^*$, and $v^* = (\tau_w/\rho)^{1/2}$ the solution is found to be the *law of the wall*

$$u^+ = K_1^{-1} \ln y^+ + C = 2.78 \ln y^+ + 3.8 \qquad\qquad (7\text{-}18)$$

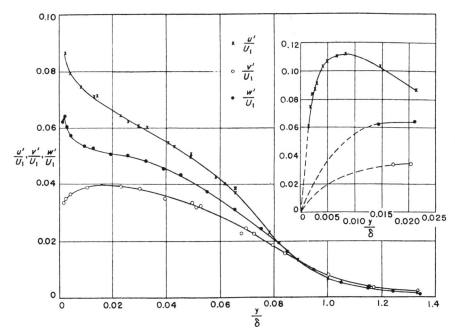

Figure 7-13 Measured rms fluctuations of the velocity components in a boundary layer along a smooth wall with constant pressure. [From P. S. Klebanoff, NACA TN 3178, 1954 [42].]

for $26 \leqslant y^+$ with $K_1 = 0.41$ being the von Karman constant. As discussed by Cantwell [43], Spalding and Coles achieved improved fits to data with more complex functions without changing the essential logarithmic dependence of u on y. Of the many fits reported by Kestin and Richardson [44], only that

$$u^+ = 2.5 \ln(1 + 0.4y^+) + 7.8\left[1 - \exp\left(-\frac{y^+}{11}\right) - \left(\frac{y^+}{11}\right)\exp(-0.33y^+)\right]$$

due to Reichardt [45] applies for all y^+. As the turbulent contribution is much larger than the molecular contribution, this region ($y^+ > 26$) is named the *fully turbulent core*. In the main part of the pipe, data show that

$$u^+ = 8.74(y^+)^{1/7} \tag{7-19}$$

for $\mathrm{Re}_D = U_{av}D/\nu \approx 10^5$. As pointed out by Schlichting [17, pp. 563–566] in his discussion of the data due to Nikuradse [46], the best fit is provided by $u^+ = C(y^+)^{1/n}$, where C and n vary with Reynolds number, as shown in Table 7-2. The power-law form is convenient for integration.

TABLE 7-2   Parameters for Power-Law Turbulent Velocity Profile in a Circular Pipe

Re = $U_{av}D/\nu$	n	C
4×10^3	6	
10^5	7	8.74
0.8×10^6	8	9.71
$\geqslant 2 \times 10^6$	10	11.5

The universal turbulent velocity profile from the work of Martinelli [47] is shown in Fig. 7-14, where it is seen that Eqs. (7-16)–(7-19) agree with data. The arbitrary division of the flow into three discontinuous zones (laminar sublayer, buffer, and turbulent core) is reasonable, although the velocity distribution is smooth throughout.

Efforts to devise a velocity distribution that is applicable to all regions of the pipe have been made. Reichardt [48] proposed

$$\frac{\varepsilon_m}{\nu} = \frac{kr_0^+}{6}\left[1 - \left(\frac{r}{r_0}\right)^2\right]\left[1 + 2\left(\frac{r}{r_0}\right)^2\right] \tag{7-20}$$

with $k = 0.4$, which, when used with a linearly varying shear stress, gives the velocity distribution applicable everywhere as

$$u^+ = 5.5 + 2.5\ln\left[y^+\frac{1.5(1 + r/r_0)}{1 + 2(r/r_0)^2}\right] \tag{7-20a}$$

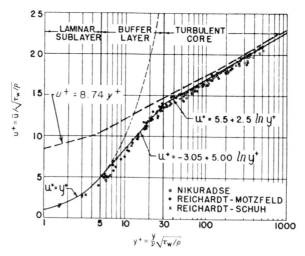

Figure 7-14 Generalized velocity distribution for turbulent flow in tubes. [From R. C. Martinelli, *Trans. ASME* **69**, 947–959 (1947) [47].]

Spalding [49] proposed, where $k = 0.0407$ and $E = 10$,

$$\frac{\varepsilon_m}{\nu} = \frac{k}{E}\left[e^{ku^+} - 1 - ku^+ - \frac{(ku^+)^2}{2!} - \frac{(ku^+)^3}{3!}\right]$$

$$y^+ = u^+ + E^{-1}\left[e^{ku^+} - 1 - ku^+ - \frac{(ku^+)^2}{2!} - \frac{(ku^+)^3}{3!} - \frac{(ku^+)^4}{4!}\right]$$

7.5 RESISTANCE FORMULAS

The flow resistance in a round duct can be deduced from the velocity distribution of Eq. (7-19). Starting with

$$u^+ = 8.74(y^+)^{1/7} \tag{7-19}$$

rearrangement gives the friction velocity as

$$v^* = u^{7/8}\left(\frac{y}{\nu}\right)^{-1/8}(8.74)^{-7/8}$$

Recalling that $v^* = (\tau_w/\rho)^{1/2}$ and solving for τ_w, one obtains

$$\tau_w = \frac{0.0225\rho u^2}{(uy/\nu)^{1/4}}$$

At the pipe center line, $u = U_{CL}$ and $y = R$ so that

$$\frac{C_f}{2} = \frac{\tau_w}{\rho U_{CL}^2} = \frac{0.0225}{(U_{CL}R/\nu)^{1/4}} \tag{7-21}$$

This relationship is accurate up to a Reynolds number of only about 10^5, as discussed in connection with Table 7-2. In terms of pressure drop, Eq. (7-21) becomes, since $U_{av} = 0.817U_{CL}$ (see Problem 7-5),

$$f = \frac{2\,\Delta P}{(L/D)\rho U_{av}^2} = \frac{0.305}{(U_{av}D/\nu)^{1/4}} \tag{7-21a}$$

Although Eq. (7-21) is of a form convenient for computation and accuracy acceptable for RE $\leq 10^5$, Prandtl's universal law of friction for a smooth pipe

$$\frac{1}{f^{1/2}} = 2 \log_{10}\left(f^{1/2}\, \frac{U_{av}D}{\nu} \right) - 0.8 \qquad (7\text{-}21b)$$

is accurate up to Re $= 3.4 \times 10^6$ and can be extrapolated far beyond that. See Section 9.8 for more convenient relationships for f.

For a turbulent boundary layer on a plate in parallel flow, it is assumed that the boundary-layer thickness can be used in place of the pipe radius and free-stream velocity can be used in place of the center-line velocity. Then the analogy to Eq. (7-21) is

$$\frac{C_f}{2} = \frac{\tau_w}{\rho U_\infty^2} = 0.0228\left(\frac{U_\infty \delta}{\nu} \right)^{-1/4} \qquad (7\text{-}22)$$

for $\mathrm{Re}_\delta = U_\infty\delta/\nu \leq 10^5$; a correction for pipe curvature has been introduced in the coefficient. Average friction coefficient formulas that are accurate for a wider range of Reynolds numbers are due to Prandtl–Schlichting as

$$\overline{C}_f = \frac{0.455}{\left[\log_{10}(U_\infty l/\nu) \right]^{2.58}}$$

accurate up to and beyond $U_\infty l/\nu = 10^9$, and von Karman–Schoenherr as

$$\frac{1}{\overline{C}_f^{1/2}} = 4.13 \log_{10}\left(\frac{\overline{C}_f U_\infty l}{\nu} \right) \qquad (7\text{-}22a)$$

based on a logarithmic velocity profile of the form of Eq. (7-18). The local friction coefficient has been correlated by Schultz-Grunow [50] as

$$C_f = \frac{0.37}{\left[\log_{10}(U_\infty x/\nu) \right]^{2.584}}$$

as discussed by Schlichting [18, p. 604], who can also be consulted for the effects of roughness.

Generally speaking, surface roughness has no effect if the roughness elements are covered by the laminar sublayer. This requires that, for the wall to be hydraulically smooth,

$$\frac{v^* k}{\nu} < 5$$

where k is the height of the roughness element.

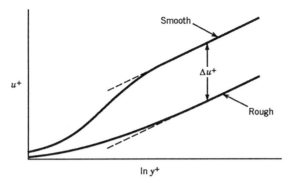

Figure 7-15 Turbulent velocity distribution near smooth and rough surfaces. [By permission from T. Cebeci and P. Bradshaw, *Physical and Computational Aspects of Convective Heat Transfer*, Springer-Verlag, 1984, p. 167 [59].]

The velocity distribution above a wall with roughness height k is

$$u^+ = K^{-1} \ln y^+ + C - \Delta u^+ \tag{7-23}$$

which is of the same form as Eq. (7-18) for a smooth wall, but shifted by the amount Δu^+ as illustrated in Fig. 7-15. The variation of Δu^+ with dimensionless roughness height $k^+ = kv^*/\nu$ is shown in Fig. 7-16. There it is seen

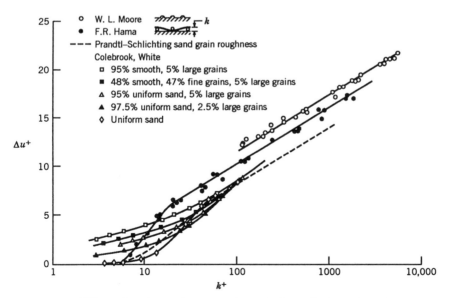

Figure 7-16 Effect of wall roughness on the velocity shift for universal turbulent velocity profiles. [By permission from F. H. Clauser, *Adv. Appl. Mech.* **4**, 1–51 (1956) [60].]

that the wall is smooth for $k^+ < 5$, transitionally rough for $5 \leqslant k^+ < 70$, and fully rough for $70 \leqslant k^+$. Krogstad [51] modified the Van Driest damping function of Eq. (7-14a) with this information; Sigal and Danberg [52] show how to account for different roughness element shapes. Christoph and Pletcher [53] used wall roughness information to predict skin friction and heat transfer from rough surfaces.

Specialized modification of the roughness elements into a system of grooves, known as riblets, aligned with the flow direction can reduce drag by as much as 8% [54]. Hot-wire anemometer measurements show turbulence intensity as low as 1% in the groove, apparently due to the damping of the spanwise vortical structures illustrated in Figs. (7-5) and (7-6). Such riblets have been used on racing sailboat hulls and airplane wings [55] as well as in pipes [56].

The observation that turbulent drag on dolphins is lower than for rigid surfaces has led to investigations on the reduction of turbulent skin friction with compliant surfaces. Reductions of up to 50% are possible [57, 58].

PROBLEMS

7-1 Consider Couette flow of two parallel walls moving with equal speeds in opposite directions. In such a case the shear stress is constant. Plot $\bar{u}/v^*$ against y/h for both laminar and turbulent flow where $v^* = (\tau_w/\rho)^{1/2}$ and h equals half the wall separation. Compare results with the measurements of Reichardt reported by Schlichting [18, p. 557]. The methods used to achieve Eqs. (7-16)–(7-18) can be applied.

7-2 Show that use of van Karman's similarity hypothesis for turbulent flow with constant shear stress gives the velocity profile as

$$u^+ = \frac{1}{K_3} \ln(K_3 y + C_1) + C_2$$

Show that if the integration constant C_1 equals zero, the result is identical to Eq. (7-18) from Prandtl's mixing-length formula.

7-3 Consider turbulent Poiseuille flow between two parallel walls with a constant axial pressure gradient so that the shear stress varies as $\tau = \tau_w y/h$, where y is the distance from the center line and h is half the wall separation.

(a) Use von Karman's similarity hypothesis to obtain the velocity profile with the condition that $\bar{u}(y = 0) = \bar{u}_{max}$ as

$$\frac{\bar{u}_{max} - \bar{u}}{v^*} = -\frac{1}{K_3} \left\{ \ln\left[1 - \left(\frac{y}{h}\right)^{1/2} \right] + \left(\frac{y}{h}\right)^{1/2} \right\}$$

where $v^* = (\tau_w/\rho)^{1/2}$. Note that $\bar{u}(y = h) \to \infty$ since molecular effects were neglected. This result is termed the *velocity-defect law*.

(b) Repeat part a with the use of Prandtl's mixing-length formula and constant shear stress. With the condition that $\bar{u}(y = h) = \bar{u}_{max}$ (y is now the distance from a wall), show that the velocity distribution is

$$\frac{\bar{u}_{max} - \bar{u}}{v^*} = \frac{1}{K_1} \ln\left(\frac{h}{y}\right)$$

(c) Equate the result for $\bar{u}$ of part b which is the *velocity-defect law* with that of Eq. (7-18) which is the *law of the wall* to derive the functional form of Prandtl's universal law of friction in Eq. (7-21b).

7-4 Show that if Deissler's relation of Eq. (7-14) is modified into $\varepsilon_m = n^2\bar{u}y$, Eq. (7-15) becomes

$$\frac{dy^+}{du^+} - n^2u^+y^+ = 1$$

and that the solution is

$$y^+ = n^{-1}\left(\frac{\pi}{2}\right)^{1/2} e^{n^2u^{+2}/2} \, \text{erf}\left(\frac{nu^+}{2^{1/2}}\right)$$

Plot this result in Fig. 7-14 and show that it fits data for $y^+ \leq 26$.

7-5 Show that if the turbulent velocity profile in a round duct of radius R is given by $u/U_{CL} = (y/R)^{1/n}$, the average velocity is related to the center-line velocity U_{CL} by

$$\frac{U_{av}}{U_{CL}} = 2n^2(n + 1)^{-1}(2n + 1)^{-1}$$

7-6 Consider a long straight pipe of 1 ft diameter through which water flows at 1 ft/sec. The pertinent water properties are $\rho = 62.4 \text{ lb}_m/\text{ft}^3$ and $\nu = 10^{-5} \text{ ft}^2/\text{sec}$. Calculate:

(a) the friction factor f for pressure drop from Eq. (7-21a),

(b) the shear stress τ_w at the wall from $\tau_w = D(P_1 - P_2)/4L$,

(c) the friction velocity $v^* = (\tau_w/\rho)^{1/2}$ (*Answer:* $v^* = 0.047$ ft/sec),

(d) the thicknesses of the laminar sublayer, the buffer zone, and the fully turbulent core (*Answer:* $\delta_1 = 10^{-3}$ ft; $\delta_b = 4 \times 10^{-3}$ ft), and

(e) whether or not the pipe can be considered to be smooth if a typical wall roughness element height k is 0.045 cm.

7-7 Estimate the extent of the transition region between laminar and turbulent flow for parallel flow of air over a flat plate with a Reynolds number based on plate total length of 10^6. *Answer*: $(x_c - x_i)/L = 0.24$.

7-8 The similarity hypothesis of von Karman for mixing-length in turbulent flow can be obtained for the velocity u in a parallel flow by considering the Taylor series in the perpendicular direction y according to Cebeci and Smith [61, p. 106]. This Taylor series is

$$u(y) = u(y_0) + (y - y_0)\frac{\partial u(y_0)}{\partial y} + (y + y_0)^2\frac{\partial^2 u(y_0)/\partial y^2}{2!} + \cdots$$

(a) Arguing that all such flows ought to be similar in shape so that scaling constants l and u_0 must exist that make the parallel velocity independent of either velocity or size of the flow field, show that the Taylor series must be expressible as

$$\frac{u}{u_0} = 1 + \left(\frac{y}{l} - \frac{y_0}{l}\right)\frac{\partial(u/u_0)}{\partial(y/l)}$$

$$+ \left(\frac{y}{l} - \frac{y_0}{l}\right)^2\frac{\partial^2(u/u_0)/\partial(y/l)^2}{2!} + \cdots$$

(b) Considering only the first three terms of the Taylor series, and arguing that to have u/u_0 similar for all y, $\partial(u/u_0)/\partial(y/l)$ must be proportional to $\partial^2(u/u_0)/\partial^2(y/l)$, show that

$$l = \left|\frac{\partial u/\partial y}{\partial^2 u/\partial y^2}\right|K_3$$

7-9 The basis for the exponential form appearing in Eq. (7-14a) due to Van Driest and Eq. (7-14) due to Deissler is to be ascertained. For this purpose, consider a flat plate in a stagnant fluid with the plate oscillating in its own plane. For laminar conditions, the adjacent fluid motion is described by

$$\frac{\partial u}{\partial t} = \nu\frac{\partial^2 u}{\partial y^2}$$

with $u(0, t) = u_0 \cos(\omega t)$ and $u(y \to \infty, t) \to 0$, where y is the perpendicular distance from the plate.

(a) Show that, with $y_s = (\nu/\omega)^{1/2}$ and $u_s = (\omega\nu)^{1/2}$, the fluid velocity varies as

$$u = u_0 \exp\left(-\frac{y}{y_s\sqrt{2}}\right)\cos\left(\omega t - \frac{yu_s}{\nu\sqrt{2}}\right)$$

(b) From the result of part a, show that if the fluid oscillates parallel to a stationary plate, the fluid velocity varies, $n = 1/y_s\sqrt{2}$, as

$$u' = u'_0[1 - \exp(-ny)]$$

where u'_0 is the velocity fluctuation far from the plate.

(c) Show that the Reynolds stress can be plausibly expressed as

$$\overline{\tau}_{yx}^{\text{turb}} = -\rho\overline{v'u'} = -\rho\overline{v'_0u'_0}[1 - \exp(-ny)]^2$$

if v'_0 and u'_0 are presumed to be similarly damped by the presence of the wall. Then, since Prandtl's mixing-length idea has

$$-\overline{v'_0u'_0} = l^2\left(\frac{\partial u}{\partial y}\right)^2$$

show that it follows that

$$\varepsilon_m = l^2[1 - \exp(-ny)]^2\left|\frac{\partial \overline{u}}{\partial y}\right|$$

Van Driest took Prandtl's relation of $l = Ky$ with $K = 0.4$, and assumed $n = (\tau_\omega/\rho)^{1/2}/\nu A^+$ with $A^+ = 26$.

7-10 Prepare a brief report of the effects of adverse pressure gradients on the transition length of boundary layers on axial flow compressor blades, consulting Walker and Gostelow [62], and on the effects of favorable pressure gradients on transition length and the heat transfer therein, consulting Schmidt and Patankar [15].

REFERENCES

1. D. R. Hofstadter, *Scientific American* **245**, 22–43 (1981).

2. J. O. Hinze, *Turbulence*, McGraw-Hill, 1975.

3. L. Burmeister, *Convective Heat Transfer*, Wiley-Interscience, 1983.

4. G. D. Catalano, R. E. Walterick, and H. E. Wright, *AIAAJ* **19**, 403–405 (1981).

5. R. J. Adrian, *Ann. Rev. Fluid Mech.* **23**, 261–304 (1991).

6. G. B. Schubauer and P. S. Klebanoff, Contributions on the mechanics of boundary layer transition, NACA TN 3489, 1955; NACA Report 1289, 1956.

7. H. W. Emmons, *J. Aeronaut. Sci.* **18**, 490–498 (1951).

8. S. K. Robinson, *Ann. Rev. Fluid Mech.* **23**, 601–639 (1991).

9. (a) C. R. Smith, A synthesized model of the near-wall behavior in turbulent boundary layers, *Proc. Eighth Symp. Turb.*, Rolla, MO, 1984; (b) M. S. Acarlar and C. R. Smith, *J. Fluid Mech.* **175**, 43–83 (1987).

10. R. F. Blackwelder and J. D. Swearingen, in S. J. Kline and N. H. Afgan, Eds., *Near Wall Turbulence* (*Proceedings Zaric Memorial Conference*), Hemisphere, 1988, pp. 268–288.

11. S. K. Robinson, *Kinematics of turbulent boundary layer structure*, Ph.D. Dissertation, Stanford University, Stanford, CA, 1990; NASA TM 103859, April 1991.

12. I. Tani, *Ann. Rev. Fluid Mech.* **1**, 169–196 (1969).

13. (a) G. B. Schubauer and H. F. Skramstad, Laminar boundary layer oscillations and stability of laminar flow, National Bureau of Standards Research Paper 1772, reprint of confidential NACA Report, April 1943 (also published as NACA Wartime Report W-8); *J. Aeronaut. Sci.* **14**, 69–78 (1947); see also NACA Report 909; (b) A. A. Hall and G. S. Hislop, Experiments on the transition of the laminar boundary layer on a flat plate, Aeronautical Research Committee Report Memorandum 1843, 1938.

14. D. Sucker and H. Brauer, *Wärme-und Stoffübertragung* **8**, 149–158 (1975).

15. R. C. Schmidt and S. V. Patankar, *ASME J. Turbomachinery* **113**, 10–26 (1991).

16. B. J. Abu-Ghannam and R. Shaw, *J. Mech. Eng. Sci.* **22** (5), 213–228 (1980).

17. O. Reynolds, *Philos. Trans. Roy. Soc.* **174**, 935–982, (1883).

18. H. Schlichting, *Boundary-Layer Theory*, McGraw-Hill, 1968.

19. I. Tosun, D. Uner, and C. Ozger, *Ind. Eng. Chem. Res.* **27**, 1955–1957 (1988).

20. A. H. P. Skelland, *Non-Newtonian Flow and Heat Transfer*, Wiley, 1967, pp. 74–172.

21. Y. I. Cho and J. P. Hartnett, *Adv. Heat Transfer* **15**, 59–141 (1982).

22. J. Pretsch, *Jb. dt. Luftfahrtforschung* **1**, 58–75 (1941).

23. W. M. Kays and M. E. Crawford, *Convective Heat Transfer and Mass Transfer* 2nd ed., McGraw-Hill, 1980, p. 163.

24. D. M. McEligot, C. W. Coon, and H. C. Perkins, *Int. J. Heat Mass Transfer.* **13**, 431–433 (1970).

25. T. Herbert, *Ann. Rev. Fluid Mech.* **20**, 487–526 (1988).

26. T. Cebeci, Ed., *Numerical and Physical Aspects of Aerodynamic Flows*, Vol. 4, Springer-Verlag, 1990, pp. 283–286.

27. R. Michel, Office Nat. Etude Recherche Aeronaut. Report 58, 1952.

28. A. M. O. Smith and N. Gamberoni, Douglas Aircraft Co., Report ES-26388, 1956.

29. P. Bradshaw, T. Cebeci, and J. Whitelaw, *Engineering Methods for Turbulent Flow*, Academic, 1981, pp. 235–241.

30. T. Cebeci and D. Egan, *AIAA J.* **27**, 870–875 (1989).

31. L. Kleiser and T. A. Zang, *Ann. Rev. Fluid Mech.* **23**, 495–537 (1991).

32. J. Boussinesq, *Mém. prés. par div. savant á l'acad. sci. Paris* **23**, 46 (1877).

33. L. Prandtl, *Zeitschrift für angewandte Mathematik und Mechanik* **5**, 136–139 (1925).

34. L. Prandtl, *Zeitschrift für angewandte Mathematik und Mechanik* **22** 241–243 (1942).

35. H. Reichardt, *VDI-Forschungsheft* **414**, 1st ed., Berlin, 1942 (2nd ed., Berlin, 1951).

36. Th. von Karman, Mechanische Ähnlichkeit und Turbulenz, *Nach. Ges. Wiss. Göttingen, Math. Phys. Klasse* **58** (1930); *Proc. Third Int. Cong. Appl. Mech., Stockholm, Part I,* 85 (1930); NACA TM 611, 1931.

37. R. B. Bird, W. E. Stewart, and E. N. Lightfoot, *Transport Phenomena*, Wiley, 1960, p. 161.

38. R. G. Deissler, Analysis of turbulent heat transfer, mass transfer, and friction in smooth tubes at high Prandtl and Schmidt numbers, NACA Report 1210, 1955.

39. E. R. Van Driest, *J. Aerospace Sci.* **23**, 1007–1011 (1956).

40. H. W. Liepmann, *American Scientist* **67**, 221–228 (1979).

41. H. J. Lugt, *American Scientist* **73** (2), 162–167 (1985).

42. P. S. Klebanoff, Characteristics of turbulence in a boundary layer with zero pressure gradient, NACA TN 3178, 1954.

43. B. J. Cantwell, *Ann. Rev. Fluid Mech.* **13**, 457–515 (1981).

44. J. Kestin and P. D. Richardson, *Int. J. Heat Mass Transfer* **6**, 147–189 (1963).

45. H. Reichardt, *Arch. Gesamte Waermetech.* **2**, 129–142 (1951).

46. J. Nikuradse, *Forsch. Arb. Ing.-Wes.*, No. 356, 1932.

47. R. C. Martinelli, *Trans. ASME* **69**, 947–959 (1947).

48. H. Reichardt, *Zeitschrift für angewandte Mathematik und Mechanik* **31**, 208–219 (1951).

49. D. B. Spalding, *Conference on International Developments in Heat Transfer, ASME, Boulder, CO, Part II,* 1961, pp. 439–446.

50. F. Schultz-Grunow, *Luftfahrtforschung* **17**, 239 (1940); NASA TM 986, 1941.

51. P.-A. Krogstad, *AIAA J.* **29**, 888–894 (1991); P. S. Granvill, *AIAA J.* **30**, 1673–1674 (1992).

52. A. Sigal and J. E. Danberg, *AIAA J.* **28**, 554–556 (1990).

53. G. H. Christoph and R. H. Pletcher, *AIAA J.* **21**, 509–515 (1983).

54. P. Vukoslavcevic, J. M. Wallace, and J.-L. Balint, *AIAA J.* **30**, 1119–1122 (1992).

55. *Scientific American* **226** (3), 107–108 (1992).

56. K. N. Liu, C. Christodoulou, O. Riccius, and D. D. Joseph, *AIAA J.* **28**, 1697–1698 (1990).

57. L. M. Weinstein and M. C. Fischer, *AIAA J.* **13**, 956–958 (1975).

58. R. D. Joslin and P. J. Morris, *AIAA J.* **30**, 332–339 (1992).

59. T. Cebeci and P. Bradshaw, *Physical and Computational Aspects of Convective Heat Transfer*, Springer-Verlag, 1984, p. 167.

60. F. H. Clauser, *Adv. Appl. Mech.* **4**, 1–51 (1956).

61. T. Cebeci and A. M. O. Smith, *Analysis of Turbulent Boundary Layers*, Academic, 1974.

62. G. J. Walker and J. P. Gostelow, *ASME J. Turbomachinery* **112**, 196–205 (1990).

8

TURBULENT BOUNDARY LAYERS

Most boundary layers become turbulent if their development is unimpeded. Although it is normal for the turbulent boundary layer to be preceded by a laminar one, such is not required. For example, a roughness element, called a *boundary-layer trip*, at the leading edge of a plate can trigger a turbulent boundary layer there. And, it is possible for transition from turbulent to laminar flow to occur under some conditions.

The $\frac{1}{7}$th power-law velocity profiles that are frequently used are not always the most accurate description. Pipe flow results show that it is more accurate to speak of $1/n$th power profiles, where $6 \leqslant n \leqslant 10$, depending on the Reynolds number. At the same time, the $\frac{1}{7}$th power-law velocity distribution is often satisfactory far from the surface as shown in Fig. 8-1.

The applicability of a logarithmic velocity distribution near the surface is shown in Fig. 8-2 to be satisfactory even for compressible flow.

8.1 VELOCITY DISTRIBUTION IN PARALLEL FLOW

Consider parallel flow of a constant-property fluid over a flat plate. The turbulent boundary layer forming on the plate as illustrated in Fig. 8-3 is to be described as to thickness and velocity distribution. The integral form of the boundary-layer equations can be used, provided shear stress includes a turbulent contribution. Thus

$$\frac{d(U^2 \delta_2)}{dx} + \delta_1 U \frac{dU}{dx} = \frac{\tau_w + \rho U v_w}{\rho} \tag{6-13}$$

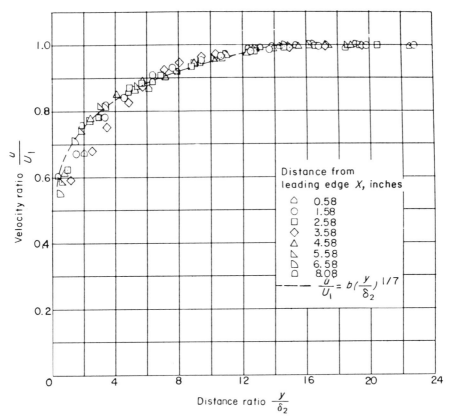

Figure 8-1 Measured velocity distribution in a turbulent boundary layer on a flat plate at zero incidence with a supersonic free stream. [From R. M. O'Donnell, NACA TN 3122, 1954 [1].]

where $\delta_1 = \int_0^\delta (1 - u/U)\, dy$ is the displacement thickness,

$$\delta_2 = \int_0^\delta (u/U)(1 - u/U)\, dy$$

is the momentum thickness, and δ is the boundary-layer thickness.

The $\frac{1}{7}$th power-law velocity distribution of Eq. (7-19)

$$u^+ = 8.74(y^+)^{1/7} \tag{7-19}$$

from pipe flow data and Eq. (7-21) with the free-stream velocity and boundary-layer thickness substituted for pipe center-line velocity and radius gives

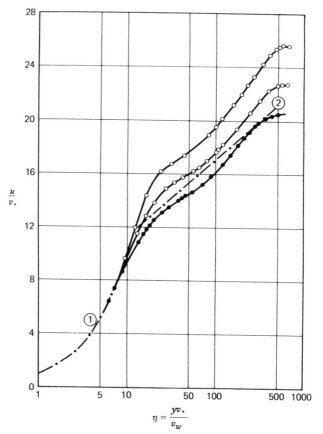

Figure 8-2 Measured velocities in a turbulent boundary layer on a channel wall with a supersonic free stream and heat transfer. Curve 1 represents the theoretical law for incompressible laminar sublayers, $u/v_* = \eta$. Curve 2 represents the theoretical law for incompressible turbulent boundary layers, $u/v_* = 5.5 + 2.5\ln(\eta)$. [From R. K. Lobb, E. M. Winkler, and J. Persh, NAVORD Report 3880, 1955 [2].]

	M_∞	$(T_{aw} - T_w)/T_{aw}$	$R_2 \times 10^{-4}$
●	5.75	0.108	1.16
◑	5.79	0.238	1.24
○	5.82	0.379	1.14

the major portion of the velocity profile in the similarity form

$$\frac{u}{U} = \left(\frac{y}{\delta}\right)^{1/7} \tag{8-1}$$

With the velocity profile of Eq. (8-1), the ratio of displacement to boundary-layer thicknesses is $\delta_1/\delta = \frac{1}{8}$ (compared to $\delta_1/\delta \approx \frac{3}{8}$ for laminar flow), and

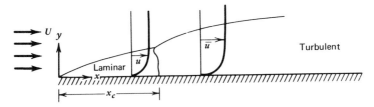

Figure 8-3 Transition from laminar to turbulent flow in a boundary layer on a flat plate.

the ratio of momentum thicknesses to boundary-layer is $\delta_2/\delta = \frac{7}{72} \approx \frac{1}{10}$ (compared to $\delta_2/\delta \approx \frac{39}{280} \sim \frac{1}{7}$ for laminar flow), the change in the shape factor $H_{12} = \delta_1/\delta_2$ from 2.6 to 1.4 is illustrated in Fig. 7-4.

Introduction of these results into Eq. (6-13) gives

$$\frac{d[U^2 7\delta/72]}{dx} = 0.0228\left(\frac{U\delta}{\nu}\right)^{-1/4} U^2 \qquad (8\text{-}2)$$

Here the free-stream velocity is constant, and Eq. (7-22), from pipe flow data, has been used to relate τ_w to U and δ. On integration, Eq. (8-2) gives

$$\delta^{5/4} = 0.2931\left(\frac{U}{\nu}\right)^{-1/4} x + c \qquad (8\text{-}3)$$

The constant of integration in Eq. (8-3) must be evaluated from knowledge of the extent of the laminar boundary-layer covering the fore part of the plate. A simple answer, first essayed by Ludwig Prandtl, is obtained by saying that the turbulent boundary layer after the transition region acts as if it had begun at the leading edge with $\delta_{\text{turb}}(x = 0) = 0$. Then $c = 0$ and

$$\frac{\delta}{x} = 0.375/\text{Re}_x^{1/5}, \qquad 5 \times 10^5 \sim \text{Re}_{x,c} \leqslant \text{Re}_x < 10^7 \qquad (8\text{-}4)$$

The turbulent boundary-layer thickness is predicted to vary as $x^{4/5}$, whereas a laminar boundary layer thickness varies as $x^{1/2}$. On physical principles, noting Eq. (6-13), the momentum thickness must vary continuously (the subject of Problem 8-3), even though Eq. (8-4) is of adequate accuracy.

The turbulent wall shear stress is calculable from Eq. (7-22) now that δ is known as, for $5 \times 10^5 \sim \text{Re}_{x,c} \leqslant \text{Re}_x < 10^7$,

$$\frac{C_f}{2} = \frac{\tau_w}{\rho U_\infty^2} = 0.0228\left(\frac{U_\infty\delta}{\nu}\right)^{-1/4} = \frac{0.0291}{\text{Re}_x^{1/5}} \qquad (8\text{-}5)$$

This is a local turbulent friction coefficient and is valid only for $x \geq x_c$. The total drag on a plate must include the contribution from the laminar boundary layer, if one exists—it might not if the boundary layer is tripped at the leading edge. Thus

$$\text{Drag} = \int_0^{x_c} \left(\frac{C_f}{2}\right)_{\text{lam}} \rho U^2 \, dx + \int_{x_c}^{x} \left(\frac{C_f}{2}\right)_{\text{turb}} \rho U^2 \, dx$$

Use of Eq. (5-17) in the first integral and Eq. (8-5) in the second integral gives

$$\frac{\bar{C}_f}{2} = \frac{0.0364}{\text{Re}_x^{1/5}} - \frac{\left(-0.66412 \, \text{Re}_{x,c}^{1/2} + 0.037 \, \text{Re}_{x,c}^{4/5}\right)}{\text{Re}_x}$$

for $5 \times 10^5 \sim \text{Re}_{x,c} \leqslant \text{Re}_x < 10^7$. If the critical Reynolds number is taken to be 5×10^5, the average friction coefficient is

$$\frac{\bar{C}_f}{2} = \frac{0.037}{\text{Re}_x^{1/5}} - \frac{850}{\text{Re}_x} \tag{8-6}$$

for $5 \times 10^5 \sim \text{Re}_{x,c} \leqslant \text{Re}_x < 10^7$.

Of interest is the thickness of the laminar sublayer δ_l whose outer edge is at

$$y^+ = \frac{v^* \delta_l}{v} = 5$$

Use of Eq. (8-4) for δ and Eq. (8-5) for τ_w results in

$$\frac{\delta_l}{\delta} = \frac{77.5}{\text{Re}_x^{7/10}} \tag{8-7}$$

Similarly, the outer edge of the buffer zone occurs at $y^+ = 26$. The buffer zone thickness δ_b then occupies the thickness $\Delta y^+ = 26 - 5 = 21$. Proceeding as for δ_l, one finds that

$$\frac{\delta_b}{\delta} = \frac{326}{\text{Re}_x^{7/10}} \tag{8-8}$$

Equations (8-7) and (8-8) show that δ_l and δ_b increase as $x^{1/10}$, nearly constant, whereas δ increases as $x^{4/5}$. Thus the turbulent core occupies increasing portions of the boundary layer at successive downstream locations.

Results for the turbulent boundary layer in the presence of a pressure gradient are attainable by the methods discussed by Schlichting [3, pp.

626–655]. In brief, the integral form of the boundary-layer equation is rearranged as

$$\frac{d\delta_2}{dx} + (H_{12} + 2)\frac{\delta_2}{U}\frac{dU}{dx} = \frac{\tau_w}{\rho U^2}$$

where $H_{12} = \delta_1/\delta_2$. The wall shear stress is related to U and δ_2 by

$$\frac{\tau_w}{\rho U^2} = \frac{\alpha}{(U\delta_2/\nu)^{1/n}}$$

Taking the flat-plate value with a $\frac{1}{7}$th power-law velocity profile for $H_{12} = \delta_1/\delta_2 = \frac{9}{7} \approx 1.3$ together with the preceding relationship for τ_w in the integral equation, one obtains

$$\delta_2\left(\frac{U\delta_2}{\nu}\right)^{1/n} = U^{-b}\left(C_1 + a\int_{x=x_c}^{x} U^b \, dx\right) \tag{8-8a}$$

where $a = \alpha(n + 1)/n$, $b = [(n + 1)(H_{12} + 2) - 1]/n$, and C_1 is a constant of integration to be determined from the laminar boundary layer at $x = x_c$. If no laminar boundary layer exists, then $C_1 = 0$ since one of U or δ_2 will be zero at $x = 0$. Since Eq. (8-5) can be expressed as

$$\frac{C_f}{2} = \frac{0.0128}{(U\delta_2/\nu)^{1/4}}$$

as was first done by Prandtl, it is reasonable to use $n = 4$ and $\alpha = 0.0128$; then $a = 0.016$ and $b = 4$. When the boundary layer does not separate from the surface [separation occurs when $\Gamma = (\delta_2/U)(dU/dx)(U\delta_2/\nu)^{1/4} \approx -0.06$, with $\Gamma > 0$ corresponding to accelerated and $\Gamma < 0$ corresponding to decelerated flow, respectively], knowledge of δ_2 essentially completes the calculation. Local wall shear stress is available by back substitution, and drag can be obtained by its integration. Such calculations are important to nozzle design and to air foils, for example. A body of revolution is the subject of Problem 8-11; three-dimensional boundary layers can be attacked by the integral method developed by Tai [4].

8.2 TEMPERATURE DISTRIBUTION IN PARALLEL FLOW

From the velocity distribution, a heat-transfer coefficient can be determined. It is assumed that thermal energy and momentum possess the same mechanisms of turbulent transport.

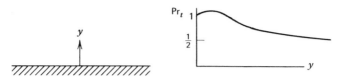

Figure 8-4   Variation of turbulent Prandtl number with distance from a solid surface.

In turbulent motions, where large clumps of molecules are moved about together, the kinetic energy associated with the motion of these clumps of fluid contributes little to the internal energy of the fluid. Hence it is assumed that neither heat nor mass are affected by interaction between these clumps in turbulent motion other than by molecular-level exchanges during a time of contact between clumps. If these molecular-level effects are neglected, the rates of turbulent transport of heat and mass must be equal.

Measurements of the turbulent Prandtl number $Pr_t = \varepsilon_m/\alpha_t$ are surveyed by Schlichting [3] and Hinze [5]. As illustrated in Fig. 8-4, near a solid surface $Pr_t \approx 1$, near the center of a duct $Pr_t \approx 0.7$, and in a free stream $Pr_t \approx \frac{1}{2}$. The $Pr_t = 1$ value corresponds to equality of mixing lengths for heat and momentum, whereas the $Pr_t = \frac{1}{2}$ value corresponds to equality of mixing lengths for heat and vorticity [5]. Efforts to develop a functional relationship between Pr_t and distance from the wall have been made by Hughmark [6] and Cebeci [7]. In liquid metals $Pr_t \sim 1.2$ for fully developed tube flow, however [7, 8].

Jischa and Rieke [9] have shown theoretically that $Pr_t = A + B(Pr + 1)/Pr$ with experiment giving $A = 0.825$ and $B = 0.0309$, a result fitting data for air well and for liquid metals as well as any other expression. Callaghan and Mason [10] measured $Pr_t \approx 1$ for nitrogen-dioxide gas (corrosive and poisonous but that undergoes the reaction $N_2O_4 \rightleftarrows 2NO_2$ at 30°F). Direct numerical simulation of turbulent flow by Antonia and Kim [11] suggests that these trends, gleaned from difficult measurements as discussed for boundary layers following Eq. (8-30c), are plausible.

Attention is now directed to the one-dimensional case where fluid flows primarily parallel to a solid wall. This corresponds to a well-developed turbulent boundary layer or to flow in a pipe well beyond the entrance region. In this way, a one-dimensional problem is posed as depicted in Fig. 8-5. The heat flux in the y direction is given by

$$\bar{q} = -k\frac{\partial \bar{T}}{\partial y} - k_t\frac{\partial \bar{T}}{\partial y} = -\rho C_p(\alpha + \alpha_t)\frac{d\bar{T}}{dy}$$

For small values of y, q is little changed from its value q_w at the wall. After

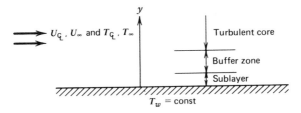

Figure 8-5 Division of a turbulent boundary layer into a laminar sublayer at the wall, an intermediate buffer zone, and an outer fully turbulent core.

rearrangement, and definition of dimensionless quantities as

$$v^* = \left(\frac{\tau_w}{\rho}\right)^{1/2}, \qquad T^+ = (\bar{T} - T_w)\left(\frac{\rho C_p}{q_w}v^*\right), \qquad y^+ = \frac{yv^*}{\nu}, \qquad u^+ = \frac{\bar{u}}{v^*}$$

one finally has

$$-1 = \left(\frac{1}{Pr} + \frac{\varepsilon_m/\nu}{Pr_t}\right)\frac{dT^+}{dy^+} \tag{8-9}$$

The velocity profile in Section 8.1 gives ε_m/ν so that Eq. (8-9) can be integrated to relate the total temperature difference to the distance required to accomplish it. The integration is divided into three portions: sublayer, $y^+ \leqslant 5$; buffer zone, $5 \leqslant y \lesssim 30$; and turbulent core, $y \gtrsim 30$. In other words,

$$\frac{T_\infty - T_w}{q_w/\rho C_p v^*} = \Delta T^+_{\text{sublayer}} + \Delta T^+_{\text{buffer zone}} + \Delta T^+_{\text{turbulent core}}$$

The laminar sublayer has molecular effects predominating over turbulence so that ε_m/ν can be neglected. Thus from Eq. (8-9)

$$\frac{dT^+}{dy^+} = -Pr, \qquad y^+ < 5$$

$$\Delta T^+_{\text{sublayer}} = T^+(y^+ = 5) - T^+(y^+ = 0) = -5\,Pr \tag{8-10}$$

In the buffer zone, where molecular effects and turbulence must both be considered, Deissler's relationship is $\varepsilon_m/\nu = n^2 u^+ y^+[1 - \exp(-n^2 u^+ y^+)]$ so that Eq. (8-9) could be numerically integrated. However, a compact analytical result is sought. An approximation, leading in this direction, is made as follows. In the buffer zone

$$\tau = \rho \nu \left(1 + \frac{\varepsilon_m}{\nu}\right) \frac{d\bar{u}}{dy}$$

Assumption that $\tau \approx \tau_w$ and rearrangement then gives the dimensionless form

$$1 = \left(1 + \frac{\varepsilon_m}{\nu}\right) \frac{du^+}{dy^+} \qquad (8\text{-}11)$$

The velocity distribution in the buffer zone is approximated by

$$u^+ \approx 5 + 5 \ln \frac{y^+}{5}$$

Insertion of this information into Eq. (8-11) gives

$$\frac{\varepsilon_m}{\nu} = \frac{y^+}{5} - 1$$

Equation (8-9) then yields, if Pr_t is constant,

$$\Delta T^+_{\text{buffer}} = -\int_5^{30} \frac{dy^+}{1/\text{Pr} + (1/\text{Pr}_t)(y^+/5 - 1)} = -5\,\text{Pr}_t \ln\left(5\frac{\text{Pr}}{\text{Pr}_t} + 1\right) \qquad (8\text{-}12)$$

In the fully turbulent core, molecular effects can be disregarded for all but liquid metals. Equation (8-9) then becomes

$$-1 = \frac{\varepsilon_m/\nu}{\text{Pr}_t} \frac{dT^+}{dy^+} \qquad (8\text{-}13a)$$

Also, Eq. (8-11) becomes

$$1 = \frac{\varepsilon_m}{\nu} \frac{du^+}{dy^+}$$

which is introduced into Eq. (8-13a) to yield

$$-\text{Pr}_t = \frac{dT^+}{du^+}$$

From this it is found that, if Pr_t is constant,

$$\Delta T_{\text{turb}}^+ = -\int_{y^+ \approx 30}^{\text{free stream}} Pr_t \, du^+ = -Pr_t[U_\infty^+ - 5(1 + \ln 6)] \quad (8\text{-}13b)$$

Combination of Eqs. (8-10), (8-12), and (8-13b) then gives the temperature difference between the wall and the free stream as

$$(T_w - T_\infty)\frac{\rho C_p v^*}{q_w} = 5\left[Pr - Pr_t + Pr_t \ln\left(\frac{5\,Pr/Pr_t + 1}{6}\right)\right] + Pr_t\,U_\infty^+ \quad (8\text{-}14)$$

A heat-transfer coefficient can now be determined from Eq. (8-14) since

$$h(T_w - T_\infty) = q_w$$

Rearrangement of this relationship gives

$$\left(\frac{T_w - T_\infty}{q_w}\right)\rho C_p v^* = \frac{\rho C_p v^*}{h} = \frac{Pr\,Re_L}{Nu_L}\,\frac{1}{U_\infty^+}$$

Equation (8-14), in conjunction with this result, gives

$$\frac{Nu_L}{Re_L\,Pr^{1/3}} = \frac{1}{U_\infty^{+2}}\frac{Pr^{2/3}}{Pr_t + 5/U_\infty^+\{Pr - Pr_t + Pr_t\ln[(5\,Pr/Pr_t + 1)/6]\}}$$

Probably attention should center near the wall where $Pr_t \approx 1$. Then

$$\frac{Nu_L}{Re_L\,Pr} = \frac{C_f}{2}\frac{1}{1 + 5(C_f/2)^{1/2}\{Pr - 1 + \ln[(5\,Pr + 1)/6]\}} \quad (8\text{-}15)$$

Equation (8-15) is not applicable to low Prandtl number fluids such as liquid metals since then molecular-level thermal conductivity effects are appreciable in the turbulent core, contrary to the assumption incorporated in Eq. (8-13). This development is attributed to von Karman [12]. The case of variable Pr_t was explored by Reichardt [13], Van Driest [14], and Rotta [15], who found no important improvement to the preceding form. Equation (10-15) is the turbulent form of Reynolds' analogy and is not as clearcut as the laminar form. Nevertheless, it is confirmed by experiment that $Nu_L/(Re_L\,Pr^{1/3}) = C_f/2$ if $Pr \approx 1$, as it is for many gases. A refined integral analysis utilizing more accurate approximations for velocity and temperature profiles near the wall confirms the results presented here [16]. Although a surface renewal model suggests [17] that $Nu_L/Re_L\,Pr^{1/2} = C_f/2$ always, measurements [18] in hypersonic flow support Reynolds' analogy in the form $Nu_L/Re_L\,Pr =$

$1.16C_f/2$ while for supersonic airflow, the coefficient of $C_f/2$ seems to vary between 1 and 1.2 [19].

If mass transfer is considered, one merely substitutes Sc for Pr and Sh for Nu in Reynolds' analogy. Whatever the coefficient of $C_f/2$ might be, Reynolds' analogy can be used to obtain mass-transfer coefficients from heat-transfer coefficients.

8.3   FLAT-PLATE HEAT-TRANSFER COEFFICIENT

For parallel flow of a Pr $\approx$ 1 fluid over a constant-temperature flat plate, the local turbulent heat-transfer coefficient from Reynolds' analogy is

$$\frac{\text{Nu}}{\text{Re Pr}} = \frac{C_f}{2} \tag{8-16}$$

With $C_f/2$ given by Eq. (8-5), Eq. (8-16) yields

$$\text{Nu}_x = 0.0291 \, \text{Re}_x^{0.8} \, \text{Pr}, \qquad 5 \times 10^5 \leqslant \text{Re}_x < 10^7 \tag{8-17}$$

Should there be a preceding laminar boundary layer, its contribution must be taken into account. From Eq. (8-17) an average turbulent heat-transfer coefficient h is obtained as

$$\bar{h} = \frac{1}{x} \int_0^x h \, dx = \frac{A}{x} \int_0^x x^{-1/5} \, dx = \tfrac{5}{4} h \tag{8-18}$$

$$\overline{\text{Nu}_x} = 0.0364 \, \text{Re}_x^{0.8} \, \text{Pr}, \qquad 5 \times 10^5 \leqslant \text{Re}_x < 10^7$$

Measurements suggest that, for many gases, the Prandtl number exponent is best taken to be 0.6 rather than 1.0.

Procedures for the calculation of heat-transfer rates in turbulent flows over nonisothermal surfaces have been devised by Spalding [20] and Kestin and co-workers [21, 22]. Measurements on turbulent boundary layers on a rough surface with blowing are available [23].

The effect of free-stream turbulence intensity is surveyed by Kestin [24]. Laminar boundary-layer heat transfer can be substantially increased if the pressure gradient is nonzero, nearly doubling at the stagnation point of a cylinder as turbulence intensity increases from 0% to 3% [25]. Traci and Wilcox [26] agree that the mechanism of heat-transfer enhancement at stagnation points is stretching of vortex lines. Blair [27] found that both the turbulent heat-transfer coefficient and friction coefficient increased by about 10% as turbulence intensity increased from 0% to 3%.

8.4   FLAT PLATE WITH UNHEATED STARTING LENGTH

The effect of an unheated starting length on the turbulent heat-transfer coefficient for constant free-stream velocity can be determined by making use of an integral method. The integral form of the turbulent boundary-layer equations can be obtained by integration in the vertical direction, but from y to δ rather than from 0 to δ as was done in Chapter 6 for laminar boundary layers.

The turbulent continuity equation of Eq. (7-11) is treated first. Integration with respect to y from y to δ gives

$$-v_\delta + v_y = \int_y^\delta \frac{\partial u}{\partial x}\, dy = \frac{\partial}{\partial x}\left[\int_y^\delta u\, dy\right] - U\frac{d\delta}{dx}$$

The turbulent x-motion equation of Eq. (7-11) is integrated next, again from y to δ. This operation gives

$$\int_y^\delta \frac{\partial u^2}{\partial x}\, dy + u_\delta v_\delta - u_y v_y = \frac{\tau_\delta - \tau_y}{\rho}$$

Now $\tau_\delta = 0$ and $u_\delta = U$, where U is the free-stream velocity. So, after employing Leibnitz' formula and the integrated form of the continuity equation, the integrated x-motion equation becomes

$$\frac{d}{dx}\left[\int_y^\delta\left(1 - \frac{u^2}{U^2}\right) dy\right] - \left(1 - \frac{u}{U}\right)\frac{v_\delta}{U} - \frac{u}{U}\frac{d}{dx}\left[\int_y^\delta\left(1 - \frac{u}{U}\right) dy\right] = \frac{\tau_y}{\rho U^2}$$

$$(8\text{-}19)$$

Use of the $\frac{1}{7}$th power-law velocity distribution of Eq. (8-1) and Eq. (6-11) to obtain $v_\delta/U = (1/8)\, d\delta/dx$ in Eq. (8-19) gives

$$\frac{\tau_y}{\rho U^2} = \frac{7}{72}\left[1 - \left(\frac{y}{\delta}\right)^{9/7}\right]\frac{d\delta}{dx}$$

Since the shear stress at $y = 0$ is just the wall shear stress τ_w, one has

$$\frac{\tau_y}{\tau_w} = 1 - \left(\frac{y}{\delta}\right)^{9/7} \tag{8-20}$$

for the vertical variation of turbulent shear. Use of the $\frac{1}{7}$th power-law velocity distribution of Eq. (8-1) and the shear stress variation of Eq. (8-20) in

$$\tau = \rho(\nu + \varepsilon_m)\frac{\partial u}{\partial y}$$

gives

$$\nu + \varepsilon_m = 7\delta\frac{\tau_w}{\rho U}\left(\frac{y}{\delta}\right)^{6/7}\left[1 - \left(\frac{y}{\delta}\right)^{9/7}\right] \qquad (8\text{-}21)$$

Attention is next directed to the temperature distribution and vertical heat flux. First, it is assumed that Pr and Pr_t equal unity so that the temperature profile is similar to the velocity profile, giving

$$\frac{T - T_w}{T_\infty - T_w} = \left(\frac{y}{\delta_T}\right)^{1/7} \qquad (8\text{-}22)$$

where δ_T is the thermal boundary-layer thickness. Then, with $\nu + \varepsilon_m = \alpha + \alpha_t$ since $Pr \approx 1 \approx Pr_t$, Eqs. (8-21) and (8-22) are substituted into

$$q_y = -\rho C_p(\alpha + \alpha_t)\frac{dT}{dy}$$

to obtain

$$\frac{q_y}{\rho C_p(T_w - T_\infty)} = \frac{\tau_w}{\rho U}\left(\frac{\delta}{\delta_T}\right)^{1/7}\left[1 - \left(\frac{y}{\delta}\right)^{9/7}\right]$$

Recognizing that $\tau_w/\rho U^2 = C_f/2$ and setting $y = 0$, one obtains

$$\frac{q_w}{\rho C_p U(T_w - T_\infty)} = \left(\frac{\delta}{\delta_T}\right)^{1/7}\frac{C_f}{2} \qquad (8\text{-}23)$$

Next the integral form of the energy equation [Eq. (6-18)]

$$\frac{\partial}{\partial x}\left[\int_0^{\delta_T} u(T - T_\infty)\, dy\right] = \frac{q_w}{\rho C_p}$$

is utilized to solve for δ/δ_T. If Eq. (8-23) is used for q_w and Eq. (8-5) for

$C_f/2$, the integral form of the energy equation is

$$\frac{d}{dx}\left(\frac{7}{72}\frac{\delta_T^{8/7}}{\delta^{1/7}}\right) = \left(\frac{\delta}{\delta_T}\right)^{1/7}\frac{0.0291}{(Ux/\nu)^{1/5}}$$

with $\delta_T \leqslant \delta$, for a flat plate with an unheated starting length x_0. Use of $\xi = \delta_T/\delta$ and Eq. (8-4) to relate δ to $x^{4/5}$ then allows the preceding equation to be recast as

$$\frac{d\xi^{9/7}}{dx} + \frac{9}{10}x^{-1}\xi^{9/7} = \frac{0.0291(9/8)(72/7)}{0.375}x^{-1}$$

from which it is found that

$$\xi^{9/7} = \frac{5}{4}\frac{(0.0291)}{0.375}\frac{72}{7} + Cx^{-9/10}$$

The constant C of integration is evaluated from the condition that $\xi(x_0) = 0$ to give

$$\xi = \left[\frac{90}{7}\left(\frac{0.0291}{0.375}\right)\right]^{7/9}\left[1 - \left(\frac{x_0}{x}\right)^{9/10}\right]^{7/9}$$

This information substituted back into Eq. (8-23) finally gives the local heat-transfer coefficient on a flat plate with an unheated starting length as

$$\frac{q_w}{\rho C_p U(T_w - T_\infty)} = \frac{Nu}{Re\,Pr^{0.6}} = \left[1 - \left(\frac{x_0}{x}\right)^{9/10}\right]^{-1/9}\frac{C_f}{2},$$

$$5 \times 10^5 \leqslant Re \leqslant 10^7$$

$$(8\text{-}24)$$

This functional form was discussed following Eq. (6-23). To better account for Pr differing from unity, the final result is taken to have $Pr^{0.6}$. Measurements [20] confirm Eq. (8-24). A sometimes more convenient approximation to Eq. (8-24) is

$$\frac{Nu}{Re\,Pr^{0.6}} = 0.0291\,Re^{-1/5}\left(1 - \frac{x_0}{x}\right)^{-0.12}$$

$$(8\text{-}25)$$

Finding the local heat flux from a plate (to a constant velocity and temperature free stream) whose surface temperature varies from these results, using Eq. (6-23) is the subject of Problem 8-9. The inverse problem of determining the surface temperature distribution that corresponds to a specified heat flux is the subject of Problem 8-10.

The turbulent velocity boundary layer on an axisymmetric body of revolution is treated in Problem 8-11. With knowledge of the local turbulent velocity boundary-layer thickness, Reynolds' analogy in the form of Eq. (8-15) or (8-18) can then be used to obtain the local heat-transfer coefficient. Refined procedures for determining the local heat-transfer coefficient have been proposed by Spalding [29] and by Ambrok [30] whose integral method, the subject of Problem 8-19, was modified by Moretti and Kays [31] and refined by Sucec and Lu [32]. In justification of the use of Reynolds' analogy, it assumes a "law of the wall" that is only weakly dependent on shear stress variation along the wall and also assumes that the ratio of shear stress to heat flux is constant in the turbulent part of the boundary layer. These assumptions are usually accurate since, for usual Prandtl number values, the principal thermal resistance is in the laminar and buffer zones.

Discussion of the effects of blowing and suction on turbulent boundary layers is available [33].

8.5   RECOVERY FACTOR FOR TURBULENT FLOW

As the boundary layer changes from its laminar state near the leading edge, for which $r = Pr^{1/2}$, r increases to a peak (~ 0.89 for air) in the laminar–turbulent transition region and then approaches a constant value predictable by

$$r = Pr^{1/3} \tag{8-26}$$

Typical experimental results are illustrated in Fig. 8-6. Additional support is provided by Seban and Doughty [35].

Wall roughness can reduce r from the 0.89 value predicted by Eq. (8-26) for air to the lower value of 0.8 in turbulent boundary layers according to Hodge et al. [36]; the direction of change is the same for laminar boundary layers.

The recovery factor for Pr appreciably differing from unity can be predicted by [17], from the original work of Chung and Thomas [37],

$$r = \begin{cases} \dfrac{4}{\pi}\left(\dfrac{Pr}{2-Pr}\right)^{1/2} \arctan\left(\dfrac{2-Pr}{Pr}\right)^{1/2}, & Pr < 2 \\[3mm] \dfrac{4}{\pi}, & Pr = 2 \\[3mm] \dfrac{2}{\pi}\left(\dfrac{Pr}{Pr-2}\right)^{1/2} \ln\left[\dfrac{1+(1-2/Pr)^{1/2}}{\left[1-(1-2/Pr)^{1/2}\right]}\right], & Pr > 2 \end{cases}$$

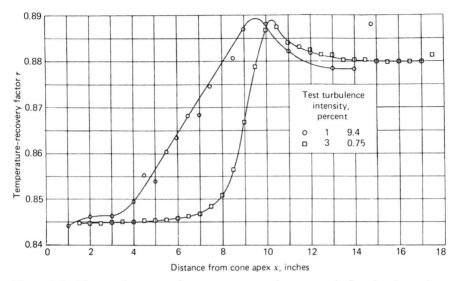

Figure 8-6 Measured recovery factor r on a cone in supersonic flow for determination of the point of laminar-to-turbulent transition. In laminar flow $r = Pr^{1/2} = 0.846$; the steep slope in the recovery factor versus distance from vertex plot indicates the location of the transition region. Curve 1 is at low turbulence intensity and curve 2 is at high turbulence intensity. [By permission from J. C. Evvard, M. Tucker, and W. C. Burgess, *J. Aeronaut. Sci.* **21**, 731–738 (1954) [34].]

8.6 FINITE-DIFFERENCE SOLUTIONS

As was remarked in Section 5.9 for laminar boundary layers, often needed details of velocity and temperature distributions can be obtained only by numerical methods. For turbulent boundary layers, accurate mathematical representation of turbulent transport of energy and momentum poses additional difficulties. A survey book by Cebeci and Bradshaw [38] can be consulted for greater detail, including discussion of the Box numerical method.

One of the difficulties is that the free outer boundary associated with boundary layers makes their velocity distribution far from the wall intermittently turbulent to an increasing degree as the free stream is approached, according to experimental observations such as those reported by Klebanoff [39] and represented in Fig. 8-7. The potential flow of the free stream penetrates the outer region of the boundary layer so that at an affected location the flow is turbulent for only a fraction γ, the *Klebanoff intermittency factor*, of the time of observation. Although this effect was apparently first noticed by Corrsin in 1943 with later investigation accomplished by Corrsin and Kistler [40], Klebanoff [39] gave the result useful for flat-plate boundary

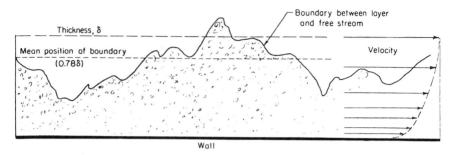

Figure 8-7 Sketch of turbulent boundary layer, showing that the potential flow of the free stream can extend far into the boundary layer at times. [From P. S. Klebanoff, NACA TN 3178, 1954 [39].]

layers of

$$\gamma \approx \frac{1}{1 + 5.5(y/\delta)^6} \tag{8-27}$$

with δ being where $u/U = 0.995$. As shown in Fig. 8-7, an appreciable portion of the flow is turbulent above the average edge of the boundary layer to $y \approx 1.2\delta$, and an appreciable portion of the flow is nonturbulent as far below as $y \approx 0.4\delta$. Of course, $\gamma = 0$ entirely outside the turbulent flow and $\gamma = 1$ entirely inside.

The utility of the intermittency factor is that, in such flows as jets and boundary layers that have an outer boundary with a nonturbulent environment, it separates turbulent transport from intermittency effects. For a turbulent boundary layer, measurements show that the eddy diffusivity is nearly constant over the outer part [39]. The effects of Mach number $0 \leqslant M \leqslant 5$ and compressibility are rendered slight in the outer part of a turbulent boundary layer [41] by the general correlation for eddy diffusivity

$$\varepsilon_{m,0} = k_2 U \delta^* \gamma \tag{8-28}$$

in which $k_2 = 0.0168$ [42], $\delta^* = \int_0^\delta (1 - u/U)\, dy$ is the displacement thickness, and γ is given by Eq. (8-27).

Close to the solid surface the effect of intermittency is negligible. There Van Driest's modification of Prandtl's mixing-length relation described by Eq. (7-14a) can be used for this inner region to give

$$\varepsilon_{m,i} = K^2 y^2 \left[1 - \exp\left(\frac{-y}{A} \right) \right]^2 \left| \frac{\partial u}{\partial y} \right| \tag{8-29}$$

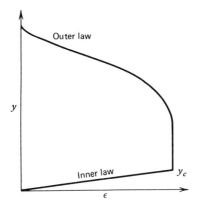

Figure 8-8   Eddy viscosity distribution across a boundary layer in a two-region model.

in which $K = 0.4$ and the generalized parameters [31, 43] are

$$A = \frac{26\nu}{v^*N}, \qquad N^2 = \frac{P^+[1 - \exp(11.8v_w^+)]}{v_w^+} + \exp(11.8v_w^+)$$

$$\text{(8-29a)}$$

$$P^+ = \frac{\nu U}{v^{*3}}\frac{dU}{dx}, \qquad v_w^+ = \frac{v_w}{v^*}$$

where v_w is the wall blowing velocity and $v^* = (\tau_w/\rho)^{1/2}$ is the friction velocity. When $v_w = 0$, $N^2 = 1 - 11.8P^+$. Equation (8-29) has been found to be accurate for compressible and incompressible turbulent flows with both mass and heat transfer (see Cebeci [43] for viscosity and density corrections).

Over the boundary-layer thickness, the eddy diffusivity is predicted by Eq. (8-29) in the inner region ($0 \leqslant y \leqslant y_c$) and by Eq. (8-28) in the outer region ($y > y_c$) with y_c the distance from the solid surface at which $\varepsilon_{m,0} = \varepsilon_{m,i}$ as illustrated in Fig. 8-8. An alternative procedure for predicting eddy diffusivity in the outer part of a turbulent boundary does not require use of the intermittency factor. Instead, Prandtl's mixing length l in

$$\varepsilon_m = l^2\left|\frac{\partial u}{\partial y}\right|$$

$$\text{(7-12)}$$

is used with the recommended correlation [44] for the nearly constant mixing length in this outer region being

$$\frac{l_0}{\delta} = 0.089, \qquad 0.6 \leqslant y/\delta$$

$$\text{(8-30a)}$$

A fit [44] to measurements [41] close to the wall is

$$\frac{l_i}{\delta} = 0.42\left[1 - \exp\left(\frac{-y^+}{26}\right)\right]\frac{y}{\delta}, \qquad 0.1 \leqslant \frac{y}{\delta} \leqslant 0.6$$

$$- 1.61489\left(\frac{y}{\delta} - 0.1\right)^2$$

$$+ 2.88355\left(\frac{y}{\delta} - 0.1\right)^3$$

$$- 1.91556\left(\frac{y}{\delta} - 0.1\right)^4 \qquad (8\text{-}30b)$$

and

$$\frac{l_i}{\delta} = 0.42\left[1 - \exp\left(\frac{-y^+}{26}\right)\right]\frac{y}{\delta}, \qquad \frac{y}{\delta} < 0.1 \qquad (8\text{-}30c)$$

Here $y^+ = y(\tau_1/\rho)^{1/2}/\nu_w$, $\tau_1 = \tau_w(1 + v_w y/\nu_w)$, and v_w is the wall blowing velocity.

Regardless of whether Eqs. (8-28) and (8-29) or Eq. (8-30) are used, the turbulent Prandtl number $\Pr_t$ is assumed to be constant at 0.9 [43, 45]; Rotta suggested that $\Pr_t = 0.9$–0.4 $(y/\delta)^2$ but the measurements of Bagheri et al. [46] in air show $\Pr_t$ about 10% higher than that—see the review by Hammon [47] for near-wall locations and $\Pr = 0.7, 7,$ and 70, showing a peak $\Pr_t \approx 1.6$ at $y^+ \approx 8$ for $\Pr = 0.7$. If $\Pr_t$ and ε_m are known, the turbulent thermal conductivity can be obtained from

$$k^t = \rho C_p \varepsilon_m / \Pr_t$$

Wedge flow with constant properties is discussed to illustrate the essential features of one of the solution procedures. Although velocities and enthalpies could be directly determined by a finite-difference procedure [36, 44, 45, 48], it is computationally more efficient to first subject the turbulent boundary-layer equations to a similarity-like coordinate transformation. This advantage is not as pronounced for turbulent boundary layers as it is for the laminar case discussed in Section 5.10, however. Proceeding as for the laminar case yields the transformed x-motion equation

$$\frac{\partial[(1 + \varepsilon_m^+)F'']}{\partial\eta} + P\left[1 - (F')^2\right] + \left(\frac{1 + P}{2}\right)FF'' = x\left[F'\frac{\partial F'}{\partial x} - F''\frac{\partial F}{\partial x}\right]$$

$$F'(x, \eta = 0) = 0$$

$$F(x, \eta = 0) = -(\nu x U)^{-1/2}\int_0^x v_w\, dx$$

$$F'(x, \eta \to \infty) \to 1 \qquad (8\text{-}31a)$$

and energy equation

$$\frac{\partial\left[(1 + \varepsilon_m^+ \,\mathrm{Pr}/\mathrm{Pr}')\theta'/\mathrm{Pr} + (1 - 1/\mathrm{Pr})(U^2/H_\infty)F'F''\right]}{\partial\eta}$$

$$= -\left(\frac{1+P}{2}\right)F\theta' + x\left[F\frac{\partial\theta}{\partial x} - \theta'\frac{\partial F}{\partial x}\right]$$

$$\theta(x, \eta = 0) = \frac{H_w - H_\infty}{H_\infty} \quad \text{or} \quad \theta'(x, \eta = 0) = -\frac{C_p q_w}{k H_\infty}$$

$$\theta(x, \eta \to \infty) \to 0 \qquad (8\text{-}31b)$$

Here quantities are as described in Section 5.10 for Eqs. (5-81) and (5-85). The remarks in Section 5.10 regarding finite-difference solutions also apply here; consult Cebeci [49] for additional details, such as the use of the Richardson extrapolation (see also Keller and Cebeci [50]) to improve accuracy and generalization to axisymmetric and variable-property cases. The dimensionless eddy diffusivity $\varepsilon_m^+ = \varepsilon_m/\nu$ for constant properties is given by the dimensionless forms of Eqs. (8-28) and (8-29) as

$$\varepsilon_{m,0}^+ = k_2\gamma\left(\frac{Ux}{\nu}\right)^{1/2}\int_0^{\eta_{\max}}(1 - F')\,d\eta, \qquad \eta > \eta_c$$

$$\varepsilon_{m,i}^+ = K^2\eta^2\left(\frac{Ux}{\nu}\right)^{1/2}\left\{1 - \exp\left[-\eta\left(\frac{Ux}{\nu}\right)^{1/4}(F'')^{1/2}\frac{N}{26}\right]\right\}^2 |F''|,$$

$$0 \leqslant \eta \leqslant \eta_c$$

where

$$\frac{1}{\gamma} = 1 + 5.5\left(\frac{\eta}{\eta_{\max}}\right)^6, \qquad P^+ = P\frac{(\nu/Ux)^{1/4}}{(F'')^{3/2}}, \qquad v_w^+ = (v_w/U)\frac{(Ux/\nu)^{1/4}}{(F'')^{1/2}}$$

where $\eta_{\max}$ is the η value at which $u/U = 0.995$, N is given in Eq. (8-29a), and η_c is the η value at which $\varepsilon_{m,0}^+ = \varepsilon_{m,i}^+$.

The eddy diffusivity representations employed are really only fits to measurements for cases of local turbulent equilibrium. So, when departure from such equilibrium is marked, the predictions based on the equilibrium assumption lose accuracy.

Heat-transfer coefficients predicted [42] by the numerical methods described here are often in good agreement with measurements [51] as shown in Fig. 8-9—St = Nu/Re Pr (St represents Stanton number).

See Cebeci and Bradshaw [52], Bradshaw, Cebeci, and Whitelaw [53], or Patankar and Spalding [54] for implementing FORTRAN programs for both external and internal flows.

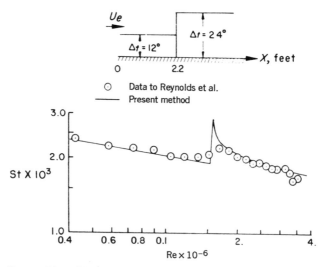

Figure 8-9 Comparison of calculated and measured Stanton numbers St = Nu/Re Pr with measurements by Reynolds et al. [51]. [From T. Cebeci, A. M. O. Smith, and G. Mosinskis, *ASME J. Heat Transfer* **92**, 133–143 (1970) [42].]

8.7 NEWER TURBULENCE MODELS

The Prandtl mixing-length model has been successfully applied to turbulent flows of the boundary-layer type (often referred to as *thin shear-layer flows*). The mixing-length hypothesis lacks universality, however, because of its assumption that turbulence is in local equilibrium, with turbulent energy locally produced and dissipated at the same rate. As an example of the computational difficulty that can be encountered, the mixing-length hypothesis predicts [see Eq. (7-12)] that the eddy diffusivity is zero at the center of a flow channel where velocity gradients vanish, which is unfortunate when heat is transferred from one channel wall to another.

One-equation models base eddy diffusivity on a quantity that is characteristic of turbulence rather than of local time-averaged velocity. The kinetic theory of gases suggests that

$$\nu \sim \text{characteristic length} \times \text{characteristic velocity}$$

In the same spirit it has been proposed that eddy diffusivity is proportional to a characteristic velocity that is $k^{1/2}$, where $k = (\overline{u'^2} + \overline{v'^2} + \overline{w'^2})/2$ is the time-averaged kinetic energy of turbulence. Thus

$$\nu^t = c'_\mu k^{1/2} L \tag{8-32}$$

which is known as the *Kolmogorov–Prandtl expression* because Kolmogorov

[55] and Prandtl [56] proposed it independently at about the same time. In one-equation models the characteristic length L is determined from empirical relations and is similar to a mixing length. The transport equation for the kinetic energy of turbulence k is determined from the Navier–Stokes equation. The procedure followed is to multiply the unsteady equation of motion for each coordinate by its corresponding unsteady velocity and then time averaging and summing the three equations. From this is subtracted the result obtained by multiplying the time-averaged equation of motion for each coordinate by its corresponding time-averaged velocity and summing. The result is

$$\frac{Dk}{Dt} = \underbrace{\frac{\partial}{\partial x_i}\left[\nu\frac{\partial k}{\partial x_i} - \overline{u'_i\left(\frac{u'_j u'_j}{2} + \frac{p'}{\varrho}\right)}\right]}_{\text{diffusive transport}} + \underbrace{\left[-\overline{u'_i u'_j}\frac{\partial u_i}{\partial x_j}\right]}_{P,\text{ production by shear}}$$

$$+ \underbrace{\left[-\beta g_i\overline{u'_i T'}\right]}_{\substack{G,\text{ buoyancy}\\ \text{production}}} - \underbrace{\left[\nu\overline{\frac{\partial u'_i}{\partial x_j}\frac{\partial u'_i}{\partial x_j}}\right]}_{\substack{\varepsilon,\text{ viscous}\\ \text{destruction}}} \qquad (8\text{-}33)$$

with primes denoting instantaneous values, overbars denoting time averaging (plain quantities are also time averaged), and employment of the Einstein convention of repeated indices (see Hinze [5, pp. 68–74] for details). For a boundary-layer flow with buoyant production G neglected, as it will be until the algebraic stress model is discussed,

$$\frac{\partial k}{\partial t} + u\frac{\partial k}{\partial x} + v\frac{\partial k}{\partial y} = \frac{\partial\left[\nu\,\partial k/\partial y - \overline{v'(p'/\rho + k')}\right]}{\partial y} - \overline{(u'v')}\frac{\partial u}{\partial y}$$

$$- \nu\overline{\left[\left(\frac{\partial u'}{\partial x}\right)^2 + \left(\frac{\partial u'}{\partial y}\right)^2 + \left(\frac{\partial v'}{\partial x}\right)^2 + \left(\frac{\partial v'}{\partial y}\right)^2\right]}$$

Equation (8-33) can be interpreted as stating that the change in k due to either unsteady events or to convective transport by the mean motion equals the diffusion of k plus the production (P) of k by turbulent stresses transferring kinetic energy from the mean flow to the turbulent motion plus the generation G by buoyant forces transferring potential energy to turbulent kinetic energy minus the dissipation ε of turbulent kinetic energy into thermal energy caused by turbulent stresses acting on the turbulent motion. Detailed study [5, pp. 68–74] reveals that the dissipation interpretation is strictly correct only for homogeneous turbulence, a situation often closely approached. Now, turbulent diffusion of k is presumed to be similar in mathematical form to molecular diffusion so that

$$-\overline{v'\left(\frac{p'}{\rho} + k'\right)} = \frac{v'}{\sigma_k}\frac{\partial k}{\partial y}$$

In the production term it is reasonable to have

$$-\overline{(u'v')} = v'\frac{\partial u}{\partial y}$$

The dissipation ε is given in terms of local turbulence quantities as

$$\varepsilon = v\sum_{i,j}\overline{\left(\frac{\partial u_j}{\partial x_i}\right)^2} = \frac{C_D k^{3/2}}{L} \tag{8-34}$$

which is dimensionally consistent. Here the dimensionless constants are commonly taken to be $\sigma_k \approx 1.0$ and $C_D \approx 0.08$. The transport equation for k is then, for highly turbulent flow,

$$\frac{Dk}{Dt} = \frac{\partial\left[(v'/\sigma_k)\partial k/\partial y\right]}{\partial y} + v'\left(\frac{\partial u}{\partial y}\right)^2 - \frac{C_D k^{3/2}}{L} \tag{8-35}$$

which is of the same form as the boundary-layer equations of motion. Equations (8-32) and (8-35) can be solved simultaneously with the boundary-layer equations of motion. Although one-equation models such as this account for the nonequilibrium character of turbulence through use of a transport equation for k, they are only marginally superior to a Prandtl mixing-length model since the characteristic length L is not adjusted to nonequilibrium turbulence except by empirical relations.

Two-equation models of turbulence determine both the turbulent kinetic energy k and the characteristic length L from transport equations. Fortunately, the increased computational complexity is compensated by a dramatic increase in universality. The best known and most extensively tested two-equation model is the so-called k–ε model, although others, such as the Wilcox–Rubesin model [57], have been developed. The k–ε model has the eddy diffusivity given by

$$v^t = f_\mu\frac{C_\mu k^2}{\varepsilon} \tag{8-36a}$$

in which L has been eliminated by combining Eqs. (8-32) and (8-34) and the turbulence stresses related to time-averaged velocities by

$$-\overline{(u_i'u_j')} = v'\left(\frac{\partial u_i}{\partial x_j} + \frac{\partial u_j}{\partial x_i}\right) - \frac{2}{3}k\delta_{ij} \tag{8-36b}$$

in which δ_{ij} is the Kronecker delta ($\delta_{ij} = 0$ for $i \neq j$ and $\delta_{ij} = 1$ for $i = j$). The transport equation for turbulent kinetic energy is given by

$$\frac{Dk}{Dt} = \frac{\partial\left[(v + v'/\sigma_k)\,\partial k/\partial x_i\right]}{\partial x_i} + k\left(\frac{P}{k} - \frac{\varepsilon}{k}\right) \tag{8-36c}$$

in which $P = -\overline{(u_i'u_j')}\,\partial u_i/\partial x_j$ is the production term and the transport equation for dissipation ε is given by

$$\frac{D\varepsilon}{Dt} = \frac{\partial\left[(v + v'/\sigma_\varepsilon)\,\partial \varepsilon/\partial x_i\right]}{\partial x_i} + \varepsilon\left(C_1\frac{P + G}{k}(1 + C_3R_f) - C_2\frac{\varepsilon}{k}\right) \tag{8-36d}$$

The turbulent Prandtl and Schmidt numbers are usually assumed to equal 0.9. Equation (8-36d) for ε is most properly viewed as being empirically of the same form as Eq. (8-36c) for k; although an exact transport equation for ε can be obtained by manipulating the Navier–Stokes equations, drastic assumptions are necessary to achieve a tractable form. The k–ε model contains five empirical constants that are determined by comparing predictions with measurements. For boundary-layer flows (see Pourahmadi and Humphrey [58] for modifications for curved channel flows and Champion and Libby [59] for stagnation streamline turbulence modifications) common values are $C_\mu = 0.09$, $\sigma_k = 1.0$, $\sigma_\varepsilon = 1.3$, $C_1 = 1.44$, $C_2 = 1.92$, and $C_3 \approx 0.8$ [60]. R_f gives the effects of buoyancy; $R_f = -G/(P + G)$ in horizontal flows, and 0 in vertical flows (see Yang and Aung [61] for natural convection). Other flows may require slightly different constants (see Rodi [62] for standard values and discussion including the effects of buoyancy).

Equations (8-36c) and (8-36d) can be integrated up to the wall through use of the wall damping function f_μ in Eq. (8-36a) to represent the effect of a wall in damping turbulent fluctuations, particularly those in the perpendicular direction. The form of f_μ is similar to that selected for the mixing length by Deissler in Eq. (7-14) and Van Driest in Eq. (7-14a). In a review, Patel et al. [60] found that of eight forms studied the best was

$$f_\mu = \left[1 - \exp(-0.0165R_y)\right]^2\left(1 + \frac{20.5}{R_T}\right)$$

where the two Reynolds numbers of turbulence are $R_y = k^{1/2}y/v$ and

$R_T = k^2/\nu\varepsilon(\approx \nu'/\nu)$. Miner et al. [63] and Speziale et al. [64] suggest improvements; So et al. [65] compared eight near-wall closure predictions, and So et al. [66] developed a near-wall closure especially for curved flows.

Alternatively, as discussed by Rodi [67], the highly turbulent transport equations for k and ε obtained by deleting the molecular momentum diffusivity ν from Eqs. (8-36c) and (8-36d) can be bridged across the viscous sublayer near the wall in what is called a wall-function approach. In this approach, at point y_c just outside the viscous sublayer it is assumed that the velocity component u_c parallel to the wall follows the law of the wall [Eq. (7-18) for smooth walls and Eq. (7-23) for rough walls] so that

$$\frac{u_c}{\upsilon^*} = K^{-1} \ln\left(\frac{y_c\upsilon^*}{\nu}\right) + 3.8 \qquad (7\text{-}18)$$

When this prediction for u versus y is matched by the solution to Eqs. (8-36c) and (8-36d), the results of the assumption that local turbulent equilibrium ensues with the turbulence production rate P equal to the dissipation rate ε are imposed as boundary conditions to give k and ε values at location y_c as

$$k_c = \frac{\upsilon^{*2}}{C_\mu^{1/2}} \qquad \text{and} \qquad \varepsilon_c = \frac{\upsilon^{*3}}{Ky_c} \qquad (8\text{-}36\text{e})$$

with $K = 0.41$ being the von Karman constant. This procedure avoids the need for small step sizes near a wall; it can also be applied to temperature distribution calculation [66] in an analogous form.

As discussed in Problem 8-23, the constant C_2 is evaluated by matching the decay of isotropic turbulence downstream from a grid. The constant C_μ is evaluated by noting, as expressed in Eq. (8-36e), that the ratio of Reynolds shear stress to turbulent kinetic energy is constant in a two-dimensional equilibrium shear flow so that

$$C_\mu = \left(\frac{\upsilon^{*2}}{k}\right)^2 = \left(\frac{-\overline{u'\upsilon'}}{k}\right)^2$$

Raithby and Schneider [68] mention that $C_\mu = 0.09$ for flow over smooth surfaces whereas $C_\mu = 0.033$ for flow over rough surfaces such as are encountered in environmental flows. As is the subject of Problem 8-25, C_1, C_μ, and K (the von Karman constant) are related.

A nonlinear $k - \varepsilon$ turbulence model developed by Speziale [69] is capable of predicting secondary flows in noncircular ducts, normal Reynolds stress differences, and highly swirling flows, all of which are beyond the capability of the standard $k - \varepsilon$ model. Applications are discussed for a straight channel by Bishnoi et al. [70] and past a backward-facing step by Thangam and Speziale [71]. Improvements can be, but not necessarily are, achieved with second-order closure models based on the Reynolds-stress transport equation, discussed by Rodi [62] as follows.

The Reynolds-stress $\overline{u_i u_j}$ transport equation, suggested by Keller and Friedmann [72] in 1924 and first derived and presented by Chou [73] in 1945, is

$$\frac{D\overline{u_i' u_j'}}{Dt} = \underbrace{-\left[\frac{\partial\left(\overline{u_L' u_i' u_j'}\right)}{\partial x_L} + \frac{1}{\varrho}\left(\frac{\partial \overline{u_j' p'}}{\partial x_i} + \frac{\partial \overline{u_i' p'}}{\partial x_j}\right)\right]}_{\text{diffusive transport}}$$

$$+ \underbrace{\left[-\overline{u_i' u_L'}\frac{\partial u_j}{\partial x_L} - \overline{u_j' u_L'}\frac{\partial u_i}{\partial x_L}\right]}_{P_{ij},\ \text{stress production}} + \underbrace{\left[-\beta\left(g_i\overline{u_j'\Phi'} + g_j\overline{u_i'\Phi'}\right)\right]}_{G_{ij},\ \text{buoyancy production}}$$

$$+ \underbrace{\left[\frac{p'}{\varrho}\left(\frac{\partial u_i'}{\partial x_j} + \frac{\partial u_j'}{\partial x_i}\right)\right]}_{\pi_{ij},\ \text{pressure strain}} - \underbrace{\left[2\nu\frac{\overline{\partial u_i'}}{\partial x_L}\frac{\partial u_j'}{\partial x_L}\right]}_{\varepsilon_{ij},\ \text{viscous dissipation}} \qquad (8\text{-}37a)$$

Noting that $k = \frac{1}{2}\overline{u_i' u_i'} = \frac{1}{2}(\overline{u_1' u_1'} + \overline{u_2' u_2'} + \overline{u_3' u_3'})$, summation of Eq. (8-37a) for $i = j = 1, 2, 3$ yields Eq. (8-36c) for k. The pressure strain term does not appear in the k-transport equation; it redistributes energy to make turbulence more isotropic.

Similarly, the transport equation for the flux $\overline{u_i T'}$ of a scalar quantity T' is

$$\frac{D\overline{u_i' T'}}{Dt} = \underbrace{-\frac{\partial}{\partial x_L}\left(\overline{u_i' u_L' T'} + \frac{1}{\varrho}\delta_{iL}\overline{p' T'}\right)}_{\text{diffusive transport}} + \underbrace{\left[-\overline{u_i' u_j'}\frac{\partial T}{\partial x_j} - \overline{u_j' T'}\frac{\partial u_i}{\partial x_j}\right]}_{\text{mean-field production}}$$

$$+ \underbrace{\left[-\beta g_i\overline{T' T'}\right]}_{\substack{\text{buoyancy} \\ \text{production}}} + \underbrace{\frac{1}{\varrho}\overline{p\frac{\partial T'}{\partial x_i}}}_{\substack{\pi_{iT'}\ \text{pressure–scalar} \\ \text{gradient production}}}$$

$$- \underbrace{(\alpha + \nu)\frac{\overline{\partial u_i'}}{\partial x_L}\frac{\partial T'}{\partial x_L}}_{\text{viscous dissipation}} \qquad (8\text{-}37b)$$

The turbulent transport equation for $\overline{T' T'}$, needed for the buoyancy term in

Eq. (8-37b), is

$$\frac{D\overline{T'T'}}{Dt} = -\frac{\partial}{\partial x_j}\underbrace{\left(\overline{u_j'T'T'}\right)}_{\text{diffusive transport}} + \underbrace{\left[-2\overline{u_j'T'}\frac{\partial T}{\partial x_j}\right]}_{\substack{P_{T'},\ \text{mean-field}\\ \text{production}}} - \underbrace{2\alpha\,\overline{\frac{\partial T'}{\partial x_j}\frac{\partial T'}{\partial x_j}}}_{\varepsilon_{T'},\ \text{dissipation}} \qquad (8\text{-}37c)$$

The algebraic stress model (ASM), a widely used alternative to the k–ε model, is a second-order closure model. Since the six components of the Reynolds stress, three components of the scalar flux, and one component of the $\overline{T'T'}$ equations require solution of ten transport equations, recourse to an algebraic approximation that retains such basic features as buoyancy effects is common since a set of algebraic equations is easier to solve than is a set of differential equations. To do this, it is assumed that the transport of $\overline{u_i'u_j'}$ is proportional to the transport of k so that

$$\frac{D\overline{u_i'u_j'}}{Dt} - \text{diffusion of } \overline{u_i'u_j'} = \frac{\overline{u_i'u_j'}}{k}\left[\frac{Dk}{Dt} - \text{diffusion of } k\right]$$

where $\overline{u_i'u_j'}/k$ is a proportionality factor with, from Eq. (8-33),

$$\frac{Dk}{Dt} - \text{diffusion of } k = P + G - \varepsilon$$

Introducing this relationship into Eq. (8-37a) gives

$$\frac{\overline{u_i'u_j'}}{k}(P + G - \varepsilon) = P_{ij} + G_{ij} + \pi_{ij} - \varepsilon_{ij} \qquad (8\text{-}38)$$

The approximation that

$$\pi_{ij} = -C_1\frac{\varepsilon}{k}\left(\overline{u_i'u_j'} - \frac{2}{3}k\delta_{ij}\right) - \gamma\left(P_{ij} - \frac{2}{3}P\delta_{ij}\right) - C_3\left(G_{ij} - \frac{2}{3}G\delta_{ij}\right)$$

is adopted. In the first term, a source for $\overline{u_i'u_i'}$ is obtained if $\overline{u_i'u_i'} > \frac{2}{3}k$ ($\overline{u_i'u_i'} \equiv \frac{2}{3}k$ in isotropic turbulence); similar remarks apply to the other two terms. The viscous dissipation term is approximated by

$$\varepsilon_{ij} = \frac{2}{3}\varepsilon\delta_{ij}$$

Use of these two approximations in Eq. (8-38) yields

$$
\overline{u_i'u_j'} = k\left[\frac{2}{3}\delta_{ij} + \frac{(1-\gamma)\left(\dfrac{P_{ij}}{\varepsilon} - \dfrac{2}{3}\dfrac{P}{\varepsilon}\delta_{ij}\right) + (1-C_3)\left(\dfrac{G_{ij}}{\varepsilon} - \dfrac{2}{3}\dfrac{G}{\varepsilon}\delta_{ij}\right)}{C_1 + \dfrac{P+\varepsilon}{\varepsilon} - 1}\right]
$$

$$(8\text{-}39)$$

Commonly used values are reported by Bergstrom [74] and by Devenport and Simpson [75] to be $C_1 = 2.2$, $\gamma = 0.55 = C_3$.

In a similar manner, the transport of the scalar $\overline{u_i'T'}$ is also assumed to be proportional to the transport of k so that

$$
\frac{D\overline{u_i'T'}}{Dt} - \text{diffusion of } \overline{u_i'T'} = \frac{\overline{u_i'T'}}{k}(P + G - \varepsilon)
$$

Use of this relationship in Eq. (8-37b) yields, neglecting viscous dissipation,

$$
\overline{u_i'T'}\left(\frac{P+G-\varepsilon}{2k}\right) = -\overline{u_i'u_j'}\frac{\partial T}{\partial x_j} - \overline{u_j'T'}\frac{\partial u_i}{\partial x_j} - \beta g_i\overline{T'T'} + \pi_{iT'} \quad (8\text{-}40)
$$

The approximation that

$$
\pi_{iT'} = -C_{1T'}\frac{\varepsilon}{k}\overline{u_i'T'} + C_{2T'}\overline{u_L'T'}\frac{\partial u_i}{\partial x_L} + C_{3T'}\beta g_i\overline{T'T'}
$$

is adopted. Use of this approximation in Eq. (8-40) yields

$$
\overline{u_i'T'} = -\frac{k}{\varepsilon}\left[\frac{\overline{u_i'u_L'}\dfrac{\partial T}{\partial x_L} + (1-C_{2T'})\overline{u_L'T'}\dfrac{\partial u_i}{\partial x_L} + (1-C_{3T'})\beta g_i\overline{T'T'}}{C_{1T'} + \dfrac{1}{2}\left(\dfrac{P+G}{\varepsilon} - 1\right)}\right]
$$

$$(8\text{-}41)$$

Bergstrom [74] has used $C_{1T'} = 3.0$, $C_{2T'} = C_{3T'} = 0.5$; Rodi [62] reports that $C_{1T'} = 3.0$, $C_{2T'} = C_{3T'} = 0.33$ among other sets by various workers.

Evaluation of $\overline{T'T'}$ requires approximation of the dissipation $\varepsilon_{T'}$ term as

$$
\varepsilon_{T'} = \frac{\varepsilon}{kR}\overline{T'T'}
$$

where $0.5 \leqslant R \leqslant 1$ is the ratio of the time scale characterizing temperature fluctuations to velocity turbulent fluctuations. Then Eq. (8-37c) becomes

$$\frac{D\overline{T'T'}}{Dt} = \frac{\partial}{\partial x_j}\left(\frac{\nu^t}{\sigma_{T'}}\frac{\partial \overline{T'T'}}{\partial x_j}\right) - 2\overline{u_j'T'}\frac{\partial T}{\partial x_j} - \frac{\varepsilon}{kR}\overline{T'T'} \qquad (8\text{-}42a)$$

Bergstrom [74] used $\sigma_{T'} = 1.4$ and $R = 0.5$. More simply, assuming equilibrium between production $P_{T'}$ and dissipation $\varepsilon_{T'}$,

$$\overline{T'T'} = -2R\overline{u_j'T'}\frac{\partial T}{\partial x_j} \qquad (8\text{-}42b)$$

The time-averaged velocity and temperature are obtained by solving, usually by computer-aided methods, the continuity equation (here written for constant-property fluid)

$$\frac{\partial u_i}{\partial x_i} = 0$$

the equation of motion

$$\frac{Du_i}{Dt} = -\frac{1}{\varrho}\frac{\partial p}{\partial x_i} + g_i + \frac{\partial}{\partial x_j}\left(\nu\frac{\partial u_i}{\partial x_j} - \overline{u_i'u_j'}\right)$$

and the energy equation

$$\frac{DT}{Dt} = \frac{\partial}{\partial x_j}\left(\alpha\frac{\partial T}{\partial x_j} - \overline{u_j'T'}\right)$$

in conjunction with Eq. (8-39) for $\overline{u_i'u_j'}$, Eq. (8-41) for $\overline{u_iT'}$, and Eq. (8-42) for $\overline{T'T'}$.

These formulations can be exercised by means of one of the commercially available computer programs as discussed in Chapter 9. In Fig. 8-10, it is demonstrated that the ASM model in one program more accurately predicted some details of a highly swirling flow than did the k–ε model. Some refined k–ε models are superior to some ASM models, however, so a trial is needed for each program.

A fast integral method for unsteady turbulent boundary layers is described by Cousteix and Houdeville [77]. It may be of acceptable accuracy and its results can guide more detailed study by the refined turbulence models discussed previously.

A recommended lucid exposition of the development of turbulence models is that by Launder and Spalding [78].

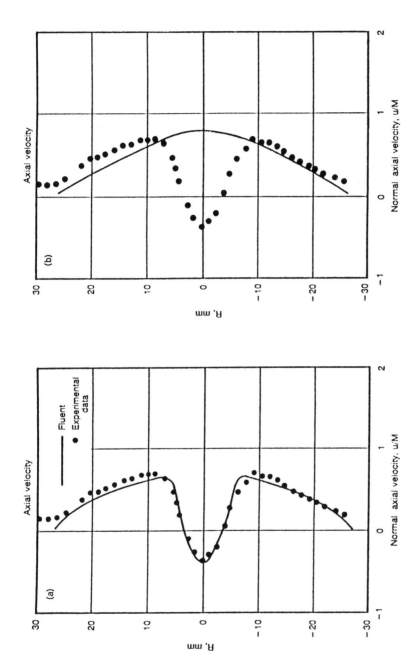

Figure 8-10 A highly swirling flow as analyzed using (*a*) the algebraic-stress turbulence model and (*b*) the *k–ε* model. (By permission from B. Hutchings and R. Iannuzzelli, *Mechanical Engineering* **109**, 72–76 (1980 [76].)

PROBLEMS

8-1 Show that for the velocity profile on a flat plate $u/U = (y/\delta)^{1/n}$, the displacement thickness $\delta_1 = \int_0^\delta (1 - u/U)\, dy$, and the momentum thickness $\delta_2 = \int_0^\delta (u/U)/(1 - u/U)\, dy$ are related to the boundary-layer thickness by

$$\frac{\delta_1}{\delta} = \frac{1}{1 + n} \quad \text{and} \quad \frac{\delta_2}{\delta} = n(1 + n)^{-1}(2 + n)^{-1}$$

8-2 Show that, if Eq. (7-22) for turbulent shear stress on a flat plate is accurate only for $Re_\delta \approx 10^5$, the range of validity of Eq. (8-4) is $5 \times 10^5 \approx Re_{x,c} < Re_x < 10^7$.

8-3 (a) Show that the integral form of the x-motion equation for constant free-stream velocity requires that the momentum thickness vary continuously as

$$\delta_2 = \int_0^x \frac{\tau_w}{\rho U^2}\, dx$$

(b) Match the laminar and turbulent momentum thickness at the transition point on a flat plate to obtain the constant of integration of Eq. (8-3) as

$$C = \frac{0.2931 x_c^{5/4}}{Re_{x,c}^{1/4}} \left(-1 + \frac{36.4}{Re_{x,c}^{3/8}} \right)$$

(c) Show from the result of part b that at typical conditions

$$\frac{\delta}{x} = \frac{0.375}{Re_x^{1/5}} \left(1 - \frac{0.732 x_c}{x} \right)^{4/5}$$

for $5 \times 10^5 \sim Re_{x,c} \leqslant Re_x < 10^7$. Compare this result with that of Eq. (8-4). Comment on the difficulty of matching laminar and turbulent momentum thicknesses at the transition point when the turbulent wall shear relation presumes "fully developed" turbulent flow.

8-4 Verify that laminar-to-turbulent transition on a flat plate in parallel flow results in an increased boundary-layer thickness at the transition point. *Answer*: $\delta_{lam}(x_c)/\delta_{turb}(x_c) = 0.24$.

8-5 Compare the laminar to the turbulent wall shear stress at the transition point for parallel flow over a flat plate. *Answer*: $\tau_{w,lam}(x_c)/\tau_{w,turb}(x_c) = 0.219$.

8-6 Compare the thickness of the laminar sublayer and the buffer zone to the turbulent boundary-layer thickness for parallel flow over a flat plate at the transition point and at a position thrice as far removed from the leading edge. *Answer*: At transition $\delta_l/\delta = 0.008$, $\delta_b/\delta = 0.033$.

8-7 Show that Eq. (8-5) can be expressed in terms of the momentum thickness δ_2 as

$$\frac{C_f}{2} = \frac{0.0128}{(U\delta_2/\nu)^{1/4}}$$

8-8 For a turbulent boundary layer of a fluid for which $\text{Pr} \neq 1$, generalize Eq. (8-17) through use of the full turbulent Reynolds analogy expressed by Eq. (8-16). From this expression for the local heat-transfer coefficient, obtain the average coefficient, the generalization of Eq. (8-18).

8-9 Consider a turbulent boundary layer of a $\text{Pr} \approx 1$ fluid on a flat plate whose surface temperature varies as

$$T_w - T_\infty = A + \sum_{n=1}^{N} B_n x^n, \qquad n = 1, 2, \ldots, N$$

with constant free-stream velocity and temperature.

(a) Show that the response to a step change in plate temperature given by Eq. (8-14) together with the integral relationship given by Eq. (6-23) gives the local heat flux as

$$q_w = \left(0.0296 k \frac{\text{Pr}^{0.4}\,\text{Re}_x^{0.8}}{x}\right)\left(A + \frac{10}{9}\sum_{n=1}^{N} n\beta_n B_n x^n\right)$$

where $\beta_n = \Gamma(8/9)\Gamma(10n/9)/\Gamma(8/9 + 10n/9)$. Here Γ denotes the gamma function.

(b) A plate surface temperature varies linearly with distance from the leading edge (assume the boundary layer to be tripped at the leading edge so that there is no laminar boundary layer of consequence) with the temperature variation over the plate length being 10% of its leading edge excess over the free-stream temperature. Evaluate the accuracy of calculating the total heat loss based on the average temperature difference. *Answer*: $\bar{q}_w/q_w(\Delta T_{av}) = 0.8$.

8-10 Consider a turbulent boundary layer of a $\text{Pr} \approx 1$ fluid on a flat plate whose local heat flux is specified. The free-stream velocity and temperature are constant.

(a) Use the results of Problem 6-9, developed for application to laminar boundary layers, to show the plate surface temperature variation to be

$$T_w(x) - T_\infty = \frac{3.31}{k} \operatorname{Pr}^{-0.6} \operatorname{Re}_x^{-0.8}$$

$$\times \int_0^x \left[1 - \left(\frac{x^*}{x} \right)^{9/10} \right]^{-8/9} q_w(x^*) \, dx^*$$

(b) Show that the local heat-transfer coefficient for the case of $q_w = $ const, using the result of part a, is

$$\frac{\mathrm{Nu}}{\operatorname{Re}_x \operatorname{Pr}^{0.6}} = \frac{0.0296}{\operatorname{Re}_x^{1/5}}$$

and give the temperature variation that accompanies it.

(c) Show that the heat-transfer coefficient of part b for a constant heat flux is 36% higher (see Problem 8-8). Check this seeming lesser sensitivity of a turbulent flow to surface temperature variation by comparing heat-transfer coefficient behavior for laminar and turbulent flow in ducts.

8-11 The turbulent boundary layer on a body of revolution for which the free-stream velocity also varies can be ascertained by application of the integral form of the x-momentum equation given in Problem 6-12 as discussed by Schlichting [3, p. 649] and first done by Millikan [79].

(a) Show that the integral form of the x-motion equation can be recast as, with $Z = U\delta_2/\nu$,

$$Z^{1/n} \frac{dZ}{dx} + Z^{(n+1)/n} \left[\frac{1}{R} \frac{dR}{dx} + (1 + H_{12}) \frac{1}{U} \frac{dU}{dx} \right] = \frac{U\alpha}{\nu}$$

if $H_{12} = \text{const} \; (\approx \frac{9}{7})$ and $\tau_w/\rho U^2 = \alpha/(U\delta_2/\nu)^{1/n}$ as was the case in the development of Eq. (8-18a).

(b) Show that the result of part a can be integrated to give

$$\left(\frac{RU^{2+H}\delta_2}{\nu} \right)^{(n+1)/n} = C + \frac{\alpha}{\nu} \frac{n+1}{n}$$

$$\times \int_{x_{\text{transition}}}^x R^{(n+1)/n} U^{2+H+(1+H)/n} \, dx$$

(c) Specialize the result of part b for the flat-plate values of the parameters—note that C can be evaluated from the laminar boundary layer at $x = x_{\text{transition}}$ and equals 0 if no laminar boundary layer exists since then one of R, U, or δ_2 is zero at $x = 0$.

8-12   Compare the thermal resistance across the laminar sublayer, the buffer zone, and the turbulent core for a turbulent boundary layer on a flat plate.

8-13   Air at a temperature of 20°F and 1 atm pressure flows along a flat surface (an idealized airfoil) at a constant velocity of 150 ft/sec. For the first 2 ft the surface is heated at a constant rate per unit of surface area; thereafter the surface is adiabatic. If the total length of the plate is 6 ft, what must the heat flux on the heated section be so that the surface temperature at the trailing edge is not below 32°F? Plot the surface temperature along the entire plate. Discuss the significance of this problem with respect to wing deicing. *Answer*: q_w = 4400 Btu/hr ft^2.

8-14   Work the previous problem but divide the heater section into two 1-ft strips, with one at the leading edge and the other 3 ft from the leading edge. What heat flux is required such that the plate surface is nowhere less than 32°F? *Answer*: q_{w1} = 5100 Btu/hr ft, q_{w2} = 2300 Btu/hr ft.

8-15   An aircraft oil cooler is to be constructed by using the skin of the wing of the cooling surface. The wing may be idealized as a flat plate over which air at 0.7 atm pressure and 25°F flows at 200 ft/sec. The leading edge of the cooler may be located 3 ft from the leading edge of the wing. The oil temperature and oil side heat-transfer resistance are such that the surface can be at approximately 130°F, uniform over the surface. How much heat can be dissipated if the cooler surface measures 2 ft × 2 ft? Would there by any substantial advantage in changing the shape to a rectangle 4 ft wide × 1 ft in the direction of flow? *Answer*: q = 8000 Btu/hr.

8-16   Consider a constant free-stream-velocity flow of air over a constant-surface-temperature plate. Let the boundary layer be initially a laminar one, but let a transition to a turbulent boundary layer take place in one case at Re_x = 300,000 and in another at Re_x = 10^6. Evaluate and plot (on log–log paper) the Stanton number St = $\text{Nu}/\text{Re Pr}$ as a function of Re_x out to Re_x = 3 × 10^6. Assume that the transition is abrupt (which is not realistic). Also, plot the Stanton number for a turbulent boundary layer originating at the leading edge of the plate. Where is the "virtual origin" of the turbulent boundary layer when there is a preceding laminar boundary layer? What is the effect of changing the transition point? How high must the Reynolds number be in order for turbulent heat-transfer coefficients to be calculated with

2% accuracy without considering the influence of the initial laminar portion of the boundary layer?

8-17 A 40-ft-diameter balloon is rising vertically upward in otherwise still air at a velocity of 10 ft/sec. When it is at 5000-ft elevation, calculate the heat-transfer coefficient over the entire upper hemispherical surface, making any assumptions that seem appropriate regarding the free-stream velocity distribution and the transition from a laminar to a turbulent boundary layer. *Answer:* $\bar{h} = 1.3$ Btu/hr ft^2 °F.

8-18 A cylindrical body 4 ft in diameter has a hemispherical cap over one end. Air flows axially along the body, with a stagnation point at the center of the cap. The air has an upstream state of 1 atm pressure, 70°F, and 200 ft/sec. Under these conditions, evaluate the local heat-transfer coefficient along the cylindrical part of the surface to a point 12 ft from the beginning of the cylindrical surface, assuming a constant temperature surface. Make appropriate assumptions about an initial laminar boundary layer and about the free-stream velocity distribution around the nose. It may be assumed that the free-stream velocity along the cylindrical portion of the body is constant at 200 ft/sec. Then calculate the heat-transfer coefficient along the same surface by idealizing the entire system as a flat plate with constant free-stream velocity from the stagnation point. On the basis of the results, discuss the influence of the nose on the boundary layer at points along the cylindrical section, and the general applicability of the constant free-stream velocity idealization. *Answer:* $h(x = 12$ ft$) = 3.5$ Btu/hr ft^2 °F.

8-19 Prediction of local heat-transfer coefficients for boundary layers over a planar two-dimensional body of arbitrary shape with variable surface temperature can be accomplished by the simple method developed by Ambrok [30] that is often in error by only about 15%. This method is to be derived.

(a) Rearrange the integral form of the energy equation into

$$\frac{d[U(T_w - T_\infty)\delta_t^*]}{dx} = \frac{q_w}{\rho C_p}$$

where x is the distance from the leading edge and

$$\delta_t^* = \int_0^{\delta_t} \frac{u}{U} \frac{T - T_\infty}{T_w - T_\infty} \, dy$$

(b) Since $q_w = h(T_w - T_\infty)$, rearrange the integral equation of part a into the form

$$[U(T_w - T_\infty)]^{-1} \frac{d[U(T_w - T_\infty)\delta_t^*]}{dx} = \frac{Nu_x}{Re_x \, Pr} \quad (19P\text{-}1)$$

(c) Recall from previous work that for flow over a flat plate of constant wall temperature, $\mathrm{Nu}_x = A\,\mathrm{Re}_x^n$ where $A = 0.33206\,\mathrm{Pr}^{1/3}$ and $n = \frac{1}{2}$ for laminar flow while $A = 0.0291\,\mathrm{Pr}^{0.6}$ and $n = \frac{4}{5}$ for turbulent flow. Show that for such a case (U, T_w, and T_∞ constant) $\mathrm{Nu}_x/\mathrm{Re}_x\,\mathrm{Pr} = A\,\mathrm{Re}_x^{n-1}/\mathrm{Pr}$ and Eq. (19P-1) gives

$$\delta_t^* = \frac{A}{\mathrm{Pr}}\frac{\nu}{nU}\,\mathrm{Re}_x^n$$

so that

$$\frac{\mathrm{Nu}_x}{\mathrm{Re}_x\,\mathrm{Pr}} = \left(n^{n-1}\frac{A}{\mathrm{Pr}}\right)^{1/n}\left(\frac{U\delta_t^*}{\nu}\right)^{1-1/n} \tag{19P-2}$$

(d) Ambrok reasoned that δ_t^* is a better indicator of local conditions than is x and that the last result of part c should apply for all cases. Then A and n would have the flat-plate values for laminar and turbulent flows. Accordingly, substitute Eq. (19P-2) into the general Eq. (19P-1) to obtain

$$\frac{U\delta_t^*}{\nu} = \left(\frac{A}{n\,\mathrm{Pr}}\right)\frac{\left[\int_0^x U(T_w - T_\infty)^{1/n}\,dx\right]^n}{\nu^n(T_w - T_\infty)}$$

(e) Substitute the result of part (d) into Eq. (19P-2) to obtain

$$\frac{\mathrm{Nu}_x}{\mathrm{Re}_x\,\mathrm{Pr}} = \frac{A}{\mathrm{Pr}}\frac{\left\{\int_0^x\left[U(T_w - T_\infty)^{1/n}/\nu\right]dx\right\}^{n-1}}{(T_w - T_\infty)^{1-1/n}}$$

(f) Show that for laminar boundary layers

$$\mathrm{Nu}_x = \frac{0.33206\,\mathrm{Pr}^{1/3}\,\mathrm{Re}_x(T_w - T_\infty)}{\left\{\int_0^x\left[U(T_w - T_\infty)^2/\nu\right]dx\right\}^{1/2}}$$

and for turbulent boundary layers

$$\mathrm{Nu}_x = \frac{0.0291\,\mathrm{Pr}^{0.6}\,\mathrm{Re}_x(T_w - T_\infty)^{1/4}}{\left\{\int_0^x\left[U(T_w - T_\infty)^{5/4}/\nu\right]dx\right\}^{1/5}}$$

See Kays and Crawford [80] for extension to high-speed flows with transpiration and Sucec [81] for extension to arbitrary, specified wall heat flux. For rocket nozzles, the review by Bartz [82] can be consulted.

8-20 A constant-temperature body immersed in a flowing stream has a free-stream velocity variation given by $U = U_0 + cx^{1/3}$. Use the result of Problem 8-19, part g to show that

$$\mathrm{Nu}_x = 0.332\,\mathrm{Pr}^{1/3}\,\mathrm{Re}_x^{1/2}\left(\frac{U_0 + cx^{1/3}}{U_0 + 3cx^{1/3}/4}\right)^{1/2}$$

for a laminar boundary layer (13.2% lower than the exact answer when $U_0 = 0$ [79]) while

$$\mathrm{Nu}_x = 0.0291\,\mathrm{Pr}^{0.6}\,\mathrm{Re}_x^{4/5}\left(\frac{U_0 + cx^{1/3}}{U_0 + 3cx^{1/3}/4}\right)^{1/5}$$

for a turbulent boundary layer. Comment on the effect of $c > 0$ and $c < 0$ on the local heat-transfer coefficient.

8-21 The velocity near the stagnation point of a constant temperature two-dimensional body immersed in a flowing stream varies as $U = cx$. Use the result of Problem 8-19, part g to show that

$$\mathrm{Nu}_x = 0.47\,\mathrm{Pr}^{1/3}\,\mathrm{Re}_x^{1/2}$$

for a laminar boundary layer (17.5% lower than the exact answer [80]) and

$$\mathrm{Nu}_x = 0.034\,\mathrm{Pr}^{0.6}\,\mathrm{Re}_x^{4/5}$$

for a turbulent boundary layer.

8-22 The leading edge of a gas turbine blade is a two-dimensional stagnation region with the external velocity given by $U = (8000\ \mathrm{s}^{-1})x$ with x in meters. Gas property values are $\mathrm{Pr} = 0.69$, $\nu = 7.7 \times 10^{-5}\ \mathrm{m}^2/\mathrm{s}$, and $k = 0.059\ \mathrm{W/m\ K}$. Use the result of Problem 8-21 to calculate the numerical values of the local heat-transfer coefficient for (a) turbulent and (b) laminar boundary layers. *Answer*: $h_{\mathrm{lam}} = 250\ \mathrm{W/m}^2$ K, $h_{\mathrm{turb}} = 4200(x/m)^{0.6}\ \mathrm{W/m}^2$ K.

8-23 In the turbulent flow downstream from a grid, diffusion and production of turbulent kinetic energy k and dissipation ε are zero.

(a) Show that Eqs. (8-36c) and (8-36d) are then, respectively,

$$U\frac{dk}{dx} = -\varepsilon \quad \text{and} \quad U\frac{d\varepsilon}{dx} = -C_2\frac{\varepsilon^2}{k}$$

where U is the stream x velocity and x is the distance downstream from the grid.

(b) Show that the solutions to the equations of part a give

$$\frac{k}{k(x=0)} = (Cx + 1)^{-1/(C_2-!)}$$

8-24 (a) Obtain the "bridging" boundary conditions of Eq. (8-36e) for ε at location y_c from the assumption that local equilibrium, the buffer region $26 \leq y^+ \leq 100$ is an equilibrium layer, gives

$$\varepsilon = P\left(= -\overline{u'v'}\frac{\partial u}{\partial y} = \nu'\frac{\partial u}{\partial y}\frac{\partial u}{\partial y} = \frac{\tau'}{\varrho}\frac{\partial u}{\partial y} = v^{*2}\frac{\partial u}{\partial y}\right)$$

with $\partial u/\partial y$ obtained from the law of the wall Eq. (7-18).

(b) Incorporate the result of part a with Eq. (8-36a) for ν' to obtain the bridging condition of Eq. (8-36e) for k.

8-25 The relationship between C_1, C_2, C_μ, and K (the von Karman constant $= 0.41$) is to be derived. To do this, consider that near the wall in a turbulent boundary layer the logarithmic velocity distribution of Eq. (7-18) gives k and ε dependence on y given by Eq. (8-36e), production P equals dissipation ε, and convection of ε is negligible. Use these in the ε-transport equation [Eq. (8-36d)] together with Eq. (8-36a) relating ν' to k and ε to show that $C_1 = C_2 - K^2/\sigma_\varepsilon C_\mu^{1/2}$.

REFERENCES

1. R. M. O'Donnell, Experimental investigation at Mach-number of 2.41 of average skin friction coefficients and velocity profiles for laminar and turbulent boundary layers and assessment of probe effects, NACA TN 3122, 1954.

2. R. K. Lobb, E. M. Winkler, and J. Persh, Experimental investigation of turbulent boundary layers in hypersonic flow, NAVORD Report 3880, 1955.

3. H. Schlichting, *Boundary-Layer Theory*, McGraw-Hill, 1968.

4. T. C. Tai, *AIAA J.* **24**, 370–376 (1986).

5. J. O. Hinze, *Turbulence*, 2nd ed., McGraw-Hill, 1975.

6. G. A. Hughmark, *AIChE J.* **17**, 902–909 (1971).

7. T. Cebeci, *ASME J. Heat Transfer* **95**, 227–234 (1973).

8. E. R. G. Eckert and R. M. Drake, *Analysis of Heat and Mass Transfer*, McGraw-Hill, 1972, p. 385.

9. U. Műeller, K. G. Rosener, and B. Schmidt, Eds., *Recent Developments in Theoretical and Experimental Fluid Mechanics*, Springer, 1979.

10. M. J. Callaghan and D. M. Mason, *Chem. Eng. Sci.* **19**, 763–774 (1964).

11. P. A. Antonia and J. Kim, *Int. J. Heat Mass Transfer* **34**, 1905–1908 (1991).

12. Th. von Karman, *Trans. ASME* **61**, 705–710 (1939).

13. H. Reichardt, *Reports of the Max-Planck-Institute für Strömungsforschung*, No. 3, 1950, pp. 1–63.

14. E. R. Van Driest, *Fifty Years of Boundary Layer Research*, Braunschweig, 1955, pp. 257–271.

15. J. C. Rotta, *Int. J. Heat Mass Transfer* **7**, 215–228 (1964).

16. L. Thomas, *ASME J. Heat Transfer* **100**, 744–746 (1978).

17. P. M. Gerhart, *AIAA J.* **13**, 966–968 (1975).

18. E. R. Kenner and T. E. Polk, *AIAA J.* **10**, 845–846 (1972).

19. E. J. Hopkins and M. Inouye, *AIAA J.* **9**, 993–1003 (1971).

20. D. B. Spalding, Heat transfer to a turbulent stream from a surface with a stepwise discontinuity in wall temperature, *International Developments in Heat Transfer* (Proceedings of the 1961–1962 Heat Transfer Conference, Boulder, CO, 1961), Part II, pp. 439–446.

21. G. O. Gardner and J. Kestin, *Int. J. Heat Mass Transfer* **6**, 289–299 (1963).

22. J. Kestin and P. D. Richardson, *Int. J. Heat Mass Transfer* **6**, 147–189 (1963).

23. R. J. Moffat, J. M. Healzer, and W. M. Kays, *ASME J. Heat Transfer* **100**, 134–142 (1978).

24. J. Kestin, *Adv. Heat Transfer* **3**, 1–32 (1966).

25. A. B. Mehendale, J. C. Han, and S. Ou, *ASME J. Heat Transfer* **113**, 843–850 (1991).

26. R. M. Traci and D. C. Wilcox, *AIAA J.* **13**, 890–896 (1975).

27. M. F. Blair, *ASME J. Heat Transfer* **105**, 33–47 (1983).

28. R. P. Taylor, P. H. Love, H. W. Coleman, and M. H. Hisni, *J. Thermophys. Heat Transfer* **4**, 121–123 (1990).

29. D. B. Spalding, Heat transfer to a turbulent stream from a surface with a stepwise discontinuity in wall temperature, *International Developments in Heat Transfer* (Proceedings of the 1961–1962 Heat Transfer Conference, Boulder, CO, 1961), Part II, pp. 439–446.

30. G. S. Ambrok, *Sov. Phys.-Tech. Phys.* **2**, 1979–1986 (1957).

31. P. M. Moretti and W. M. Kays, *Int. J. Heat Mass Transfer* **8**, 1187–1201 (1965).

32. J. Sucec and Y. Lu, *ASME J. Heat Transfer* **112**, 905–912 (1990).

33. S. S. Kutateladze and A. I. Leontiev, in A. E. Bergles, Ed., *Heat Transfer, Mass Transfer, and Friction in Turbulent Boundary Layers*, Hemisphere, 1990.

34. J. C. Evvard, M. Tucker, and W. C. Burgess, *J. Aeronaut. Sci.* **21**, 731–738 (1954); NACA TN 3100, 1954.

35. R. A. Seban and D. L. Doughty, *Trans. ASME* **78**, 217–223 (1956).

36. B. K. Hodge, R. P. Taylor, and H. W. Coleman, *AIAA J.* **24**, 1560–1561 (1986).

37. B. T. F. Chung and L. C. Thomas, *ASME J. Heat Transfer* **95**, 562–564 (1973).

38. T. Cebeci and P. Bradshaw, *Physical and Computational Aspects of Convective Heat Transfer*, Springer-Verlag, 1984.

39. P. S. Klebanoff, Characteristics of turbulence in a boundary layer with zero pressure gradient, NACA TN 3178, 1954.

40. S. Corrsin and A. L. Kistler, The free-stream boundaries of turbulent flows, NACA TN 3133, 1954.

41. G. Maise and H. McDonald, *AIAA J.* **6**, 73–80 (1968).

42. T. Cebeci, A. M. O. Smith, and G. Mosinskis, *ASME J. Heat Transfer* **92**, 133–143 (1970).

43. T. Cebeci, *AIAA J.* **9**, 1091–1097 (1971).

44. R. H. Pletcher, On a solution for turbulent boundary layer flows with heat transfer, pressure gradients, and wall blowing and suction, *Heat Transfer 1970, Proceedings Fourth International Heat Transfer Conference*, Vol. 2, Elsevier, 1970, Paper FC 2.9.

45. R. H. Pletcher, *AIAA J.* **10**, 245–246 (1972).

46. N. Bagheri, C. J. Strataridakis, and B. R. White, *AIAA J.* **30**, 33–43 (1992).

47. G. P. Hammond, *AIAA J.* **23**, 1668–1669 (1985).

48. R. H. Pletcher, *AIAA J.* **7**, 305–311 (1969).

49. T. Cebeci, *Handbook of Numerical Heat Transfer*, Wiley, 1988, pp. 117–154.

50. H. B. Keller and T. Cebeci, *AIAA J.* **10**, 1191–1199 (1970).

51. W. C. Reynolds, W. M. Kays, and S. J. Kline, Heat transfer in the turbulent incompressible boundary layer—III: arbitrary wall temperature and heat flux, NASA Memo 12-3-58W, December 1958.

52. T. Cebeci and P. Bradshaw, *Momentum Transfer in Boundary Layers*, Hemisphere, 1977.

53. P. Bradshaw, T. Cebeci, and J. Whitelaw, *Engineering Calculational Methods for Turbulent Flow*, Academic, 1981.

54. S. V. Patankar and D. B. Spalding, *Heat and Mass Transfer in Boundary Layers*, 2nd ed., Intertext, 1970.

55. A. N. Kolmogorov, *Izv. Akad. Nauk. SSR Ser. Fiz. Vi.* (1–2), 1942, 56–58 (1942), English translation Imperial College, Mechanical Engineering Department, Report ON/6, 1968.

56. L. Prandtl, Über ein neues Formel-system für die ausgebildete Turbulenz, *Nach. Akad. Wiss. Göttingen, Math.-Phys. Klasse*, 1945, p. 6.

57. D. C. Wilcox and M. W. Rubesin, Progress in turbulence modelling for complex flowfields, NASA TP 1517, 1980.

58. F. Pourahmadi and J. A. C. Humphrey, *AIAA J.* **21**, 1365–1371 (1983).

59. M. Champion and P. A. Libby, *AAIA J.* **28**, 1525–1526 (1990).

60. V. C. Patel, W. Rodi, and G. Scheuerer, *AIAA J.* **23**, 1308–1319 (1985).

61. R. J. Yang and W. Aung, *Natural Convection*, Hemisphere, 1985, pp. 259–300.

62. W. Rodi, *Turbulence Models and Their Application in Hydraulics*, 2nd ed., Book Publication of the International Association for Hydraulic Research, Delft, The Netherlands, 1984.

63. E. W. Miner, T. F. Swean, R. A. Handler, and R. I. Leighton, Evaluation of wall-damping models by comparison with direct simulations of turbulent channel flow, *Numerical Methods in Laminar and Turbulent Flow* (Proc. Seventh Int. Conf., Swansea, July 11–15, 1989), Vol. 6, Part 1, pp. 273–284.

64. C. G. Speziale, R. Abid, and E. C. Anderson, *AIAA J.* **30**, 324–331, 1145 (1992); R. M. C. So, H. S. Zhang, and C. G. Speziale, *AIAA J.* **29**, 2069–2076 (1991); S. Thangam, R. Abid, and C. G. Speziale, *AIAA J.* **30**, 552–554 (1992).

65. R. M. C. So, Y. G. Lai, H. S. Zhang, and B. C. Hwang, *AIAA J.* **29**, 1819–1835 (1991).

66. R. M. S. So, Y. G. Lai, H. S. Zhang, and B. C. Hwang, *AIAA J.* **29**, 1202–1213 (1991).

67. W. Rodi, *AIAA J.* **20**, 872–879 (1982).

68. G. D. Raithby and G. E. Schneider, in W. J. Minkowycz, E. M. Sparrow, G. E. Schneider, and R. H. Pletcher, Eds., *Handbook of Numerical Heat Transfer*, Wiley, 1988, p. 283.

69. C. G. Speziale, *J. Fluid Mech.* **178**, 459–475, 1987.

70. P. K. Bishnoi, U. Ghia, and K. N. Ghia, Prediction of normal Reynolds stresses with nonlinear k–ε model of turbulence, *Advances in Numerical Simulation of Turbulent Flows* (Proc. First ASME/JSME Fluids Engineering Conf., Portland, Oregon, June 23–27, 1991), 1991, ASME FED-Vol. 117, pp. 25–34.

71. S. Thangam and C. G. Speziale, *AIAA J.* **30**, 1314–1320 (1992).

72. Keller and A. Friedmann, Differentialgleichungen für die turbulente Bewegung einer kompressibeln-Flussigkeiten, *Proc. First Int. Cong. Appl. Mech.*, Delft, 1924, pp. 395–405.

73. P. Y. Chou, *Quart. J. Appl. Math.* **3** (1), 38–45 (1945).

74. D. J. Bergstrom, Modelling the buoyancy production of dissipation in a plane turbulent plume, *Advances in Numerical Simulation of turbulent flows* (Proc. First ASME/JSME Fluids Engineering Conf., Portland, Oregon, June 23–27, 1991), ASME FED-Vol. 117, 1991, pp. 89–96.

75. W. J. Devenport and R. L. Simpson, *AIAA J.* **30**, 873–881 (1992).

76. B. Hutchings and R. Iannuzzelli, *Mechanical Engineering* **109**, 72–76 (1987).

77. J. Cousteix and R. Houdeville, in S. Klien and N. Afgan, Eds., *Near-Wall Turbulence*, Hemisphere, 1988, pp. 782–799.

78. B. E. Launder and D. B. Spalding, *Mathematical Models of Turbulence*, Academic, 1972.

79. C. B. Millikan, *Trans. ASME J. Appl. Mech.* **54**, 29–43 (1932); Paper PAM-54-3.

80. W. M. Kays and M. E. Crawford, *Convective Heat and Mass Transfer*, McGraw-Hill, 1980, p. 139.

81. J. Sucec, *Int. J. Heat Mass Transfer* **32**, 1189–1192 (1989).

82. D. R. Bartz, *Adv. Heat Transfer* **2**, 1–108 (1965).

9

TURBULENT FLOW IN DUCTS

Much of the early understanding of turbulent flow came from experimental study of flow in circular pipes as discussed in Chapter 7. These conclusions were then applied to exterior turbulent boundary layers in Chapter 8.

Many applications require information concerning turbulent flow and heat transfer for flow internal to a duct. No new essential difficulty is encountered in such a flow, because the basic information came from it.

The final theory developed can be considered to be a correlation of a large body of measurements. Its utility is to impart a general understanding of turbulent transport phenomena and allow new areas to be explored by numerical simulation without extensive and expensive experimentation as is reported by Pollard and Martinuzzi [1] in a comparison of turbulence models for prediction of turbulent pipe flow.

9.1 FULLY DEVELOPED PROFILE IN CIRCULAR DUCT

As discussed in Chapter 7, measured velocity profiles for fully developed turbulent flow in smooth circular ducts are well fitted in the central region by

$$u^+ = C(y^+)^{1/n} \tag{7-19}$$

with C and n somewhat dependent on the Reynolds number $\mathrm{Re}_D = U_{av}D/\nu$, where D is the duct diameter. As before, y is distance from the wall whereas $u^+ = \bar{u}/v^*$, $y^+ = yv^*/\nu$, and $v^* = (\tau_w/\rho)^{1/2}$. As shown in Table 7-2 $C = 8.74$ and $n = 7$ for the typical case of $\mathrm{Re} \approx 10^5$.

In the buffer zone, the velocity profile is approximated by

$$u^+ = 5 + 5\ln\left(\frac{y^+}{5}\right), \qquad 5 \leqslant y^+ \leqslant 30 \tag{7-17}$$

while, nearer the wall in the laminar sublayer,

$$u^+ = y^+, \qquad 0 \leqslant y^+ \leqslant 5 \tag{7-16}$$

The shear stress is nearly constant in the region near the wall ($5 \leqslant y^+ \leqslant 30$), but in the central region to which the $1/n$th power-law velocity profile applies the shear stress varies linearly as

$$\tau = \tau_w\left(1 - \frac{y}{r_0}\right)$$

For prediction of temperature profiles, and the heat-transfer coefficient that follows from them, it will be necessary to know the ratio of eddy diffusivity of momentum to kinematic viscosity ε_m/ν. Knowledge of the velocity profile and the shear stress variation is sufficient to yield ε_m/ν since

$$\tau = (\varepsilon_m + \nu)\frac{du}{dy}$$

Accounting for the linear variation of τ and employing dimensionless quantities allows this equation to be expressed as

$$\frac{\varepsilon_m}{\nu} = -1 + \frac{(1 - y^+/r_0^+)}{du^+/dy^+} \tag{9-1}$$

Substantial error can be incurred by employing velocity profiles that, although they give a good fit on the average to velocity over a range of y, might give a poor fit to velocity gradient. For the laminar sublayer, the linear velocity profile of Eq. (7-16) used in Eq. (9-1) results in

$$\frac{\varepsilon_m}{\nu} \approx 0, \qquad 0 \leqslant y^+ \leqslant 5 \tag{9-2}$$

This is a reasonable result since molecular effects should outweigh turbulent effects there.

However, the laminar sublayer is not entirely free of turbulent effects as shown in Fig. 7-13. A refined variation that smoothly merges with the result due to Granville [2] for the buffer zone is

$$\frac{\varepsilon_m}{\nu} = Ky^+\left\{1 - \exp\left[-\left(\frac{y^+(1 + ap^+)}{24}\right)^2\right]\right\}\left[\frac{\tau}{\tau_w}\right]$$

where $K = 0.41$ is the von Karman constant, $p^+ = (v\, dp/dx)/\varrho v^{*3}$ is a dimensionless pressure gradient, and $a = 14.5$ for $p^+ > 0$ and 18 for $p^+ < 0$ fits the laminar-sublayer variation of ε_m/v as y^3 demonstrated by Chapman and Kuhn [3].

For the buffer zone, the logarithmic velocity profile of Eq. (7-17) used in Eq. (9-1) yields

$$\frac{\varepsilon_m}{v} \approx \frac{y^+ - 5}{5}, \qquad 5 \leq y^+ \leq 30 \qquad (9\text{-}3)$$

This is also a reasonable result since at the laminar sublayer–buffer zone interface there is a match. For the fully turbulent core of the pipe, the $\frac{1}{7}$th power-law velocity profile of Eq. (7-19) results in

$$\frac{\varepsilon_m}{v} = -1 + 0.8\left(1 - \frac{y}{r_0}\right)(y^+)^{6/7} \qquad (9\text{-}4)$$

In comparison, the more exact velocity profile of Eq. (7-20a) due to Reichardt for the central region of the pipe leads to Eq. (7-20).

The predicted variation of ε_m/v is displayed in Fig. 9-1, where the mismatch at the outer edge of the buffer zone is evident. The desire for

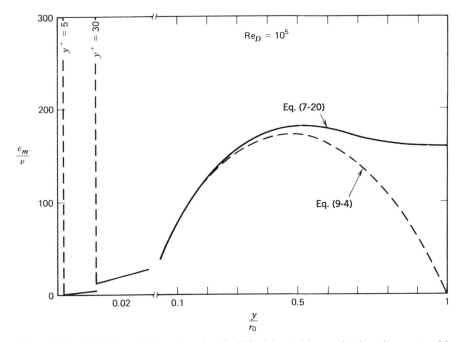

Figure 9-1   Variation of the ratio of eddy diffusivity to kinematic viscosity ε_m/v with distance from the wall in a circular duct.

closed-form solutions motivates the discontinuous representations used here. Most striking is the erroneous prediction from the simpler velocity profiles that ε_m/ν falls to low values near the pipe axis. The more accurate prediction of Reichardt from Eq. (7-20) for ε_m/ν to be large in the central region testifies to the vigorous mixing in turbulence. Since the effective turbulent viscosity is about two orders of magnitude greater than the molecular value, the central core nearly behaves as a solid slug sliding down the pipe on a thin lubricating film whose thickness is about that of the laminar sublayer.

9-2   FULLY DEVELOPED TEMPERATURE PROFILE IN CIRCULAR DUCT FOR CONSTANT HEAT FLUX

The fully developed temperature profile is desired for a circular duct subjected to a constant wall heat flux q_w. From this temperature distribution the local heat-transfer coefficient can be determined as it was for laminar flow in Chapter 4, except that the effective viscosity and thermal conductivity are now space dependent. This situation is important because it yields the turbulent heat-transfer coefficient for the long ducts that commonly occur and because the results are close to those for the constant wall temperature case.

The energy equation in cylindrical coordinates is

$$\rho C_p u \frac{\partial T}{\partial z} = -\frac{1}{r} \frac{\partial (r q_r)}{\partial r}$$

subject to

$$\frac{\partial T(r = r_0, z)}{\partial r} = -\frac{q_w}{k}$$

$$\frac{\partial T(r = 0, z)}{\partial r} = 0 \quad \text{or} \quad q_r(r = 0, z) = 0 \tag{9-5}$$

Here the z direction is down the pipe and interest is confined to regions far removed from the start of the heated region and the pipe inlet. In this fully developed region the temperature must increase linearly in the z direction as discussed in Chapter 4; according to Eq. (4-37c), $\partial T/\partial z = dT_{\text{mixing cup}}/dz =$ const.

Integration of Eq. (9-5) with respect to y, where $y = r_0 - r$, gives

$$C_1 + \rho C_p \frac{dT_m}{dz} \int_0^y u(r_0 - y)\, dy = (r_0 - y) q_r$$

The constant of integration is evaluated from the condition at $y = r_0$ to give

$$-\rho C_p \frac{dT_m}{dz} \int_y^{r_0} u(r_0 - y)\, dy = (r_0 - y) q_r \tag{9-6}$$

The radial heat flux is related to temperature gradient by

$$q_r = -\rho C_p (\varepsilon_H + \alpha) \frac{\partial T}{\partial r}$$

$$= \rho C_p \left(\frac{\varepsilon_m}{\mathrm{Pr}_t} + \frac{\nu}{\mathrm{Pr}} \right) \frac{\partial T}{\partial y}$$

Insertion of this expression into Eq. (9-6) gives, with the realization that $\rho C_p\, dT_m/dz = -2q_w/r_0 U_{av}$ where q_w is positive if leaving the fluid,

$$\frac{dT}{dy} = \frac{2q_w}{\rho C_p r_0} \frac{\int_y^{r_0} (u/U_{av})(r_0 - y)\, dy}{(\varepsilon_m/\mathrm{Pr}_t + \nu/\mathrm{Pr})(r_0 - y)}$$

One further integration gives, where T_w is the yet unknown wall temperature, $v^* = (\tau_w/\rho)^{1/2}$, $T^+ = (T - T_w)(\rho C_p v^*/q_w)$, $y^+ = yv^*/\nu$, and $u^+ = u/v^*$,

$$T^+ = 2 \int_0^{y^+} \left[\frac{\int_{y/r_0}^1 (u/U_{av})(1 - y'/r_0)\, d(y'/r_0)}{[(\varepsilon_m/\nu)/\mathrm{Pr}_t + 1/\mathrm{Pr}](1 - y/r_0)} \right] dy^+ \tag{9-7}$$

Integration is simplified by the assumption that $u/U_{av} \approx 1$. Then

$$T^+ = \int_0^{y^+} \frac{(1 - y^+/r_0^+)\, dy^+}{(\varepsilon_m/\nu)/\mathrm{Pr}_t + 1/\mathrm{Pr}} \tag{9-8}$$

The temperature distribution will be obtained by integration of Eq. (9-8) in three parts, considering Pr_t to be constant. Then

$$\Delta T^+ = \Delta T^+_{\text{sublayer}} + \Delta T^+_{\text{buffer zone}} + \Delta T^+_{\text{turbulent core}}$$

The laminar sublayer is considered first. Here $\varepsilon_m/\nu \approx 0$ from Eq. (9-2) and $1 - y/r_0 \approx 1$ so that

$$\Delta T^+_{\text{sublayer}} = \int_0^5 \mathrm{Pr}\, dy^+ = 5\, \mathrm{Pr}$$

In the buffer zone $\varepsilon_m/\nu = (y^+ - 5)/5$ from Eq. (9-3) and $1 - y/r_0 \approx 1$ so that

$$\Delta T^+_{\text{buffer}} = \int_5^{30} \frac{5\,\text{Pr}_t\,dy^+}{y^+ - 5(1 - \text{Pr}_t/\text{Pr})} = 5\,\text{Pr}_t\,\ln\left(1 + \frac{5\,\text{Pr}}{\text{Pr}_t}\right)$$

The turbulent core has $\varepsilon_m/\nu = (1 - y/r_0)y^+/2.5$ from Eq. (7-20a) evaluated at $r \approx r_0$ so that

$$\Delta T^+_{\text{core}} = 2.5\,\text{Pr}_t\int_{30}^{r_0^+} \frac{(1 - y^+/r_0^+)\,dy^+}{(1 - y^+/r_0^+)y^+ + 2.5\,\text{Pr}_t/\text{Pr}} \approx 2.5\,\text{Pr}_t\,\ln\left(\frac{r_0^+}{30}\right)$$

Neglect of $1/\text{Pr}$ makes the result inaccurate for liquid metals where $\text{Pr} \ll 1$. Realizing that

$$r_0^+ = \frac{r_0 \upsilon^*}{\nu} = \frac{r_0}{\nu}\left(\frac{\tau_w}{\rho}\right)^{1/2} = \frac{r_0}{\nu}\left(\frac{C_f}{2}U_{\text{av}}^2\right)^{1/2} = \frac{\text{Re}_D}{2}\left(\frac{C_f}{2}\right)^{1/2}$$

and adding the three temperature drops yields the temperature difference between the center line and the wall as

$$T_{\text{CL}} - T_w = \frac{q_w}{\rho C_p U_{\text{av}}(C_f/2)^{1/2}}\left\{5\,\text{Pr} + 5\,\text{Pr}_t\,\ln(1 + 5\,\text{Pr}/\text{Pr}_t)\right.$$

$$\left. + 2.5\,\text{Pr}_t\,\ln\left[\frac{\text{Re}_D(C_f/2)^{1/2}}{60}\right]\right\} \quad (9\text{-}9)$$

in which it is conventional to take $\text{Pr}_t = 1$.

From Eq. (9-9) the heat-transfer coefficient can be determined since

$$q_w = h(T_m - T_w)$$

Rearrangement gives

$$\text{Nu}_D = \frac{hD}{k} = \frac{q_w D}{k(T_{\text{CL}} - T_w)}\frac{T_{\text{CL}} - T_w}{T_m - T_w} \quad (9\text{-}10)$$

The first term on the right-hand side of Eq. (9-10) is available from Eq. (9-9), but the second term remains to be calculated. The velocity profile is represented by $u/U_{\text{CL}} = (y/r_0)^{1/7}$. For $\text{Pr} \approx 1$, the temperature distribution must be of the same form so that

$$\frac{T - T_w}{T_{CL} - T_w} = \left(\frac{y}{r_0}\right)^{1/7}$$

The $\frac{1}{7}$th power-law profiles used in the mixing-cup temperature formula give

$$\frac{T_m - T_w}{T_{CL} - T_w} = \frac{2\int_0^{r_0}(u/U_{CL})[(T - T_w)/(T_{CL} - T_w)]r\,dr}{2\int_0^{r_0}(u/U_{CL})r\,dr} = \frac{5}{6} \quad (9\text{-}11)$$

Substitution of Eqs. (9-9) and (9-11) into Eq. (9-10) gives, with $Pr_t = 1$,

$$\frac{Nu_D}{Re_D\,Pr} = \frac{6(C_f/2)^{1/2}}{25\left\{Pr + \ln(1 + 5\,Pr) + 0.5\ln\left[Re_D(C_f/2)^{1/2}/60\right]\right\}} \quad (9\text{-}12)$$

Equation (9-12) follows the essential developments of Martinelli [4] and Boelter [5], who included the Reynolds number dependence of the mixing-cup temperature and allowed for application to liquid metals ($Pr \ll 1$) by including molecular effects in the turbulent core. A form of the Reynolds analogy for constant wall heat flux in the fully developed flow region of a smooth round duct developed by Petukhov and Popov [6, 7] is accurate within 15% for $0.7 < Pr < 50$ according to Webb [8]. Essentially, they used Reichardt's velocity and eddy diffusivity distributions to achieve (for rectangular ducts, too [9])

$$\frac{Nu_D}{Re_D\,Pr} = \frac{C_f/2}{1.07 + 12.7(Pr^{2/3} - 1)(C_f/2)^{1/2}} \quad (9\text{-}13)$$

where the friction factor $f = -(2D/\rho U_{av}^2)\,dp/dx$ is calculated from the Prandtl–Karman equation [Eq. (7-21b)] which applies only to long smooth circular ducts—$f = 4C_f$.

As outlined by Webb [8], a variety of correlations of the general form

$$\frac{Nu_D}{Re_D\,Pr} = \frac{C_f/2}{C_1 + C_2F(Pr)(C_f/2)^{1/2}}$$

have been proposed with varying $C_{1,2}$ values and functions of Pr, $F(Pr)$; usually $C_1 \approx 1$ and $F(Pr = 1) = 0$.

It is instructive to examine the Reynolds analogy of Eq. (9-12). First, note that in turbulent flow Nu depends on Re and Pr separately in contrast to its constant value for laminar flow. Second, note the temperature distributions in turbulent flow displayed in Fig. 9-2. At large Pr, such as characterize water

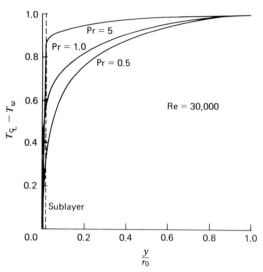

Figure 9-2 Effect of Prandtl number on temperature distribution in turbulent flow.

and oils, for instance, the temperature variation occurs mostly in the laminar sublayer and buffer zone, where molecular effects are important. At small Pr, such as characterize liquid metals, the temperature variation occurs in the turbulent core to a major extent as a result of the large thermal conductivity that makes molecular effects important even in the central region of the pipe. Molecular effects predominate everywhere when Pr is small; not only is the temperature profile similar to that for laminar flow, but the thermal entry length and the response to changes in wall temperature are also similar to the laminar case. At large Pr, the thermal entry length is short and there is rapid response to change in wall temperature, a result of rapid diffusion of heat through the fluid once it has penetrated the sublayers where the major thermal resistance occurs.

An analysis by Deissler [10] has several important features, including applicability to large Prandtl numbers. Deissler assumed a turbulent Prandtl number Pr_t of unity and utilized the eddy diffusivity of Eq. (7-14) near the wall to obtain

$$u^+ = \int_0^{y^+} \frac{dy^+}{1 + n^2 u^+ y^+ \left[1 - \exp(-n^2 u^+ y^+)\right]}, \qquad y^+ \leqslant 26 \quad (7\text{-}14)$$

$$T^+ = \int_0^{y^+} \frac{dy^+}{1/Pr + n^2 u^+ y^+ \left[1 - \exp(-n^2 u^+ y^+)\right]}, \qquad y^+ \leqslant 26 \quad (9\text{-}14)$$

Far from the wall ($y^+ \geqslant 26$), he used

$$u^+ = 13 + \frac{1}{0.36} \ln\left(\frac{y^+}{26}\right) \tag{9-15}$$

to fit velocity data. Because molecular effects are negligible far ($y^+ \geqslant 26$) from the wall at large Pr, so that velocity and temperature profiles are the same, the same relationship was used for T^+. Numerical integration was used to solve Eqs. (7-15), (9-14), and (9-15), and the Nusselt number results closely agree with measurements for $0.7 \leqslant$ Pr. Of interest here is the case where Pr $\to \infty$. Setting $u^+ = y^+$, since the temperature drop will occur in the laminar sublayer at large Pr, and retaining only the first two terms of the denominator series expansion gives Eq. (9-14) as

$$T^+ = \int_0^{y^+} \frac{dy^+}{1/\mathrm{Pr} + n^4(y^+)^4}$$

Evaluation of the integral at $y^+ = \infty$ gives

$$T^+(y^+ = \infty) = \frac{\pi}{2^{3/2}n} \mathrm{Pr}^{3/4}$$

where $n = 0.124$. Now T^+ ($y^+ = \infty$) is essentially $T^+_{\text{mixing cup}}$ because T^+ is nearly constant when Pr $\to \infty$. Recognizing that

$$\mathrm{Nu}_D = \frac{2r_0^+ \, \mathrm{Pr}}{T_m^+}$$

one then finds that

$$\lim_{\mathrm{Pr} \to \infty} \mathrm{Nu}_D = \frac{32^{1/2}}{\pi} n r_0^+ \, \mathrm{Pr}^{1/4} = \frac{2^{3/2}n}{\pi} \mathrm{Re}_D \, \mathrm{Pr}^{1/4}\left(\frac{C_f}{2}\right)^{1/2}$$

Here $C_f = 2\tau_w/\rho U_{\text{av}}^2$. Deissler also explored entrance region and temperature-dependent viscosity effects. A similar study was executed by Sparrow et al. [11], who obtained good agreement with Deissler's work.

Subsequently, an analysis was performed by Kays and Leung [12] which is applicable for Prandtl numbers ranging from the low values of liquid metals to the high values of oils. They used Deissler's expression [Eq. (7-14)] for the eddy diffusivity of momentum near the wall and took the turbulent Prandtl number to be unity there. In the turbulent core, Reichardt's Eq. (7-20) was used for the eddy diffusivity of momentum; the turbulent Prandtl number was not taken to be unity there. Instead, an idea suggested by Jenkins [13] and

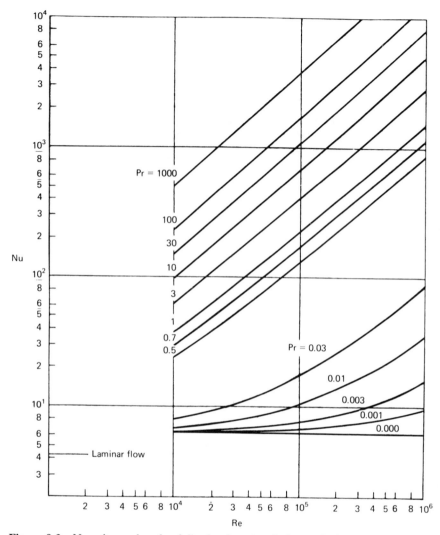

Figure 9-3   Nusselt number for fully developed turbulent velocity and temperature profiles in a circular duct with constant wall heat flux. [From W. M. Kays and E. Y. Leung, *Int. J. Heat Mass Transfer* **6**, 537–557 (1963) [12].]

refined by Deissler [14] (cogently discussed by Jakob [15]) was adopted in a modified form to ascertain the dependence of Pr_t on Pr. Jenkins, as discussed by Hinze [16] in connection with an early version of the idea due to Burgers [17], postulated that an eddy can be thought of as a solid sphere that starts its flight with a uniform temperature and exchanges heat with its environment along the way. The results shown in Fig. 9-3 were obtained by numerical procedures. The slopes of the curves in Fig. 9-3 are nearly equal on a log–log

TABLE 9-1   Circumferential Heat Flux Functions S_1 for Fully Developed Turbulent Flow in a Circular Tube with Constant Heat Rate [18]

Pr \ Re	Laminar	10^4	3×10^4	10^5	3×10^5	10^6
0	1.000	1.000	1.000	1.000	1.000	1.000
0.001		1.000	1.000	0.999	0.974	0.901
0.003		0.999	0.994	0.957	0.831	0.473
0.01		0.991	0.952	0.733	0.409	0.161
0.03		0.923	0.699	0.348	0.145	0.0535
0.7		0.121	0.0490	0.0180	0.00721	0.00275
3		0.0448	0.0178	0.00629	0.00246	0.000902
10		0.0239	0.00931	0.00322	0.00123	0.000438
30		0.0151	0.00582	0.00199	0.00751	0.000166
100		0.00994	0.00383	0.00130	0.000486	0.000166
1000		0.00513	0.00198	0.000667	0.000248	0.0000841

plot for $Pr \geq 1$, justifying the conventional correlation of turbulent data by $Nu_D = C\,Re_D^a\,Pr^b$. The same result does not hold for $Pr \ll 1$, where the slopes change considerably; there a correlation of the form $Nu_D = C_1 + C_2(Re_D\,Pr)^n$ is successful.

Because ε_H and ε_m are little affected by thermal boundary conditions, the ε_H and ε_m obtained for the constant-wall-heat-flux case can be applied to calculate the effects of other thermal boundary conditions. Sometimes the heat flux is nonuniform around the circular duct, as in a solar collector coolant tube. General solutions have been developed by Reynolds [18] of which only those applicable to a cosine variation, $q_w(\phi) = q_{w_0}(1 + b\cos\phi)$, are presented here. For this case the local Nusselt number is given by

$$\frac{Nu_D(\phi)}{Nu_D} = \frac{1 + b\cos\phi}{1 + (S_1 b\,Nu_D/2)\cos\phi}$$

The S_1 parameter is given in Table 9-1 as a function of Reynolds and Prandtl number, and Nu_D is the Nusselt number for a uniform peripheral temperature from Fig. 9-3.

9.3   FULLY DEVELOPED NUSSELT NUMBER FOR OTHER GEOMETRIES WITH SPECIFIED WALL HEAT FLUX

Nusselt numbers for fully developed turbulent flow in annular ducts of constant wall heat flux calculated by Kays and Leung [12] agree with measurements for air, $Pr \approx 1$, and are believed to be accurate for $Pr \gg 1$ and $Pr \ll 1$. Results for the limiting case of parallel planes, $r_i/r_0 = 1$, are

TABLE 9-2   Nusselt Numbers for Fully Developed Turbulent Flow Between Parallel Planes; Constant Heat Rate; One Side Heated and the Other Side Insulated [12]

Re Pr	Laminar	10^4	3×10^4	10^5	3×10^5	10^6
0.0	5.385	5.70	5.78	5.80	5.80	5.80
0.001		5.70	5.78	5.80	5.88	6.23
0.003		5.70	5.80	5.90	6.32	8.62
0.01		5.80	5.92	6.70	9.80	21.5
0.03		6.10	6.90	11.0	23.0	61.2
0.5		22.5	47.8	120	290	780
0.7		27.8	61.2	155	378	1,030
1.0		35.0	76.8	197	486	1,340
3.0		60.8	142	380	966	2,700
10.0		101	214	680	1,760	5,080
30.0		147	367	1,030	2,720	8,000
100.0		210	514	1,520	4,030	12,000
1000.0		390	997	2,880	7,650	23,000

shown in Table 9-2. The parallel-plane results give good estimates of Nu for annuli of $r^* > 0.5$. Sutherland and Kays [20] ascertained the effect of an arbitrary variation of heat flux around the periphery of either surface. The characteristic length for the Nusselt and Reynolds numbers is the hydraulic diameter in Tables 9-2–9-5.

The effect of eccentricity of tube center lines in an annulus was studied analytically by Deissler and Taylor [21], who also give the friction factor for eccentric tubes, and experimentally by Leung et al. [22]. The experimental results in Table 9-3, in which the peripheral conduction through duct walls

TABLE 9-3   Effect of Eccentricity on Turbulent-Flow Heat Transfer in Circular-Tube Annuli (Experimental Data [20])[a]

Radius Ratio	e^*	$\dfrac{Nu_{ii,\max}}{Nu_{ii,\mathrm{conc}}}$	$\dfrac{Nu_{ii,\min}}{Nu_{ii,\mathrm{conc}}}$	$\dfrac{Nu_{00,\max}}{Nu_{00,\mathrm{conc}}}$	$\dfrac{Nu_{00,\min}}{Nu_{00,\mathrm{conc}}}$
0.255	0.27	0.99	0.97	1.02	0.93
	0.50	0.94	0.92	0.98	0.86
	0.77	0.92	0.88	0.93	0.77
0.500	0.54	0.96	0.87	1.01	0.78
	0.77	0.87	0.67	0.88	0.62

Source: By permission from W. M. Kays and M. E. Crawford, *Convective Heat and Mass Transfer*, McGraw-Hill, 1980 [19].

[a]*Notes*: $e^* = e/(r_0 - r_i)$ where e is the eccentricity of the tube center lines. $Nu_{ii,\mathrm{conc}}$ and $Nu_{00,\mathrm{conc}}$ refer to the Nusselt numbers for the concentric annulus at the same Reynolds number and Prandtl number.

that has a smoothing effect on Nu was made negligible, suggest that eccentricity decreases the average Nusselt number.

The effect of peripheral conduction through the wall was studied by Baughn [23].

Generally speaking, fully developed turbulent heat-transfer coefficients in noncircular ducts are accurately predicted by use of the hydraulic diameter D_h as the characteristic dimension in the circular duct predictive equations just as for pressure drop calculations. This approach is accurate for $\text{Pr} \geq 0.5$ (it can break down in such geometries as a triangle where the corners can be filled with laminar flow even though the central portion is filled with turbulent flow) but is inaccurate at low Prandtl numbers. As pointed out in Section 9.2, the physical basis for this Prandtl number effect is that at high Pr the major thermal resistance occurs in the thin sublayers next to the wall, which are little affected by the duct shape. At low Pr the thermal resistance of the central turbulent region is an important part of the total thermal resistance, and this region is strongly affected by the duct shape. Turbulent flow at low Pr is, therefore, similar to laminar flow in that neither admits the use of a hydraulic diameter to perfectly correlate results for different geometries.

9.4 FULLY DEVELOPED NUSSELT NUMBER IN CIRCULAR DUCT FOR CONSTANT SURFACE TEMPERATURE

The case of fully developed turbulent flow in a long smooth circular duct of constant wall temperature occurs frequently. It is necessary to obtain the heat-transfer coefficient for this case and, additionally, to be aware of the difference caused by a constant wall temperature rather than a constant wall heat-flux boundary condition.

The iterative method utilized for the constant wall temperature case with laminar flow as discussed in connection with Eq. (4-41) can also be used for turbulent flow. Seban and Shimazaki [24] used such a procedure. Alternatively, the solution for the thermal entry region with fully developed flow such as was developed by Sleicher and Tribus [25] can be used to determine the constant wall temperature limiting Nusselt numbers. From their results the ratio of constant wall heat flux to constant wall temperature Nusselt numbers is as displayed in Fig. 9-4.

The constant wall heat flux and constant wall temperature results are similar for most gases and liquids ($\text{Pr} \geqslant 0.5$). It is only for the low Prandtl number fluids, such as liquid metals, that a significant difference is found. In all cases the difference diminishes with increasing Reynolds number.

The physical basis for these results is as mentioned in connection with Fig. 9-2. As the Reynolds number and the turbulence level in the central core increase, molecular effects in the central core make the thermal resistance there a lesser part of the total; so, at high Reynolds numbers the ratio of the two Nusselt numbers approaches unity for all values of the Prandtl number

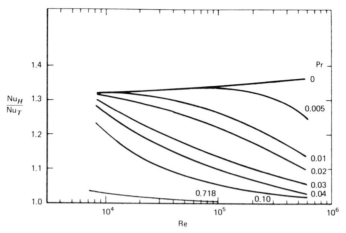

Figure 9-4 Ratio of the Nusselt number for constant wall heat flux to that for constant wall temperature for fully developed turbulent flow in a circular duct. [By permission from C. A. Sleicher and M. Tribus, *Heat Transfer and Fluid Mechanics Institute*, Stanford University Press, Stanford, CA, 1956, pp. 59–78 [25].]

but does so more slowly for low Prandtl number. As a consequence, the Nusselt number for a large Prandtl number fluid is insensitive to axial variation of either wall heat flux or wall temperature.

9.5 CORRELATING EQUATIONS FOR FULLY DEVELOPED FLOW IN ROUND DUCTS

Correlating equations that represent experimentally or analytically obtained heat-transfer coefficients for long round ducts have been reviewed by Sleicher and Rouse [26], who recommend, for $0.1 < \mathrm{Pr} < 10^5$ and $10^4 < \mathrm{Re}_D < 10^6$ and no more than 10% error,

$$\mathrm{Nu}_{D,m} = 5 + 0.015\,\mathrm{Re}_{D,f}^a\,\mathrm{Pr}_w^b \qquad (9\text{-}16a)$$

where

$$a = 0.88 - \frac{0.24}{4 + \mathrm{Pr}_w} \qquad \text{and} \qquad b = \tfrac{1}{3} + 0.5\exp(-0.6\,\mathrm{Pr}_w)$$

and the subscripts m, f and w designate property evaluation at the mixing-cup, film $[T_f = (T_w + T_m)/2]$, and wall temperatures, respectively. For gases

$(0.6 < \text{Pr} < 0.9)$, an approximation is

$$\text{Nu}_{D,m} = 5 + 0.012\,\text{Re}_{D,f}^{0.83}(\text{Pr}_w + 0.29) \tag{9-16b}$$

For viscous fluids $(\text{Pr} > 50)$, Eq. (9-16a) is approximated by

$$\text{Nu}_{D,m} = 0.015\,\text{Re}_{D,f}^{0.88}\,\text{Pr}_w^{1/3} \tag{9-16c}$$

For liquid metals $(\text{Pr} < 0.1)$, with 10% or less error,

$$\text{Nu}_{D,m} = 4.8 + 0.0156\,\text{Re}_{D,f}^{0.85}\,\text{Pr}_w^{0.93}, \qquad T_w = \text{const} \tag{9-16d}$$

$$= 6.3 + 0.0167\,\text{Re}_{D,f}^{0.85}\,\text{Pr}_w^{0.93}, \qquad q_w = \text{const} \tag{9-16e}$$

Additional discussion is given by Kays and Crawford [19], Notter and Sleicher [27], and Sleicher et al. [28].

The Lyon–Martinelli equation [29] for liquid metals

$$\text{Nu}_D = 7 + 0.025(\text{Re}_D\,\text{Pr})^{0.8}, \qquad q_w = \text{const}$$

gives results that are 30% to 70% too high, possibly due to the nonwetting of some solids by some liquid metals and the effects of impurities. Information on liquid metals is given in the handbook by Lyon [30] and in the review by Stein [31]. Since flow in the first portion of the entrance region of a duct behaves much like a boundary layer on a flat plate, the result of Problem 5-13 applies as

$$\lim_{\text{Pr}\to 0}\,\text{Nu}_x = \left(\frac{\text{Re}_x\,\text{Pr}}{\pi}\right)^{1/2}$$

A common correlation for long ducts is the Dittus–Boelter equation [32]

$$\text{Nu}_D = 0.023\,\text{Re}_D^{0.8}\,\text{Pr}^n \tag{9-16f}$$

where $n = 0.4$ for heating and $n = 0.3$ for cooling, $10^4 < \text{Re}_D < 1.2 \times 10^5$, $0.7 < \text{Pr} < 120$, and when the wall temperature does not exceed the fluid mixing-cup temperature by more than 10°F for liquids or 100°F for gases; it can be 20% high for gases and as much as 40% low for water at high Reynolds number [26]. Fluid properties are evaluated at the average local film temperature, $T_f = (T_m + T_w)/2$. The Sieder–Tate equation [33] for fluids of temperature-sensitive viscosity is

$$\text{Nu}_{D,m} = 0.023\,\text{Re}_{D,m}^{0.8}\,\text{Pr}_m^{1/3}\left(\frac{\mu_m}{\mu_w}\right)^{0.14}$$

and is intended for $Re_D > 10^4$ and $0.7 \leqslant Pr \leqslant 16,700$. Fluid properties are evaluated at the local mean temperature except for the last term in which viscosity is evaluated at the mean fluid temperature T_m and the wall temperature T_w.

9.6 THERMAL ENTRY LENGTH

When heating begins in a duct after the velocity profile is fully developed, the general trends observed in any thermal entry region for laminar flow would be expected; Nusselt numbers would be large at the start of the heated section and would asymptotically approach the limiting value for long ducts farther down stream.

Analyses for a circular duct have been executed for a constant wall surface temperature by Sleicher and Tribus [25] and for a constant wall heat flux by Sparrow et al. [11]. Their solutions use the same procedures as for laminar flow but now applied to Eq. (9-5) with the turbulent contributions to thermal transport discussed in Section 9.2. The resulting eigenvalues and eigenfunctions were numerically determined and can be used in an evaluation of entrance-region effects as for laminar flow. Some of the results collected by Kays [19] that can be used in Eq. (4-48a) for the constant wall temperature case are given in Table 9-4. Equation (4-48a) gives the Nusselt number based on log–mean-temperature difference (LMTD) as

$$\overline{Nu_D} = \frac{\lambda_0^2}{2} + \frac{Re_D \, Pr}{(x/D)} \ln \frac{\lambda_0^2}{8G_0} \tag{4-48a}$$

Recognition that $\lambda_0^2/2 = \overline{Nu_D}(x = \infty)$ and rearrangement then gives

$$\frac{\overline{Nu_D}(x)}{\overline{Nu_D}(x = \infty)} = 1 + \frac{Re_D \, Pr}{(x/D)} \frac{2}{\lambda_0^2} \ln \frac{\lambda_0^2}{8G_0}$$

Substitution of values from Table 9-4 for $Re_D = 5 \times 10^4$ and $Pr = 0.7$ into this equation gives

$$\frac{\overline{Nu_D}(x)}{\overline{Nu_D}(x = \infty)} = 1 + \frac{2}{x/D}$$

TABLE 9-4 Some Function Values for the Thermal Entry Length in Turbulent Flow Through a Circular Tube [19]

Pr	Re_D	λ_0^2	G_0
0.01	5×10^4	11.7	1.11
	10^5	13.2	1.3
	2×10^5	16.9	1.7
0.7	5×10^4	235	28.6
	10^5	400	49.0

Or, more generally,

$$\frac{\overline{Nu}_D(x)}{\overline{Nu}_D(x = \infty)} = 1 + C_1\left(\frac{D}{x}\right)^{C_2}$$

Such a procedure is commonly used but, from the results shown in Table 9-4, C_1 and C_2 depend on both Re_D and Pr. The general behavior for local

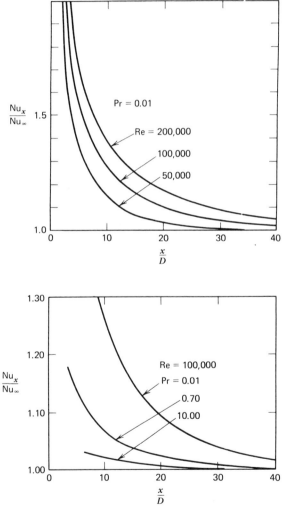

Figure 9-5 Nusselt numbers in the thermal entry region of a circular duct for constant wall heat flux and turbulent flow with (*a*) Pr = 0.01 at various Reynolds numbers, (*b*) Re = 10^5 at various Prandtl numbers, and (*c*) Pr = 0.7 at various Reynolds numbers. [By permission from W. M. Kays and M. E. Crawford, *Convective Heat and Mass Transfer*, McGraw-Hill, 1980 [19].]

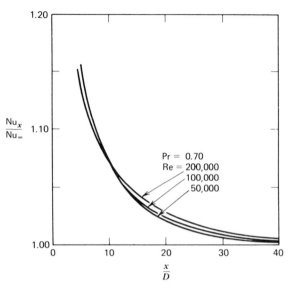

Figure 9-5   *(Continued)*

Nusselt numbers is displayed in Fig. 9-5 for constant wall heat flux. There it is seen that the entrance-region effect is pronounced at small Pr, but is small for large Pr. Also noteworthy is the near independence of entrance-region effects on Re_D for $Pr \approx 1$. For an abrupt contraction of the inlet Molki and Sparrow [34] gave $C_1 = 23.99\,Re^{-0.23}$ and $C_2 = 0.815 - 2.08 \times 10^{-6}\,Re$ for $7 \times 10^3 \leq Re \leq 10^5$ and $x/D \geq 2$.

In many applications the velocity and temperature profiles develop simultaneously, or nearly so as in heat exchangers where a coolant enters tubes through a sudden contraction from a header. Primary reliance is placed on measurements because of the complexities of the entrance conditions of interest (consult Hartnett [35], Latzko [36], Deissler [37], and Boelter et al. [38]). Modifications for the average heat-transfer coefficient [5], on the basis of experiments with fluids such as air, are

$$\frac{\bar{h}}{h_\infty} = 1 + \frac{C}{L/D}, \qquad \frac{L_c}{D} < \frac{L}{D} < 60$$

$$= 1.11\frac{Re_D^{1/5}}{(L/D)^{4/5}}, \qquad \frac{L}{D} < \frac{L_c}{D}$$

where h_∞ is the heat-transfer coefficient for an infinitely long pipe and $L_c/D = 0.625\,Re_D^{1/4}$ is the number of diameters required for the friction factor to become constant. Values of C for selected inlet configurations and for $26{,}000 < Re_D < 56{,}000$ are given in Table 9-5. The value of C for a fully

TABLE 9-5   Values of C for Various Inlets [5]

Inlet Configuration	C
Fully developed velocity profile	1.4
Bell-mouthed with screen	1.4
Calming section	
$\quad L/D = 11.2$	1.4
$\quad L/D = 2.8$	3.0
45° sharp bend	5.0
Abrupt contraction	6.0
180° round bend	6.0
90° right-angle bend	7.0

developed velocity profile is 1.4, agreeing substantially with the analytically derived 2.0. The entrance-region effects would be greater for low-Prandtl number fluids and lesser for high-Prandtl number fluids [39]. The scatter of measurements does not warrant distinguishing between constant wall temperature and constant wall heat flux conditions in the entrance-region correction.

9.7   AXIALLY VARYING SURFACE TEMPERATURE AND HEAT FLUX

When either wall surface temperature or wall heat flux vary along the axis of a duct, the inconstancy of the heat-transfer coefficient can be accounted for in the same manner as was done for laminar flow in Chapter 4. For this purpose, the eigenvalues and constants in Table 9-4 (for more extensive tabulations, consult references [11, 19, 25]) are needed.

Cooling of a nuclear reactor with its high heat flux in the center and low heat flux at inlet and exit by use of a liquid metal coolant requires variable heat-flux calculations, whereas use of a gas or pressurized water allows a Nusselt number based on constant heat flux—the local heat flux must be used to calculate local temperature differences.

9.8   EFFECT OF SURFACE ROUGHNESS

Previous discussions have assumed the wall to be smooth. Manufacturing and operating conditions often lead to rough walls. Usually the roughness is small relative to the duct diameter, however.

Rough walls usually have little effect on laminar flow. In turbulent flow the roughness elements are found to have little effect if they are small enough to be immersed in the laminar sublayer. When the roughness elements protrude above the laminar sublayer into the turbulent flow, on the other hand, the

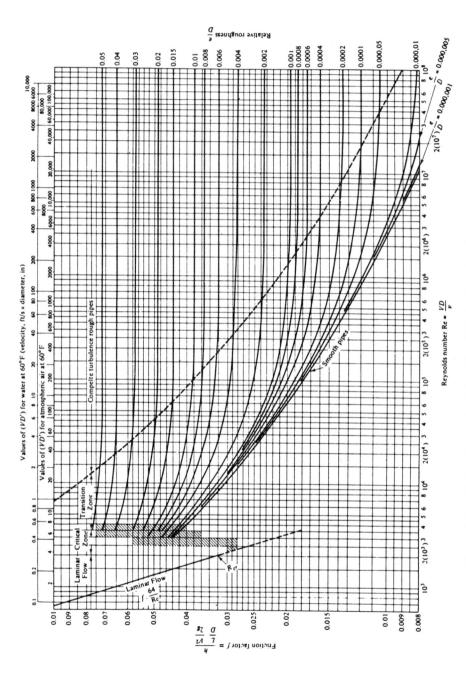

Figure 9-6 Friction factor $f = (D/L)(2/\rho U_{av}^2)\, \Delta p$ for pipe flow versus Reynolds number $Re_D = U_{av} D/\nu$. [From L. F. Moody, *Trans. ASME* **66**, 671–684 (1944) [40].]

TABLE 9-6   Roughness of Commercial Pipes

Pipe Type	e (roughness), in.
Drawn tubing	6×10^{-5}
Brass, lead, glass, spun cement	3×10^{-4}
Commercial steel, wrought iron	1.8×10^{-3}
Cast iron (asphalt dipped)	4.8×10^{-3}
Galvanized iron	6×10^{-3}
Wood stave	$0.72–3.6 \times 10^{-2}$
Cast iron (uncoated)	1.02×10^{-2}
Concrete	$1.2–12 \times 10^{-2}$
Riveted steel	$3.2–36 \times 10^{-2}$

head loss H_f is caused primarily by pressure drag on the roughness elements rather than by viscous shear. As pressure drag on an object is nearly a constant fraction of the square of the velocity, the measured independence of friction coefficient f for head loss ($f = 2g_c H_f D/LU_{av}^2$) from the Reynolds number is plausible on physical grounds.

Measurements on pipes with commercial roughnesses are summarized by Moody [40] in the Moody diagram in Fig. 9-6. Typical values of the average height e of a roughness element are given in Table 9-6; it is seen in Fig. 9-6 that the relative roughness e/D is of greatest significance. For commercially rough pipes the correlation of Colebrook [41]

$$\frac{1}{f^{1/2}} = 1.74 - 2\log_{10}\left(2\frac{e}{D} + \frac{18.7}{\text{Re}_D f^{1/2}}\right)$$

can be used in the turbulent range. For sand-roughened pipes, use the correlation given by Nedderman and Shearer [42] since commercial roughness does not show the dip in the transition region exhibited by sand roughness.

For both laminar and turbulent flow Churchill [43] gives

$$\frac{f}{8} = \left[\left(\frac{8}{\text{Re}_D}\right)^{12} + \frac{1}{(A+B)^{3/2}}\right]^{1/12}$$

where

$$A = \left[2.457\ln\left\{\frac{1}{(7/\text{Re}_D)^{0.9} + 0.27e/D}\right\}\right]^{16}, \qquad B = \left(\frac{37,530}{\text{Re}_D}\right)^{16}$$

Both of these relationships give the friction factor explicitly when the volumetric flow rate Q is specified, but still require a trial-and-error solution procedure when head loss/length is specified. This difficulty was circumvented by Swamee and Jain [44] who give, with H_f the head loss,

$$\frac{1}{f^{1/2}} = 2\log_{10}\left[\frac{e}{3.7D} + \frac{5.74}{\text{Re}_D^{0.9}}\right],$$

$$Q = -2.22\left(g_c H_f D^5\right)^{1/2}\log_{10}\left[\frac{e}{3.7D} + \frac{1.78\nu}{\left(g_c H_f D^3/L\right)^{1/2}}\right]$$

with less than 1% error for $10^{-6} \leqslant \varepsilon/D \leqslant 10^{-2}$ and $5 \times 10^3 \leqslant \text{Re}_D \leqslant 10^8$ and

$$D = 0.66\left(\frac{LQ^2}{g_c H_f}\right)^{19/100}\left[\varepsilon^{5/4} + \frac{\nu}{Q}\left(\frac{LQ^2}{g_c H_f}\right)^{9/20}\right]^{1/25}$$

with less than 2% error for $10^{-6} \leqslant \varepsilon/D \leqslant 10^{-2}$ and $3 \times 10^3 \leqslant \text{Re}_D \leqslant 3 \times 10^8$.

The head losses caused by fittings, such as valves and joints, can be significant, although they are usually referred to as *minor losses*. The loss coefficients $k_L = 2H_f/\rho U_{av}^2$ for some common valves and fittings are given in Table 9-7. If fittings are closer together than the 40–50 diameters required to achieve fully developed flow, the actual loss coefficients will be smaller. Little is reported regarding the heat-transfer coefficients downstream from these components, but they must be correspondingly increased above their values for fully developed conditions. See Joshi and Shah [45] for a survey.

The effect of pipe roughness on the heat-transfer coefficient is usually less than on the friction factor for head loss. The Reynolds analogy, which suggests that h is proportional to f, is predicated on shear providing flow resistance; this is not the case when roughness elements protrude above the laminar sublayer. A useful parameter for expressing the effect of roughness on heat transfer is $k^+ = e/\delta_L$, where e is the roughness height and δ_L is the laminar sublayer thickness. As illustrated in Fig. 9-1, the laminar sublayer thickness is given by $y^+ \approx 5$, or

$$\frac{\delta_L v^*}{\nu} \approx 5$$

Rearrangement in terms of the friction factor for pressure drop gives

$$k^+ = \frac{e}{\delta_L} = \text{Re}_D \frac{e}{D}\left(\frac{f}{8}\right)^{1/2}$$

TABLE 9-7   Loss Coefficients for Various Fittings

Fitting	k_L (Loss Coefficient)
Angle valve, fully open	3.1–5.0
Ball check valve, fully open	4.5–7.0
Gate valve, fully open	0.19
Globe valve, fully open	10.0
Swing check valve, fully open	2.3–3.5
Regular-radius elbow,	
Screwed	0.9
Flanged	0.3
Long-radius elbow	
Screwed	0.6
Flanged	0.23
Close return bend, screwed	2.2
Flanged return bend, two elbows	
Regular radius	0.38
Long radius	0.25
Standard tee, screwed	
Flow through run	0.6
Flow through side	1.8

where constants have been ignored. The departure from Reynolds' analogy is expressed by the effectiveness parameter η, where

$$\eta = \frac{\mathrm{Nu}/f}{\mathrm{Nu}_s/f_s} = \log_{10}\left[\frac{\mathrm{Pr}^{0.33}}{(k^+)^{0.243}}\right] - 0.32 \times 10^{-3}k^+ \log_{10}(\mathrm{Pr}) + 1.25$$

and the subscript s refers to smooth-wall quantities. Burck [46] correlated the measurements of Nunner [47] and of Dipprey and Sabersky [48], for $\mathrm{Pr} \approx 1$,

$$\frac{\mathrm{Nu}}{\mathrm{Nu}_s} = \left(\frac{f}{f_s}\right)^{1/2}$$

On the physical grounds that velocity fluctuations are increased by roughness but temperature fluctuations are not with $\tau_w \sim \overline{u'v'}$ and $q_w \sim \overline{v'T'}$, as pointed out by Christoph and Pletcher [49]. The magnitude of expected roughness effects is illustrated in Problem 9-4. Variable-property results are reported by Wassel and Mills [50].

It is expected that roughness would have the greatest effect on a high-Prandtl number fluid since its thermal resistance is primarily in the sublayer. A low-Prandtl number fluid such as a liquid metal, where molecular effects predominate throughout, would be little affected by wall roughness, on the other hand.

Measurements have been reported by Webb et al. [51] on the effect of discrete roughness elements. Spiraled internal fins inclined at an angle with the tube axis cause increase in heat-transfer coefficients and pressure drop [52, 53]. Kreith and Bohn [54] provide examples of the use of the performance ranking method developed by Soland et al. [55] for enhanced-surface heat exchangers.

9.9   CURVED DUCTS

In curved pipes there is a secondary flow, as there is for turbulent flow in straight ducts of irregular cross section due to anisotropy of turbulent stresses as illustrated by Demuren [56], because fluid particles near the pipe axis have a larger velocity and are acted on by a larger centrifugal force than are slower particles near the walls. The resultant secondary flow is directed outward in the center and inward (toward the center of curvature of the pipe bend) near the wall as shown in Fig. 9-7.

Curvature has a stronger effect on pressure drop in laminar than in turbulent flow. For laminar flow, the analyses by Dean [57] and Adler [58] suggest that dimensionless Dean number De controls where

$$\text{De} = \frac{\text{Re}_D}{2} \left(\frac{D}{D_c} \right)^{1/2}$$

with D the pipe diameter, D_c the diameter of the pipe curvature, and $\text{Re}_D = U_{av}D/v$. The results of computational fluid dynamics (CFD) calculations are correlated by the relationship

$$\frac{f}{f_{\text{straight}}} = \begin{cases} 0.513(2\,\text{De})^{0.216}, & (2\,\text{De})^2 < 12{,}900 \\ 0.318(2\,\text{De})^{0.326}\,[1 - 6.13 \times 10^4(2\,\text{De})^{0.5} \\ \qquad + 1.08 \times 10^{-5}(2\,\text{De}) \\ \qquad + 2.82 \times 10^{-6}(2\,\text{De})^{1.5}], & (2\,\text{De})^2 > 20{,}500 \end{cases}$$

For the laminar heat-transfer coefficient augmented by curvature-induced

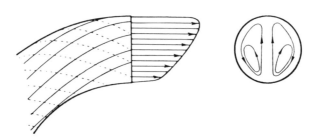

Figure 9-7   Flow pattern in a curved pipe.

secondary flow, they give

$$\frac{Nu}{Nu_{straight}} = \begin{cases} 0.518(4\,De^2\,Pr)^{0.147}, & 4\,De^2\,Pr < 1.44 \times 10^5 \\ 0.278(4\,De^2\,Pr)^{0.222}, & 1.44 \times 10^5 < 4\,De^2\,Pr < 10^7 \end{cases}$$

$$(9\text{-}17)$$

which are complemented by those of Futagami and Aoyama [60] for a coiled tube that account for buoyancy effects. Numerical procedures have also been applied by Yee et al. [61].

In turbulent flow the Dean number does not correlate measurements well. Ito [62] gave the correlation

$$\frac{f}{f_{straight}} = \left[Re_D \left(\frac{D}{D_c}\right)^2 \right]^{1/20}, \qquad Re_D \left(\frac{D}{D_c}\right)^2 > 6 \qquad (9\text{-}18)$$

He also found that the critical value of the Reynolds number above which fully turbulent flow exists in a pipe is given by

$$Re_{D,\,critical} = 2 \times 10^4 \left(\frac{D}{D_c}\right)^{0.32}$$

Seban and McLaughlin [63] found that Reynolds' analogy holds for turbulent flow in curved pipes so that

$$\frac{Nu_D}{Re_D\,Pr^{0.4}} = \frac{f}{8}$$

with f found from Eq. (9-18). Consult Mori and Nakayama [64] for detailed discussion and general predictive equations regarding heat transfer and pressure drop for laminar and turbulent flow in curved pipes for constant-temperature and constant-heat-flux boundary conditions.

A related situation is that of a duct rotating about its central axis. Such effects are of interest when, for example, rotating electrical generator windings are cooled convectively according to Metzger and Afgan [65] and Yang [66].

9.10 COMPUTATIONAL FLUID DYNAMICS

The situations for which solutions are wanted are complicated by both geometry and boundary condition. In the filling of multiple molds in a plastic injection molding machine, the transient viscous motion of the molten plastic in each mold is wanted with temperature-dependent viscosity and freezing of the plastic at the chilled mold wall taken into account. In a combustor, the

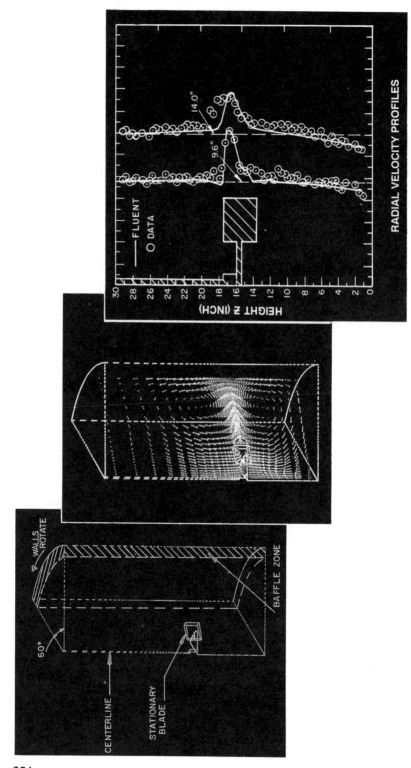

Figure 9-8 Three-dimensional simulation of turbulent flow in a mixing tank with a Rushton impeller. [By permission from Fluent Inc. 1991]

swirling gas reacts with vaporizing liquid droplets at high temperature and gives rise to a wall heat flux whose distribution is to be determined. Gas flow over an airplane wing–fuselage assembly or a rotating fan blade results in lift and drag which is to be computed. The efficacy of a scheme for removing heat from electronic equipment might need to be evaluated. Or, the effectiveness of a ventilation hood in removing paint droplets from a paint-spray booth might be wanted. Many of these applications require a three-dimensional representation as shown in Fig. 9-8.

The desired details for such applications are usually beyond the capability of all but numerical techniques. With them velocity, temperature, and mass fraction distributions can be obtained and displayed with greater economy than is possible with only measurement and flow visualization techniques. Thus computer programs for these purposes are in commercial use although development continues [67, 68]. They and the general numerical solution procedures and solution display techniques have come to be called computational fluid dynamics (CFD). Although the applications extend beyond merely determining velocity distributions, the view that such other quantities as energy are just "along for the ride" is close enough to the truth that CFD broadly refers to all uses.

The capabilities of some commercially available CFD codes were reviewed by Hutchings and Iannuzzelli [69, 71], Wolfe [72], and O'Connor [73]. The Fluent Inc. CFD programs [74–77] include FLUENT based on the control-volume method with a renormalization-group-based anisotropic k–ϵ turbulence model for evaluation of k–ε constants in a range (near-wall, curved geometry, flow separation, flow curvature, liquid metal, etc.) of conditions, NEKTON based on the spectral finite-element method (see Orzag [78] and Canuto et al. [79] and finite-element discussions by Pepper and Heinrich [80]), and RAMPANT with a solution mesh adaptively altered to cluster solution points in regions of rapid changes. Other commercially available CFD programs include CFD 2000 by Adaptive Research Corporation, FIDAP by Fluid Dynamics International, microCOMPACT by Innovative Research Inc., FLOW-3D by Flow Science Inc., FIRE by AVL List GmbH, FLOTRAN by Compuflo, INCA by Amtec Engineering, and Phoenics by Cham of North America Inc.

Augmenting these whole-field solvers, so-called because they produce solutions at many points interior to the solution domain (whether or not solutions are wanted there), are boundary-element (see Brebbia and Dominguez [81, 82] and Kinnas and Hsin [83] for propellers), panel (see Chow [84]), and vortex-lattice methods (see Kerwin and Lee [85]) that produce solutions only on domain boundaries, often the only points of interest; solutions at interior points can then be obtained from these boundary values. Some hybrid techniques, such as Chen's [86] finite-analytic technique, reduce the number of solution points required for accuracy by using information from exact solutions near each solution point. The statistical Monte Carlo method described by Haji-Sheikh [87] offers the advantage of obtaining a numerical solution at only one point.

A Finite-Volume Method—SIMPLE

Numerical simulation of flow is commonly accomplished by utilizing a version of the semi-implicit method for pressure-linked equations (SIMPLE) devised by Patankar and Spalding [75], who drew upon earlier work by Harlow and Welch [88] with the marker-and-cell (MAC) method, Chorin [89], and Amsden and Harlow [90]. It is around a version of the SIMPLE method that the often-used FLUENT program is organized. The use of conservation principles applied to a control volume in the derivation of the method has led to its description as a finite-volume, rather than finite-difference or finite-element, method. The closeness of the physical principles to the final mathematical forms needed for numerical calculation makes the SIMPLE method a natural one for a first CFD exposition. It is also well suited here, since it was originally derived for a flow with a preferred direction, as for flow down a duct.

The SIMPLE method is illustrated with a two-dimensional derivation applied to the control volume shown in Fig. 9-9. The shaded control volume surrounds the primary point p and the surfaces are at the halfway point to adjacent primary points N, S, E, and W (denoting directions); the midpoints are at locations n, s, e, and w. The control volume is not necessarily square,

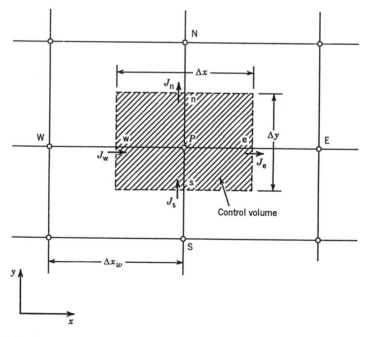

Figure 9-9 Control volume for the two-dimensional situation. [By permission from S. V. Patankar, *Numerical Heat Transfer and Fluid Flow*, Taylor and Frances, Washington, D.C., 1980 [91].]

and a face dimension (Δx or Δy) is not necessarily divided into two equal parts by location n, s, e, or w. The primary points are not necessarily equally spaced, either. A flux J of a scalar quantity crosses each control-volume surface. Not shown is a volumetric distributed source S of the scalar.

Application of the conservation principle for ϕ, as discussed in Chapter 2, to the control volume gives

$$\sum \text{ rate of } \phi_{\text{incoming}} - \sum \text{ rate of } \phi_{\text{outgoing}} + \text{ rate of } \phi_{\text{generated}} = \text{ rate of } \phi_{\text{stored}}$$

$$J_{\text{w}} \, \Delta y + J_{\text{s}} \, \Delta x - J_{\text{e}} \, \Delta y - J_{\text{n}} \, \Delta x + \int_A S \, dA \approx \frac{(\rho_P \phi_P - \rho_P^o \phi_P^o) \, \Delta x \, \Delta y}{\Delta t}$$

$$\approx (S_c + S_P \phi_P) \, \Delta x \, \Delta y \quad (9\text{-}19)$$

The superscript o denotes evaluation at the start of the time interval that starts at t and is of duration Δt. All other values are unknown at the end $t + \Delta t$ of the time interval. The source term S has been approximated as shown with a linearized possible dependence on the value of the scalar at primary point P. Here flux approximations such as $\int J \, dy \approx J_{\text{w}} \, \Delta y$ over the surface at w have also been employed. The conservation of mass principle applied to the control volume similarly gives, with F being the mass flux,

$$F_{\text{w}} \, \Delta y + F_{\text{s}} \, \Delta x - F_{\text{e}} \, \Delta y - F_{\text{n}} \, \Delta x \approx \frac{(\rho_P - \rho_P^o) \, \Delta x \, \Delta y}{\Delta t} \quad (9\text{-}20)$$

Multiplication of Eq. (9-20) by ϕ_P followed by subtraction from Eq. (9-19) results in

$$\frac{\rho_P^o \, \Delta x \, \Delta y}{\Delta t} (\phi_P - \phi_P^o) + (J_{\text{e}} - F_{\text{e}} \phi_P) \, \Delta y - (J_{\text{w}} - F_{\text{w}} \phi_P) \, \Delta y$$

$$+ (J_{\text{n}} - F_{\text{n}} \phi_P) \, \Delta x - (J_{\text{s}} - F_{\text{s}} \phi_P) \, \Delta x = (S_c + S_P \phi_P) \, \Delta x \, \Delta y \quad (9\text{-}21)$$

The flux J is, as discussed in Chapter 2, the sum of convective transport $\rho u \phi$ and diffusion $-\Gamma \, d\phi/dx$, giving

$$J_{\text{w}} = \rho_{\text{w}} u_{\text{w}} \phi_{\text{w}} - \Gamma \left(\frac{d\phi}{dx} \right)_{\text{w}}$$

and

$$F_{\text{w}} = \rho_{\text{w}} u_{\text{w}}$$

However, the scalar is to be determined only at the primary point P, not at the surface point w. Approximations for ϕ_{w} and $(d\phi/dx)_{\text{w}}$ must be made thoughtfully or else physically unrealistic predictions can be made. For example, if the linear, central-difference approximations

$$\phi_w = \frac{\phi_W + \phi_P}{2}, \qquad \left(\frac{d\phi}{dx}\right)_w = \frac{\phi_P - \phi_W}{\Delta x_W}$$

$$\phi_e = \frac{\phi_E + \phi_P}{2}, \qquad \left(\frac{d\phi}{dx}\right)_e = \frac{\phi_E - \phi_P}{\Delta x_E}$$

are first employed, then

$$J_w - F_w\phi_P = \left(D_W + \frac{F_w}{2}\right)(\phi_W - \phi_P)$$

$$J_e - F_e\phi_P = \left(D_E - \frac{F_e}{2}\right)(\phi_P - \phi_E)$$

letting $F_w = \rho_w u_w$ and $D_W = \Gamma/\Delta x_W$. These approximations used in Eq. (9-21) yield

$$a_P\phi_P = a_E\phi_E + a_W\phi_W + a_N\phi_N + a_S\phi_S + b \qquad (9\text{-}22)$$

where

$$a_E = \left(D_E - \frac{F_e}{2}\right)\Delta y, \qquad a_W = \left(D_W + \frac{F_w}{2}\right)\Delta y$$

$$a_N = \left(D_N - \frac{F_n}{2}\right)\Delta x, \qquad a_S = \left(D_S + \frac{F_s}{2}\right)\Delta x$$

$$b = \left(\frac{\rho_P^o \phi_P^o}{\Delta t} + S_c\right)\Delta x\,\Delta y$$

$$a_P = \left(\frac{\rho_P^o}{\Delta t} - S_P\right)\Delta x\,\Delta y + a_E + a_W + a_N + a_S$$

The form of Eq. (9-22) is familiar from application of finite-difference methods to heat conduction, giving the scalar value at a primary point as the weighted sum of the values at neighboring primary points. As required for physical plausibility, the sum $\Sigma_i a_i/a_P$ of the weighting factors is unity. In this case a_E and a_N can be negative when the convective flux is sufficiently large that $F_e > 2D_E$, for example. To see that this is physically unrealistic, consider the case in which $\phi_E = 1$ with $\phi_W = 0 = \phi_N = \phi_S$ and all F are equal while all D are equal subject, also, to $F = 4D$. Then Eq. (9-22) is

$$\phi_P = -\frac{1}{4}\phi_E^1 + \frac{3}{4}\phi_W^0 - \frac{1}{4}\phi_N^0 + \frac{3}{4}\phi_S^0$$

predicting that $\phi_P = -\frac{1}{4}$. The mathematical cause of this unrealistic prediction, unrealistic because ϕ_P should be positive since all of its neighbors are, is that one or more of the coefficients became negative (note that, since velocities can be either positive or negative, the other set of coefficients might be the ones to become negative). This can be avoided, of course, by restricting the control-volume size to ensure that positive coefficients ensue, or requiring that $a_i \geq 0$ so that

$$\left(D_i \pm \frac{F_i}{2} \right) \Delta x_i \geq 0$$

In general, this requires

$$\left| \frac{\rho_i u_i \, \Delta x_I}{\mu} \frac{\nu}{\alpha_\Gamma} \right| \leq 2$$

where $\alpha_\Gamma = \Gamma/\rho C_p$ is the diffusivity and the term inside the absolute value signs is the cell Peclet number $\text{Pe} = F/D$ (see Section 4.3). The linear, central-difference approximations are stable only if each cell has the cell Peclet number $\text{Pe}_i = F_i/D_i \leq 2$; when convection predominates over diffusion, many small cells are needed.

A second approximation is the upwind scheme for the convection; the diffusion is still approximated by the previously used linear, central-difference scheme. Then

$$F_w \phi_w = \begin{cases} F_w \phi_W, & F_w > 0 \\ F_w \phi_P, & F_w < 0 \end{cases} = \phi_W (F_w, 0)_{\max} - \phi_P (-F_w, 0)_{\max},$$

$$\left(\frac{d\phi}{dx} \right)_w = \frac{\phi_P - \phi_W}{\Delta x_W}$$

$$F_e \phi_e = \begin{cases} F_e \phi_P, & F_e > 0 \\ F_e \phi_E, & F_e < 0 \end{cases} = \phi_P (F_e, 0)_{\max} - \phi_E (-F_e, 0)_{\max},$$

$$\left(\frac{d\phi}{dx} \right)_e = \frac{\phi_E - \phi_P}{\Delta x_E}$$

and so forth. With these approximations the coefficients for Eq. (9-22) are

$$a_E = D_E + (-F_e, 0)_{\max}, \qquad a_W = D_W + (F_w, 0)_{\max}$$
$$a_N = D_N + (-F_n, 0)_{\max}, \qquad a_S = D_S + (F_s, 0)_{\max}$$

which are always positive as required for physically realistic predictions to come from Eq. (9-22). The function $(\ )_{\max}$ is the maximum of the items listed in parentheses.

A third approximation that leads to a so-called hybrid scheme is constructed from an exact solution to the one-dimensional problem of the form treated in Problem 3-18

$$\rho_w u_w \frac{\partial \phi}{\partial x} = \Gamma \frac{d^2 \phi}{dx^2}, \qquad \phi(0) = \phi_w \qquad \text{and} \qquad \phi(\Delta x_w) = \phi_P$$

whose solution is

$$\frac{\phi - \phi_w}{\phi_P - \phi_w} = \frac{\exp\left(\dfrac{\text{Pe}_w \, x}{\Delta x_w}\right) - 1}{\exp(\text{Pe}_w) - 1}$$

in which the Peclet number $\text{Pe}_w = \rho_w u_w \Delta x_w / \Gamma = F_e/D_w$ can take on any value. From this exact one-dimensional solution it is found that

$$\phi_w = \frac{\phi_w \exp(\text{Pe}_w/2) + \phi_P}{\exp(\text{Pe}_w/2) + 1}, \quad \cdot \quad \left(\frac{d\phi}{dx}\right)_w = (\phi_P - \phi_w) \frac{\text{Pe}_w}{\Delta x_w} \frac{\exp(\text{Pe}_w/2)}{\exp(\text{Pe}_w) - 1}$$

with similar expressions for the other three directions. Use of this approximation in Eq. (9-21) gives the coefficients of Eq. (9-22) as

$$a_E = D_E \frac{F_e/D_E}{\exp(F_e/D_E) - 1}, \qquad a_w = D_w \frac{(F_w/D_w) \exp(F_w/D_w)}{\exp(F_w/D_w) - 1}$$

$$a_N = D_N \frac{F_n/D_n}{\exp(F_n/D_N) - 1}, \qquad a_S = D_S \frac{(F_s/D_S) \exp(F_s/D_S)}{\exp(F_s/D_S) - 1}$$

Because the exponentials are expensive to compute, a power-law representation

$$\frac{a_E}{D_E} = \begin{cases} -\dfrac{F_e}{D_E}, & \dfrac{F_e}{D_E} < -10 \\[2mm] \left(1 + \dfrac{0.1F_e}{D_E}\right)^5 - \dfrac{F_e}{D_E}, & -10 \leqslant \dfrac{F_e}{D_E} < 0 \\[2mm] \left(1 - \dfrac{0.1F_e}{D_E}\right)^5, & 0 < \dfrac{F_e}{D_E} \leqslant 10 \\[2mm] 0, & \dfrac{F_e}{D_E} > -10 \end{cases}$$

$$= \left[0, \left(1 - 0.1 \left|\frac{F_e}{D_E}\right|\right)^5\right]_{\max} + \left[0, -\frac{F_e}{D_E}\right]_{\max}$$

has been recommended [91, 92]. Similarly,

$$\frac{a_w}{D_W} = \left[0, \left(1 - 0.1\left|\frac{F_w}{D_W}\right|\right)^5\right]_{\text{max}} + \left[0, \frac{F_w}{D_W}\right]_{\text{max}}$$

and so forth.

All such scalar quantities as temperature, mass fraction, and pressure are calculated according to Eq. (9-22) at the primary (grid, mesh) points of the grid (mesh) shown in Fig. 9-9. However, the velocity (a vector) components are not yet known and must be computed. To do this, a second grid, staggered with respect to the first one shown in Fig. 9-9, is used as illustrated in Fig. 9-10. Such a staggered grid enables the velocity components $[u_w, u_e, u_n,$ and u_s on the faces of the control volume in Fig. 9-9, needed in Eq. (9-22) for computation of scalar quantities] to be ascertained directly rather than by interpolation. To understand the potential difficulty, consider application of the x-momentum principle to the control volume shown in Fig. 9-9 which gives

$$\sum F_x + x \text{ momentum}_{\text{in}} - x \text{ momentum}_{\text{out}} = x \text{ momentum}_{\text{stored}}$$

$$(p_w - p_e)\,\Delta y + \Delta y\,\rho_w u_w\left(\frac{u_w}{g_c}\right) - \Delta y\,\rho_e u_e\left(\frac{u_e}{g_c}\right) = \frac{(\rho_P u_P - \rho_P^o u_P^o)\,\Delta x\,\Delta y}{\Delta t}$$

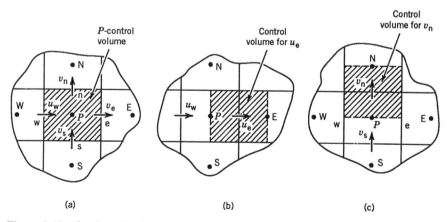

Figure 9-10 Section of a Cartesian grid showing placement of control-volume boundaries. [By permission from J. P. Van Doormaal and G. D. Rahthby, *Numerical Heat Transfer* 7, 147–163 (1984) [93].]

Linear interpolation gives the pressure difference as

$$p_w - p_e \approx \frac{p_W + p_P}{2} - \frac{p_P + p_E}{2} = \frac{p_W - p_E}{2}$$

In other words, this approximation leads to the x-momentum equation depending on the pressure difference between two alternate grid points, rather than two adjacent ones. A similar procedure applied for the y direction leads to the y-momentum equation depending on

$$p_s - p_n \approx \frac{p_S - p_N}{2}$$

The result is that a checkerboard pressure distribution such as shown in Fig. 9-11 would have the same effect on velocity distribution as would a uniform pressure, and this is not right on physical grounds. The same result is obtained when the conservation of mass principle is applied to the control volume of Fig. 9-9, as was done to achieve Eq. (9-20) rewritten here in the form appropriate to constant density

$$(u_w - u_e)\,\Delta y + (u_s - u_n)\,\Delta x = 0$$

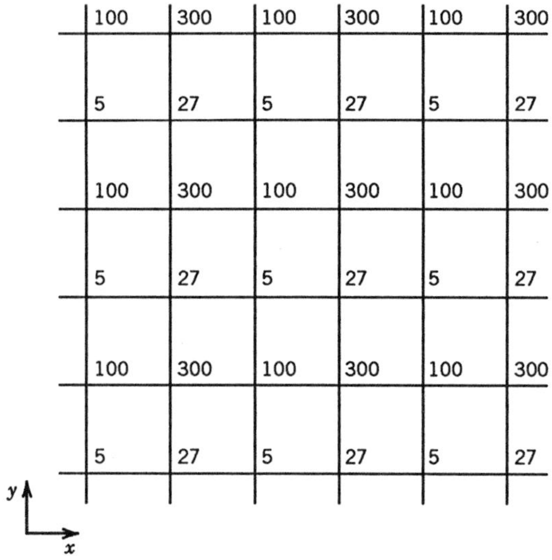

Figure 9-11 Checkerboard pressure field. [By permission from S. V. Patankar, *Numerical Heat Transfer and Fluid Flow*, Taylor and Francis, Washington, D.C., 1980 [91].]

A checkerboard velocity distribution could, erroneously, be taken as the solution for a case in which velocity ought to be uniform. This, again, is because the linear interpolation approximation employed leads to continuity equation dependence on velocity difference between two alternate, rather than adjacent, grid points.

Use of the staggered grid eliminates the possibility of physically unrealistic numerical solutions while simple approximations are employed. The x-momentum principle applied to the control volume for u_e in Fig. 9-10b leads to

$$(p_P - p_E)\,\Delta y + (J_P - J_E)\,\Delta x + (J_{se} - J_{ne})\,\Delta y + (S_c + S_p u_E)\,\Delta x\,\Delta y$$
$$= \frac{(\rho_e u_e - \rho_e^o u_e^o)\,\Delta x\,\Delta y}{\Delta t} \qquad (9\text{-}23)$$

The source term $S_c + S_p u_e$ represents the effects of a body force. In addition, the flux J_P of x momentum is the sum of a convective and a diffusive part as

$$J = \rho u u - \mu\frac{\partial u}{\partial x}$$

and u has taken the place of the scalar ϕ. The conservation of mass principle applied to the same control volume leads to

$$(F_P - F_E)\,\Delta x + (F_{se} - F_{ne})\,\Delta y = \frac{(\rho_e - \rho_e^o)\,\Delta x\,\Delta y}{\Delta t} \qquad (9\text{-}24)$$

Multiplication of Eq. (9-24) by u_e followed by subtraction from Eq. (9-23) leads to

$$\frac{\rho_P^o\,\Delta x\,\Delta y}{\Delta t}(u_e - u_e^o)$$
$$= (J_P - F_P u_e)\,\Delta y - (J_E - F_E u_e)\,\Delta y + (J_{se} - F_{se} u_e)\,\Delta x$$
$$- (J_{ne} - F_{ne} u_e)\,\Delta x + (S_c + S_p u_e)\,\Delta x\,\Delta y + (p_P - p_E)\,\Delta y \quad (9\text{-}25)$$

The result of discretizing Eq. (9-25) for x momentum will be of the same form as Eq. (9-22), the result of discretizing Eq. (9-21) for a scalar. The coefficients will, however, differ with the details depending on the discretization approximations employed. Linear interpolations and central-difference approximations for gradients give

$$J_P - F_P u_e = \rho_P u_P u_P - \mu \left(\frac{du}{dx} \right)_P - \rho_P u_P u_e$$

$$= \rho_P u_P (u_P - u_e) - \mu \left(\frac{du}{dx} \right)_P$$

$$\approx \left(\frac{\rho_e u_e + \rho_w u_w}{2} \right) \left(\frac{u_e + u_w}{2} - u_e \right) - \mu \frac{u_e - u_w}{\Delta x}$$

$$J_{se} - F_{se} u_e = \rho_{s,se} v_{s,se} u_{e,se} - \mu \left(\frac{du}{dy} \right)_{se} - \rho_{s,se} v_{s,se} u_e$$

$$= \rho_{s,se} v_{s,se} (u_{e,se} - u_e) - \mu \left(\frac{du}{dy} \right)_{se}$$

$$\approx \left(\frac{\rho_s v_s + \rho_{s,e} v_{s,e}}{2} \right) \left(\frac{u_e + u_{e,s}}{2} - u_e \right) - \mu \frac{u_e - u_{e,s}}{\Delta y_s}$$

with similar linear approximations applied to the remaining terms. Insertion of these approximations into Eq. (9-25) and rearrangement leads to

$$A_e u_e = A_E u_{e,E} + A_W u_{e,W} + A_N u_{e,N} + A_S u_{e,S} + b + (p_P - p_E) \Delta y$$

where $u_{e,W} = u_w$ as it was formerly designated and

$$A_E = \left[\frac{\mu}{\Delta x_E} - \frac{(\rho_e u_e + \rho_{ee} u_{ee})/2}{2} \right] \Delta y$$

$$A_W = \left[\frac{\mu}{\Delta x_W} + \frac{(\rho_e u_e + \rho_w u_w)/2}{2} \right] \Delta y$$

$$A_N = \left[\frac{\mu}{\Delta y_N} - \frac{(\rho_n v_n + \rho_{n,E} v_{n,E})/2}{2} \right] \Delta x$$

$$A_S = \left[\frac{\mu}{\Delta y_s} + \frac{(\rho_s v_s + \rho_{s,E} v_{s,E})/2}{2} \right] \Delta x$$

$$b = \left(\frac{\rho_P^o u_e^o}{\Delta t} + S_c \right) \Delta x \Delta y$$

$$A_e = \left(\frac{\rho_P^o}{\Delta t} - S_P \right) \Delta x \Delta y + A_E + A_W + A_N + A_S$$

Other approximations, as discussed in connection with Eq. (9-22), would give slightly different coefficients. Application of the y-momentum principle to the control volume staggered in the y direction as shown in Fig. 9-10 similarly leads to the equation for v_n, similar in form to that

$$A_e u_e = \sum_n A_{nb} u_{nb} + b + (p_P - p_E)\,\Delta y \qquad (9\text{-}26)$$

for u_e, of

$$A_n v_n = \sum_n A'_{nb} v_{nb} + b' + (p_P - p_N)\,\Delta x \qquad (9\text{-}27)$$

in which the subscript b refers to neighboring values on the staggered grid for velocity components.

Equations (9-26) and (9-27) for the velocity components at the E and N faces of the P-control volume in Fig. 9-9 can be used to determine corrections u'_e, v'_n to initial velocities u^*_e, v^*_n if correcting p'_P, p'_N, p'_E are available to initial pressure p^*_P, p^*_N, p^*_E. Then

$$u'_e = u_e - u^*_e \qquad \text{and} \qquad p'_P = p_P - p^*_E$$

Multiplication by A_e gives

$$A_e u'_e = A_e u_e - A_e u^*_e$$

Use of Eq. (9-26) on the right-hand side then gives

$$u'_e = \sum_n \frac{A_{nb}(u_{eb} - u^*_{eb})}{A_e} + \frac{(p'_P - p'_E)\,\Delta y}{A_e} \qquad (9\text{-}28a)$$

Neglecting the summation, which will be zero when the correct solution is obtained, gives

$$u'_e = d_e(p'_P - p'_E) \qquad (9\text{-}28b)$$

where $d_e = \Delta y / A_e$. Similarly,

$$v'_n = d_n(p'_P - p'_N) \qquad (9\text{-}29)$$

Thus

$$u_e = u^*_e + d_e(p'_P - p'_E), \qquad v_n = v^*_n + d_n(p'_P - p'_N) \qquad (9\text{-}30)$$

To obtain an equation for determination of the pressure corrections, attention is directed back to Eq. (9-20). This was the result of applying the conservation of mass principle to the P-control volume shown in Fig. 9-10a.

Rewritten, it is

$$(\rho_w u_w - \rho_e u_e)\,\Delta y + (\rho_s v_s - \rho_n v_n)\,\Delta x = \frac{(\rho_P - \rho_P^o)\,\Delta x\,\Delta y}{\Delta t}$$

Use of Eq. (9-30) for the velocity components leads to

$$a_P' p_P' = a_E' p_E' + a_W' p_W' + a_N' p_N' + a_S' p_S' + b' = \sum_n a_{nb}' p_{nb} + b' \quad (9\text{-}31)$$

where

$$a_E' = \rho_e d_e\,\Delta y, \qquad a_W' = \rho_w d_w\,\Delta y, \qquad a_N' = \rho_n d_n\,\Delta x, \qquad a_S' = \rho_s d_s\,\Delta x$$

$$b' = \frac{\rho_P^o - \rho_P}{\Delta t}\,\Delta x\,\Delta y + (\rho_w u_w^* - \rho_e u_e^*)\,\Delta y + (\rho_s v_s^* - \rho_n v_n^*)\,\Delta x$$

$$a_P' = a_E' + a_W' + a_N' + a_S' + b'$$

Note that b' represents a "mass source" which will be zero when the initial velocities (u_w^*, etc.) satisfy conservation of mass for the P-control volume; no pressure corrections are needed then.

The SIMPLE Algorithm

The SIMPLE algorithm, or solution procedure, consists of seven steps.

1. Specify an initial pressure p^* distribution.
2. Evaluate the coefficients of the x- and y-momentum equations [Eqs. (9-26) and (9-27)] and solve to obtain u^* and v^*. An iterative solution procedure, using underrelaxation to ensure convergence, can be employed as illustrated by considering u_e^* as an example, for the ith iteration

$$u_{e,i}^* = \left(\sum_n \frac{A_{nb} u_{nb}^*}{A_e} + \frac{b}{A_e} + d_e(p_P^* - p_E^*) \right)_{i-1} \quad (9\text{-}26)$$

The correction to the previous iteration is $u_{e,i}^* - u_{e,i-1}^*$. The value for the ith iteration is then

$$u_{e,i}^* = u_{e,i-1}^* + \alpha_u(u_{e,i}^* - u_{e,i-1}^*)$$

where α_u is a relaxation factor that accounts for inaccuracy in evaluating the correction. Overrelaxation ($\alpha_u > 1$) speeds convergence when the correction is too small. Here, though, underrelaxation ($\alpha_u \approx 0.5$) is usually needed to ensure convergence. The underrelaxed version of

Eq. (9-26) follows as

$$u_{e,i}^* = \alpha_u \frac{\left[\sum_n A_{nb} u_{nb}^* + b + (p_P^* - p_E^*)\,\Delta y\right]_{i-1}}{A_e} + (1 - \alpha_u)u_{e,i-1}^*$$

Equation (9-27) for v^* is similarly modified. This requirement to simultaneously solve algebraic equations is a consequence of the implicit formulation.

3. Evaluate the coefficients of the pressure equation Eq. (9-31) and solve for the pressure correction p' distribution, using underrelaxation as discussed for step 2.

4. Correct the pressure p distribution according to

$$p = p^* + \alpha_p p'$$

where α_p is an underrelaxation factor ($\alpha_p \approx 0.8$).

Underrelaxation is needed because the summation $\sum_n A_{nb}(u_{eb} - u_{eb}^*)$ was neglected in Eq. (9-28), so only the portion α_p of the correction p' is used.

5. Correct the velocities u, v from their initial, starred, values according to Eq. (9-30).

6. Solve Eq. (9-22), underrelaxed ($a_\phi \approx 0.5$) as described in step 2, for scalar ϕ (temperature, mass fraction, turbulence quantities, etc.). If the flow is not influenced by ϕ, it is better to wait until velocity and pressure calculations are finished to calculate ϕ.

7. Use the corrected pressure p distribution as an initial one, and go back to step 2. Repeat this sequence until a converged solution is obtained.

Improvements to SIMPLE

The SIMPLE algorithm can be improved by calculating pressure corrections more accurately. Repetition of the steps that led to Eq. (9-28) begins the derivation of one improvement. First, the x- and y-momentum equations are written as

$$u_e = \underline{u_e} + d_e(p_P - p_E) \tag{9-26a}$$

$$v_n = \underline{v_n} + d_n(p_P - p_N) \tag{9-27a}$$

where the pseudovelocities $\underline{u_e}$, $\underline{v_n}$ are

$$\underline{u_e} = \frac{\sum_n A_{nb} u_{nb} + b}{A_e}, \qquad \underline{v_n} = \frac{\sum A'_{nb} v_{nb} + b'}{A_n}$$

Comparison with Eq. (9-30) shows that the pseudovelocities u_e, v_n take the place of initial velocities u^*, v^* and pressure p takes the place of pressure correction p'. Thus, either by repetition of the steps that led to Eq. (9-31) for the pressure correction p' or by inspection, it follows that the equation for calculating a new pressure p from a specified velocity distribution is

$$a'_P p_P = a'_E p_E + a'_W p_W + a'_N p_N + a'_S p_S + b'' \tag{9-32}$$

where the a' are as for Eq. (9-31) and b'' differs from b' only in having the pseudovelocities u, v used in place of the starred velocities u^*, v^*.

The algorithm based on this improved pressure calculation is called the SIMPLE-Revised (SIMPLER) algorithm as described by Patankar [91, 92]. The sequence of eight SIMPLER steps differs from the SIMPLE algorithm in starting with a guessed velocity distribution (it is easier to make a good guess for velocity than for pressure).

1. Specify an initial velocity distribution.
2. Evaluate the coefficients of the momentum equations [Eqs. (9-26a) and (9-27a)] and the pseudovelocities u, v.
3. Evaluate the coefficients of the pressure equation [Eq. (9-32)] and solve it to obtain the pressure distribution p.
4. Treat this pressure distribution as p^* and solve the momentum equations [Eqs. (9-26) and (9-27)] to obtain u^*, v^*.
5. Solve for the pressure correction p' distribution from Eq. (9-31).
6. Correct the velocity by use of Eq. (9-30).
7. Solve Eq. (9-22) for other scalars ϕ.
8. Go to step 2 and repeat this sequence until a converged solution is obtained.

An alternative procedure, called SIMPLE-Consistent (SIMPLEC), for more accurately calculating the pressure correction has been devised by Van Doormaal and Raithby [93] that is preferred by some. In it, $\sum_n A_{nb} u'_e$ is subtracted from both sides of the x-momentum equation [Eq. (9-28a)] to get

$$\left(A_e - \sum A_{nb}\right) u'_e = \sum_n A_{nb}(u'_{nb} - u'_e) + (p'_P - p'_E)\,\Delta y \tag{9-33}$$

Again neglecting the summation on the right-hand side of this equation, a consistent approximation from which the C in SIMPLEC comes, replacing u' by $u - u^*$ then puts Eq. (9-33) in the form

$$u_e = u^*_e + d_e(p'_P - p'_E) \tag{9-34}$$

where $d_e = \Delta y / (A_e - \Sigma_n A_{nb})$. Similar expressions apply to the other velocity components. See the original paper [93] for treatment of nonrectangular control volumes.

The steps in SIMPLEC are identical to those in SIMPLE except for two:

1. The new d_e are calculated as defined for Eq. (9-34) rather than as for Eq. (9-28b), and the new d_e are used in the coefficients of the p' equation [Eq. (9-31)] and the velocity correction equation [Eq. (9-34)].

2. The improved pressure is obtained from

$$p = p^* + p'$$

in which no underrelaxation ($\alpha_p < 1$) is used.

The CELS method of Raithby and colleagues, extended by Davidson [94], is said to be more efficient and robust than SIMPLEC.

The SIMPLEX algorithm described by Raithby and Schneider [95, p. 276] provides an even better calculation of the d_e.

A further improvement to the SIMPLE algorithm was devised by Miller and Schmidt [96] that eliminates the need for a staggered grid. This improvement, substantially easing programming and implementation for the SIMPLE algorithm since the programmer need keep track of only one grid, is made possible by more accurately coupling the pressure and velocity distributions. This procedure has been employed in the FLUENT program.

The complex geometry that can constrain the flow often does not conform to an orthogonal coordinate system, of which the rectangular and the cylindrical are common examples. Accommodation to such an irregular geometry within the structure of a SIMPLE-type of algorithm can be accomplished by the use of body-fitted coordinates as described by Dvinsky [97, 98] for the FLUENT/BFC program. Control-volume finite-element formulations discussed by Baliga and Patankar [99] to handle irregular geometries also utilize a SIMPLE-type algorithm to solve the resultant equations.

The quadratic-upstream-interpolation-for-convective kinematics (QUICK) method described by Leonard [100] is often used to evaluate the convective flux out of an interface of a control volume by fitting a parabola between two upstream nodes and one downstream node containing the interface. QUICK gives more accurate results than either an upwind or hybrid method when flow is not perpendicular to control-volume faces. Discussion of QUICK, QUICKEST, the newest version SHARP [104], and the tridiagonal-matrix-algorithm (TDMA) technique for solving the simultaneous algebraic equations encountered is available [95, 99, 101]. Briefly, the one-dimensional

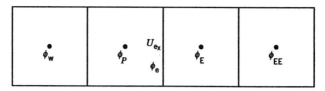

Figure 9-12 Diagram of nodes for the general variable ϕ used in the QUICK scheme (x direction).

QUICK scheme is illustrated in Fig. 9-12, showing that

$$
\phi_e = \begin{cases} \dfrac{\phi_P + \phi_E}{2} - \dfrac{\phi_W - 2\phi_P + \phi_E}{8}, & u_e > 0 \\[3mm] \dfrac{\phi_P + \phi_E}{2} - \dfrac{\Phi_{EE} - 2\phi_E + \phi_P}{8}, & u_e < 0 \end{cases}
$$

Three-Dimensional Form of SIMPLE

The general flow down a duct is three dimensional so three-dimensional forms of the momentum, continuity, and scalar conservation equations must be solved. The pressure at a point in the duct exerts an influence on the upstream flow; this is said to be an elliptic flow since the describing equations are elliptic in form—the possible forms are elliptic (usually for equilibrium problems), parabolic (usually for unsteady diffusion problems), and hyperbolic (usually for wave propagation problems).

A fully elliptic procedure would entail storing all dependent flow variables throughout the solution domain. To reduce storage requirements, thereby enabling a finer solution grid to be utilized, the semielliptic procedure of storing only the pressure at every node is adopted; all other quantities are stored at only two adjacent planes at any time. Since the pressure is then computed at only one cross-sectional plane at a time, a number of iterative passes must be made in the flow direction (one plane at a time) before a converged pressure distribution in the flow direction as well as in the crossflow direction is achieved; regions of reversed flow (recirculation) require special treatment.

At each cross-sectional plane the simultaneous finite-difference equations are solved (usually by TDMA) as discussed earlier. The velocities are corrected from the pressure corrections as before for the crossflow components u, v of velocity in a cross-sectional plane. For the z component w of velocity down the duct, the equations are of the same form

$$
w_d = w_d^* + d_d(p_P' - p_D')
$$

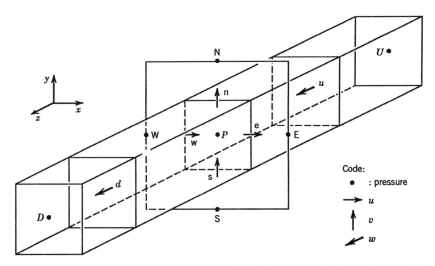

Figure 9-13 Three-dimensional staggered grid showing control volume for nonveloc-ity properties such as pressure. [From R. W. Johnson, Turbulent convecting flow in a square duct with a 180° bend: an experimental and numerical study, Ph.D. Dissertation, Faculty of Technology, University of Manchester, 1984 [102].]

for example, with the subscripts d and D referring to the downstream face of a control volume and the subscripts u and U referring to the upstream face as illustrated in Fig. 9-13.

Initially it can be assumed that $p'_D = 0 = p'_U$. A "bulk pressure" correction due to Pratap [101] and discussed by Patankar and Spalding [75] can be made by substituting into the total mass flow rate $\dot{m}$ relationship

$$\dot{m} = \sum_x \sum_y \rho w_P \, \Delta x \, \Delta y$$

the expression for w_P of

$$w_P = w_P^* + d_P(p'_P - p'_D) \tag{9-35}$$

to obtain

$$p'_P - p'_D = \frac{\dot{m} - \sum_x \sum_y \rho w_P^* \, \Delta x \, \Delta y}{\sum_x \sum_y \rho \, d_P \, \Delta x \, \Delta y}$$

After this correction is made to p'_D, the w-velocity distribution is readjusted according to Eq. (9-35).

Additionally, the correction p'_P can be added to upstream planes as soon as it is computed. In this way, fewer iterative passes in the streamwise direction are required to achieve convergence. In this scheme, the upstream pressure p_U at the adjacent upstream plane is corrected according to

$$p_{U,\text{new}} = p_{U,\text{old}} + \beta p'_P$$

The parameter β is

$$\beta = \frac{\Delta x \, \Delta y \, d_{P,\text{w}}}{\Delta x \, \Delta y (d_{P,\text{w}} + d_{U,\text{w}}) + \Delta x \, \Delta z (d_{N,v} + d_{U,v}) + \Delta y \, \Delta z (d_{E,u} + d_{U,u})}$$

and is derived by eliminating the mass flow rate $\dot{m}_P$ imbalance, created at the upstream plane when p'_P was applied to p^*_P in the plane of interest, just as was done to derive the equation for the pressure correction p'. Of course, the correction to pressure in the adjacent upstream plane creates an imbalance in the adjacent plane upstream to it, and a pressure correction should be applied to it in the same manner, and so on. Thus the presure correction for the nth upstream plane is

$$p'_n = \beta_1, \beta_2, \ldots \beta_n p'_P$$

where $n = 6$ has been found to balance reduction in streamwise iteration and increase in computational complexity for each iteration. Detailed description of a three-dimensional form of the SIMPLE algorithm is provided by Johnson [102].

Methods for incompressible flows generally use pressure as a main dependent variable and use a staggered grid to prevent a checkerboard pressure pattern from being accepted as a solution. Since pressure changes are always finite, a calculation procedure can be developed that is applicable for all Mach numbers, including the compressible range. Such an extension for boundary-fitted, nonorthogonal coordinates was developed by Karki and Patankar [103].

PROBLEMS

9-1   Perform the calculations on which Fig. 9-1 is based.

(a) Show that wall shear stress τ_w and pressure drop Δp are related by

$$\tau_w = \frac{\Delta p}{4} \frac{D}{L}$$

(b) Introduce the relationship of Eq. (7-21a) into the result of part a for a smooth wall and rearrange to find

$$r_0^+ = \frac{r_0 v^*}{\nu} = 0.09944 \, \text{Re}_D^{7/8}$$

(c) Recognizing that $y^+ = r_0^+ y/r_0$, plot ε_m/ν against y/r_0 for $\mathrm{Re}_D = 10^5$. In the central region of the pipe show the results for the $\frac{1}{7}$th power-law velocity distribution from Eq. (9-4), the approximation to Reichardt's velocity distribution from Eq. (7-20a) evaluated at $r \approx r_0$, and Reichardt's velocity distribution from Eq. (7-20).

9-2 (a) Put the Reynolds analogy of Eq. (9-12) into the form

$$\frac{\mathrm{Nu}}{\mathrm{Re}_D\,\mathrm{Pr}}$$

$$= \frac{C_f/2}{0.848 + 2.86\big(1 + 1.46\{\mathrm{Pr} - 1 + \ln[(1 + 5\,\mathrm{Pr})/6]\}\big)(C_f/2)^{1/2}}$$

(b) Compare the prediction of part a with Eq. (9-13) for water ($\mathrm{Pr} = 5$) for Re_D in the range 10^4–10^6 for a long, smooth circular pipe. Comment on the apparent ability of the simplified theory of Eq. (9-12) to explain the physical processes in this situation.

(c) Compare the results of part a against those of the common Dittus–Boelter correlation

$$\mathrm{Nu}_D = 0.023\,\mathrm{Re}_D^{0.8}\,\mathrm{Pr}^{1/3}$$

9-3 Consider a long, smooth round pipe around whose periphery the heat flux varies according to $q_w(\phi) = q_{w0}(1 + b\cos\phi)$. For $\mathrm{Re}_D = 10^5$ and $b = 0.2$, determine the maximum and minimum value of $\mathrm{Nu}_D(\phi)$ and the percentage deviation of the average value of $\mathrm{Nu}_D(\phi)$ from Nu_D for a constant wall heat flux for:

(a) $\mathrm{Pr} = 0.01$, characteristic of liquid metals.

(b) $\mathrm{Pr} = 0.7$, characteristic of gases. *Answer*: $\mathrm{Nu}(\phi = 0)/\mathrm{Nu}_D = 0.91$, $\mathrm{Nu}(\phi = \pi/\mathrm{Nu}_D = 1.18$.

(c) $\mathrm{Pr} = 3.0$, characteristic of water.

Comment on the accuracy of using Nu_D for constant wall heat flux as an estimate of the average Nu_D in a varying wall heat flux situation.

9-4 Consider fully developed flow of air at $\mathrm{Re}_D = 10^5$ in a round pipe of 1 in. diameter and made of commercial steel.

(a) From Fig. 9-6 and Table 9-6, verify that $f/f_s = 1.41$.

(b) From Fig. 9-7 verify that $h/h_s = 1.27$. Compare this result against $h/h_s = (f/f_s)^{1/2}$.

(c) Comment on the relative effect of roughness on pressure drop and heat transfer.

9-5 Consider fully developed flow at 25 ft/sec mean velocity in a circular tube of 1 in. diameter with constant heat rate per unit length. Evaluate the heat-transfer coefficient for the following cases and discuss the reasons for the differences:

(a) Air, 200°F, 1 atm pressure.

(b) Hydrogen gas, 200°F, 1 atm pressure.

(c) Liquid water, 100°F,

(d) Air, 200°F, 10 atm pressure. *Answer*: $h = 33.6$ Btu/hr ft^2 °F.

9-6 Consider a 0.5-in-i.d. tube 6 ft long wound by an electric resistance heating element. The function of the tube is to heat an organic fluid from 50 to 150°F. The mass rate of the fluid is 1000 lb$_m$/hr, and the following average properties may be treated as constant: Pr = 10, $\rho = 47$ lb$_m$/ft^3, $C_p = 0.5$ Btu/lb$_m$ °F, $k = 0.079$ Btu/hr ft °F, and $\mu = 1.6$ lb$_m$/hr ft. Plot both tube surface temperature and fluid mean temperature as a function of tube length.

9-7 Liquid potassium flows in a 1-in.-i.d. tube at a mean velocity of 8 ft/sec and a mean temperature of 1200°F. The tube is heated at a constant rate per unit of length, but the heat flux varies around the periphery of the tube in a sinusoidal manner, with the maximum heat flux twice the minimum heat flux. If the maximum surface temperature is 1500°F, evaluate the axial mean temperature gradient and plot temperature around the periphery of the tube.

9-8 Air at 70°F and 1 atm pressure flows at a Reynolds number of 50,000 in a 1-in.-i.d. circular tube. The tube wall is insulated for the first 30 in., but for the next 50 in. the tube surface temperature is constant at 100°F. Then it abruptly increases to 130°F and remains constant for another 50 in. Plot the heat-transfer coefficient as a function of axial distance. Does the abrupt increase in surface temperature to 130°F cause a significant change in h over the heated length? In heat exchanger theory, a mean h with respect to tube length is generally employed. In this case, how much does the mean differ from the asymptotic value of h?

9-9 Starting with the constant-surface-temperature, thermal-entry-length solutions for a circular tube, calculate and compare the Nusselt numbers for constant surface temperature and for a linearly varying surface temperature, very long tubes, a Reynolds number of 50,000, and a Prandtl number of 0.01. Repeat for a Prandtl number of 0.7 and discuss the reasons for the differences noted.

9-10 Specifications for the cooling tubes in a pressurized-water nuclear-power reactor are: *tube configuration*—concentric-circular tube annulus, with heating from the inner tube (containing the uranium fuel), and the outer tube surface having no heat flux; *tube dimensions*—inner tube diameter 1 in., outer tube diameter 2 in., tube length 15 ft; *water* *temperatures*—inlet 525°F, outlet 575°F; *water mean velocity*—3 ft/sec; *axial heat flux distribution*—$\dot{q}''/\dot{q}''_{max} = [1 + 2\sin(\pi x/L)]/3$. Plot heat flux, mean water temperature, and inner and outer tube surface temperature as a function of x. Assume that h is independent of x and the value is that for fully developed constant heat rate. Justify these assumptions. Is it possible to avoid boiling with these specifications? What would be the effect of local boiling at the highest-temperature parts of the system?

9-11   Consider fully developed turbulent flow in a circular tube with heat transfer to or from the fluid at a constant rate per unit of tube length. There is internal heat generation (perhaps from nuclear reaction) at a rate S, Btu/hr ft³. For Re = 50,000 and Pr = 4, evaluate the Nusselt number as a function of the pertinent parameters. The heat-transfer coefficient is defined in the usual manner on the basis of the heat flux at the surface, the surface temperature, and the mixed mean fluid temperature.

9-12   Water flows in a heat exchanger tube 3 ft long and 1 in. diameter at a velocity of 15 ft/sec. The tube wall temperature is constant at 210°F as a result of condensing steam on the outside. If the inlet temperature is 60°F, what is the exit temperature? *Answer*: T_{out} = 95°F.

9-13   Mercury at an average temperature of 200°F flows through a $\frac{1}{2}$-in.-diameter tube at the rate of 10^4 lb$_m$/hr. Calculate the average heat-transfer coefficient. *Answer*: h = 1600 Btu/hr ft² °F.

9-14   Power generation in a nuclear reactor is principally limited by the ability to transfer heat into a coolant. A solid-fuel reactor is to be cooled by a fluid flowing inside 0.25-in.-diameter stainless-steel tubes. If the tube wall temperature is 600°F, compare the relative merits of the use of water or liquid sodium as the coolant. In each case the velocity is 15 ft/sec and the fluid inlet temperature is 400°F. *Answer*: $q_{H_2O}/q_{Na} \approx 1/100$ for short tubes, 3 for long tubes.

9-15   The Reynolds analogy between the coefficients of heat transfer and wall shear can be suggested on the basis of the following simple analysis.

(a) Divide the expression for shear stress

$$\tau = \rho(\nu + \varepsilon_m)\frac{du}{dy}$$

by the expression for heat flux

$$q = \rho C_p(\alpha + \varepsilon_H)\frac{dT}{dy}$$

to obtain

$$\frac{\tau C_p}{q} = \frac{\nu + \varepsilon_m}{\alpha + \varepsilon_H} \frac{du}{dT}$$

(b) Assume that the molecular and turbulent Prandtl numbers are unity so that the velocity and temperature profiles are nearly the same and $du/Dt \approx$ const. Replace the left-hand side by wall values and then integrate to obtain

$$\frac{\tau_w C_p}{q_w}(T_{max} - T_{wall}) = (U_{max} - U_{wall})$$

(c) Set $\tau_w = (C_f/2)\rho U_{av}^2$ and $q_w = h(T_{mixing\ cup} - T_{wall})$ to obtain

$$\frac{h}{\rho C_p U_{av}} = \frac{C_f}{2}\left[\frac{U_{av}}{U_{max} - U_{wall}} \frac{T_{max} - T_{wall}}{T_{mixing\ cup} - T_{wall}}\right]$$

which, since the bracketed term must be nearly unity, can be rephrased as $\mathrm{Nu}/\mathrm{Re}\ \mathrm{Pr} \approx C_f/2$ and constitutes Reynolds' analogy.

9-16 The form of the Dittus–Boelter evaluation is to be suggested by appealing to Reynolds' analogy.

(a) Into Reynolds' analogy, substitute Eq. (7-21) to obtain

$$\mathrm{Nu}_D = (\mathrm{const})\ \mathrm{Re}_D^{3/4}\ \mathrm{Pr}^{1/3}$$

(b) Discuss the factors involved in the equation of the constant.

9-17 Use a computer program to solve the two-dimensional problem of turbulent flow and heat transfer sketched in Fig. 9P-17 and described as follows.

(a) Determine the numerical values of the pressure drops between the inlet and the outlet for the right-angle duct with entrance turbulent intensity levels of 1% and 10%.

(b) Compare the numerical values of the pressure drops for part a with those for a 5-m length in the midregion of a very long duct and with those that account for entrance and right-angle bend losses. Comment briefly.

(c) Determine the numerical values of the heat-transfer coefficients at locations A and B for a 310-K inlet fluid temperature and a 300-K wall temperature. Compare these results with those that would be obtained far from the inlet of a very long duct.

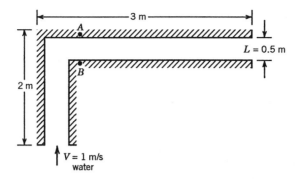

Figure 9P-17

REFERENCES

1. A. Martinuzzi and A. Pollard, *AIAA J.* **27**, 29–36 (1989); A. Pollard and R. Martinuzzi, *AIAA J.* **27**, 1714–1721 (1989).

2. P. S. Granville, A near-wall eddy viscosity formula for turbulent boundary layers in pressure gradients, *Forum on Turbulent Flows*–1989 (Proc. Third Joint ASCE/ASME Mechanics Conf., La Jolla, CA, July 9–12, 1989), ASME FED-Vol. 76, 1989, pp. 25–28.

3. D. R. Chapman and G. D. Kuhn, *J. Fluid Mech.* **170**, 265–292 (1986).

4. R. C. Martinelli, *Trans. ASME* **69**, 947–959 (1947).

5. L. M. K. Boelter, R. C. Martinelli, and F. Jonassen, *Trans. ASME* **63**, 447–455 (1941).

6. B. S. Petukhov and V. N. Popov, *Teplofiz. Vysok. Temperatur (High Temperature Heat Physics)* **1** (1), 69–83 (1963).

7. B. S. Petukhov, *Adv. Heat Transfer* **6**, 503–564 (1970).

8. R. L. Webb, *Wärme-und Stoffübertragung* **1**, 197–204 (1971).

9. F. D. Haynes and G. D. Ashton, *ASME J. Heat Transfer* **102**, 384–386 (1980).

10. R. G. Deissler, Analysis of turbulent heat transfer, mass transfer, and friction in smooth tubes at high Prandtl and Schmidt numbers, NACA Report 1210, 1955.

11. E. M. Sparrow, T. M. Hallman, and R. Siegel, *Appl. Sci. Res.* **A7**, 37–52 (1957).

12. W. M. Kays and E. Y. Leung, *Int. J. Heat Mass Transfer* **6**, 537–557 (1963).

13. R. Jenkins, *Heat Transfer and Fluid Mechanics Institute*, Stanford University Press, Stanford, CA, 1951, pp. 147–158.

14. R. G. Deissler, Analysis of fully developed turbulent heat transfer at low Peclet numbers in smooth tubes with application to liquid metals, NACA Research Memo E52F05, 1955.

15. M. Jacob, *Heat Transfer*, Vol. 2, Wiley, 1957, pp. 509–529.

16. J. O. Hinze, *Turbulence*, McGraw-Hill, 1975, pp. 386–391.

17. J. M. Burgers, lecture notes, California Institute of Technology, Pasadena, CA, 1951.

18. W. C. Reynolds, *Int. J. Heat Mass Transfer* **6**, 445–454 (1963).

19. W. M. Kays and M. E. Crawford, *Convective Heat and Mass Transfer*, McGraw-Hill, 1980.

20. W. A. Sutherland and W. M. Kays, *Int. J. Heat Mass. Transfer* **7**, 1187–1194 (1964).

21. R. G. Deissler and M. F. Taylor, Analysis of fully developed turbulent heat transfer and flow in an annulus with various eccentricities, NACA TN 3451, Washington, DC, 1955.

22. E. Y. Leung, W. M. Kays, and W. C. Reynolds, Report AHT-4, Department of Mechanical Engineering, Stanford University, Stanford, CA, April 15, 1962.

23. J. W. Baughn, *ASME J. Heat Transfer* **100**, 537–539 (1978).

24. R. A. Seban and T. T. Shimazaki, *Trans. ASME* **73**, 803–809 (1951).

25. C. A. Sleicher and M. Tribus, *Heat Transfer and Fluid Mechanics Institute*, Stanford University Press, Stanford, CA, 1956, pp. 59–78.

26. C. A. Sleicher and M. W. Rouse, *Int. J. Heat Mass Transfer* **18**, 677–683 (1975).

27. R. H. Notter and C. A. Sleicher, *Chem. Eng. Sci.* **27**, 2073–2093 (1972).

28. C. A. Sleicher, A. S. Awad, and R. H. Notter, *Int. J. Heat Mass Transfer* **16**, 1565–1575 (1973).

29. R. N. Lyon, *Chem. Eng. Prog.* **47**, 75–79 (1951).

30. R. N. Lyon, Ed., *Liquid Metals Handbook*, 3rd ed., U.S. Atomic Energy Commission and Department of the Navy, Washington, DC, 1952.

31. R. Stein, *Adv. Heat Transfer* **3**, 101–174 (1966).

32. F. W. Dittus and L. M. K. Boelter, University of California, Berkeley, CA, Publications in Engineering **2**, 443 (1930).

33. E. N. Sieder and G. E. Tate, *Ind. Eng. Chem.* **28**, 1429–1435 (1936).

34. M. Molki and E. M. Sparrow, *ASME J. Heat Transfer* **108**, 482–484 (1986).

35. J. P. Harnett, *Trans. ASME* **77**, 1211–1234 (1955).

36. H. Latzko, *Zeitschrift für angewandte Mathematik und Mechanik* **1**, 268–190 (1921).

37. R. G. Deissler, *Trans. ASME* **77**, 1221–1234 (1955); NACA TN 3016.

38. L. M. K. Boelter, G. Young, and H. W. Iverson, An investigation of aircraft heaters—XXVII. Distribution of heat transfer rate in the entrance section of a circular tube, NACA TN 1451, 1948.

39. S. Faggiani and F. Gori, *ASME J. Heat Transfer* **102**, 292–296 (1980).

40. L. F. Moody, *Trans. ASME* **66**, 671–684 (1944).

41. C. F. Colebrook, *J. Inst. Civ. Eng.* **11**, 133–156 (1938/1939).

42. R. M. Nedderman and G. J. Shearer, *Chem. Eng. Sci.* **19**, 423–428 (1964).

43. S. W. Churchill, *Chem. Eng.* **84** (24), 91–92 (1977).

44. P. K. Swamee and A. K. Jain, *J. Hydraulic Division of the ASCE* **102** (HY5), 657–664 (May 1976).

45. S. D. Joshi and R. K. Shah, in S. Kakac, R. K. Shah, and W. Aung, Eds., *Handbook of Single-Phase Convective Heat Transfer*, Wiley-Interscience, 1987, pp. 10-1–10-32.

46. E. Burck, *Wärme-und Stoffübertragung* **2**, 87–98 (1969).

47. W. Nunner, *Verein Deutscher Ingenieure—Forschungsheft*, Nu. 4551, 1956.

48. D. F. Dipprey and R. H. Sabersky, *Int. J. Heat Mass Transfer* **6**, 329–353 (1963).

49. G. H. Christoph and R. H. Pletcher, *AIAA J.* **21**, 509–515 (1983).

50. A. T. Wassel and A. G. Mills, *ASME J. Heat Transfer* **101**, 469–474 (1979).

51. R. L. Webb, E. R. G. Eckert, and R. J. Goldstein, *Int. J. Heat Mass Transfer* **14**, 601–617 (1971).

52. T. C. Carnavos, *Heat Transfer Eng.* **1**, 32–37 (1980).

53. R. L. Webb and M. J. Scott, *ASME J. Heat Transfer* **102**, 38–43 (1980).

54. F. Kreith and M. S. Bohn, *Principles of Heat Transfer*, 4th ed., Harper and Row, 1986, pp. 621–629.

55. J. G. Soland, W. W. Mack, and W. M. Rohsenow, *ASME J. Heat Transfer* **100**, 514–519 (1978).

56. A. M. Demuren, *AIAA J.* **29**, 531–537 (1991).

57. W. R. Dean, *Philos. Mag.* **4**, 208 (1927); **5**, 673 (1928).

58. W. Adler, *Zeitschrift für angewandte Mathematik und Mechanik* **14**, 257–275 (1934).

59. G. J. Hwang and C.-H. Chao, *ASME J. Heat Transfer* **113**, 48–55 (1991).

60. K. Futagami and Y. Aoyama, *Int. J. Heat Mass Transfer* **31**, 387–396 (1988).

61. G. Yee, R. Chilukuri, and J. A. C. Humphrey, *ASME J. Heat Transfer* **102**, 285–291 (1980).

62. H. Ito, *ASME J. Basic Eng.* **81**, 123–124 (1959).

63. R. A. Seban and E. F. McLaughlin, *Int. J. Heat Mass Transfer* **6**, 87–95 (1963).

64. Y. Mori and W. Nakayama, *Int. J. Heat Mass Transfer* **10**, 681–695 (1967); **10**, 37–59 (1967); **8**, 67–82 (1965).

65. D. E. Metzger and N. H. Afgan, *Heat and Mass Transfer in Rotating Machinery*, Hemisphere, 1984.

66. W.-J. Yang, *Heat Transfer and Fluid Flow in Rotating Machinery*, Hemisphere, 1986.

67. J. P. Boris, *Ann. Rev. Fluid Mech.* **21**, 345–385 (1989).

68. *Aerospace America* **30** (1–2) (1992).

69. B. Hutchings and R. Iannuzzelli, *Mech. Eng.* **109**, 72–76 (May 1987).

70. B. Hutchings and R. Iannuzzelli, *Mech. Eng.* **109** (6), 54–58 (1987).

71. R. Iannuzzelli and B. Hutchings, *Mech. Eng.* **109** (7), 60–63 (1987).

72. A. Wolfe, *Mech. Eng.* **113** (1), 48–54 (1991).

73. L. O'Connor, *Mech. Eng.* **114** (5), 44–50 (1992).

74. A. D. Gossman, W. M. Pun, A. K. Runchal, D. B. Spalding, and M. Wolfshtein, *Heat and Mass Transfer in Recirculating Flows*, Academic, 1969.

75. S. V. Patankar and D. B. Spalding, *Int. J. Heat Mass Transfer* **5**, 1787–1806 (1972).

76. S. V. Patankar, *Numerical Heat Transfer and Fluid Flow*, Francis and Taylor, 1980.

77. S. V. Patankar, *Computation of Conduction and Duct Flow Heat Transfer*, Innovative Research Inc., 7846 Ithaca Lane N., Maple Grove, MN 55369-8549, 1990.

78. S. Orzag, *J. Comput. Phys.* **37**, 70–92 (1980).

79. C. Canuto, M. Y. Hussaini, A. Quateroni, and T. A. Zang, *Spectral Methods in Fluid Dynamics*, Springer-Verlag, 1988.

80. D. Pepper and J. Heinrich, *The Finite Element Method: Basic Concepts and Applications*, Hemisphere, 1992.

81. C. A. Brebbia, Ed., *Boundary Elements X: Heat Transfer, Fluid Flow and Electrical Applications*, Springer-Verlag, 1988.

82. C. A. Brebbia and J. Dominguez, *Boundary Elements: An Introductory Course*, McGraw-Hill, 1989.

83. S. A. Kinnas and C.-Y. Hsin, *AIAA J.* **30**, 688–696 (1992).

84. C.-Y. Chow, *An Introduction to Computational Fluid Mechanics*, Seminole Publishing Co., P.O. Box 3315, High-Mar Station, Boulder, CO 80307, 1983.

85. J. E. Kerwin and C.-S. Lee, *Trans. Soc. Naval Arch. Marine Eng.* **86**, 218–253 (1978).

86. C.-J. Chen, in W. J. Minkowycz, E. M. Sparrow, G. E. Schneider, and R. H. Pletcher, Eds., *Handbook of Numerical Heat Transfer*, Wiley, 1988, pp. 723–746.

87. A. Haji-Sheikh, in W. J. Minkowycz, E. M. Sparrow, G. E. Schneider, and R. H. Pletcher, Eds., *Handbook of Numerical Heat Transfer*, Wiley, 1988, pp. 673–722.

88. F. H. Harlow and J. E. Welch, *Phys. Fluids* **8**, 2182–2189 (1965).

89. A. J. Chorin, *Math. Comput.* **22**, 745–762 (1968).

90. A. A. Amsden and F. H. Harlow, The SMAC method: a numerical technique for calculating incompressible fluid flows, Los Alamos Scientific Laboratory Report LA-4370, 1970, Los Alamos, NM.

91. S. V. Patankar, *Numerical Heat Transfer and Fluid Flow*, Hemisphere, 1980.

92. S. V. Patankar, *Numerical Heat Transfer* **4**, 409–425 (1981).

93. J. P. Van Doormaal and G. D. Raithby, *Numerical Heat Transfer* **7**, 147–163 (1984).

94. L. Davidson, *Numerical Heat Transfer* **18**, 129–147 (1990).

95. G. D. Raithby and G. E. Schneider, in W. J. Minkowycz, E. M. Sparrow, G. E. Schneider, and R. H. Pletcher, Eds., *Handbook of Numerical Heat Transfer*, Wiley, 1988, pp. 241–291.

96. T. F. Miller and F. W. Schmidt, *Numerical Heat Transfer* **14**, 213–233 (1988).

97. A. S. Dvinsky, in C. Taylor, W. G. Habashi, and M. M. Hafez, Eds., *Numerical Methods in Laminar and Turbulent Flow* **5** (1), 1987, pp. 137–148.

98. A. S. Dvinsky, A new framework for grid generation, *Proc. AIAA/ASME/SIAM/APS First National Fluid Dynamics Congress (July 25–28, 1988, Cincinnati, OH)*, AIAA, 1988, pp. 480–483.

99. B. R. Baliga and S. V. Patankar, in W. J. Minkowycz, E. M. Sparrow, G. E. Schneider, and R. H. Pletcher, Eds., *Handbook of Numerical Heat Transfer*, Wiley, 1988, pp. 421–461.

100. B. P. Leonard, *Comp. Meths. Appl. Eng.* **19**, 59–98 (1979).

101. V. S. Pratap and D. B. Spalding, *Aeronaut. Quart.* **26**, 219–228 (1975); *A calculation procedure for partially-parabolic flow situations*, Imperial College, Mechanical Engineering Department, Report HTS/75/19, 1975.

102. R. W. Johnson, *Numerical Heat Transfer* **13**, 205–228 (1988); *Turbulent convecting flow in a square duct with a 180° bend: an experimental and numerical study*, Ph.D. Dissertation, Faculty of Technology, University of Manchester, 1984.

103. K. C. Karki and S. V. Patankar, *AIAA J.* **27**, 1167–1174 (1989).

104. B. P. Leonard and H. S. Niknafs, *Computers & Fluids* **19**, 144–154 (1991).

10

NATURAL CONVECTION

In all the previous examples the origin of fluid motion lay outside the heat- or mass-transfer problem, and there was no need to be concerned about it. However, in natural convection fluid motion is caused by density variations, resulting from temperature distributions in the case of heat transfer, which are acted on by local gravitational and centrifugal forces. As might be expected, velocities are usually small since the forces are usually small.

As with forced convection, natural convection is conveniently divided into internal flows, in which the fluid is enclosed by solid boundaries as in a package of electronic equipment, and external flows, in which the fluid extends indefinitely from the solid surface from which transport occurs as in a porous medium near a waste disposal site. In addition, a further classification as to the laminar or turbulent nature of the flow is also made.

10.1 LAMINAR NATURAL CONVECTION FROM A CONSTANT-TEMPERATURE VERTICAL PLATE IN AN INFINITE FLUID—EXACT SOLUTION

Studies are initiated by considering a vertical flat plate immersed in an infinitely large body of fluid. The fluid properties are constant (except insofar as density variations are important in the body force terms of the equations of motion), flow is laminar, the plate and the fluid are at constant temperature, and steady state prevails. For this situation, illustrated in Fig. 10-1, the

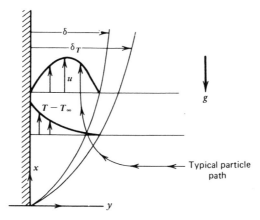

Figure 10-1 Laminar boundary layer in natural convection on a hot vertical flat plate.

boundary-layer equations and boundary conditions are

Continuity $\qquad \dfrac{\partial u}{\partial x} + \dfrac{\partial v}{\partial y} = 0$

x Motion $\qquad u\dfrac{\partial u}{\partial x} + v\dfrac{\partial u}{\partial y} = \nu_\infty \dfrac{\partial^2 u}{\partial y^2} + \dfrac{B_x - \partial p/\partial x}{\rho_\infty}$

y Motion $\qquad \dfrac{\partial p}{\partial y} = 0$

Energy $\qquad u\dfrac{\partial T}{\partial x} + v\dfrac{\partial T}{\partial y} = \alpha_\infty \dfrac{\partial^2 T}{\partial y^2}$

$\qquad$ At $y = 0 \qquad u = 0 = v \qquad$ and $\qquad T = T_w$

$\qquad$ At $y = \infty \qquad u = 0 \qquad$ and $\qquad T = T_\infty$

Because the expected velocities are small and temperature differences are not, viscous dissipation is ignored as is the possibility of fluid cooling as it rises and expands in the lower pressure existing at higher positions, although this is an important phenomenon in meteorology; see Gebhart [1] for studies of both effects.

The pressure gradient and body force must be evaluated. The body force B_x is the result of gravity; thus $B_x = -\rho g$. Because $\partial p/\partial y = 0$, $\partial p/\partial x$ can be evaluated at a large distance from the plate where the equations of motion show that for a stagnant fluid ($u = 0$), $\partial p/\partial x = -\rho_\infty g$. As a result, affecting only the x-motion equation,

$$\frac{1}{\rho_\infty}\left(B_x - \frac{\partial p}{\partial x}\right) = (\rho_\infty - \rho)\frac{g}{\rho_\infty}$$

A similarity solution is sought first in order to establish the exact solution of the laminar boundary-layer model. These exact results are then compared against measurements. Unfortunately, the basic physical idea on which a similarity solution is built for a forced convection boundary layer [$\delta \sim t^{1/2} \sim (x/U)^{1/2}$] lacks a specified characteristic velocity U. As a result, a more mathematical approach is required along the lines suggested in Appendix D. The upshot of it is that the similarity variable is $\eta = yc/x^{1/4}$ and the stream function is $\psi = 4\nu cx^{3/4}F(\eta)$, where $c^4 = (g|\rho_\infty - \rho_w|)/(4\nu_\infty^2\rho_\infty)$, with $u = 4\nu c^2 x^{1/2}F'(\eta)$ and $v = (\nu c/x^{1/4})[\eta F' - 3F]$. The integral treatment presented later motivates these functional forms that were first advanced by Pohlhausen (in cooperation with Schmidt and Beckmann [2]—early partially successful analyses were proposed by Oberbeck [3] in 1879 and Lorenz [4] in 1881). Note that if a problem requiring a nonzero normal velocity at the wall is to be considered, that velocity must vary as $x^{-1/4}$. The boundary-layer equations are then given in similarity form as

$$F''' + 3FF'' - 2(F')^2 + \frac{\rho_\infty - \rho}{|\rho_\infty - \rho_w|} = 0 \qquad (10\text{-}1a)$$

$$\theta'' + 3\,\mathrm{Pr}\,F\theta' = 0 \qquad (10\text{-}1b)$$

with boundary conditions of

$$F'(0) = F'(\infty) = \theta(\infty) = 0 \qquad (10\text{-}1c)$$

$$F(0) = -\frac{v_w x^{1/4}}{3\nu c} \qquad (10\text{-}1d)$$

$$\theta(0) = 1 \qquad (10\text{-}1e)$$

where $\theta = (T - T_\infty)/(T_w - T_\infty)$, v_w is a possible wall "blowing" velocity (here assumed to be zero), and a prime denotes an ordinary derivative with respect to η. It can be seen that the $(\rho_\infty - \rho)/|\rho_\infty - \rho_w|$ term acts as a forcing function for velocity; if that term is zero, Eq. (10-1a) would be satisfied by a constant that means that then there is no motion.

There are a number of ways for the forcing function to be generated. In a mass-transfer situation in which water, for example, evaporates from a wetted surface into stagnant air, the fluid density is somewhat dependent on the mass fraction of water vapor in the air. Since water vapor is lighter than air, the effect would be the same as if the surface were hotter than the air—there is an upward convection. In combined heat and mass transfer, it is possible for these two body forces to oppose one another.

Attention is here restricted to density dependence on temperature. From the definition of the coefficient β of thermal expansion

$$\beta = -\frac{1}{\rho}\frac{\partial \rho}{\partial T}\bigg|_{p=\text{const}}$$

in conjunction with the series expansion

$$\rho = \rho_\infty + \frac{\partial \rho}{\partial T}(T - T_\infty) + \frac{\partial^2 \rho}{\partial T^2}\frac{(T - T_\infty)^2}{2!} + \cdots$$

it is found that

$$\rho_\infty \beta (T - T_\infty) \approx \rho_\infty - \rho$$

The differential equations then are coupled, requiring simultaneous solution, as

$$F''' + 3FF'' - 2(F')^2 + \theta = 0 \tag{10-2a}$$

$$\theta'' + 3\,\text{Pr}\,F\theta' = 0 \tag{10-2b}$$

with boundary conditions of

$$F'(0) = F'(\infty) = \theta(\infty) = 0 \tag{10-2c}$$

$$F(0) = -\frac{v_w x^{1/4}}{3\nu c} \tag{10-2d}$$

$$\theta(0) = 1 \tag{10-2e}$$

When "blowing" at the wall is absent, $v_w = 0$. As with forced convection laminar boundary layers, the essence of the problem solution is to find the two unknown quantities—$F''(0)$ and $\theta'(0)$. Because the equations are coupled, the velocity and temperature profiles depend on two separate parameters: the Prandtl number Pr and the Grashof number $\text{Gr} = g\beta|T_w - T_\infty|x^3/v_\infty^2$.

Refined numerical solutions to Eq. (10-2) obtained by Ostrach [5] are displayed in Fig. 10-2. The velocity is seen to achieve a maximum near the plate and, like the temperature, to approach ambient conditions far from the plate. At large Pr the thermal boundary layer is thinner than the velocity boundary layer. Tabular presentation of the numerical solution requires several tables, of which only one for Pr = 1 is given in Table 10-1 [5]. Note there that a fluid particle far from the plate approaches the plate perpendicularly as sketched in Fig. 10-1, since as $\eta \to \infty$, $u = (2v/x)\,\text{Gr}_x^{1/2}\,F' \to 0$ and $v = (v/x)(\text{Gr}_x/4)^{1/4}(\eta F' - 3F) \to -1.556(v/x)(\text{Gr}_x/4)^{1/4}$; the shape of the leading edge should not matter greatly.

Local heat-transfer coefficients are available from these solutions since

$$q_w = -k\frac{\partial T(y=0)}{\partial y} = h(T_w - T_\infty)$$

or, in terms of similarity variables,

$$-k(T_w - T_\infty)\frac{\partial\theta(\eta=0)}{\partial\eta}\frac{\partial\eta}{\partial y} = h(T_w - T_\infty) \qquad h = -\theta'(0)\frac{kc}{x^{1/4}}$$

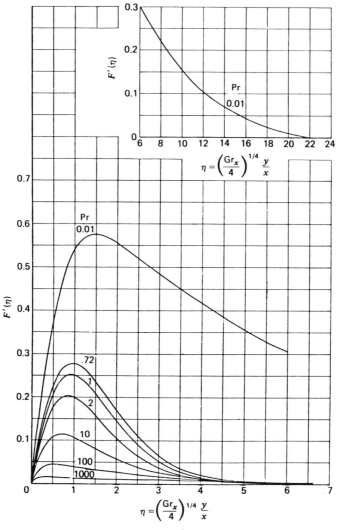

Figure 10-2   Calculated dimensionless profiles in the laminar boundary layer on a hot vertical flat plate in natural convection of (a) velocity and (b) temperature. [From S. Ostrach, NACA Report 1111, 1953 [5].]

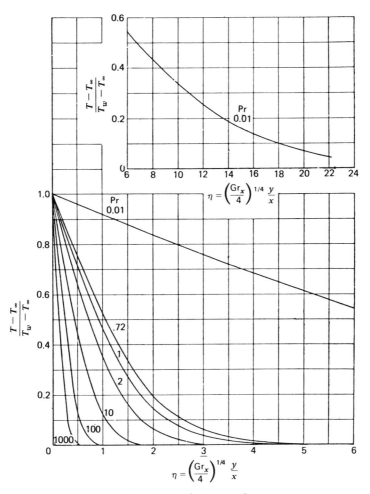

Figure 10-2   (*Continued*)

The average heat-transfer coefficient is obtainable from

$$q_{w_{total}} = \bar{h}(T_w - T_\infty)x = \int_0^x h(T_w - T_\infty)\, dx$$

Inasmuch as $h = Bx^{-1/4}$, integration of h with subsequent division by x gives

$$\bar{h} = \frac{\int_0^x Bx^{-1/4}\, dx}{x} = \frac{4}{3}Bx^{-1/4} = \frac{4}{3}h$$

TABLE 10-1 Functions F and θ and Derivatives for Pr = 1 [5]

η	F	F'	F''	θ	θ'
0	0.0000	0.0000	0.6421	1.0000	-0.5671
0.1	0.0030	0.0593	0.5450	0.9433	-0.5669
0.2	0.0115	0.1092	0.4540	0.8867	-0.5657
0.3	0.0246	0.1503	0.3694	0.8302	-0.5627
0.4	0.0413	0.1833	0.2916	0.7742	-0.5572
0.5	0.0610	0.2089	0.2208	0.7189	-0.5488
0.6	0.0829	0.2277	0.1572	0.6645	-0.5371
0.7	0.1064	0.2406	0.1008	0.6116	-0.5221
0.8	0.1308	0.2481	0.0516	0.5602	-0.5038
0.9	0.1558	0.2511	0.0093	0.5109	-0.4826
1.0	0.1809	0.2502	-0.0263	0.4638	-0.4589
1.1	0.2058	0.2461	-0.0557	0.4192	-0.4330
1.2	0.2300	0.2393	-0.0793	0.3772	-0.4056
1.3	0.2535	0.2304	-0.0975	0.3381	-0.3772
1.4	0.2761	0.2199	-0.1110	0.3018	-0.3484
1.5	0.2975	0.2083	-0.1203	0.2684	-0.3197
1.6	0.3177	0.1960	-0.1260	0.2379	-0.2915
1.7	0.3367	0.1832	-0.1287	0.2101	-0.2642
1.8	0.3543	0.1708	-0.1288	0.1850	-0.2382
1.9	0.3707	0.1575	-0.1268	0.1624	-0.2136
2.0	0.3859	0.1450	-0.1233	0.1422	-0.1907
2.1	0.3997	0.1239	-0.1185	0.1242	-0.1695
2.2	0.4125	0.1213	-0.1127	0.1082	-0.1501
2.3	0.4240	0.1104	-0.1064	0.0941	-0.1324
2.4	0.4346	0.1001	-0.0997	0.0817	-0.1164
2.5	0.4441	0.0904	-0.0928	0.0708	-0.1020
2.6	0.4527	0.0815	-0.0859	0.0613	-0.0892
2.7	0.4604	0.0733	-0.0791	0.0529	-0.0777
2.8	0.4673	0.0657	-0.0725	0.0457	-0.0676
2.9	0.4736	0.0588	-0.0662	0.0392	-0.0587
3.0	0.4791	0.0524	-0.0602	0.0339	-0.0509
3.1	0.4841	0.0467	-0.0546	0.0291	-0.0441
3.2	0.4885	0.0415	-0.0493	0.0250	-0.0381
3.3	0.4924	0.0368	-0.0444	0.0215	-0.0329
3.4	0.4959	0.0326	-0.0399	0.0185	-0.0283
3.6	0.5016	0.0254	-0.0321	0.0136	-0.0210
3.8	0.5061	0.0197	-0.0255	0.0009	-0.0155
4.0	0.5096	0.0151	-0.0202	0.0072	-0.0115
4.2	0.5122	0.0116	-0.0158	0.0053	-0.0084
4.4	0.5143	0.0087	-0.0124	0.0038	-0.0062
4.6	0.5158	0.0066	-0.0096	0.0027	-0.0045
4.8	0.5169	0.0049	-0.0075	0.0020	-0.0034
5.0	0.5177	0.0035	-0.0057	0.0014	-0.0024
5.2	0.5183	0.0025	-0.0044	0.0010	-0.0018
5.5	0.5189	0.0014	-0.0029	0.0006	-0.0011
6.0	0.5194	0.0004	-0.0014	0.0002	-0.0005
6.25	0.5194	0.0000	-0.0010	0.0000	-0.0004

It follows that the average Nusselt number $\overline{\mathrm{Nu}}_x$ is

$$\overline{\mathrm{Nu}}_x = \frac{\overline{h}x}{k} = -\frac{4\theta'(0)}{3\sqrt{2}}\,\mathrm{Gr}_x^{1/4} \tag{10-3}$$

In similar fashion the local shear stress on the plate is found as

$$\tau_w = \mu\frac{\partial u(\,y=0)}{\partial y} = \mu\frac{\partial u}{\partial\eta}\frac{\partial\eta}{\partial y} = \frac{\sqrt{2}\,\rho_\infty v_\infty^2}{x^2}F''(0)\,\mathrm{Gr}_x^{3/4} \tag{10-4}$$

Hence $\tau_w \sim x^{1/4}$ and it is seen that, in contrast to a forced convection laminar boundary layer, wall shear stress increases with distance from the leading edge. The average shear stress is found from

$$\overline{\tau}_w = x^{-1}\int_0^x \tau_w\,dx$$

to be

$$\overline{\tau}_w = \frac{4\tau_w}{5} = \frac{4\sqrt{2}}{5}\frac{\rho_\infty v_\infty^2}{x^2}F''(0)\,\mathrm{Gr}_x^{3/4} \tag{10-5}$$

Values of $\theta'(0)$ and $F''(0)$ are given in Table 10-2 for various values of Prandtl number [5, 6]. Also included are values of A for use in the representation for the average Nusselt number

$$\overline{\mathrm{Nu}} = \frac{\overline{h}x}{k} = A(\mathrm{Gr}_x\,\mathrm{Pr})^{1/4} = A\,\mathrm{Ra}_x^{1/4} \tag{10-6}$$

with, according to LeFevre [7],

$$A^4 \approx 0.4\,\mathrm{Pr}/(1 + 2\,\mathrm{Pr}^{1/2} + 2\,\mathrm{Pr}) \tag{10-7}$$

where the Rayleigh number Ra is defined as the product of the Grashof and Prandtl numbers $\mathrm{Ra} = \mathrm{Gr}\,\mathrm{Pr}$. Use of the Rayleigh number in Eq. (10-6) takes advantage of the fact that for the creeping flow expected under normal conditions, inertial effects are negligible and the Nusselt number depends solely on the Rayleigh number rather than on Gr and Pr separately; this is the topic of Problem 10-4. It is seen in Table 10-2 that when $\mathrm{Pr} \to \infty$,

$$\overline{\mathrm{Nu}} = 0.67(\mathrm{Ra}_x)^{1/4} \tag{10-8}$$

At small Prandtl numbers it can be argued that viscosity cannot be important since, as suggested in Fig. 10-2b, the fluid velocity near the plate then tends

TABLE 10-2 $\theta'(0)$ and $F''(0)$ versus Pr [5, 6]

Pr	A	$\theta'(0)$	$F''(0)$
0	$0.800564\,\mathrm{Pr}^{1/4}$	$0.849126\,\mathrm{Pr}^{1/2}$	
0.01	0.240279	0.080592	0.9862
0.03	0.308	0.136	
0.09	0.377	0.219	
0.5	0.496	0.442	
0.72	0.516492	0.50463	0.676
0.733	0.517508	0.50789	0.6741
1.0	0.534705	0.56714	0.6421
1.5	0.555059	0.651534	
2.0	0.568033	0.716483	0.5713
3.5	0.589916	0.855821	
5.0	0.601463	0.953956	
7.0	0.611035	1.05418	
10	0.619	1.168	0.4192
100	0.653349	2.1914	0.2517
1,000	0.665	3.97	0.1450
10,000	0.668574	7.0913	
∞	0.670327	$0.710989\,\mathrm{Pr}^{1/4}$	

Source: By permission from A. J. Ede, *Adv. Heat Transfer* **4**, 1–64 (1967) [6]; © 1967, Academic.

to be imposed by events far from the plate and to resemble slug flow. This is a consequence of the high thermal conductivity that is associated with a low Pr that allows the wall temperature excess to penetrate far into the fluid. For viscosity to be unimportant, it is necessary that $\mathrm{Nu} = f(\mathrm{Gr}\,\mathrm{Pr}^2)$, and this is confirmed in Table 10-2 to be the case. So, for $\mathrm{Pr} \to 0$,

$$\overline{\mathrm{Nu}} = 0.8(\mathrm{Gr}\,\mathrm{Pr}^2)^{1/4} \qquad (10\text{-}9)$$

Problem 10-5 deals in greater detail with this limiting case. Calculations for small Pr were made by Sparrow and Gregg [8], and the limiting cases of $\mathrm{Pr} \to 0$ and $\mathrm{Pr} \to \infty$ were treated by LeFevre [7].

The agreement of the boundary-layer predictions with temperature and velocity measurements is good, as shown in Fig. 10-3, but measured values of the average heat-transfer coefficient are higher than the boundary-layer predictions. A comparison for air made by Ede [6] is shown in Fig. 10-4 in which the solid line represents the boundary-layer prediction [from Table 10-2 and Eq. (10-6)], and the points represent measurements. Since uncontrolled air currents, the commonest experimental error, lead to increased heat transfer, the deviation is generally understandable. Additionally, some of the deviation at high Rayleigh numbers is probably due to the development of turbulence, seen to become important at $\mathrm{Ra}_L \approx 10^9$, and that at low Rayleigh numbers may be due to inaccuracy of the boundary-layer assump-

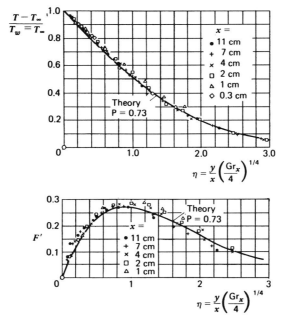

Figure 10-3 Measured dimensionless profiles in the laminar boundary layer on a hot vertical plate in natural convection of (*a*) velocity and (*b*) temperature; *x* is distance from lower edge of plate. Measurements by E. Schmidt and W. Beckmann [2]. [By permission from H. Schlichting, *Boundary-Layer Theory*, 6th ed., McGraw-Hill, 1968.]

tions for thick boundary layers. The 1935–36 data reported by McAdams [9] agree substantially with the data points of Fig. 10-4. For purposes of rapid estimation when $\Pr \approx 1$, the relationship

$$\overline{\mathrm{Nu}} = \frac{\overline{h}x}{k} = 0.555(\mathrm{Ra}_x)^{1/4}, \qquad 10 \leqslant \mathrm{Ra}_x \leqslant 10^9 \qquad (10\text{-}10)$$

can be used [10], which was originally intended for air.

All properties should be evaluated at the film temperature T_f where

$$T_f = \frac{T_w + T_\infty}{2} \qquad (10\text{-}11)$$

for use in such constant-property correlations as Eq. (10-7). A study of the effect of variable properties on the vertical flat-plate problem was conducted by Sparrow and Gregg [11]. They found that Nusselt number correlations obtained from constant-property solutions give good accuracy when properties are all evaluated at T_f from Eq. (10-11) but that slightly better accuracy

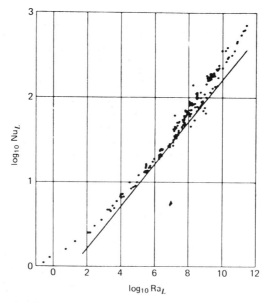

Figure 10-4 Measured average Nusselt number versus Rayleigh number for natural convection from vertical plates to air. [By permission from A. J. Ede, *Adv. Heat Transfer* **4**, 1–64 (1967) [6]; © 1967, Academic.]

results from evaluating all properties at a reference temperature T_r given by

$$T_r = T_w - 0.38(T_w - T_\infty) \tag{10-12}$$

for gases and for liquid mercury. The velocity distribution and, hence, the wall shear stress are affected more than the Nusselt number by temperature-dependent properties [10, 11]. The rapid variation of liquid viscosity with temperature, often exponential, appears to be best taken into account in the manner suggested in Section 4.9 for forced convection. For such fluids Fujii et al. [12] suggest, according to the survey of Kakac et al. [13], for constant temperature

$$\mathrm{Nu}_{L,\infty} \left(\frac{\nu_w}{\nu_\infty} \right)^{0.21} = \begin{cases} 0.6\,\mathrm{Ra}_{L,\infty}^{1/4}, & \mathrm{Ra}_{L,\infty} \leqslant 10^{10} \\ 0.13\,\mathrm{Ra}_{L,\infty}^{1/3}, & \mathrm{Ra}_{L,\infty} \geqslant 10^{10} \end{cases} \tag{10-13}$$

where the subscripts w and ∞ refer to wall and ambient conditions, respectively.

The effect of suction or blowing on the rate of heat transfer in laminar natural convection from a vertical flat plate has been studied [12, 13]. This effect can be important in mass-transfer applications, for example.

10.2   LAMINAR NATURAL CONVECTION FROM A CONSTANT-TEMPERATURE VERTICAL PLATE IN AN INFINITE FLUID—APPROXIMATE SOLUTION

The problem described in Section 10.1 can be approached by an integral method. The describing boundary-layer equations can be put in an integral form in the manner previously explored in Chapter 6. Application of Eqs. (6-12) and (6-18) yields

$$\frac{\partial\left[\int_0^\delta u^2\, dy\right]}{\partial x} - g\beta\int_0^\delta (T - T_\infty)\, dy = -\frac{\tau_w}{\rho} \tag{10-14a}$$

for the x-motion equation and

$$\frac{\partial\left[\int_0^\delta u(T - T_\infty)\, dy\right]}{\partial x} = \frac{q_w}{\rho C_p} \tag{10-14b}$$

for the energy equation. Here $\tau_w = \mu\,\partial u(y = 0)/\partial y$, $q_w = -k\,\partial T(y = 0)/\partial y$, and the thermal δ_T and velocity δ boundary-layer thicknesses have been assumed to be equal for simplicity. The integral method was apparently first applied to natural convection by Squire [14]. Later Merk and Prins [15] treated natural convection from vertical flat plate, vertical cone, horizontal cylinder, and sphere geometries.

The exact solutions illustrated in Fig. 10-2 show that $\delta \approx \delta_T$ only if $Pr \approx 1$ —the cases in which $\delta \neq \delta_T$ were treated by the integral method by Sugawara and Michiyoshi [16a]. The results obtained for $\delta \approx \delta_T$ will, nevertheless, show the effect of Prandtl number, suggest the manner in which data might be correlated, and be surprisingly accurate.

A polynomial is assumed to describe the velocity distribution as

$$u = a + b\eta + c\eta^2 + d\eta^3 \tag{10-15a}$$

where $\eta = y/\delta$ and can be subjected to the five conditions:

$$\text{At } y = 0 \qquad u = 0 \tag{10-15b}$$

and

$$g\beta(T_w - T_\infty) + \nu\frac{\partial^2 u}{\partial y^2} = 0 \tag{10-15c}$$

$$\text{At } y = \delta \qquad u = 0 \tag{10-15d}$$

$$\frac{\partial u}{\partial y} = 0 \tag{10-15e}$$

and

$$\frac{\partial^2 u}{\partial y^2} = 0 \tag{10-15f}$$

Equations (10-15b) and (10-15d) are the no-slip condition at the plate and the stagnant fluid pool at the edge of the boundary layer, respectively. Equation (10-15e) acknowledges the absence of a shear stress τ in the fluid at the edge of the boundary layer, whereas Eqs. (10-15c) and (10-15f) are the boundary-layer x-motion equation evaluated at the wall and the edge of the boundary layer, respectively. Only the three conditions with the lowest-ordered derivatives and the condition involving velocity's second derivative at the wall are used here. It is then found that

$$u = u_1 \eta (1 - \eta)^2 \qquad (10\text{-}16)$$

where

$$u_1 = \frac{g\beta(T_w - T_\infty)\delta^2}{4\nu}$$

Similarly, a polynomial describes the temperature distribution as

$$\frac{T - T_\infty}{T_w - T_\infty} = A + B\eta + C\eta^2 \qquad (10\text{-}17\text{a})$$

and can be subjected to the five conditions:

$$\text{At } y = 0 \qquad T = T_w \qquad (10\text{-}17\text{b})$$

and

$$\frac{\partial^2 T}{\partial y^2} = 0 \qquad (10\text{-}17\text{c})$$

$$\text{At } y = \delta \qquad T = T_\infty \qquad (10\text{-}17\text{d})$$

$$\frac{\partial T}{\partial y} = 0 \qquad (10\text{-}17\text{e})$$

and

$$\frac{\partial^2 T}{\partial y^2} = 0 \qquad (10\text{-}17\text{f})$$

Equations (10-17b) and (10-17d) are the requirement that the fluid take on the wall and ambient values at the plate and edge of the boundary layer, respectively. Equation (10-17e) acknowledges that there is no heat flux at the edge of the boundary layer, whereas Eqs. (10-17c) and (10-17f) come from the boundary-layer energy equation evaluated at the wall and the edge of the boundary layer, respectively. The three conditions involving temperature's lowest-ordered derivatives are employed to obtain

$$\frac{T - T_\infty}{T_w - T_\infty} = (1 - \eta)^2 \qquad (10\text{-}18)$$

Substitution of the profiles of Eqs. (10-16) and (10-18) into the integral relationships of Eq. (10-14) gives

$$\frac{d\left(u_1^2\delta/105\right)}{dx} = \left[\frac{g\beta(T_w - T_\infty)}{3}\right]\delta - \frac{u_1\nu}{\delta} \tag{10-19a}$$

$$\frac{d(u_1\delta/30)}{dx} = \frac{2\alpha}{\delta} \tag{10-19b}$$

Keep in mind that u_1 must be assumed to be unrelated to δ in order to avoid having two equations in only one unknown. This situation is the result of assuming $\delta = \delta_T$. With the insight that $u_1 = c_1 x^m$ and $\delta = c_2 x^n$, Eq. (10-19) gives

$$(2m + n)c_1^2 c_2^2 x^{2m+2n-1} = 35\left[g\beta(T_w - T_\infty)\right]c_2^2 x^{2n} - 105\nu c_1 x^m \tag{10-20a}$$

and

$$(m + n)c_1 c_2^2 x^{m+2n-1} = 60\alpha \tag{10-20b}$$

The only way for these two relationships to be true for all values of x is to have all exponents of x be equal. This requires that

$$2m + 2n - 1 = 2n = m \qquad \text{and} \qquad m + 2n - 1 = 0$$

which is satisfied by $m = \frac{1}{2}$ and $n = \frac{1}{4}$. Thus

$$u_1 = c_1 x^{1/2} \tag{10-21a}$$

$$\delta = c_2 x^{1/4} \tag{10-21b}$$

where, as a result of substitution of the m and n values back into Eq. (10-20),

$$c_1 = 4\left(\frac{5}{3}\right)^{1/2}\frac{\left[g\beta(T_w - T_\infty)\right]^{1/2}}{(\text{Pr} + 20/21)^{1/2}} \tag{10-21c}$$

$$c_2 = \frac{4(15/16)^{1/4}(1 + 20/21\,\text{Pr})^{1/4}}{\left[g\beta(T_w - T_\infty)\,\text{Pr}/\nu^2\right]^{1/4}} \tag{10-21d}$$

The integral method results give $\eta = y/\delta = y/(c_2 x^{1/4})$, according to Eq. (10-21b), which suggests that dimensionless velocity and temperature profiles solely depend on $\eta = cy/x^{1/4}$, the similarity variable used to achieve exact solutions. In other words, the integral method can suggest the proper form of a similarity transformation to be used to obtain an exact solution. Note also that the integral method suggests that $u \sim x^{1/2}$ just as does the similarity transformation.

The local heat-transfer coefficient is obtained as

$$h(T_w - T_\infty) = -k\frac{\partial T(y = 0)}{\partial y} = \frac{2k(T_w - T_\infty)}{\delta}$$

From this the local Nusselt number follows as

$$\frac{\mathrm{Nu}_x}{(\mathrm{Ra}_x)^{1/4}} = 0.51\left(1 + \frac{20}{21\,\mathrm{Pr}}\right)^{-1/4}$$

where $\mathrm{Nu}_x = hx/k$ and $\mathrm{Ra}_x = [g\beta(T_w - T_\infty)x^3/\nu^2]\,\mathrm{Pr}$. The average Nusselt number comes from this result, since $\overline{\mathrm{Nu}} = 4\,\mathrm{Nu}_x/3$, as

$$\frac{\overline{\mathrm{Nu}_x}}{(\mathrm{Ra}_x)^{1/4}} = \frac{0.68\,\mathrm{Pr}^{1/4}}{(\mathrm{Pr} + 20/21)^{1/4}} \tag{10-22}$$

A refined correlation [17] is

$$\overline{\mathrm{Nu}} = 0.68 + \frac{0.67\,\mathrm{Ra}_x^{1/4}}{\left[1 + (0.492/\mathrm{Pr})^{9/16}\right]^{4/9}}$$

For $\mathrm{Pr} = 0.73$ as it is for air, Eq. (10-22) gives

$$\frac{\overline{\mathrm{Nu}_x}}{\mathrm{Ra}_x^{1/4}} = 0.55$$

which is only 1% below the 0.555 value of the recommended Eq. (10-11). When $\mathrm{Pr} \to \infty$, Eq. (10-22) gives

$$\frac{\overline{\mathrm{Nu}_x}}{\mathrm{Ra}_x^{1/4}} = 0.68$$

which is within 1% of the exact Eq. (10-8). When $\mathrm{Pr} \to 0$, Eq. (10-22) predicts

$$\frac{\overline{\mathrm{Nu}_x}}{(\mathrm{Gr}_x\,\mathrm{Pr}^2)^{1/4}} = 0.69$$

which, although agreeing in form with Eq. (10-9), is 14% low.

The failure of the boundary-layer description of this natural convection problem at low values of the Rayleigh number is a result of the thick boundary layer that then occurs. This is demonstrated by Eqs. (10-21b) and (10-21d), which can be combined to give the local boundary-layer thickness as

$$\frac{\delta}{x} = 4\left(\frac{15}{16}\right)^{1/4} \frac{(1 + 20/21\,\mathrm{Pr})^{1/4}}{\mathrm{Ra}_x^{1/4}}$$

It can also be seen that the boundary-layer thickens as Pr decreases, holding Ra_x constant.

The local shear stress τ_w at the plate is given by

$$\tau_w = \mu\frac{\partial u(y = 0)}{\partial y} = \left(\frac{80}{27}\right)^{1/4}\mathrm{Gr}_x^{3/4}\frac{\rho\nu^2}{x^2}\frac{\mathrm{Pr}^{1/4}}{(\mathrm{Pr} + 20/21)^{3/4}} \quad (10\text{-}23)$$

For Pr $= 1$, the integral method prediction is 13% below the exact Eq. (10-5).

The maximum velocity is found by this integral method to occur at $y_{\max}/\delta = \frac{1}{3}$ as

$$u_{\max}\left[g\beta(T_w - T_\infty)\right]^{-1/2}x^{-1/2} = \frac{16}{27}\left(\frac{5}{3}\right)^{1/2}\bigg/\left(\mathrm{Pr} + \frac{20}{21}\right)^{1/2}$$

which is 9% above the exact answer from Table 10-1 for Pr $= 1$.

These comparisons indicate that the integral method is good for heat-transfer-coefficient predictions. Its predictions of such interior details as the velocity distribution are less accurate but still useful.

10.3 LAMINAR NATURAL CONVECTION FROM A CONSTANT-HEAT-FLUX PLATE IN AN INFINITE FLUID

If surface temperature variation influences the heat-transfer coefficient for laminar natural convection from a plate in an infinite fluid, attention must be given to accurate description of surface temperature. On the other hand, if surface temperature variation is not influential, no great care need be exercised. Fortunately, the latter case holds as will be seen.

A similarity analysis for constant heat flux q_w from the plate and constant properties is made. The boundary-layer description is provided by Eq. (10-1), but with the single change that at the plate the heat flux is constant so that

$$-k\frac{\partial T(y = 0)}{\partial y} = q_w$$

As set forth by Sparrow and Gregg [16], the similarity solution proceeds by selecting a stream function ψ such that $u = \partial\psi/\partial y$ and $v = -\partial\psi/\partial x$ so that the continuity equation is automatically satisfied. Then it is found that a similarity solution is possible if

$$\psi = c_2 x^{4/5} F(\eta)$$

where

$$\eta = \frac{c_1 y}{x^{1/5}}, \qquad c_1^5 = \frac{g\beta q_w}{5k\nu^2}, \qquad \text{and} \qquad c_2^5 = \frac{5^4 g\beta q_w \nu^3}{k}$$

Thus

$$u = c_1 c_2 x^{3/5} F'(\eta) \qquad \text{and} \qquad v = c_2 \frac{\eta F'(\eta) - 4F(\eta)}{5x^{1/5}}$$

Then, with $\theta = c_1(T_\infty - T)/(x^{1/5} q_w/k)$, the x-motion and energy equations become, with a prime denoting a derivative with respect to η,

$$F''' - 3(F')^2 + 4FF'' - \theta = 0 \qquad (10\text{-}24a)$$
$$\theta'' + \Pr(4\theta'F - \theta F') = 0 \qquad (10\text{-}24b)$$

subject to the boundary conditions:

$$F(\eta = 0) = 0 = F'(\eta = 0), \qquad \theta'(\eta = 0) = 1, \qquad \text{and}$$
$$\theta(\eta \to \infty) = 0 = F'(\eta \to \infty)$$

The temperature excess $T_w - T_\infty$ at the wall from Eqs. (10-24) is

$$T_w(x) - T_\infty = c_1^{-1} \frac{q_w}{k} \theta(0) x^{1/5} = -\left(\frac{5\nu^2 q_w^4 x}{k^4 g\beta}\right)^{1/5} \theta(0)$$

This relationship can be rephrased as

$$T_w(x) - T_\infty = \left[T_w(L) - T_\infty\right]\left(\frac{x}{L}\right)^{1/5}$$

It is seen that the wall temperature excess increases as the $\frac{1}{5}$th power of distance from the leading edge and is proportional to $q_w^{4/5}$. The local heat-transfer coefficient is found in the usual way to be

$$h = \frac{q_w}{T_w - T_\infty} = -\frac{\left[(k^4 g\beta q_w)/(5\nu^2 x)\right]^{1/5}}{\theta(0)}$$

varying as $q_w^{1/5}$ and $x^{-1/5}$, in contrast to the $x^{-1/4}$ variation found for

TABLE 10-3 **Nusselt Number Dependence on Boundary Condition for Laminar Natural Convection from a Vertical Plate [16b]**

Pr	$\overline{\mathrm{Nu}}_{q_w}/\overline{\mathrm{Nu}}_{T_w}$ Mean Temperature	Halfway Temperature	$\theta(0)$	$F''(0)$
0.1	1.08	1.02	−2.7507	1.6434
1	1.07	1.015	−1.3574	0.72196
10	1.06	1.01	−0.76746	0.30639
100	1.05	1	−0.46566	0.12620

constant wall temperature. The local Nusselt number then is

$$\mathrm{Nu}_x = \frac{hx}{k} = -\frac{(\mathrm{Gr}_x^*/5)^{1/5}}{\theta(0)}$$

where $\mathrm{Gr}_x^* = g\beta(q_w x/k)x^3/\nu^2$ is a modified Grashof number with $q_w x/k$ playing the role of a temperature difference. To calculate the average heat-transfer coefficient $\overline{h}$, the temperature difference can be specified as either the mean temperature excess $L^{-1}\int_0^x(T_w - T_\infty)\,dx$ or the temperature halfway along the plate. As shown in Table 10-3 there is little difference between the heat-transfer coefficients for constant-heat-flux and constant-temperature conditions, particularly if the driving temperature excess is evaluated at the halfway point on the plate.

Note that the modified Grashof number $\mathrm{Gr}_x^* = g\beta(q_w x/k)x^3/\nu^2$ is related to the conventional Grashof number $\mathrm{Gr}_x = g\beta(T_w - T_\infty)x^3/\nu^2$ by

$$\mathrm{Gr}_x^* = \mathrm{Gr}_x\,\mathrm{Nu}_x$$

since $h(T_w - T_\infty) = q_w$. If it is presumed that the criterion for laminar flow remains $\mathrm{Gr}_x\,\mathrm{Pr} \leqslant 10^9$ as it was for the constant wall temperature case, and if the integral method result of Problem 10-6, $\mathrm{Nu}_x = 0.62[\mathrm{Pr}^2\,\mathrm{Gr}_x^*/(\mathrm{Pr} + 0.8)]^{1/5}$, is used, it is found that the criterion for laminar flow in the constant-heat-flux case is [18]

$$\mathrm{Gr}_x^* \leqslant \left(\frac{0.916\,\mathrm{Pr}}{\mathrm{Pr} + 0.8}\right)^{1/4} 10^{11}$$

Application of the integral method to the case where heat flux (or surface temperature) is specified and varies over the plate surface was studied by Sparrow [19] and is the subject of Problem 10-6. Similarity solutions for plate temperature variations of the form $T_w - T_\infty = Nx^n$ and $T_w - T_\infty = Me^{mx}$ were obtained by Sparrow and Gregg [20]. The integral method for specified

q_w was generalized for natural convection by Tribus [21] and then further extended by Bobco [22]; the results are usually within 10% of the exact solutions—for $q_w \sim x^n$, it is found that $T_w - T_\infty \sim x^{(4n+1)/5}$ as in exact solutions and

$$\mathrm{Nu}_x = \frac{(7\,\mathrm{Pr}/27)^{1/4}(n+1)^{1/2}\,\mathrm{Ra}_x^{1/4}}{[(1+35\,\mathrm{Pr}/12)(n+1)+4/3]^{1/4}}$$

For liquids such as water and oil whose viscosity is exponentially sensitive to temperature, Fujii [12] suggests that the correlation for constant wall heat flux is

$$\mathrm{Nu}_{x,\infty}\left(\frac{\nu_w}{\nu_\infty}\right)^{0.17} = \begin{cases} 0.62(\mathrm{Gr}_{x,\infty}\,\mathrm{Pr}_\infty)^{1/5}, & \mathrm{Gr}_{x,\infty}\,\mathrm{Pr}_\infty \le 2\times 10^{13} \\ 0.055(\mathrm{Gr}_{x,\infty}\,\mathrm{Pr}_\infty)^{2/7}, & \mathrm{Gr}_{x,\infty}\,\mathrm{Pr}_\infty \ge 2\times 10^{13} \end{cases}$$

according to the survey of Kakac et al. [13].

10.4 LAMINAR NATURAL CONVECTION FROM A CONSTANT-TEMPERATURE VERTICAL CYLINDER IN AN INFINITE FLUID

Laminar natural convection from a constant-temperature vertical cylinder of diameter D and length L deviates from that for a vertical plate if D/L is sufficiently small for curvature to be significant.

A similarity solution of the boundary-layer equations in cylindrical coordinates

$$\frac{\partial(ru)}{\partial x} + \frac{\partial(rv)}{\partial r} = 0$$

$$u\frac{\partial u}{\partial x} + v\frac{\partial u}{\partial r} = g\beta(T - T_\infty) + \frac{\nu}{r}\frac{\partial(r\,\partial u/\partial r)}{\partial r}$$

$$u\frac{\partial T}{\partial x} + v\frac{\partial T}{\partial r} = \frac{\alpha}{r}\frac{\partial(r\,\partial T/\partial r)}{\partial r}$$

was executed by Sparrow and Gregg [23]. It is of interest here not only because of its useful results, but also because it illustrates a perturbation technique. A stream function ψ is defined such that $ru = \partial\psi/\partial r$ and $rv = -\partial\psi/\partial x$ and a coordinate transformation is employed with $\eta = c_1(r^2 - R^2)/x^{1/4}$ and $\xi = c_2 x^{1/4}$. Then, with $\theta = (T - T_\infty)/(T_w - T_\infty)$ and $f = c_3\psi/x^{1/4}$, it is found that $u = (2c_1/c_3)\partial f/\partial\eta$ and $v = -[\xi(f + \xi\,\partial f/\partial\xi) - \eta\xi\,\partial f/\partial\eta]/(c_2 c_3 4rx)$. The boundary-layer equations of motion and energy then become

$$\xi\left(\frac{\partial f}{\partial \eta}\frac{\partial^2 f}{\partial \xi\,\partial \eta} - \frac{\partial f}{\partial \xi}\frac{\partial^2 f}{\partial \eta^2}\right) - f\frac{\partial^2 f}{\partial \eta^2} = \xi^2\frac{\partial}{\partial \eta}\left[(1 + \xi\eta)\frac{\partial^2 f}{\partial \eta^2}\right] + \xi^4\theta$$

$$\xi\left(\frac{\partial\theta}{\partial \xi}\frac{\partial f}{\partial \eta} - \frac{\partial\theta}{\partial \eta}\frac{\partial f}{\partial \xi}\right) - f\frac{\partial\theta}{\partial \eta} = \frac{\xi^2}{\mathrm{Pr}}\frac{\partial}{\partial \eta}\left[(1 + \xi\eta)\frac{\partial\theta}{\partial \eta}\right]$$

which are subject to the boundary conditions:

$$\text{At } \eta = 0 \qquad f = 0, \qquad \frac{\partial f}{\partial \eta} = 0, \qquad \frac{\partial f}{\partial \xi} = 0, \qquad \text{and} \qquad \theta = 1$$

$$\text{At } \eta = \infty \qquad \frac{\partial f}{\partial \eta} = 0 \qquad \text{and} \qquad \theta = 0$$

Here,

$$c_1 = \left[\frac{g\beta(T_w - T_\infty)R^3}{\nu^2}\right]^{1/4}\frac{R^{-7/4}}{2^{3/2}}, \qquad c_2 = \left[\frac{g\beta(T_w - T_\infty)R^3}{\nu^2}\right]^{-1/4}\frac{2^{3/2}}{R^{1/4}}$$

$$c_3 = \left[\frac{g\beta(T_w - T_\infty)R^3}{\nu^2}\right]^{-3/4}\frac{2^{3/2}}{\nu R^{3/4}}$$

At this point it is recognized that ξ is small if R is large, and interest is concentrated on that case. Therefore, f and θ are expanded in series as

$$f(\xi, \nu) = \xi^2\left[f_0(\eta) + \xi f_1(\eta) + \xi^2 f_2(\eta) + \cdots\right]$$

$$\theta(\xi, \eta) = \theta_0(\eta) + \xi\theta_1(\eta) + \xi^2\theta_2(\eta) + \cdots$$

These series are then substituted into the motion and energy equations with the result that, with a prime denoting a derivative with respect to η,

$$\xi^4\left[f_0''' + 3f_0 f_0'' - 2(f_0')^2 + \theta_0\right]$$

$$+ \xi^5[f_1''' + f_0'' + \eta f_0''' - 5f_0'f_1' + 4f_0''f_1 + 3f_1''f_0 + \theta_1] + \cdots = 0$$

$$\xi^2[\theta_0'' + 3\,\mathrm{Pr}\,f_0\theta_0']$$

$$+ \xi^3[\theta_1'' + \theta_0' + \eta\theta_0'' - \mathrm{Pr}(f_0'\theta_1 - 4f_1\theta_0' - 3f_0\theta_1')] + \cdots = 0$$

TABLE 10-4   Nusselt Number Dependence on Vertical Cylinder Diameter [23]

$Ra_L^{1/4} D/L$	$\overline{Nu}_{L,cyl}/\overline{Nu}_{L,FP}{}^a$	
	Pr = 0.72	Pr = 1.0
100	1.02	1.02
30	1.06	1.05
10	1.17	1.16
6	1.27	1.26

[a]FP = flat plate.

and the boundary conditions:

$$\text{At } \eta = 0 \qquad f_0 = f_1 = \cdots = 0, \qquad f_0' = f_1' = \cdots = 0, \qquad \theta_0 = 1,$$
$$\text{and} \qquad \theta_1 = \theta_2 = \cdots = 0$$
$$\text{At } \eta = \infty \qquad f_0' = f_1' = \cdots = 0 \qquad \text{and} \qquad \theta_1 = \theta_2 = \cdots = 0$$

Each coefficient of ξ^n must equal zero if the left-hand side of these two equations is to be satisfied for all ξ. Hence the solutions for f_0 and θ_0 (the plate results) can be obtained first. Then f_1 and θ_1 can be determined numerically, by using f_0 and θ_0 as known functions. By proceeding in this way, as many terms of the series can be evaluated as accuracy requires.

The local heat-transfer coefficient is expressed in terms of these solutions as

$$Nu_x = \left[-\theta_0'(0) \left(\frac{Gr_x}{4} \right)^{1/4} \right] \left[1 + \frac{\xi \theta_1'(0)}{\theta_0'(0)} + \frac{\xi^2 \theta_2'(0)}{\theta_0'(0)} + \cdots \right]$$

in which the first bracketed term is the local Nusselt number for a vertical flat plate. Numerical results for Pr = 0.72 and 1 are displayed in Table 10-4. There it is seen that curvature effects make the average Nusselt number greater for the cylinder than the plate. Plate results are accurate within 5% for Pr = 1 if $Ra_L^{1/4} D/L \geq 33$—this corresponds to $\xi = 0.15$, so the series solution encounters no convergence difficulties. These results were extended for $0.01 \leqslant Pr \leqslant 100$ by Cebeci [24].

Essentially the same results were obtained by LeFevre and Ede [25] who applied an integral method to the boundary-layer equations to get

$$\overline{Nu}_{x,cyl} = \frac{4}{3} \left[\frac{(7/5) Pr}{(20 + 21 Pr)} \right]^{1/4} Ra_x^{1/4} + \frac{(4/35)(272 + 315 Pr)}{64 + 63 Pr} \frac{L}{D} + \cdots$$

It is seen again that the integral method offers the advantage over an exact method of an explicit formula. The first term is the plate result of Eq. (10-22). Rearrangement then gives

$$\frac{\overline{Nu_{L,cyl}}}{\overline{Nu_{L,fp}}} = 1 + K\frac{L}{D\,Ra_L^{1/4}} + \cdots \tag{10-25}$$

where $K = \frac{3}{7}[(Pr + 272/315)/(Pr + 64/63)][15(Pr + 20/21)/Pr]^{1/4}$, showing that the curvature effect tends to increase the average Nusselt number above the value it would have for a plate. Turbulent natural convection about a vertical cylinder was investigated with an integral method by Na and Chiou [26].

Additional information on this special topic is given by Kyte et al. [27].

For isothermal vertical cylinders the findings of Nagendra et al. [28] for vertical cylinders with a constant heat flux are

$$\overline{Nu_D} = \begin{cases} 0.6\left(Ra_D\dfrac{D}{L}\right)^{0.25}, & Ra_D\dfrac{D}{L} \geqslant 10^4 \\[2ex] 1.37\left(Ra_D\dfrac{D}{L}\right)^{0.16}, & 0.05 \leqslant Ra_D\dfrac{D}{L} \leqslant 10^4 \\[2ex] 0.93\left(Ra_D\dfrac{D}{L}\right)^{0.05}, & Ra_D\dfrac{D}{L} \leqslant 0.05 \end{cases}$$

in which the average temperature difference is used in the Rayleigh number Ra. These results differ by less than 5% from the isothermal case.

10.5 THERMAL PLUMES

The thermal plume that rises from a heated object into the fluid in which it is immersed is of technical importance. Fires, electronic components, and cooling tower exhausts are examples of sources of plumes. The rate at which ambient fluid is entrained and the variation of plume temperature with distance from the source is of interest; the heat-transfer rate is understood to be specified. Jaluria [29] provides an overview.

Line Source

For a line source as shown in Fig. 10-5 the rate at which thermal energy is convected by the plume must equal that Q provided by the plume source.

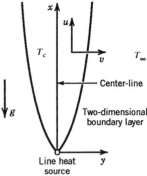

Figure 10-5 Thermal plume from a line heat source. [By permission from Y. Jaluria, in S. Kakac, W. Aung, and R. Viskanta, Eds., *Natural Convection*, Frances and Taylor, Washington, D.C., 1985, pp. 51–74 [29].]

Hence

$$Q = \int_{-\infty}^{\infty} \rho C_p u (T - T_\infty) \, dy \qquad (10\text{-}26a)$$

The boundary-layer equations describing the laminar flow are

Continuity $\qquad \dfrac{\partial u}{\partial x} + \dfrac{\partial v}{\partial y} = 0$

x Motion $\qquad u\dfrac{\partial u}{\partial x} + v\dfrac{\partial u}{\partial y} = g\beta(T - T_\infty) + v\dfrac{\partial^2 u}{\partial y^2}$

Energy $\qquad u\dfrac{\partial T}{\partial x} + v\dfrac{\partial T}{\partial y} = \alpha\dfrac{\partial^2 T}{\partial y^2}$

These equations are subjected to

$$T(x,0) = T_{\text{center}}(x), \qquad \frac{\partial T(x,0)}{\partial y} = 0, \qquad v(x,0) = 0, \qquad \frac{\partial u(x,0)}{\partial y} = 0$$

$$T(x, y \to \infty) \to T_\infty, \qquad u(x, y \to \infty) \to 0$$

A similarity solution is possible by letting the similarity variable η, stream function ψ, and dimensionless temperature θ be

$$\eta = \frac{y}{x}\left(\frac{\text{Gr}_x}{4}\right)^{1/4}, \qquad \Psi = 4v\left(\frac{\text{Gr}_x}{4}\right)^{1/4} f(\eta), \qquad \theta = \frac{T - T_\infty}{T_{\text{center}} - T_\infty}$$

Here

$$Gr_x = g\beta(T_{center} - T_\infty)\frac{x^3}{\nu^2}$$

$$u = \frac{\partial \Psi}{\partial y} = \left[4\left(\frac{g\beta Q}{C_p I}\right)^2 \Big/ \mu\rho\right]^{1/5} x^{1/5} f'$$

$$v = -\frac{\partial \Psi}{\partial x} = \left[\frac{64\mu^2 g\beta Q}{C_p I\rho^{5/2}}\right]^{1/5}\left[\frac{x^{-2/5}}{5}\right][3f - 2\eta f']$$

and

$$T_{center} - T_\infty = Nx^n$$

Then the similarity form of Eq. (10-26a) is

$$Q = 4\mu C_p N\left(\frac{g\beta Nx^{5n+3}}{4\nu^2}\right)^{1/4} I \qquad (10\text{-}26b)$$

Since Q is constant, it must be that $n = -\frac{3}{5}$; $I = \int_{-\infty}^{\infty} f'\theta \, d\eta$. Then the describing equations and boundary conditions are

$$f''' + \tfrac{12}{5}ff'' - \tfrac{4}{5}(f')^2 + \theta = 0 \qquad (10\text{-}27a)$$

$$\theta'' + \tfrac{12}{5}\Pr(f\theta)' = 0 \qquad (10\text{-}27b)$$

$$\theta(0) = 1, \qquad \theta'(0) = 0, \qquad f''(0) = 0,$$

$$f(0) = 0, \qquad \text{and} \qquad f'(\infty) = 0 \qquad (10\text{-}27c)$$

The condition $\theta(\infty) = 0$ is automatically satisfied since Eq. (10-27b) can be integrated twice to obtain

$$\theta = \exp\left[-\frac{12}{5}\Pr \int_0^\eta f \, d\eta\right]$$

which tends to zero as η becomes large since f there is large and positive.

TABLE 10-5   Two-Dimensional Plume Parameters [30]

Pr	$f'(0)$	$(\eta_e)_v$	$(\eta_e)_t$	I	J
0.01	0.9751	14.6	—	—	—
0.1	0.8408	9.3	11.0	3.090	4.316
0.7	0.6618	4.1	3.9	1.245	1.896
1.0	0.6265	3.8	3.2	1.053	1.685
2.0	0.5590	3.7	2.2	0.756	1.393
6.7	0.4480	4.1	1.2	0.407	1.094
100.0	0.2505	5.6	0.4	—	—

From the solutions to Eqs. (10-27) the quantities of major interest can be obtained. The temperature-excess parameter N is obtained from Eq. (10-26b) as $N = (Q^4/64g\beta\rho^2\mu^2 C_p^4 I^4)^{1/5}$. In like manner the mass-flow rate $\dot{m}$ in the plume is

$$\dot{m} = \int_{-\infty}^{\infty} \rho u \, dy = \left(\frac{64g\rho^2\mu^2 Qx^3}{C_p I} \right)^{1/5} J$$

in which $J = \int_{-\infty}^{\infty} f' \, d\eta = 2f(\infty)$. The solutions due to Gebhart [30] are given in Table 10-5 where it is seen that the edge locations $(\eta_e)_v$ and $(\eta_e)_t$ at which $f'/f'(0) = \frac{1}{100} = \theta$, respectively, are provided. From these η_e values the spread δ of the plume can be ascertained.

Measurements far from the source are difficult to make, the plume often continuously swaying; Bejan [31] discusses this swaying from the point of view of a buckling criterion. It is seen that the center-line velocity increases as $x^{1/5}$, the center-line temperature excess decays as $x^{-3/5}$, the plume width increases as $x^{2/5}$, and the entrained mass-flow rate increases as $x^{3/5}$. Closed-form solutions are reported by Jaluria [29]; he also discusses the turbulent plume from a line source (the plume will eventually become turbulent).

Point Source

For a point source of heat as illustrated in Fig. 10-6, the treatment is parallel to that for a line source. The rate of thermal energy convected in the plume is

$$Q = \int_0^{\infty} \rho C_p u (T - T_\infty) 2\pi r \, dr \qquad (10\text{-}28a)$$

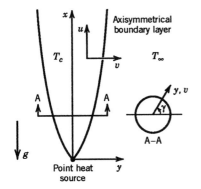

Figure 10-6 Thermal plume from a point source. [By permission from Y. Jaluria, in S. Kakac, W. Aung, and R. Viskanta, Eds., *Natural Convection*, Frances and Taylor, Washington, D.C., 1985, pp. 51–74 [29].]

The describing equations for laminar flow

Continuity
$$\frac{\partial u}{\partial x} + \frac{1}{r}\frac{\partial (rv_r)}{\partial r} = 0$$

x Motion
$$u\frac{\partial u}{\partial x} + v_r\frac{\partial u}{\partial r} = \frac{\nu}{r}\frac{\partial}{\partial r}\left(r\frac{\partial u}{\partial r}\right) + g\beta(T - T_\infty)$$

Energy
$$u\frac{\partial T}{\partial x} + v_r\frac{\partial T}{\partial r} = \frac{\alpha}{r}\frac{\partial}{\partial r}\left(r\frac{\partial T}{\partial r}\right)$$

are subjected to the boundary conditions

$$T(x,0) = T_{\text{center}}(x), \qquad \frac{\partial T(x,0)}{\partial r} = 0, \qquad v_r(x,0) = 0, \qquad \frac{\partial u(x,0)}{\partial r} = 0$$

$$T(x, r \to \infty) \to T_\infty, \qquad u(x, r \to \infty) \to 0$$

A similarity solution proceeds by letting

$$\eta = \frac{r}{x}\,\mathrm{Gr}_x^{1/4}, \qquad \Psi = \nu x f(\eta), \qquad \text{and} \qquad \theta = \frac{T - T_\infty}{T_{\text{center}} - T_\infty}$$

Here

$$\mathrm{Gr}_x = g\beta(T_{\text{center}} - T_\infty)\frac{x^3}{\nu^2}$$

$$u = r^{-1}\frac{\partial \Psi}{\partial r} = \nu\,\mathrm{Gr}_x^{1/2}\frac{f'}{x\eta}$$

$$v = -r^{-1}\frac{\partial \Psi}{\partial x} = \nu\,\mathrm{Gr}_x^{1/4}\left(\frac{f'}{2} - \frac{f}{\eta}\right)\Big/x$$

and

$$T_{center} - T_\infty = Nx^n$$

The similarity form of Eq. (10-28a) then is

$$Q = 2\pi\rho C_p \nu N x^{n+1} I \tag{10-28b}$$

Since Q is constant, it is required that $n = -1$; $I = \int_0^\infty f'\theta\, d\eta$. The describing equations then become

$$\frac{f'''}{\eta} + \frac{(f-1)}{\eta} + \left(\frac{f'}{\eta}\right)' + \theta = 0 \tag{10-29a}$$

$$(\eta\theta' + \Pr f\theta)' = 0 \tag{10-29b}$$

and their boundary conditions take the form

$$\theta(0) = 1, \qquad \theta'(0) = 0, \qquad \left(\frac{f'}{2} - \frac{f}{\eta}\right)_{\eta=0} = 0$$

$$\left(\frac{f'}{\eta}\right)'_{\eta=0} = 0, \qquad \theta(\infty) = 0, \qquad \left(\frac{f'}{\eta}\right)_{\eta=\infty} = 0$$

As for the line source case, $\theta(\infty) = 0$ is automatically satisfied due to the integrated similarity form of the energy equation. Numerical solutions due to Mollendorf and Gebhart [32] are

$$\Pr = 0.7; \quad f(\infty) = 7.91, \quad I = 2.074, \quad (\eta_e)_v \approx 9, \quad (\eta_e)_t \approx 6$$

$$\Pr = 7.0; \quad f(\infty) = 3.08, \quad I = 0.2497, \quad (\eta_e)_v \approx 7, \quad (\eta_e)_t \approx 3$$

Since the effect of any plume source is that of a point source when sufficiently far removed, the turbulent plume from a point source is of importance. In terms of the average axial velocity $\bar{u}$ and temperature $\bar{T}$, the entrained mass-flow rate $\dot{m}$ varies with axial distance x from the buoyancy source as

$$\frac{d\dot{m}}{dx} = \frac{d(\pi R^2 \rho \bar{u})}{dx}$$

The approximation due to Morton [33] is that $d\dot{m}/dx = 2\pi R\rho\bar{u}\hat{\alpha}$. Hence

$$\frac{d(R^2\bar{u})}{dx} = 2\hat{\alpha}R\bar{u} \tag{10-30}$$

The axial momentum equation is

$$\frac{d\left(\pi R^2 \rho \overline{u}\,\overline{u}\right)}{dx} = \pi R^2 g\beta\left(\overline{T} - T_\infty\right) \tag{10-31}$$

and the energy equation is

$$Q = \pi R^2 \rho C_p \overline{u}\left(\overline{T} - T_\infty\right) \tag{10-32}$$

As given by Turner [34], the solutions to Eqs. (10-30)–(10-32) are

$$R = \frac{6}{5}\hat{\alpha}x \quad \text{and} \quad \overline{u} = \left[\frac{25}{48\pi}\frac{g\beta Q}{\hat{\alpha}^2 \rho C_p}\right]^{1/3} x^{-1/3}$$

It is seen that the spreading angle R/x of the plume is constant. Measurements by Rouse et al. [35] suggest $\hat{\alpha} = 0.12$. Use of the Gaussian profiles

$$u = u_{\text{center}} \exp\left(\frac{-r^2}{R^2}\right), \qquad T - T_\infty = (T_{\text{center}} - T_\infty)\exp\left(\frac{-r^2}{\lambda^2 R^2}\right)$$

in Eqs. (10-30)–(10-32) gives, with $F = (1 + \lambda^2)g\beta Q/\pi\rho C_p$,

$$R = \frac{6}{5}\hat{\alpha}x, \quad u_{\text{center}} = \left(\frac{25}{24}\frac{F}{\hat{\alpha}^2}\right)^{1/3} x^{-1/3},$$

$$T_{\text{center}} - T_\infty = \frac{5F}{6\hat{\alpha}\lambda^2 g\beta}\left(\frac{5}{9\hat{\alpha}F}\right)^{1/3} x^{-5/3}$$

for which Yokoi [36] and Zukoski et al. [37] report $\hat{\alpha} = 0.1042$ and $\lambda = 1.15$. Comparison of these with the previous results shows little difference; application of numerical methods based on such turbulent closures as the k-ε and ASM to model the plume–plane interaction as illustrated in Fig. 10-7 is pertinent to optimal packaging of electronic components.

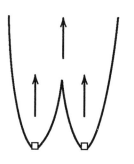

Figure 10-7 Interactions of plumes. [By permission from Y. Jaluria, in S. Kakac, W. Aung, and R. Viskanta, Eds., *Natural Convection*, Frances and Taylor, Washington, D.C., 1985, pp. 51–74 [29].]

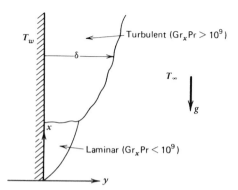

Figure 10-8 Transition from laminar to turbulent boundary-layer flow in natural convection on a hot vertical flat plate.

10.6 TURBULENT NATURAL CONVECTION FROM A VERTICAL PLATE OF CONSTANT TEMPERATURE

Reference to Fig. 10-4 shows that the laminar relationships of previous sections do not accurately predict the heat-transfer coefficient when the Rayleigh number $Ra_x = Gr_x\,Pr$ becomes large. This change in behavior is caused by the boundary layer next to the surface undergoing a transition from laminar to turbulent flow. Data suggest that a vertical plate can be taken to be covered by a turbulent boundary layer if $Ra_L \geq 10^9$ [6]. However, turbulence begins gradually over a transition length whose distance from the plate leading edge is influenced by external disturbances, just as for forced convection.

The turbulence fundamentals studied in Chapter 7 are incorporated into the integral method for the constant-property case as was first done by Eckert and Jackson [38]. The physical situation is as sketched in Fig. 10-8. The integral formulation of the boundary-layer equations is Eq. (10-14) where $\delta_T = \delta$—this assumption is justified only by the accuracy of the final results. Selection of appropriate velocity and temperature distributions is guided by the insight that vigorous turbulent mixing should yield a large velocity U_1 near the wall. This would correspond to a turbulent boundary layer, parallel to a flat plate, of external velocity U_1. Thus, the $\frac{1}{7}$th power-law velocity distribution of Chapter 8 is tentatively adopted as

$$u = U_1 \left(\frac{y}{\delta}\right)^{1/7}$$

But this form cannot stand unchanged since, although it correctly predicts $u(y = 0) = 0$, it does not satisfy the other requirements:

At $y = 0$ $g\beta(T_w - T_\infty) + \nu\dfrac{\partial^2 u}{\partial y^2} = 0$

At $y = \delta$ $u = 0,\qquad \dfrac{\partial u}{\partial y} = 0,\qquad \dfrac{\partial^2 u}{\partial y^2} = 0,\qquad$ and $\qquad \dfrac{\partial^3 u}{\partial y^3} = 0$

Hence it is finally assumed that

$$u = U_1[a + b\eta + c\eta^2 + d\eta^3 + e\eta^4]\eta^{1/7}$$

with $\eta = y/\delta$. Recognition that $a = 1$ if the $\frac{1}{7}$th power law holds near the wall and imposition of the four conditions at $y = \delta$ then give

$$u = U_1(1 - \eta)^4\eta^{1/7} \qquad (10\text{-}33)$$

Equation (10-33) has $u_{max}/U_1 = 0.613$ at $\eta = \frac{1}{29}$. This suggests that U_1 can be interpreted as nearly the maximum velocity in the boundary layer. Further, since the maximum velocity occurs near the plate, the wall shear stress and heat flux must behave nearly as for forced convection with a free-stream velocity of U_1. The temperature profile near the wall must be similar to the velocity profile under turbulent conditions, $T - T_w \sim u$. The conditions imposed on the temperature distribution are

$$\text{At } y = 0 \qquad T = T_w, \qquad \frac{\partial^2 T}{\partial y^2} = 0$$

$$\text{At } y = \delta \qquad T = T_\infty, \qquad \frac{\partial T}{\partial y} = 0, \qquad \frac{\partial^2 T}{\partial y^2} = 0$$

Since only the requirement that $T(y = 0) = T_w$ can be met simply

$$T - T_\infty = (T_w - T_\infty)(1 - \eta^{1/7}) \qquad (10\text{-}34)$$

The turbulent wall heat flux q_w and shear stress τ_w can be obtained by appeal to the forced convection turbulence information developed in Chapter 7 in which a $\frac{1}{7}$th power-law velocity profile gave

$$\frac{\tau_w}{\rho U_\infty^2} = \frac{C_f}{2} = 0.0228\left(\frac{U_\infty\delta}{\nu}\right)^{-1/4} \qquad (7\text{-}22)$$

If it is presumed that U_1 of Eq. (10-33) plays the role of U_∞, it follows that

$$\frac{\tau_w}{\rho} = 0.0228U_1^2\left(\frac{\nu}{U_1\delta}\right)^{1/4} \qquad (10\text{-}35)$$

From Reynolds' analogy it is known that

$$\frac{\text{Nu}}{\text{Re Pr}^{1/3}} = \frac{C_f}{2}$$

Since $h = q_w/(T_w - T_\infty)$, $C_f/2 = \tau_w/\rho U_1^2$, and $\mathrm{Re} = U_1 L/\nu$

$$\frac{q_w}{\rho C_p} = (T_w - T_\infty) \mathrm{Pr}^{-2/3} \frac{\tau_w}{\rho U_1} = 0.0228 \, \mathrm{Pr}^{-2/3}(T_w - T_\infty)U_1 \left(\frac{\nu}{U_1 \delta}\right)^{1/4}$$

$$(10\text{-}36)$$

The integrals required in the integral equations in Eq. (10-14) are

$$\int_0^\delta u^2 \, dy = I_1 \, \delta U_1^2 \qquad \text{with} \quad I_1 = 0.052315,$$

$$\int_0^\delta (T - T_\infty) \, dy = I_2(T_w - T_\infty)\delta$$

with $I_2 = \frac{1}{8}$ and $\int_0^\delta u(T - T_\infty) \, dy = I_3 U_1(T_w - T_\infty)\delta$ with $I_3 = 0.03663$. The integral equations [Eq. (10-14)] are then

$$I_1 \frac{d(\delta U_1^2)}{dx} = g\beta(T_w - T_\infty) \, \delta I_2 - 0.0228 U_1^2 \left(\frac{\nu}{U_1 \delta}\right)^{1/4}$$

$$(T_w - T_\infty)I_3 \frac{d(\delta U_1)}{dx} = 0.0228 \, \mathrm{Pr}^{-2/3}(T_w - T_\infty)U_1 \left(\frac{\nu}{U_1 \delta}\right)^{1/4}$$

To solve these two simultaneous equations, assume that $U_1 = Ax^m$ and $\delta = Bx^n$. This assumed solution substituted into the differential equations gives

$$(2m + n) A^2 x^{2m+n-1} = g\beta(T_w - T_\infty) \frac{I_2}{I_1} x^n - 0.0228 \frac{A^{7/4} \nu^{1/4}}{I_1 B^{5/4}} x^{(7m-n)/4}$$

$$(m + n) x^{m+n-1} = 0.0228 \frac{\nu^{1/4}}{I_3 \, \mathrm{Pr}^{2/3} \, A^{1/4} B^{5/4}} x^{(3m-n)/4}$$

These two relationships can only be true for all values of x if all exponents of x are equal. Hence

$$2m + n - 1 = n = \frac{7m - n}{4} \qquad \text{and} \qquad m + n - 1 = \frac{3m - n}{4}$$

These conditions are satisfied by $m = \frac{1}{2}$ and $n = \frac{7}{10}$. Since the exponents are now known, the coefficients are found to be

$$A^2 = \frac{10g\beta(T_w - T_\infty)I_2/17I_1}{1 + 12I_3\,\mathrm{Pr}^{2/3}/17I_1}$$

$$B = \left(0.0228\frac{0.89}{I_3}\right)^{4/5}\nu^{1/5}\,\mathrm{Pr}^{-8/15}\left[\frac{1 + 12I_3\,\mathrm{Pr}^{2/3}/17I_1}{g\beta(T_w - T_\infty)I_2/I_1}\right]^{1/10}$$

The local heat-transfer coefficient can now be computed from Eq. (10-36), by recognizing that $h = q_w/(T_w - T_\infty)$, as

$$h = 0.02979k\,\mathrm{Pr}^{1/15}\left\{\frac{\left[g\beta(T_w - T_\infty)\,\mathrm{Pr}/\nu^2\right]^{2/5}}{(1 + 0.494\,\mathrm{Pr}^{2/3})^{2/5}}\right\}x^{1/5} \qquad (10\text{-}37)$$

In turbulent flow h increases as $x^{1/5}$, in contrast with the laminar flow finding that h decreases as $x^{-1/4}$. Evidently, turbulent mixing becomes more intense as x increases.

If it is presumed that the plate is covered by a turbulent boundary layer, the average coefficient $\bar{h}$ is found from Eq. (10-37) to be $\bar{h} = 5h/6$. In terms of dimensionless groups,

$$\overline{\mathrm{Nu}}_x = 0.0248\frac{\mathrm{Pr}^{1/15}}{(1 + 0.494\,\mathrm{Pr}^{2/3})^{2/5}}(\mathrm{Ra}_x)^{2/5} \qquad (10\text{-}38)$$

which is to be applied only if $\mathrm{Ra}_x \geqslant 10^9$. For air ($\mathrm{Pr} = 0.72$), this becomes

$$\overline{\mathrm{Nu}}_x = 0.0212(\mathrm{Ra}_x)^{2/5}$$

in which the coefficient is only 1% different from the experimental coefficient of 0.021. Application of these results to determination of the shear stress at the wall is the subject of Problem 10-8.

The agreement of Eq. (10-38) with measurements is satisfactory. These measurements were made on plates of human size, however, so there is still room to question the ability of Eq. (10-38) to be extrapolated to very large Rayleigh numbers. As reviewed by Ede [6], a number of alternative suggestions have been put forth. It was suggested that when turbulence occurs over all of the plate, h should be constant, which requires $\mathrm{Nu}_x \sim \mathrm{Gr}_x^{1/3} f(\mathrm{Pr})$. Another suggestion is that in fully developed turbulence, h should be independent of viscosity, which requires that $\mathrm{Nu}_x \sim f(\mathrm{Gr}_x\,\mathrm{Pr}^2)$; yet another suggestion is that thermal conductivity should be of no influence, which, taken together with the result for independence from viscosity, requires that $\mathrm{Nu}_x \sim (\mathrm{Gr}_x\,\mathrm{Pr}^2)^{1/2}$. The $\frac{2}{5}$ exponent of Gr_x in Eq. (10-38) lies between the suggested values of $\frac{1}{3}$ and $\frac{1}{2}$ but is open to criticism because it utilizes a forced convection wall shear stress relationship whose accuracy is restricted to a range of Reynolds numbers.

Turbulent analyses such as these have been refined by Bayley [39]. He found that for the Prandtl numbers typical of liquid metals and for $10^{10} \leqslant \mathrm{Gr}_x \leqslant 10^{15}$,

$$\overline{\mathrm{Nu}} = 0.08\,\mathrm{Gr}_x^{1/4}$$

with the plate at constant temperature. For a fluid with $\mathrm{Pr} \approx 1$ such as air, he found

$$\overline{\mathrm{Nu}} = 0.1\,\mathrm{Ra}^{1/3}, \qquad 2 \times 10^9 \leqslant \mathrm{Ra} \leqslant 10^{12}$$
$$= 0.183\,\mathrm{Ra}^{0.31}, \qquad 2 \times 10^9 \leqslant \mathrm{Ra} \leqslant 10^{15} \qquad (10\text{-}39)$$

which agrees well with measurements [40]. The exponent of the Rayleigh number is uncertain even for $\mathrm{Pr} \approx 1$, ranging from 0.4 in Eq. (10-38) to 0.31 in Eq. (10-40) with $\frac{1}{3}$ in Eq. (10-39) as an intermediate value. An exponent of $\frac{1}{3}$ is appealing since it makes the choice of a characteristic length immaterial as far as h is concerned and shows h to be constant along the plate. Available measurements are not adequate to resolve the question. A refined correlation [17], applicable for all Ra, is

$$\overline{\mathrm{Nu}}_x^{1/2} = 0.825 + \frac{0.387\,\mathrm{Ra}_x^{1/6}}{\left[1 + (0.492/\mathrm{Pr})^{9/16}\right]^{8/27}} \qquad (10\text{-}40)$$

Numerical simulation of turbulent natural convection along a vertical plate with both a k–ε and an algebraic mixing-length model (along the lines discussed in Chapter 8) was accomplished by To and Humphrey [41]. Excellent agreement between prediction and measurement, especially that of Siebers et al. [42], was obtained, primarily proving that the numerical simulation was correctly done and could be applied to the more complex geometries and flow conditions encountered in vented and unvented enclosures (see also Yang and Lloyd [43]).

Siegel [44] obtained a solution for the case of a prescribed uniform heat flux that gives the local coefficient as

$$\mathrm{Nu}_x = \frac{0.080\,\mathrm{Pr}^{1/3}\,\mathrm{Gr}_x^{*2/7}}{(1 + 0.444\,\mathrm{Pr}^{2/3})^{2/7}}$$

which is in general agreement with the measurements of Vliet and Liu [45] for water who suggested $\mathrm{Nu}_x = 0.568(\mathrm{Pr}\,\mathrm{Gr}_x^*)^{0.22}$ for $2 \times 10^{13} < \mathrm{Gr}_x^*\,\mathrm{Pr} < 10^{16}$—the exponent of Gr_x^* is not important to a calculation of Nu_x over a restricted Gr_x^* range since the coefficient is selected properly, but it is important to the question of whether h is increasing, constant, or decreasing along the plate. Lemlich and Vardi [46] extended this analysis to the case where the body force is proportional to x, as it would be in a centrifuge, to find

$$\mathrm{Nu}_x = \frac{0.0729\,\mathrm{Pr}^{1/3}\,\mathrm{Gr}_x^{*\,2/7}}{\left(1 + 0.316\,\mathrm{Pr}^{2/3}\right)^{2/7}}$$

In these two preceding expressions, $\mathrm{Gr}_x^* = g\beta(q_w x/k)x^3/\nu^2$ as in the two that follow. They found in the case of laminar flow that

$$\mathrm{Nu}_x = \frac{0.546\,\mathrm{Pr}^{1/4}(\mathrm{Gr}_x^*\,\mathrm{Pr})^{1/4}}{(\mathrm{Pr} + 1.143)^{1/4}} \quad \text{and} \quad \mathrm{Nu}_x = \frac{0.616\,\mathrm{Pr}^{1/4}(\mathrm{Gr}_x^*\,\mathrm{Pr})^{1/4}}{(\mathrm{Pr} + 1.143)^{1/4}}$$

for uniform plate temperature and constant heat flux, respectively.

10.7   COMBINED NATURAL AND FORCED CONVECTION

Even a small externally forced flow can play an appreciable role in fluid motion near the heated surface when natural convection is still dominant. When the externally forced flow is large, forced convection effects are dominant but natural convection effects can be appreciable. In the absence of detailed information concerning the heat-transfer coefficient in a situation of combined natural and forced convection, a design procedure suggested by McAdams [9] is to calculate the coefficient for forced convection and for natural convection separately; then the larger value is adopted.

Sometimes the demarcation between the purely forced, the combined, and the purely natural regimes is given in terms of the dimensionless group $\mathrm{Gr}/\mathrm{Re}^2$. To make the importance of $\mathrm{Gr}/\mathrm{Re}^2$ plausible on physical grounds, an elementary consideration of the ratio of buoyancy to inertial forces F_b/F_i is essayed in a mixed natural and forced convection case. The buoyancy force is roughly given by $F_b \sim g\,\Delta\rho\,L^3$ where $\Delta\rho$ is the density differential in a cube of side L. The inertial force on a surface of side L in a flowing stream of normal velocity U is roughly given by $F_i \sim \rho U^2 L^2$. Thus

$$\frac{F_b}{F_i} \sim \frac{g(\Delta\rho/\rho)L^3/\nu^2}{(UL/\nu)^2} = \frac{\mathrm{Gr}}{\mathrm{Re}^2}$$

When $\mathrm{Gr}/\mathrm{Re}^2$ is of the order of unity, both natural and forced convection are important. On the other hand, natural convection is dominant when $\mathrm{Gr}/\mathrm{Re}^2$ is large, whereas forced convection is dominant when $\mathrm{Gr}/\mathrm{Re}^2$ is small.

Natural convection can act in a direction that is antiparallel, parallel, or perpendicular to that of the forced convection. The numerous studies, mostly for the laminar state, were surveyed by Churchill [47, 48].

For natural convection acting in a direction antiparallel to that of the forced convection, the resultant Nusselt number Nu is related to that Nu_F from purely forced convection correlations and that Nu_N from purely natural convection correlations by

$$\text{Nu}^n = \text{Nu}_F^n - \text{Nu}_N^n \tag{10-41}$$

where $n = 3$ is suggested for both the laminar and the turbulent state. The predicted null in Nu is a flaw; measurements show a nonzero value.

When the directions of action of natural and forced convection are parallel, the spirit of the suggested correlation of Eq. (10-41) is maintained for the turbulent case as

$$\text{Nu}^n = \text{Nu}_F^n + \text{Nu}_N^n \tag{10-42}$$

where $n = 3$, again. A flaw in Eq. (10-42) is that it does not account for the fact that buoyant effects can delay the transition from laminar to turbulent flow; Nu can decrease below Nu_F for small Nu_N, contrary to the prediction of Eq. (10-42). For the laminar state of flow the suggested correlation is

$$(\text{Nu} - \text{Nu}_0)^3 = -\left\{ \frac{A_F \, \text{Re}^{1/2} \, \text{Pr}^{1/3}}{\left[1 + (C_F/\text{Pr})^{2/3} \right]^{1/4}} \right\}^3 + \left\{ \frac{A_N \, \text{Ra}^{1/4}}{\left[1 + (C_N/\text{Pr})^{9/16} \right]^{4/9}} \right\}^3 \tag{10-43}$$

where A_F, A_N, C_F, and C_N values are given in Table 10-6. Use the distance x from the leading edge or forward stagnation point as the characteristic length to obtain a local Nu value; use the characteristic length in Table 10-6 to obtain a mean Nu value.

When the directions of action of the natural and forced convection are perpendicular, little guidance is offered for the turbulent case beyond that of taking the larger of the h predicted for purely forced and purely natural convection. For the laminar case the suggestions are only a bit more detailed. Churchill [47] suggests for cylinders and spheres

$$(\text{Nu} - \text{Nu}_0)^4 = \text{Nu}_F^4 + \text{Nu}_N^4 \tag{10-44}$$

TABLE 10-6   Constants and Characteristic Lengths for Mixed Convection [47]

	l	A_F	A_N	C_F	C_N	Nu_0
Vertical plate						
Local						
Uniform T	x	0.339	0.503	0.0468	0.492	0.5
Uniform q'	x	0.464	0.563	0.0205	0.437	0.5
Mean						
Uniform T	x	0.677	0.670	0.0468	0.492	0.5
Uniform q'	$x/2$	0.656	0.669	0.0205	0.437	0.5
Horizontal cylinder						
Mean						
Uniform T	πD	1.08	0.690	0.412	0.559	1.0
Uniform q'	πD		0.694	0.442		
Sphere						
Mean						
Uniform T	$\pi D/2$	0.69	0.659			
General approximate						
values		0.67^a	0.67^b	0.45^c	0.45	

[a] Except for Nu on plates and $\overline{\text{Nu}}$ on cylinders.
[b] Except for Nu on plates.
[c] Except for plates.

and for horizontal plates with a laminar boundary layer

$$(Nu - Nu_0)^{7/2} = \left\{ A_F \frac{Re^{1/2}\, Pr^{1/3}}{\left[1 + (C_F/Pr)^{2/3}\right]^{1/4}} \right\}^{7/2}$$

$$\pm \left\{ \frac{A'_N\, Ra^{1/5}}{\left[1 + (C'_N/Pr)^{9/16}\right]^{16/45}} \right\}^{7/2} \qquad (10\text{-}45)$$

with A'_N and C'_N taken from Table 10-7 and the positive sign used when either the fluid on the upper side is heated or the fluid on the lower side is cooled (use the negative sign otherwise).

Metais and Eckert [49] reviewed experimental information concerning forced, mixed, and natural convection in horizontal and vertical tubes. The various regimes for vertical tubes are shown in Fig. 10-9; Fig. 10-10 applies to horizontal tubes. In both figures the Reynolds number is based on diameter, and the Grashof number is based on pipe diameter and difference between wall and bulk fluid temperatures. The demarcations of the forced and natural convection regimes were established at the conditions under which the actual heat flux deviated by less than 10% from the value predicted for either forced

TABLE 10-7 Constants for Free Convection from a Horizontal Plate [47]

	A'_N	C'_N
Local coefficient		
Uniform T	0.456	0.312
Uniform q'	0.625	(0.30)
Mean coefficient		
Uniform T	0.760	0.312
Uniform q'	0.760	(0.30)*

* Postulated value.

or natural convection acting singly. In Fig. 10-9 data for forced convection both aiding and opposing natural convection were included as were data for uniform wall temperature and uniform heat flux; Fig. 10-10 includes only uniform-wall-temperature data. In Fig. 10-10 the Graetz number is Gz = $\mathrm{Re}_D \mathrm{Pr}\, D/L$. These Metais–Eckert charts are for Pr $\geqslant$ 1; for liquid metals consult the experimental study by Buhr et al. [50].

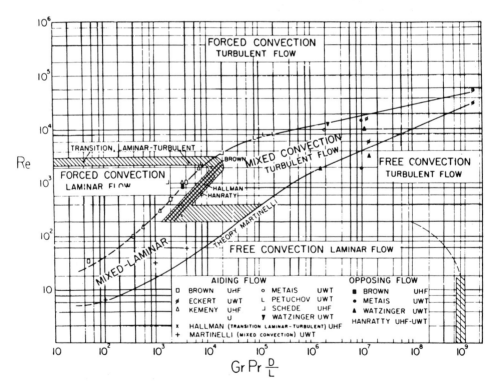

Figure 10-9 Regimes of free, forced, and mixed convection for flow through vertical tubes (UHF, uniform heat flux; UWT, uniform wall temperature). [From B. Metais and E. R. G. Eckert, *ASME J. Heat Transfer* **86**, 295–296 (1964) [49].]

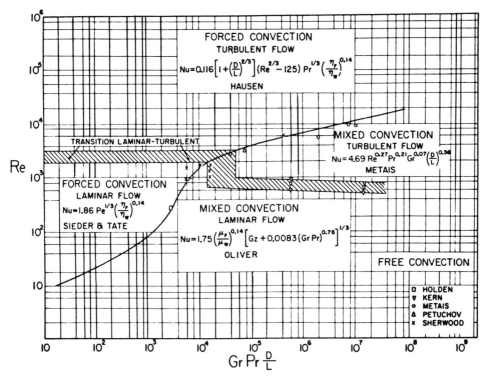

Figure 10-10 Regimes of free, forced, and mixed convection for flow through horizontal tubes. [From B. Metais and E. R. G. Eckert, *ASME J. Heat Transfer* **86**, 295–296 (1964) [49].]

As surveyed by Churchill [47] for vertical channels with fully developed forced and natural convection acting parallel and with constant wall temperature,

$$\text{Nu}^3 = \text{Nu}_F^3 + \text{Nu}_N^3$$

where

$$\text{Nu}_F = 3.657 \quad \text{and} \quad \text{Nu}_N = \frac{0.75(\text{Ra } D/L)^{1/4}}{\left[1 + (0.492/\text{Pr})^{9/16}\right]^{4/9}}$$

The log–mean-temperature difference is to be used for both Ra and heat flux. In the entrance region of a round tube,

$$\text{Nu}_{F,\,\text{AM}} = 1.75 F_1 \,\text{Gz}^{1/3} \quad \text{and} \quad \text{Nu}_{N,\,\text{AM}} = 0.729 F_1 F_2^{1/3} (\text{Ra } D/L)^{0.28}$$

$$(10\text{-}46)$$

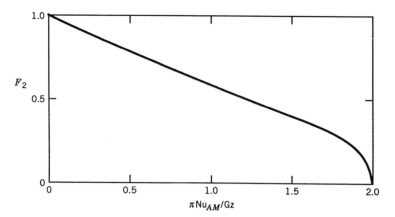

Figure 10-11 Correction factor F_2 for use in Eq. (10-46) [47].

where

$$Gz = \dot{m}\frac{C_p}{kZ}, \qquad F_1 = Z \ln\left(\frac{2+Z}{2-Z}\right)$$

with

$$Z = \frac{2(T_\theta - T_i)}{2T_w - T_\theta - T_i} = \frac{\pi \, \text{Nu}_{AM}}{Gz}$$

Ra is based on the temperature difference $T_w - T_i$ between the wall and the inlet, and F_2 is shown in Fig. 10-11. The AM subscript denotes the use of the arithmetic–mean-temperature difference $(2T_w - T_i - T_o)/2$ for heat flux. When the wall heat flux is constant with fully developed forced and natural convection acting parallel, the limiting cases

$$\text{Nu}_F = \tfrac{48}{11} \qquad \text{and} \qquad \text{Nu}_N = 0.846 \, \text{Ra*}^{1/4}$$

with $\text{Ra*} = g\beta\rho^2 C_p D^4 (dT_m/dx)/\mu k$ are to be used in

$$\text{Nu}^6 = \text{Nu}_F^6 + \text{Nu}_N^6$$

but with no recommendation for the entrance region. When laminar natural convection opposes laminar forced convection in a vertical tube, it is suggested that

$$\text{Nu}^3 = \text{Nu}_F^3 - \text{Nu}_N^3$$

although the null prediction is likely to be a flaw. For turbulent fully developed flow in a vertical tube, it is suggested that

$$\text{Nu}_F = \frac{0.0357\,\text{Re}\,\text{Pr}^{1/3}}{\ln(\text{Re}/7)[1 + \text{Pr}^{-4/5}]^{5/6}}$$

and

$$\text{Nu}_N = \frac{0.15\,\text{Ra}^{1/3}}{\left[1 + (0.492/\text{Pr})^{9/16}\right]^{16/27}}$$

be used in

$$\text{Nu}^3 = \text{Nu}_F^3 - \text{Nu}_N^3$$

with, contrary to the laminar case, the plus sign used for the antiparallel case (the reduced velocity near the wall increases turbulent core mixing due to a higher velocity in the core) and the minus sign used for the parallel case (the increased velocity near the wall decreases turbulent core mixing due to a lower velocity in the core). For the inlet region of a horizontal tube with laminar conditions and constant wall temperature,

$$\text{Nu}^6 = 362\,\text{Gz}^2 + \frac{0.0905\,\text{Ra}^{3/2}}{\left[1 + (0.492/\text{Pr})^{9/16}\right]^{8/3}}$$

with properties evaluated for Gz at the bulk temperature of the fluid, while those for Nu and Ra are evaluated at the wall temperature but with the log–mean-temperature difference used in Ra. For the inlet region of a horizontal tube with laminar conditions and constant wall heat flux q_w,

$$\text{Nu}^6 = \text{Nu}_F^6 + \text{Nu}_N^6$$

where

$$\text{Nu}_F = 5.364\left[1 + (\text{Gz}/55)^{10/9}\right]^{3/10} - 1$$

and

$$\text{Nu}_N = \frac{0.847\,\text{Ra}^{*0.177}}{\left[1 + (0.492/\text{Pr})^{9/16}\right]^{0.315}}$$

and

$$\text{Ra}^* = \frac{g\rho^2 C_p \beta q_w D^4}{\mu k^2}$$

When wall thermal conductivity and wall thickness δ_w are such as to affect the wall temperature, use of the parameter $f = (hD/k_w)(D/\delta_w)$ in

$$\mathrm{Nu}_F = \frac{0.378\ \mathrm{Ra}^{0.28}\ \mathrm{Pr}^{0.05}}{f^{0.12}}$$

is suggested as an alternative for Nu_F. For fully developed laminar, constant wall heat flux conditions in a horizontal duct when wall conduction is a factor,

$$\mathrm{Nu}^2 = (4.36)^2 + \left(\frac{0.055\ \mathrm{Ra}^{0.4}\ \mathrm{Pr}^{0.14}}{f^{0.1}}\right)^2$$

For a heated cone with vertical axis spinning in air, Kreith [51], in a survey of convection heat transfer in rotating systems, concludes that natural convection has less than 5% influence on the flow forced by the rotation if $\mathrm{Gr}_x/\mathrm{Re}_x^2 < 0.052$, where $\mathrm{Gr}_x = (g\beta\,\Delta T x^3/\nu^2)\cos\alpha$, where α is the cone half-angle and $\mathrm{Re}_x = x^2\omega\sin(\alpha/\nu)$ with ω cone rotational speed.

10.8 EMPIRICAL RELATIONS FOR NATURAL CONVECTION

A large amount of analytical and experimental information concerning natural convection is available for both laminar and turbulent flow. Most of this information pertains to the steady state, but a substantial amount of work has also been done [6] on both the transient state and the quasi-steady state [52].

The results of Siegel [53] for a vertical plate provide a basis for estimating the time t_D required for departure from the conduction regime and the time t_∞ for steady conditions to be achieved after a step change in wall temperature at a distance x above the leading edge;

$$t_D = 1.8(1.5 + \mathrm{Pr})^{1/2}\left[\frac{x}{g\beta\,\Delta T}\right]^{1/2}$$

$$t_\infty = 5.24(0.952 + \mathrm{Pr})^{1/2}\left[\frac{x}{g\beta\,\Delta T}\right]^{1/2}$$

More extensive discussion of transients is provided by Raithby and Hollands [54].

For the steady state, it is convenient to have a summary of results additional to those previously presented.

The survey of Raithby and Hollands [54] augments, in detail and cases, those presented here. In many instances the accurate range of correlations of measured Nusselt numbers Nu_m has been extended by blending the Nusselt numbers Nu_1, Nu_2, and Nu_3 for the limiting stagnant, laminar, and turbulent conditions, respectively. As explained by Churchill and Usagi [55], this has

$$Nu_m^n = C_1 Nu_1^n + C_2 Nu_2^n + C_3 Nu_3^n$$

where n is an integer selected to minimize an integrated error.

Flat Plate at Small Angle from Vertical

For a flat plate inclined at an angle θ from the vertical, Vliet recommends [56] that the component of gravity parallel to the heated surface be used (but only for the laminar case) in vertical plate correlations for $0° < \theta < 60°$. All fluid properties are to be evaluated at the film temperature, as usual. The plate inclination affects the laminar–turbulent transition as roughly given by

$$Gr\,Pr_{trans} = 3 \times 10^5 \exp[0.1368(90° - \theta)]$$

with transition beginning at about $\frac{1}{10}$ this value and ending at about 10 times this value. For nearly horizontal plates, Raithby and Hollands [54] suggest that the greater of the Nusselt number calculated as described previously and for a horizontal plate be used.

In turbulent heat transfer Raithby and Hollands suggest, θ is the angle from the vertical,

$$Nu = C\,Ra^{1/3}$$

where

$$C = \begin{cases} C_a \cos^{1/3}(\theta), & -90° \leqslant \theta \leqslant \arctan(C_a/C_b)^3 \\ C_b \sin^{1/3}(\theta), & \arctan(C_a/C_b)^3 \leqslant \theta \leqslant 90° \end{cases}$$

with $C_a = 0.13\,Pr^{0.22}/(1 + 0.61\,Pr^{0.81})^{0.42}$ and $C_b = 0.14$ for $Pr < 100$ ($C_b = 0.15$ for $Pr > 1,000$). For upward-facing heated plates or downward-facing cooled plates, $0° \leqslant \theta \leqslant 90°$; for downward-facing heated plates or upward-facing cooled plates, $-90° \leqslant \theta \leqslant 0°$.

TABLE 10-8   Parameters for Natural Convection from Horizontal Plates [9]

Case	Type of Flow	Ra Range	C	n
1	Hot surface up or cool surface down			
	Laminar	10^5–2×10^7	0.54	$\frac{1}{4}$
	Turbulent	2×10^7–3×10^{10}	0.14	$\frac{1}{3}$
2	Hot surface down or cool surface up			
	Laminar	3×10^5–3×10^{10}	0.27	$\frac{1}{4}$

Horizontal Plates

For a horizontal plate, the correlations to measurements are recommended by McAdams [9] to be of the form $\overline{Nu} = C\,Ra^n$. The characteristic length L for a square is the side of the square; for a rectangle it is the arithmetic average of the long and short sides, and for a circular disk it is 0.9 times the disk diameter. The values of C and n are cited in Table 10-8—note that in case 2 there is no turbulent case cited.

The foregoing recommendations for C and n correlate measurements fairly well. However, Raithby and Hollands [54] found that the prediction of Fujii et al. [57] for a horizontal downward-facing heated (or an upward-facing cooled) strip

$$Nu = \frac{0.527\,Ra^{1/5}}{\left[1 + (1.9/Pr)^{9/10}\right]^{2/9}}$$

applies to many other plate shapes (squares, circles, etc.) if the characteristic length L for the Rayleigh and Nusselt numbers is $L = $ plate area/plate perimeter. For an upward-facing heated (or a downward-facing cooled) plate

$$Nu^{10} = Nu_L^{10} + Nu_T^{10}$$

where, see Table 10-9 for C_1 and with C_2 as defined for inclined plates,

$$Nu_L = \frac{1.4}{\ln(1 + 1.4/Nu_1)}, \qquad Nu_1 = 0.835\,C_1\,Ra^{1/4}, \qquad Nu_T = C_2\,Ra^{1/3}$$

For Laminar natural convection from an upward-facing circular disk in an infinite medium, but with an adiabatic lowerface, Yovanovich and Jafarpur [96] give

$$Nu_{\sqrt{A}} = 2\sqrt{\pi} + 2^{1/8}F\,Ra_{\sqrt{A}}^{1/4}$$

with $F = 0.67/[1 + (0.5/Pr)^{9/16}]^{4/9}$ while for the same case but with the disk embedded in an infinite adiabatic plane it is

$$Nu_{\sqrt{A}} = 4/\sqrt{\pi} + \pi^{-1/4}F Ra_{\sqrt{A}}^{1/4}$$

All laminar natural convection correlations for inclined surfaces of any planform should lay between these two bounds.

Horizontal Cylinders

Natural convection from a single horizontal cylinder has a heat-transfer coefficient predicted [54] by, using cylinder diameter as the characteristic length,

$$Nu^{3.3} = Nu_L^{3.3} + Nu_T^{3.3}, \qquad 10^{-10} < Ra < 10^{10}$$

where, see Table 10-9 for C_1 and C_3,

$$Nu_L = \frac{2f}{\ln(1 + 2f/Nu_1)}, \qquad f = 1 - \frac{0.13}{(Nu_1)^{0.16}}$$

$$Nu_1 = 0.772 C_1 Ra^{1/4}, \qquad Nu_T = C_3 Ra^{1/3}$$

following the studies of Morgan [58] and Hesse and Sparrow [59]. See Morgan [60] or Ali-Arabi and Salman [61] for inclined cylinders.

Spheres and Miscellaneous Shapes

The heat-transfer coefficient from a sphere to air is predicted [54] by

$$Nu^6 = Nu_L^6 + Nu_T^6, \qquad 10 \leqslant Ra \leqslant 2 \times 10^9$$

where, see Table 10-9 for C_1 and C_4 values and use the characteristic length as the sphere diameter,

$$Nu_L = 2 + 0.878 C_1 Ra^{1/4}, \qquad Nu_T = C_4 Ra^{1/3}$$

King's generalized correlation [62], mostly for air and water, shown in Fig. 10-12 can be applied to such miscellaneous shapes as blocks, spheres, horizontal cylinders, and vertical plates if high accuracy is not required. The characteristic dimension L to be used is

$$\frac{1}{L} = \frac{1}{L_{\text{vertical}}} + \frac{1}{L_{\text{horizontal}}} \qquad (10\text{-}47)$$

TABLE 10-9 C Values for Various Prandtl Numbers

Pr	0.01	0.1	0.71	2	6	50	100	2000
C_1	0.242	0.387	0.515	0.568	0.608	0.65	0.656	0.668
C_3	0.077	0.09	0.103	0.108	0.109	0.100	0.097	0.088
C_4	0.074	0.088	0.104	0.110	0.111	0.101	0.097	0.086

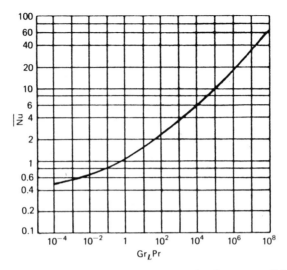

Figure 10-12 King's generalized correlation for miscellaneous solid shapes. [From W. J. King, *Mech. Eng.* **54**, 347–353 (1932) [62].]

For a horizontal cylinder, L is diameter. L_v might be the height of a vertical cylinder and L_h might be its diameter; in the turbulent range the exact value of L is immaterial as a result of the $\frac{1}{3}$ slope of the curve, and in the laminar range it is still largely immaterial since the curve slope then is about $\frac{1}{4}$. Although each geometric shape might fall on a curve slightly different from the average one, this correlation is adequate for most purposes. In the case of a vertical plate, for example, its finite width might allow a three-dimensional flow pattern whose effects are roughly taken into account by use of L from Eq. (10-47) in the correlations previously given for an infinitely wide vertical plate.

A refined procedure for estimating natural convection heat-transfer coefficients from three-dimensional bodies of arbitrary shape was devised by Hassani and Hollands [63]. The characteristic length for the Nusselt number is $L = A^{1/2}$ where A is the heat-transfer surface area of the body while for the Rayleigh number it is $H = (Z_f \bar{P}^2)^{1/3}$ where Z_f is the total height of the body and $\bar{P} = Z_f^{-1} \int_0^{Z_f} P(z)\,dz$ is the average perimeter $P(z)$ of the body. They give the Nusselt number Nu_L as

$$\mathrm{Nu}_L^n = \left[\left(C_l\,\mathrm{Ra}_H^{1/4} \right)^m + \left(C_t\,\mathrm{Ra}_H^{1/3} \right)^m \right]^{n/m} + \left(\mathrm{Nu}_{C,L} \right)^n$$

Here the exponents are evaluated from

$$n = \left[1, 1.26 - \frac{2 - L/L_m}{9\left(1 - 4.79 v^{2/3}/A\right)^{1/2}} \right]_{\max}$$

$$m = 2.5 + 12\exp\left(-13|C_t\,\mathrm{Ra}_H^{1/12} - 0.5| \right)$$

with v the body volume and L_m the longest straight line passing through the body ($n = 1.07$ is usually within 6%; for example a sphere of radius R has $L/L_m = \pi^{1/2}$, $A/v^{2/3} = 36\pi$, $n = 1$, $Z_f = 2R$, and $\bar{P} = \pi^2 R/2$). The coefficients are obtained from

$$C_l = \frac{0.671}{\left[1 + (0.492/\text{Pr})^{9/16}\right]^{4/9}}$$

$$\frac{HC_t}{L} = 0.098 - 0.065\frac{A_h}{A} + 0.008Z_f\frac{\bar{P}}{A}$$

with A_h either the horizontal downward-facing surface area of a heated body or the horizontal upward-facing surface area of a cooled body. Usually, $\text{Nu}_{C,L} \approx 3.51$. The subscripts C, l, and t refer to the conducting layer that controls at vanishingly small Rayleigh number, the laminar boundary layer that envelopes the body at small Ra, and the turbulent boundary layer that envelopes the body at large Ra, respectively.

For laminar conditions, Yovanovich [64] found equal or better accuracy with the simpler, again $L = A^{1/2}$,

$$\text{Nu}_L = 3.47 + 0.51\,\text{Ra}_L^{1/4}, \qquad 0 \leqslant \text{Ra}_L \leqslant 10^8$$

Enclosed Spheres and Cylinders

Natural convection inside a spherical cavity of diameter D is correlated by Kreith [65] by the empirical relation

$$\overline{\text{Nu}}_D = 0.59\,\text{Ra}_D^{1/4}, \qquad 10^4 \leqslant \text{Ra}_D \leqslant 10^9$$

$$= 0.13\,\text{Ra}_D^{1/3}, \qquad 10^9 \leqslant \text{Ra}_D \leqslant 10^{12}$$

For concentric spheres of outer diameter D_o and inner diameter D_i, Raithby and Hollands [54] give

$$\text{Nu} = 1.16C_1\left(\frac{L}{D_i}\right)^{1/4}\frac{\text{Ra}_L^{1/4}}{\left[(D_i/D_o)^{3/5} + (D_o/D_i)^{4/5}\right]^{5/4}}$$

which fits the data of Scanlan et al. [66] for $1.3 \times 10^3 \leqslant \text{Ra} \leqslant 6 \times 10^8$, $5 \leqslant \text{Pr} \leqslant 4000$, $1.25 \leqslant D_o/D_i \leqslant 2.5$. C_1 is obtained from Table 10-9,

$$L = \frac{(D_o - D_i)}{2} \qquad \text{and} \qquad \text{Nu} = \frac{qL}{\pi D_i D_o \,\Delta T k}$$

Natural convection in a cylindrical cavity of length L and diameter D was found by Evans and Stefany [67] to be correlated by

$$\overline{Nu}_D = 0.55 \, Ra_L^{1/4}, \qquad 0.75 \leqslant \frac{L}{D} \leqslant 2$$

In the cylindrical annulus formed by two long and concentric cylinders,

$$Nu = \left[1, 0.603 C_1 \frac{Ra_L^{1/4} \ln(D_o/D_i)}{\left[(L/D_i)^{3/5} + (L/D_o)^{3/5} \right]^{5/4}} \right]_{max}$$

where C_1 is obtained from Table 10-9, $L = (D_o/D_i)/2$, and $Nu = q \ln(D_o/D_i)/2\pi \, \Delta T k$.

Surveys of natural convection by Raithby and Hollands [54, 68] give additional details.

Enclosed Spaces Between Planes

Natural convection in the enclosure between two planes, each of constant temperature, tilted at an angle θ from the horizontal has been studied for many years. Recent interest in solar collector applications has led to refined correlations, most of which envision air as the fluid. The separation of the planes is L, the height of the enclosure along the tilt is H, and the width of the enclosure perpendicular to the tilt is W.

For $Pr \approx 0.7$ and $W/H \geqslant 8$ and subject to $H/L > 5$ and $0° \leqslant \theta \leqslant 60°$, Raithby and Hollands [54] recommend

$$Nu_L = 1 + 1.44 \left[1 - \frac{1708}{Ra_L \cos \theta} \right]^{\bullet} \left[1 - \frac{1708(\sin 1.8 \, \theta)^{1.6}}{Ra_L \cos \theta} \right]$$

$$+ \left[\left(\frac{Ra_L \cos \theta}{5830} \right)^{1/3} - 1 \right]^{\bullet}$$

El Sherbiny et al. [69a] recommend for $\theta = 60°$

$$Nu_L = \left\{ \left[1 + \frac{(0.0936 \, Ra^{0.314})^7}{1 + G} \right]^{1/7} , \left(0.1044 + 0.175 \frac{L}{H} \right) Ra_L^{0.283} \right\}_{max}$$

where $G = 0.5/[1 + (Ra_L/3165)^{20.6}]^{0.1}$ and for $\theta = 90°$

$$Nu_L = \left\{ \left[1 + \left(\frac{0.104\,Ra_L^{0.293}}{1 + (6310/Ra_L)^{1.36}} \right)^3 \right]^{1/3}, 0.242\left(\frac{Ra_L L}{H} \right)^{0.273}, 0.0605\,Ra_L^{1/3} \right\}_{max}$$

Then for $60° \leqslant \theta \leqslant 90°$, linear interpolation is recommended as

$$Nu_L = \frac{(90° - \theta)\,Nu_{L,60} + (\theta - 60°)\,Nu_{L,90}}{30°}$$

As explained by Raithby and Hollands, in the vertical ($\theta = 90°$) cavity there is flow for a Newtonian fluid at any Ra. The flow is small and parallel to the planes at low Ra, and Nu ≈ 1 since heat flows between the planes by conduction. At higher Ra, the flow has several possibilities. For $H/L \leqslant 40$, a laminar boundary layer forms on each plane with a nearly stationary core between them [the core is nearly isothermal horizontally but with a vertical gradient of $(T_{hot} - T_{cold})/2H$]; at sufficiently large Ra, after a transition range turbulent boundary layers form on the planes. For $H/L \geqslant 40$, the flow in the conduction regime becomes unstable at Ra in excess of a critical Rayleigh number Ra_c that depends on Pr ($Ra_c/Pr \approx 7300$ for Pr $\leqslant 12.7$ according to Korpela [70b]); for Pr > 12.7 the instability results in unsteady vertically traveling waves while for Pr < 12.7 the instability results in stationary horizontal-axis rolls. The turbulent boundary layers that are on the planes for these two cases are separated by a core of horizontally uniform temperature and a vertical temperature gradient of $0.36\,(T_{hot} - T_{cold})/H$.

For horizontal planes the recommended correlation is

$$Nu = 1 + \left[1 - \frac{Ra_c}{Ra} \right]^{\bullet} \left[k_1 + 2\left(\frac{Ra^{1/3}}{k_2} \right)^{1 - \ln(Ra^{1/3}/k_2)} \right]$$

$$+ \left[\left(\frac{Ra}{5380} \right)^{1/3} - 1 \right]^{\bullet} \left(1 - \exp\left\{ -0.95\left[\left(\frac{Ra}{Ra_c} \right)^{1/3} - 1 \right] \right\} \right)$$

where $[x]^{\bullet} = (|x| + x)/2$, $k_1 = 1.44/[1 + 0.018/Pr + 0.00136/Pr^2]$, $k_2 = 75\exp(1.5/Pr^{1/2})$, and Ra_c is the critical Rayleigh number for the geometry and boundary conditions encountered ($Ra_c = 1708$ for infinite horizontal planes, but see Raithby and Hollands for a collection of Ra_c for various conditions).

Ostrach's [70] survey on natural convection in fluids heated from below is recommended for detailed information and references to work prior to 1957.

With regard to horizontal planes heated from below, one of the first physical descriptions of the convective motion that can result when a dense fluid overlays a lighter fluid was given by Thomson [71] in 1882. In 1900 Bénard's [72] more complete description of the flow pattern—hexagonal cells with flow ascending in the center and descending along the sides (sometimes, however, descending in the center and ascending along the sides)—has resulted in these cells being termed *Bénard cells*. The first analysis was accomplished in 1916 by Rayleigh [73], who showed that motion will first occur at $Ra_L = 27\pi^4/4 = 657$, where L is the plane separation, if the top and bottom surfaces of the fluid are free. This analysis was confirmed and extended in 1926 by Jeffreys [74], who found that the critical Rayleigh number for the onset of motion is $Ra_L = 1108$ if the top fluid surface is free and the bottom surface is rigid. When both surfaces are rigid, the critical Rayleigh number is 1708 and turbulence first appears at $Ra_L \approx 5830$ [70]; flow becomes fully turbulent at $Ra_L \approx 45,000$ for air but at lower values for water. At very small temperature differences or very small plane separations, flow is columnar—a few randomly distributed and migratory columns of fluid ascend—rather than cellular [70]; the cellular pattern emerges only as plane separation or temperature difference exceeds a critical value. General discussion including the effects of externally imposed cross flow in a variety of configurations is given by Turner [75]. More recently, the influence of surface tension when one surface is free has been elucidated (see Velarde and Normand [76]). The influence of surface tension is traceable to its temperature dependence, and it is this influence that is responsible for the generally (but not always) hexagonal convection cells originally described by Bénard. In the absence of surface tension a container with one free surface is filled with rolls, each of width equal to the container depth—roll axes are parallel to the short side of a rectangular container and form concentric rings for a cylindrical container. Accurate description of a system with one free surface requires consideration of the Marangoni number—the same as the Rayleigh number but with the buoyant force replaced by the surface tension force. A fuller discussion of surface tension and the Marangoni number is given in Chapter 12 in connection with condensation. The hexagonal flow pattern occurs when the Marangoni number exceeds a critical value; other flow patterns exist prior to that. The roll and hexagon flow patterns are illustrated in Figs. 10-13 and 10-14, respectively. Thus the Raleigh number should not solely be relied on to predict the temperature difference required to initiate flow. As discussed by Velarde and Normand [70], the stability of natural convective systems is more accurately determined by the so-called Landau theory, devised in about 1937, which accounts for many terms in the series expansions considered—the older Rayleigh theory accounts for only the first term. Instructions for the construction of a natural convection demonstration apparatus driven by temperature differences and by the difference in density between saline and fresh water are given by Stong [77, 78] and Walker [80].

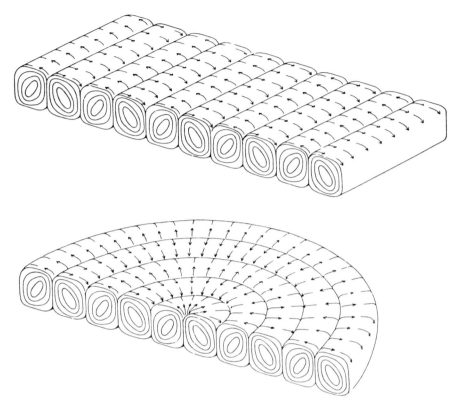

Figure 10-13 Roll-shaped cells are a stable configuration in convection driven by buoyancy forces rather than by surface tension. The fundamental unit of the pattern consists of two rolls that rotate in opposite directions; the width of this unit is twice the depth of the fluid layer. The plan form of the pattern depends strongly on the boundaries of the layer. In a rectangular container the rolls are parallel to the shorter sides; in a circular container they form concentric rings. A stable roll pattern is usually observed only when the fluid has no free surface. [By permission from M. G. Velarde and C. Normand, *Scientific American* **243** (1), 92–108 (1980) [76]; © 1980 by Scientific American, Inc., all rights reserved.]

The related natural convection that occurs when a heat-generating fluid is enclosed in a cavity whose walls are cooled has also been studied [54]. Among the practical situations of importance are those of nuclear reactor safety in which molten radioactive liquid pools must be cooled.

Natural convection in freezing or melting water has some peculiar characteristics because the maximum density of water is at 4°C. Experiments for horizontal ice layers have been made [80], as have experiments on freezing paraffin [81].

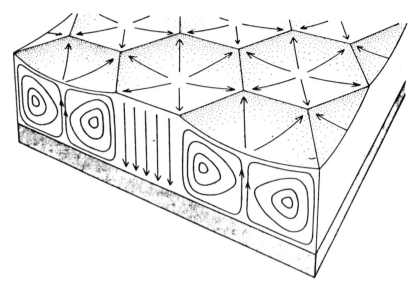

Figure 10-14 Tessellation of the surface with hexagonal cells is a characteristic feature of convection driven by a gradient in surface tension. Where the tension is greatest the surface becomes puckered, so that its area is reduced. Over the ascending plume in the center of each cell the surface is depressed; the fluid must flow uphill before descending at the edge of the cell. [By permission from M. G. Velarde and C. Normand, *Scientific American* **243** (1), 92–108 (1980) [76]; © 1980 by Scientific American, Inc., all rights reserved.]

Optimal Fin Array for Natural Convection

The ability of a finned surface to reject heat by natural convection is influenced by fin spacing and by fin dimensions. If the fins are too close together, the fluid is trapped and little heat is transferred. On the other hand, if the fins are too far apart, the surface area is so small that, again, little heat is transferred. Thus it is evident that an optimal space between fins exists. Similarly, the temperature distribution along the fin affects the heat-transfer coefficient and the fin dimensions also have optimal values.

Several common fin geometries are illustrated in Fig. 10-15. For the correlations recommended in the survey by Raithby and Hollands [54], the nomenclature in Fig. 10-15 is employed; it is also assumed that the fluid is air, that all properties are evaluated at the average temperature $(T_w + T_\infty)/2$, and that the fins are nearly isothermal.

For vertical rectangular fins as in Fig. 10-15a for $0.6 < \mathrm{Ra} < 100$, $\mathrm{Pr} = 0.71$, $0.33 < W/S < 4$, and $10.6 < H/S < 42$,

$$\mathrm{Nu} = \frac{\mathrm{Ra}}{\Psi}\left\{1 - \exp\left[-\frac{\Psi}{(2\,\mathrm{Ra})^{3/4}}\right]\right\}$$

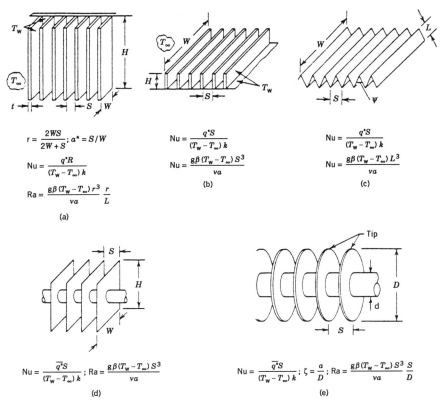

Figure 10-15 Common fin geometries for cooling by natural convection. [By permission from G. D. Raithby and K. G. T. Hollands, in W. Rohsenow, J. Hartnett, and E. Ganic, Eds., *Handbook of Heat Transfer Fundamentals*, 2nd ed., McGraw-Hill, 1985, pp. 6-1–6-94 [54].]

with

$$\Psi = \frac{24[1 - 0.483\exp(-0.17/\alpha^*)]}{\left\{(1 + \alpha^*/2)\left[1 + \{1 - \exp(-0.83\alpha^*)\}\{9.14\alpha^{*1/2}\exp(S^*) - 0.61\}\right]\right\}^3}$$

$$S^* = -4.65S \ (S \text{ in cm}), \qquad -11.8S \ (S \text{ in in.})$$

As suggested by Van De Pol and Tierney [82], the heat-transfer coefficient decreases after spacing decreases below a critical value S_1 due to trapping. If more fins are added in the freed-up space, the heat-transfer rate increases, however, due to the increased surface area. As fin spacing continues to decrease below a second critical value S_2, the heat-transfer rate decreases thereafter. For long fins, $\alpha^* \ll 1$, $S_1 \approx 60$ and $S_2 \approx 600$. Bar-Cohen [83] found the optimal fin length W^*, space S^* between fins, fin thickness t^*, and heat rejection rate q^* from a fin surface to be given by

$$W^* = 2.89 \left(\frac{k_{fin}}{k_{fluid}} \right)^{1/2} P, \qquad S^* = t^* = 2.89P$$

$$\frac{q}{H(S + t)(T_w - T_\infty)} = 0.217 \frac{(k_{fin} k_{fluid})^{1/2}}{P}$$

where

$$P^4 = \frac{H \nu^2}{[g\beta(T_w - T_\infty) \text{Pr}]}$$

For horizontal rectangular fins as in Fig. 10-15b,

$$\text{Nu}^{-2} = \left(\frac{\text{Ra}}{1500} \right)^{-2} + (0.081 \, \text{Ra}^{0.39})^{-2}$$

for $2 \times 10^2 < \text{Ra} < 6 \times 10^5$, $\text{Pr} = 0.71$, $0.026 \leqslant H/W \leqslant 0.19$, and $0.016 \leqslant S/W \leqslant 0.2$ as suggested by Jones and Smith [84]. When H/W is large, horizontal inflow through the ends is important, and the ends should not be blocked.

For square isothermal fins on a horizontal cylinder as in Fig. 10-15d, $0.2 < \text{Ra} < 4 \times 10^4$ and $\text{Pr} = 0.71$, as suggested by Elenbaas [85],

$$\text{Nu}^{2.7} = \left(\frac{\text{Ra}^{0.89}}{18} \right)^{2.7} + (0.62 \, \text{Ra}^{1/4})^{2.7}$$

10.9 NATURAL CONVECTION IN A POROUS MEDIUM

Natural convection in a porous medium is usually slow since the intimate contact between the porous medium and the fluid enables shearing forces to oppose body and pressure forces with small apparent velocities. In the cases discussed here, it is assumed that the porous medium is saturated with the fluid as air saturates fiberglass insulation in a building wall; a partially saturated porous medium can be encountered in a petroleum reservoir or an unconfined layer of water-bearing porous rock, for example. Greater detail than provided here is available in surveys [86–89].

Vertical Plate in an Infinite Porous Medium

For a constant-temperature vertical plate immersed in a saturated porous medium as illustrated in Fig. 10-1 with the porous medium taking the place of the continuous fluid, the boundary-layer form of Eqs. (2-74), (2-81), and (2-84) is

$$\text{Continuity} \qquad \frac{\partial u}{\partial x} + \frac{\partial v}{\partial y} = 0 \qquad (10\text{-}48a)$$

$$x \text{ Motion} \qquad 0 = -\frac{\partial p}{\partial x} - \rho g - \frac{\mu}{K} u \qquad (10\text{-}48b)$$

$$y \text{ Motion} \qquad 0 = -\frac{\partial p}{\partial y} - \frac{\mu}{K} v \qquad (10\text{-}48c)$$

$$\text{Energy} \qquad \rho C_p \left(u \frac{\partial T}{\partial x} + v \frac{\partial T}{\partial y} \right) = k_e \frac{\partial^2 T}{\partial y^2} \qquad (10\text{-}48d)$$

subject to the conditions

$$v(y = 0) = 0, \qquad u(y \to \infty) \to 0$$
$$T(y = 0) = T_w, \qquad T(y \to \infty) \to T_\infty$$

The inertial terms as well as both the Brinkman extension and the Forchheimer extension are neglected in the equations of motion since the velocities are expected to be small. Because the Brinkman extension $(\mu_e/\phi)/\partial^2 u/\partial y^2$ from Eq. (2-81) is neglected it will not be possible to satisfy a zero-slip condition at the plate.

Evaluation of the pressure gradient is different from the continuous-fluid case of Section 10.1 because $\partial p/\partial y$ in the y-motion equation is not zero even in a boundary-layer framework. Fortunately, pressure can be eliminated by differentiating the x-motion equation with respect to x and the y-motion equation with respect to x and then subtracting. This gives

$$\frac{\partial^2 p}{\partial y\,\partial x} - \frac{\partial^2 p}{\partial x\,\partial y} = -g\frac{\partial \rho}{\partial y} - \frac{\mu}{K}\frac{\partial u}{\partial y} + \frac{\mu}{K}\frac{\partial v}{\partial x}$$

from which it is seen that

$$\frac{\partial u}{\partial y} - \frac{\partial v}{\partial x} = -\frac{Kg}{\mu}\frac{\partial \rho}{\partial y}$$

The coefficient of thermal expansion β gives

$$-\frac{\partial \rho}{\partial y} = \rho\left(-\frac{1}{\rho}\frac{\partial \rho}{\partial T}\right)\frac{\partial T}{\partial y} = \rho\beta\frac{\partial T}{\partial y}$$

so that the combination of the x- and y-motion equations is

$$\frac{\partial u}{\partial y} - \frac{\partial v}{\partial x} = \frac{Kg\beta}{\mu}\frac{\partial T}{\partial y} \qquad (10\text{-}49)$$

An integral solution will be obtained first to ascertain the form of the solution. The assumed temperature and velocity profiles, $\eta = y/\delta$,

$$u = a + b\eta \qquad \text{and} \qquad \frac{T - T_\infty}{T_w - T_\infty} = A + B\eta + C\eta^2$$

are subjected to the conditions

At $\eta = 0$
$$\frac{\partial u}{\partial y} = \frac{Kg\beta}{\mu}\frac{\partial T}{\partial y} \qquad (10\text{-}50\text{a})$$

$$T = T_w \qquad (10\text{-}50\text{b})$$

$$\frac{\partial^2 T}{\partial y^2} = 0 \qquad (10\text{-}50\text{c})$$

At $\eta = 1$
$$u = 0 \qquad (10\text{-}50\text{d})$$

$$T = T_\infty \qquad (10\text{-}50\text{e})$$

$$\frac{\partial T}{\partial y} = 0 \qquad (10\text{-}50\text{f})$$

Equation (10-50a) is the combination of the x- and y-motion equations applied at the impermeable constant-temperature plate, and Eq. (10-50c) is the energy equation applied at the plate. The zero heat flux at the outer edge of the boundary layer is expressed by Eq. (10-50f). The stagnant state far from the plate is expressed by Eq. (10-50d) while the specified plate and ambient temperatures are reflected in Eqs. (10-50b) and (10-50e), respectively. Neglecting the second-order condition of Eq. (10-50c) for temperature gives

$$u = 2\frac{Kg\rho\beta\,\Delta T}{\mu}(1 - \eta) \qquad \text{and} \qquad \frac{T - T_\infty}{T_w - T_\infty} = (1 - \eta)^2 \quad (10\text{-}51\text{a})$$

This velocity profile has an apparent slip velocity at the plate, a consequence of neglecting $\partial^2 u/\partial y^2$ in the x-motion equation. These profiles can now be used in the integral form of the energy equation [Eq. (10-48d)], obtained as explained in Chapter 6,

$$\frac{\partial}{\partial x}\left[\int_0^\delta u(T - T_\infty)\, dy\right] = -\alpha\frac{\partial T(0)}{\partial y}$$

(10-51b)

to obtain

$$\delta = \left(\frac{8\mu\alpha x}{Kg\rho\beta\,\Delta T}\right)^{1/2}$$

The local heat-transfer coefficient h is found in terms of the Rayleigh number Ra from use of the temperature profile and the solution for δ in

$$q_w = h(T_w - T_\infty) = -k\frac{\partial T(0)}{\partial y} = \frac{2k(T_w - T_\infty)}{\delta}$$

to be, the Rayleigh number is $\mathrm{Ra}_x = Kg\rho\beta\,\Delta Tx/\mu\alpha$,

$$\mathrm{Nu}_x = \frac{hx}{k} = \frac{\mathrm{Ra}_x^{1/2}}{\sqrt{2}}$$

The variable $\eta = y/\delta$ is seen to be of the form $\eta = Cy/x^{1/2}$ where $C^2 = C_1^2 Kg\rho\beta\,\Delta T/\mu\alpha$. A similarity solution can be formed from this information. As before, the continuity equation is automatically satisfied by selecting a stream function $\psi = F(x)f(\eta)$ such that

$$u = \frac{\partial\Psi}{\partial y} \quad\text{and}\quad v = -\frac{\partial\Psi}{\partial x}$$

so that a similarity transformation gives

$$u = \frac{\partial\Psi}{\partial x}\frac{\partial x^0}{\partial y} + \frac{\partial\Psi}{\partial\eta}\frac{\partial\eta^{c/x^{1/2}}}{\partial y} = F(x)cx^{-1/2}f'$$

Comparison of this result with the integral method velocity profile of Eq. (10-51a) shows that

$$F(x) = C_1\alpha\left(\frac{Kg\rho\beta\,\Delta Tx}{\mu\alpha}\right)^{1/2} = \alpha Cx^{1/2}$$

Hence

$$\Psi = \alpha C x^{1/2} f(\eta)$$

and

$$u = \frac{Kg\rho\beta \,\Delta T}{\mu} f' = \frac{\alpha C^2}{C_1^2} f'$$

Similarly,

$$v = -\frac{\partial \Psi}{\partial x}\frac{\partial y}{\partial x}^{1} - \frac{\partial \Psi}{\partial \eta}\frac{\partial \eta}{\partial x}^{-\eta/2x} = \frac{\alpha C}{2x^{1/2}}(\eta f' - f)$$

and it is seen that $v(x \to \infty) \to 0$ and, from the y-motion equation, $\partial p(x \to \infty)/\partial y \to 0$ as for the continuous-fluid case. The energy equation [Eq. (10-48d)] becomes, with $(T - T_\infty)/(T_w - T_\infty) = \theta(\eta)$,

$$\frac{\alpha C^2}{C_1^2} f'\left(\frac{\partial \theta}{\partial x}^{0}\frac{\partial x}{\partial x}^{1} + \frac{\partial \theta}{\partial \eta}\frac{\partial \eta}{\partial x}^{-\eta/2x}\right) + \frac{\alpha C}{2x^2}(\eta f' - f)\left(\frac{\partial \theta}{\partial x}\frac{\partial x}{\partial y}^{0}\right.$$

$$\left. + \frac{\partial \theta}{\partial \eta}\frac{\partial \eta}{\partial y}^{c/x^{1/2}}\right) = \alpha\left(\frac{\partial \eta}{\partial y}\right)^2 \frac{\partial^2 \theta}{\partial \eta^2}$$

$$\eta f' \theta'\left(1 - \frac{1}{C_1^2}\right) - f\theta' = 2\theta''$$

A simplification results if $C_1 = 1$. Then the energy equation is

$$2\theta'' + f\theta' = 0$$

The combination of the x- and y-motion equations [Eq. (10-49)] similarly becomes

$$\left[\frac{\partial u}{\partial x}\frac{\partial x}{\partial y}^{0} + \frac{\partial u}{\partial \eta}\frac{\partial \eta}{\partial y}^{c/x^{1/2}}\right] - \left[\frac{\partial v}{\partial x}\frac{\partial x}{\partial x}^{1} + \frac{\partial v}{\partial \eta}\frac{\partial \eta}{\partial x}^{c/x^{1/2}}\right]$$

$$= \frac{\alpha C^2}{C_1^2}\left(\frac{\partial \theta}{\partial x}^{0}\frac{\partial x}{\partial y}^{0} + \frac{\partial \theta}{\partial \eta}\frac{\partial \eta}{\partial y}^{c/x^{1/2}}\right)$$

which reduces to, noting that $C^2 x = \mathrm{Ra}_x$ if $C_1 = 1$,

$$f'' - (f - \eta f' - \eta^2 f'')/4\,\mathrm{Ra}_x = \theta'$$

The second term on the left-hand side does not have a similarity form inasmuch as it depends on x rather than η. But in the limit as $\mathrm{Ra}_x \to \infty$, the boundary layer becomes a thin shear layer and the equations to be solved have the similarity forms of

$$2\theta'' + f\theta' = 0 \quad \text{and} \quad f'' = \theta' \tag{10-52}$$

subject to the boundary conditions

$f(0) = 0$ (impermeable plate), $\qquad f(\infty) = 0$ (stagnant medium)

$\theta(0) = 1$ (isothermal plate), $\qquad \theta(\infty) = 0$ (isothermal medium)

Numerical integration of Eqs. (10-52) requires determination of the missing initial conditions $f'(0)$ and $\theta'(0)$. Then the local Nusselt number for the thin-shear-layer limit follows as

$$h(T_w - T_\infty) = -k\frac{\partial T(0)}{\partial y} = -k(T_w - T_\infty)\frac{\partial \eta}{\partial y}\theta'(0)$$

$$\mathrm{Nu}_x = \frac{hx}{k} = \mathrm{Ra}_x^{1/2}[-\theta'(0)]$$

The boundary-layer thickness δ is the value of η at which $\theta = 1/100$. Thus

$$\delta = \frac{\eta_T x^{1/2}}{C}$$

This similarity procedure applied to the more general case of a power-law variation of plate temperature

$$T_w = T_\infty + Ax^\lambda$$

with the stream function ψ and similarity variable η as

$$\Psi = \alpha\,\mathrm{Ra}_x^{1/2}\,f(\eta) \quad \text{and} \quad \eta = \frac{\mathrm{Ra}_x^{1/2}\,y}{x}$$

where $\mathrm{Ra}_x = Kg\rho\beta Ax^{\lambda+1}/\mu\alpha$ and letting $\theta = (T - T_\infty)/Ax^n$ gives

$$2\theta'' + (1 + \lambda)f\theta' - \lambda f'\theta = 0 \tag{10-53a}$$

and

$$f'' = \theta' \tag{10-53b}$$

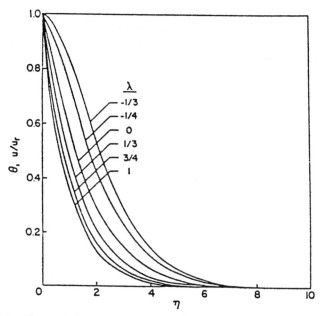

Figure 10-16 Dimensionless temperature and vertical velocity versus the similarity variable for natural convection in a porous medium adjacent to a vertical heated surface. [By permission from P. Cheng and W. J. Minkowycz, *J. Geophys. Res.* **82**, 2040–2044 (1977) [90].]

subject to

$$f(0) = 0, \qquad f'(\infty) = 0$$
$$\theta(0) = 1, \qquad \theta(\infty) = 0$$

Numerical solutions due to Cheng and Minkowycz [90] are shown in Fig. 10-16 with $\theta'(0)$ and η_T values listed in Table 10-10. Integration of Eq. (10-53b) with consideration of the boundary conditions reveals that $f' = \theta$; the distribution of both $f' = u/[Kg\rho\beta Ax^\lambda/\mu]$ and θ is shown in Fig. 10-16. The local heat flux q_w at the plate is

$$q_w = k \left(\frac{Kg\rho\beta A^3 x^{3\lambda-1}}{\mu} \right)^{1/2} [-\theta'(0)]$$

with $\lambda = 1/3$ for constant heat flux; the local Nusselt number Nu_x is

$$\mathrm{Nu}_x = [-\theta'(0)] \, \mathrm{Ra}_x^{1/2}$$

TABLE 10-10   Values of η_T and $-\theta'(0)$ for Different λ [90]

λ	η_T	$-\theta(0)$
$-\frac{1}{3}$	7.2	0
$-\frac{1}{4}$	6.9	0.162
0	6.3	0.444
$\frac{1}{4}$	5.7	0.630
$\frac{1}{3}$	5.5	0.678
$\frac{1}{2}$	5.3	0.761
$\frac{3}{4}$	4.9	0.892
1	4.6	1.001

and the boundary-layer thickness δ is

$$\frac{\delta}{x} = \frac{\eta_T}{\mathrm{Ra}_x^{1/2}}$$

Horizontal Cylinder and Sphere in an Infinite Porous Medium

For an isothermal horizontal cylinder of diameter D in an infinite porous medium, the average Nusselt number is

$$\overline{\mathrm{Nu}}_D = 0.565\,\mathrm{Ra}_D^{1/2}, \qquad \mathrm{Ra}_D = \frac{Kg\rho\beta\,\Delta TD}{\mu\alpha}$$

For an isothermal sphere of radius D, it is

$$\overline{\mathrm{Nu}}_D = 0.362\,\mathrm{Ra}_D^{1/2}$$

Such additional details as the mass flow rate in the plume above these objects can be ascertained from the fuller treatment provided by Cheng [86].

Plumes

The plume rising above a line source of heat releasing heat per unit length at the rate Q immersed in an infinite porous medium has a temperature excess given by

$$T - T_\infty = \left[\frac{QB^2}{6\rho C_p\alpha\,\mathrm{Ra}_x^{1/3}}\right]\tanh\left(\frac{B\eta}{6}\right)$$

where $B^3 = \frac{9}{2}$, $\mathrm{Ra}_x = Kg\beta Qx/\alpha^2\mu C_p$, x is the vertical distance above the line source, T_∞ is the ambient temperature, and $\eta = \mathrm{Ra}_x^{1/3}\,y/x$. The stream function ψ from which the vertical velocity component $u = \partial\psi/\partial y$ can be obtained is

$$\Psi = \alpha \, \mathrm{Ra}_x^{1/3} \, B \tanh\left(\frac{B\eta}{6}\right)$$

according to Yih [91] and in the survey of Cheng [86].

Above a point source of heat with a heat release rate Q, the temperature excess and the vertical velocity at a height x above the heat source is given by Bejan [31, pp. 376–380] as

$$\frac{(T - T_\infty) kx}{Q} = \frac{ux}{\alpha \, \mathrm{Ra}} = \frac{2C^2}{1 + (C\eta/2)^2}$$

Here $C = 0.141$, $\mathrm{Ra} = Kg\beta Q/\alpha\nu k$, $\eta = \mathrm{Ra}^{1/2} r/x$, and r is the radial distance from the vertical axis through the sphere center.

A physical situation similar to a plume is a heated plate immersed in a porous medium of infinite extent in which there is a parallel, uniform externally forced velocity U. As mentioned before, the predicted velocity in the porous medium can be uniform, apparently slipping along the plate, if the Brinkman extension is neglected in the equations of motion as is consistent with a creeping-flow assumption. The boundary-layer description of the problem then has only the energy equation as

$$U\frac{\partial T}{\partial x} = \alpha\frac{\partial^2 T}{\partial y^2}, \qquad x \geqslant 0, \qquad y \geqslant 0$$

$$T(x,0) = T_w \qquad \text{and} \qquad T(x, y \to \infty) \to T_\infty$$

This is of the form of a transient conduction problem in a semi-infinite domain. The solution (see Carslaw and Jaeger [92]) is

$$\frac{T - T_w}{T_\infty - T_w} = \mathrm{erf}\left(\frac{Uy^2}{4\alpha x}\right)^{1/2}$$

with x the distance from the plate leading edge and y the distance from the plate into the porous medium. The local Nusselt number Nu_x is

$$\mathrm{Nu}_x = \frac{hx}{k} = 0.564 \, \mathrm{Pe}_x^{1/2}$$

where the Peclet number is $\mathrm{Pe}_x = Ux/\alpha$.

Enclosures

Similar to the case for a continuous fluid, a horizontal layer of saturated porous medium of thickness L heated from below will have natural convec-

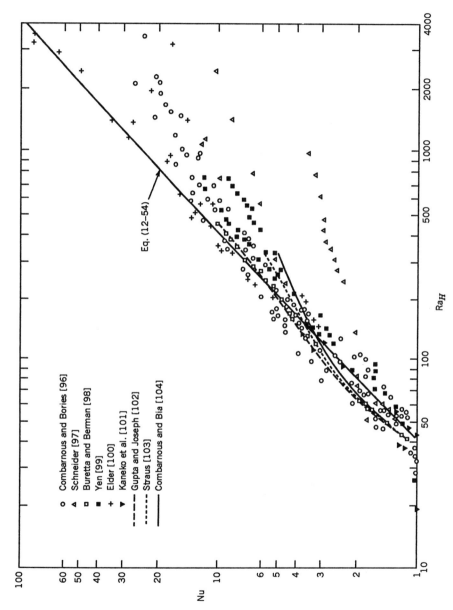

Figure 10-17 Heat-transfer measurements in a porous layer heated from below. [By permission from P. Cheng, *Adv. Heat Transfer* **14**, 1–105 (1979) [93].]

Eq. (12–54)

Ra_H

Nu

o Combarnous and Bories [96]
△ Schneider [97]
□ Buretta and Berman [98]
■ Yen [99]
+ Elder [100]
▼ Kaneko et al. [101]
 ‒ ‒ ‒ Gupta and Joseph [102]
 ······ Straus [103]
 —— Combarnous and Bia [104]

443

tion cells if the Rayleigh number $\mathrm{Ra}_L = Kg\beta\,\Delta TL/\nu\alpha$ exceeds a critical value $\mathrm{Ra}_c = 4\pi^2 = 39.5$ (see Appendix E for a stability analysis). After Ra_c is exceeded, the Nusselt number is given by

$$\mathrm{Nu}_L = \frac{hL}{k} = \begin{cases} 1, & \mathrm{Ra}_L < 40 \\ \mathrm{Ra}_L/40, & \mathrm{Ra}_L \geqslant 40 \end{cases} \qquad (10\text{-}54)$$

as shown in Fig. 10-17. The data shown in Fig. 10-17 and those of Kladias and Prasad [94] differ from prediction when the thermal conductivity of the solid greatly exceeds that of the fluid.

Mixed forced–natural convection with volumetric heat generation studied by Hadim and Burmeister [95] exemplifies the variety of information that annually increases.

PROBLEMS

10-1 Show that the similarity transformation in Section 10.1 puts the laminar boundary-layer equations and boundary conditions into the similarity form of Eq. (10-1).

10-2 Laminar natural convection Nusselt numbers can be large. To illustrate this, calculate the Nusselt number for a vertical flat plate in room air—take the Grashof number to be near its upper allowed value for laminar flow, $\mathrm{Gr}_L \leqslant 10^9$. The result should be $\mathrm{Nu}_L \sim 10^2$.

10-3 Natural convection is important to cooling of gas turbine rotor blades where a fluid circulates through internal blade passages. Sometimes the maximum fluid velocity in natural convection is then surprisingly large because centrifugal force, rather than gravity, acts as the body force.

(a) Show for air with $g = 10^6$ ft/sec^2, $L = 0.25$ ft, and $T_w - T_\infty = 100°$F that $U_{\max}$ is of the order of 10^2 ft/sec for a flat-plate geometry. The body force can be approximated as a constant.

(b) Repeat part a for $g = 32.2$ ft/sec^2 and comment on the statement that velocities are usually small in natural convection.

10-4 Velocities in natural convection from an impermeable vertical plate are usually small. In such a case the convective (inertia) terms in the equations of motion would be negligible, leaving body forces to be entirely countered by viscous shear forces, and "creeping motion" would exist.

(a) Show that the boundary-layer equation of x motion with convective terms deleted becomes

$$0 = \nu_\infty \frac{\partial^2 u}{\partial y^2} + g\frac{\rho_\infty - \rho}{\rho_\infty}$$

and that the other boundary-layer equations and boundary conditions are unchanged from those given just prior to Eq. (10-1).

(b) Use a stream function $\Psi = 4\alpha(\mathrm{Ra}_x/4)^{1/4}F(\eta)$ such that $u = \partial\Psi/\partial y$ and $v = -\partial\Psi/\partial x$ and a similarity variable $\eta = (\mathrm{Ra}_x/4)^{1/4}(y/x)$ to put the boundary-layer equations and boundary conditions into the similarity form

$$F''' + \theta = 0, \qquad \theta'' + 3F\theta' = 0$$
$$F'(0) = F(0) = F'(\infty) = \theta(\infty) = 0, \qquad \theta(0) = 1$$

where

$$u = \frac{2\alpha}{x}\mathrm{Ra}_x^{1/2}\,F', \qquad v = \frac{\alpha}{x}\left(\frac{\mathrm{Ra}_x}{4}\right)^{1/4}(\eta F' - 3F)$$

(c) Show that the similarity formulation in part b gives the local Nusselt number and shear stress as

$$\mathrm{Nu}_x = \frac{hx}{k} = -\theta'(0)\left(\frac{\mathrm{Ra}_x}{4}\right)^{1/4}$$

$$\tau_w = \left(4\,\mathrm{Ra}_x^3\right)^{1/4}\left(\frac{\rho\nu^2}{x^2\,\mathrm{Pr}}\right)F''(0)$$

Hence, in "creeping motion," Nu_x depends on the Gr Pr product rather than the Gr and Pr separately.

(d) Use the separation of variables technique outlined in Appendix C to obtain the similarity transformation applied in part b.

10-5 Just as inertial effects are negligible at the low velocities encountered for $\mathrm{Pr} \to \infty$ in natural convection from a vertical plate, so might viscous effects be negligible at the higher velocities that might be encountered for $\mathrm{Pr} \to 0$. To see this:

(a) Render the boundary-layer equations dimensionless by letting $x' = x/L$, $y' = y/L$, $u' = u/u_r$, $v' = v/u_r$, and $T' = (T - T_w)/(T_w - T_\infty)$ to obtain

$$\frac{\partial u'}{\partial x'} + \frac{\partial v'}{\partial y'} = 0 \tag{5P-1}$$

$$u'\frac{\partial u'}{\partial x'} + v'\frac{\partial u'}{\partial y'} = \frac{\nu}{Lu_r}\frac{\partial^2 u'}{\partial y'^2} + \left[\frac{gL|\rho_\infty - \rho_w|}{\rho_\infty u_r^2}\right]\frac{\rho_\infty - \rho}{|\rho_\infty - \rho_w|} \tag{5P-2}$$

$$u'\frac{\partial T'}{\partial x'} + v'\frac{\partial T'}{\partial y'} = \frac{\nu}{\mathrm{Pr}\,Lu_r}\frac{\partial^2 T'}{\partial y'^2} \tag{5P-3}$$

Here L is a constant reference length and u_r is a constant reference velocity.

(b) Select $u_r = \nu / \mathrm{Pr}\, L$ to put Eqs. (5P-1)–(5P-3) into the form

$$\frac{\partial u'}{\partial x'} + \frac{\partial v'}{\partial y'} = 0 \tag{5P-4}$$

$$u' \frac{\partial u'}{\partial x'} + v' \frac{\partial u'}{\partial y'} = \mathrm{Pr}\, \frac{\partial^2 u'}{\partial y'^2} + \mathrm{Gr}_L \, \mathrm{Pr}^2 \, \frac{\rho_\infty - \rho}{|\rho_\infty - \rho_w|} \tag{5P-5}$$

$$u' \frac{\partial T'}{\partial x'} + v' \frac{\partial T'}{\partial y'} = \frac{\partial^2 T'}{\partial y'^2} \tag{5P-6}$$

where

$$\mathrm{Gr}_L = g |\rho_\infty - \rho_w| \frac{L^3}{\rho \nu^2}$$

(c) Verify from the result of part b that (1) viscous effects are negligible and $\mathrm{Gr}\,\mathrm{Pr}^2$ is the controlling parameter when $\mathrm{Pr} \to 0$ and (2) inertial effects are negligible and $\mathrm{Gr}\,\mathrm{Pr}$ is the controlling parameter when $\mathrm{Pr} \to \infty$.

10-6 An integral method is to be applied to free convection from a vertical plate, immersed in a constant-property fluid, whose surface heat flux q_w is constant [19].

(a) Show that use of the assumed profiles

$$T - T_\infty = \frac{q_w \delta}{2k} (1 - \eta)^2, \qquad u = u_1 \eta (1 - \eta)^2$$

where $\eta = y/\delta$, δ is the common thermal and velocity boundary-layer thickness, and u_1 is a reference velocity to be determined in Eq. (10-14) gives

$$\frac{d(W^2 D/105)}{dX} = \frac{D^2}{6} - \frac{W}{D}, \qquad \frac{d(WD^2/30)}{dX} = \frac{2}{\mathrm{Pr}}$$

Here $W = u_1 (g\beta q_w \nu^2 / k)^{-1/4}$, $D = \delta (g\beta q_w / k\nu^2)^{1/4}$, and $X = x(g\beta q_w / k\nu^2)^{1/4}$.

(b) Solve these two differential equations to achieve

$$W = \left(\frac{6000}{\mathrm{Pr}} \right)^{1/5} \left(\mathrm{Pr} + \frac{4}{5} \right)^{-2/5} X^{3/5}$$

$$D = \left(\frac{360}{\mathrm{Pr}^2} \right)^{1/5} \left(\mathrm{Pr} + \frac{4}{5} \right)^{1/5} X^{1/5}$$

(c) From the result of part b show that

$$T_w - T_\infty = 1.622 \frac{q_w x}{k} \left(\frac{Pr + 0.8}{Pr^2 \, Gr_x^*} \right)^{1/5}$$

$$h_x = \frac{q_w}{T_w - T_\infty} = 0.62 \frac{k}{x} \left(\frac{Pr^2 \, Gr_x^*}{Pr + 0.8} \right)^{1/5}$$

where $Gr_x^* = (g\beta x^3 / \nu^2)(q_w x / k)$ is a modified Grashof number. Note that T_w varies as $x^{1/5}$ as predicted by the exact solution.

(d) From the results of part c show that the average Nusselt number is $\overline{Nu}_x = 0.775[Pr^2 \, Gr_x^* / (Pr + 0.8)]^{1/5}$.

(e) Recognizing that $Gr_x^* = Gr_x \, Nu_x$, rearrange the result of part d into the form

$$\frac{\overline{Nu}_x}{Ra_x^{1/4}} = \frac{0.683 \, Pr^{1/4}}{(Pr + 0.8)^{1/4}}$$

and compare with Eq. (10-22) for the case of constant wall temperature.

(f) Use the result of part d and the corresponding integral method result for a constant wall temperature as expressed by Eq. (10-22) to compare with the exact results given in Table 10-3.

10-7 Show by use of Eq. (10-25) that the Nusselt number for natural convection from a vertical cylinder of constant wall temperature immersed in an infinite fluid first exceeds the flat-plate value by 5% when $Ra_L^{1/4} D/L \approx 33$ for $Pr = 1$.

10-8 Employ the results found by the integral method to show that the shear stress at the wall in turbulent natural convection from a constant-temperature vertical plate immersed in a constant property fluid has:

(a) A local value given by

$$\tau_w = \frac{0.06686 \rho \nu^2}{x^2} \, Pr^{-23/30} (1 + 0.494 \, Pr^{2/3})^{-9/10} \, Ra_x^{9/10}$$

(b) An average value, assuming the plate to be wholly covered by a turbulent boundary layer, of $\overline{\tau}_w = 5\tau_w/12$. Or

$$\overline{\tau}_w = 0.02786 \frac{\rho \nu^2}{x^2} \, Pr^{-23/30} (1 + 0.494 \, Pr^{2/3})^{-9/10} \, Ra_x^{9/10}$$

10-9 Nuclear waste buried 500 m below the surface of the Earth in a water-saturated sandstone rock (see Tables 2-1 and 2-2 for permeability and porosity) formation releases heat at the rate of 500 W. Use information from Section 10.9 to estimate a numerical value for the maximum velocity in the plume. From this result, estimate the time required for a fluid particle to rise to the surface after contacting the waste.

10-10 Show by numerical example that the laminar natural convection heat-transfer coefficient from a horizontal cylinder can be computed by use of $\pi D/2$ in a vertical plate correlation. In other words, let the characteristic dimension for a horizontal cylinder be the maximum distance traveled by a fluid particle, and use an average gravitational acceleration tangent to the cylinder surface. *Answer*: $\overline{Nu}_D = 0.443\,Ra_D^{1/4}$.

10-11 Demonstrate that the thermal resistance offered by vertical air layers (e.g., building construction) approaches a lower limit—nearly attained with a thickness of 2 in.—as the air gap becomes large. Does the increase in thermal resistance obtained by dividing a thick vertical air gap by vertical partitions increase proportionally to the number of partitions? Also, establish that the thermal resistance for horizontal air layers with a plane separation of 2–8 in. is equal to the maximum thermal resistance for a vertical air layer (at about 2 in.).

10-12 Compare the predictions for air in horizontal and vertical enclosures between two planes.

10-13 When installed, a storm window is 1 in. away from the inside window and is 2 ft high. If the inside window is at 65°F and the outside window is at 35°F, estimate the convection heat-transfer rate across the air gap between the two windows. *Answer*: $q \approx 10$ Btu/hr ft^2.

10-14 Estimate the average heat-transfer coefficient for natural convection between a cubical instrumentation package that is 1 ft on a side at 120°F and a pool of water at 80°F. *Answer*: $h \approx 110$ Btu/hr ft^2 °F.

10-15 A short solid vertical cylinder 6 in. high and 6 in. in diameter is at 500°F and cools by natural convection in 100°F air. Estimate the average heat-transfer coefficient. *Answer*: $h \approx 1.8$ Btu/hr ft^2 °F, $h_r = 2.9$ Btu/hr ft^2 °F.

10-16 Water in a large tank at 60°F is heated by a 1-in.-diameter vertical pipe at 200°F. The water depth is 3 ft. Estimate the heat-transfer rate by natural convection from the pipe to the water. *Answer*: $q \approx 35,000$ Btu/hr.

10-17 A 6-in.-diameter sphere at 700°F cools by natural convection in 100°F air after heat treatment. Estimate the heat-transfer coefficient for natural convection and estimate the magnitude of the forced convection velocity that would be required to appreciably affect the heat-transfer coefficient. *Answer*: $h \approx 0.6$ Btu/hr ft^2 °F, $h_r = 0.5$ Btu/hr ft^2 °F.

10-18 A 1-ft-diameter spherical container holds electronic instruments near the bottom of a lake whose water is at 60°F. The sphere surface is maintained at constant temperature by the 1020 W dissipated by the electronics. Estimate the sphere surface temperature. *Answer*: $T \approx$ 70°F.

10-19 Water flows in forced convection past a vertical flat plate of 1 ft height so that Re $= 5 \times 10^4$. If the water temperature is 60°F, what is the highest plate temperature that can be used without natural convection effects influencing the heat-transfer coefficient by more than 5%? *Answer*: $T \approx 80$°F.

10-20 The side of a small laboratory furnace can be idealized as a vertical plate 2 ft high and 8 ft wide. The furnace sides are at 100°F and the surrounding air is at 70°F.

(a) Estimate the heat loss from the sides by natural convection.

(b) Estimate the lowest vertical forced velocity that would cause the heat-transfer coefficient to depart noticeably from its natural convection value. *Answer*: (a) $q \approx 3800$ Btu/hr; (b) $V \approx 1$ ft/sec.

10-21 Air at 100°F flows at Re $= 10^2$ in a horizontal tube that is 5 ft long of 1 in. diameter and of 500°F wall temperature. Is the flow purely forced convection? *Answer*: Pure forced convection.

10-22 Water at 60°F flows at 10 ft/sec perpendicular to a 1-in.-diameter horizontal tube that is at 200°F. The water velocity is 10 ft/sec. Are natural convection effects important to a determination of the heat-transfer coefficient?

10-23 Air at 140°F flows through an 8-in.-diameter duct that is 15 ft long, and the duct wall temperature is 80°F. Estimate the flow velocity at which natural convection effects will become important with possible fluid stratification. *Answer*: $V \approx 0.08$ ft/sec.

10-24 Determine the water pressure necessary to prevent boiling of a horizontal heating element 2 ft long and 1 in. in diameter dissipating electrical energy at the rate of 4500 W in 140°F water. *Answer*: 20 psia.

10-25 A $\frac{1}{32}$-in.-diameter wire is horizontal in 60°F water. If the wire temperature is 140°F, what is the energy loss by natural convection from the wire? *Answer*: 80 W/ft.

10-26 The top of a stove approximates a square plate of 3-ft side. Estimate the heat-transfer rate by natural convection from the stove top if its temperature is 110°F while the room temperature is 70°F. *Answer*: $q = 300$ Btu/hr.

10-27   A small resistance heater in the shape of a square plate of 2 in. side is to be designed for the bottom of a fish tank. The tank water must be kept at 75°F while the heater surface temperature must not exceed 85°F. Specify the wattage of the heater. *Answer*: 3 W.

10-28   Square tiles of 3 in. side are heated in an oven at 1100°F. The tiles are at 900°F. Compute the natural convection heat-transfer coefficient when the tiles rest horizontally on the oven floor. *Answer*: $h = 0.5$ Btu/hr ft^2 °F.

10-29   A block of ice is of 2 ft side and is suspended in a stagnant room with its top horizontal. The ice is at 20°F, and the room is at 80°F. Estimate the natural convective heat-transfer coefficient for the block on the (a) top, (b) bottom, and (c) average surface. *Answer*: (a) $h \approx 0.3$ Btu/hr ft^2 °F; (b) $h \approx 1$ Btu/hr ft^2 °F.

10-30   Circular sheets of plastic of 1 ft diameter and at 40°F are heated by floating on the surface of 160°F water. Estimate the heat-transfer coefficient between the sheet and the water. *Answer*: $h = 200$ Btu/hr ft^2 °F.

10-31   Consider a vertical isothermal surface to which rectangular aluminum fins are attached as shown in Fig. 10-15a. The fins have a height L of 0.13 m, and the isothermal surface is 100°C hotter than the surrounding air. Show that under these conditions (a) the optimal fin length is 60 cm, the optimal space between fins and the optimal fin thickness are both 0.6 cm, and that the optimal heat rejection rate from the vertical isothermal surface is 260 W/m^2 K, (b) the optimal fin dimensions are not sensitive to temperature differences, varying only with its $\frac{1}{4}$th power, and (c) the ratio of the optimal finned heat-rejection rate q to the heat-rejection rate from the unfinned vertical surface q_{uf} is 40.

REFERENCES

1. B. Gebhart, in S. Kakac, W. Aung, and R. Viskanta, Eds., *Natural Convection*, Hemisphere, 1985, pp. 3–35.
2. E. Schmidt and W. Beckmann, *Forsch-Ing.-Wes.* **1**, 391 (1930).
3. A. Oberbeck, *Annalen der Physik und Chemie* **7**, 271–292 (1879).
4. L. Lorenz, *Annalen der Physik und Chemie* **13**, 582–606 (1881).

5. S. Ostrach, An analysis of laminar free-convection flow and heat transfer about a plate parallel to the direction of the generating body force, NACA Report 1111, 1953.

6. A. J. Ede, *Adv. Heat Transfer* **4**, 1–64 (1967).

7. E. J. LeFevre, Laminar free convection from a vertical plane surface, Mechanical Engineering Research Laboratory, Heat 113 (Great Britain), 1956; *Proc. Ninth International Congress on Applied Mechanics, Brussels*, Vol. 4, 1956, p. 168.

8. E. M. Sparrow and J. L. Gregg, Details of exact low Prandtl number boundary layer solutions for forced and for free convection, NASA Memo 2-27-59 E, 1959.

9. W. H. McAdams, *Heat Transmission*, 3rd ed., McGraw-Hill, 1954.

10. J. Gryzagoridis, *Int. J. Heat Mass Transfer* **14**, 162–164 (1971).

11. E. M. Sparrow and J. L. Gregg, *Trans. ASME* **80**, 879–886 (1958).

12. T. Fujii, M. Takeuchi, M. Fujii, K. Suzaki, and H. Uehara, *Int. J. Heat Mass Transfer* **13**, 753–787 (1970).

13. S. Kakac, O. E. Atesoglu, and Y. Yener, in S. Kakac, W. Aung, and R. Viskanta, Eds., *Natural Convection*, Hemisphere, 1985, pp. 729–773.

14. H. B. Squire, in S. Goldstein, Ed., *Modern Developments in Fluid Dynamics*, Vol. II, Oxford University Press, 1938, pp. 801–810.

15. H. J. Merk and J. A. Prins, *Appl. Sci. Res.* **A4**, 11–24, 195–206, 207–221 (1954).

16. (a) S. Sugawara and I. Michiyoshi, *Trans. JSME* **17**, 109 (1951); (b) E. M. Sparrow and J. L. Gregg, *Trans. ASME* **78**, 435–440 (1956).

17. S. W. Churchill and H. H. S. Chu, *Int. J. Heat Mass Transfer* **18**, 1323–1329 (1975).

18. G. C. Vliet, *ASME J. Heat Transfer* **91**, 511–516 (1969).

19. E. M. Sparrow, Laminar free convection on a vertical plate with prescribed non-uniform wall heat flux or prescribed non-uniform wall temperature, NACA TN 3508, 1955.

20. E. M. Sparrow and J. L. Gregg, *Trans. ASME* **80**, 379–386 (1958).

21. M. Tribus, *Trans. ASME* **80**, 1180–1181 (1958).

22. R. P. Bobco, *J. Aerospace Sci.* **26**, 846–847 (1959).

23. E. M. Sparrow and J. L. Gregg, *Trans. ASME* **78**, 1823–1829 (1956).

24. T. Cebeci, *Fifth International Heat Transfer Conference*, Vol. 3, Sect. NC 1.4, 1975, pp. 15–19.

25. E. J. LeFevre and A. J. Ede, *Proc. Ninth International Congress on Applied Mechanics, Brussels*, Vol. 4, 1957, pp. 175–183.

26. T. Y. Na and J. P. Chiou, *Wärme-und Stoffübertragung* **14**, 157–164 (1980).

27. J. R. Kyte, A. J. Madden, and E. L. Piret, *Chem. Eng. Prog.* **49**, 653–662 (1953).

28. H. R. Nagendra, M. A. Tirunarayanan, and A. Ramachandran, *ASME J. Heat Transfer* **92**, 191–194 (1970).

29. Y. Jaluria, in S. Kakac, W. Aung, and R. Viskanta, Eds., *Natural Convection*, Taylor and Francis, 1985, pp. 51–74.

30. B. Gebhart, L. Pera, and A. W. Schorr, *Int. J. Heat Mass Transfer* **13**, 161–171 (1970).

31. A. Bejan, *Convection Heat Transfer*, Wiley-Interscience, 1984.

32. J. C. Mollendorf and B. Gebhart, *Proc. Fifth International Heat Transfer Conference, Tokyo*, Vol. 5, 1974, pp. 10–14.

33. B. R. Morton, *J. Fluid Mech.* **2**, 127–144 (1957).

34. J. S. Turner, *Buoyancy Effects in Fluids*, Cambridge University Press, 1973.

35. H. Rouse, Y. S. Yih, and H. W. Humphreys, *Tellus* **4**, 201–210 (1952).

36. S. Yokoi, *Bulletin of the Journal of the Association of Fire Science Engineering* **5**, 53–59 (1956).

37. E. E. Zukoski, T. Kubota, and B. Centegen, Entrainment in fire plumes, National Bureau of Standards, Report NBS-6CR-80-294, 1980.

38. E. R. G. Eckert and T. Jackson, Analysis of turbulent free convection boundary layer on a flat plate, NACA Report 1015, 1951.

39. F. J. Bayley, *Institution of Mechanical Engineers (London) Proceeding* **169**, 361–370 (1955).

40. C. Y. Warner and V. S. Arpaci, *Int. J. Heat Mass Transfer* **11**, 397–406 (1968).

41. W. M. To and J. A. C. Humphrey, *Int. J. Heat Mass Transfer* **29**, 573–592 (1986).

42. D. L. Siebers, R. J. Moffat, and R. G. Schwind, *ASME J. Heat Transfer* **107**, 124–132 (1985).

43. K. T. Yang and J. R. Lloyd, in S. Kakac, W. Aung, and R. Viskanta, Eds., *Natural Convection*, Hemisphere, 1985, pp. 303–329.

44. R. Siegel, General Electric Company Technical Information Service Report R54GL89, 1954.

45. G. C. Vliet and C. K. Liu, *ASME J. Heat Transfer* **91**, 517–531 (1969).

46. R. Lemlich and J. Vardi, *ASME J. Heat Transfer* **86**, 562–563 (1964).

47. S. W. Churchill, *Heat Exchanger Design Handbook*, Taylor and Francis, 1986, pp. 2.5.9-1–2.5.9-7.

48. S. W. Churchill, *AIChE J.* **23**, 10–16 (1977).

49. B. Metais and E. R. G. Eckert, *ASME J. Heat Transfer* **86**, 295–296 (1964).

50. H. O. Buhr, E. A. Horsten, and A. D. Carr, *ASME J. Heat Transfer* **96**, 152–158 (1974).

51. F. Kreith, *Adv. Heat Transfer* **5**, 129–251 (1968).

52. Y. Kamotani, A. Prasad, and S. Ostrach, *AIAA J.* **19**, 511–516 (1981).

53. R. Siegel, *Trans. ASME* **89**, 347–359 (1958).

54. G. D. Raithby and K. G. T. Hollands, in W. Rohsenow, J. Hartnett, and E. Ganic, Eds., *Handbook of Heat Transfer Fundamentals*, 2nd ed., McGraw-Hill, 1985, pp. 6-1–6-94.

55. S. W. Churchill and R. Usagi, *AIChE J.* **18**, 1121–1128 (1972).

56. G. C. Vliet, *ASME J. Heat Transfer* **91**, 511–516 (1969).

57. T. Fujii, H. Honda, and I. Morioka, *Int. J. Heat Mass Transfer* **16**, 611–627 (1973).

58. V. T. Morgan, *Adv. Heat Transfer* **11**, 199–264 (1975).

59. G. Hesse and E. M. Sparrow, *Int. J. Heat Mass Transfer* **17**, 796–798 (1974).

60. V. T. Morgan, *Adv. Heat Transfer* **11**, 199–264 (1975).

61. M. Ali-Arabi and Y. K. Salman, *Int. J. Heat Mass Transfer* **23**, 45–51 (1980).

62. W. J. King, *Mech. Eng.* **54**, 347–353 (1932).

63. A. V. Hassani and K. G. T. Hollands, *ASME J. Heat Transfer* **111**, 363–371 (1989).

64. M. M. Yovanovich, On the effect of shape, aspect ratio, and orientation upon natural convection from isothermal bodies of complex shape, *Convective Transport* (ASME Winter Annual Meeting, December 3–18, 1987, Boston, MA), ASME HTD-Vol. 82.

65. F. Kreith, *ASME J. Heat Transfer* **92**, 307–332 (1970).

66. J. A. Scanlan, E. H. Bishop, and R. E. Powe, *Int. J. Heat Mass Transfer* **19**, 1127–1134 (1976).

67. L. B. Evans and N. E. Stefany, An experimental study of transient heat transfer to liquids in cylindrical enclosures, AIChE Paper 4, Heat Transfer Conference, Los Angeles, August 1965.

68. G. D. Raithby and K. G. T. Hollands, *Adv. Heat Transfer* **11**, 266–315 (1975).

69. (a) S. M. El Sherbiny, G. D. Raithby, and K. G. T. Hollands, *ASME J. Heat Transfer* **104**, 96–102 (1982); (b) S. E. Korpela, D. Gozum, and C. B. Baxi, *Int. J. Heat Mass Transfer* **16**, 1683–1690 (1973).

70. S. Ostrach, *Trans. ASME* **79**, 299–305 (1957).

71. J. Thomson, *Proc. Glasgow Philos. Soc.* **13**, 469 (1882).

72. H. Bénard, *Revue genérale des Sciences Pures et Appliqees* **11**, 1261–1271, 1309–1328 (1900).

73. Lord Rayleigh, *Philos. Mag. J. Sci.* **32**, 529–546 (1916).

74. H. Jeffreys, *Philos. Mag. J. Sci.* **2**, 833–844 (1926); *Proc. Roy. Soc. London, Ser. A* **113**, 195–208 (1928).

75. J. S. Turner, *Buoyancy Effects in Fluids*, Cambridge University Press, 1973.

76. M. G. Velarde and C. Normand, *Scientific American* **243** (1), 92–108 (1980).

77. C. L. Stong, *Scientific American* **216** (1), 124–128 (1967).

78. C. L. Stong, *Scientific American* **223** (3), 221–234 (1970).

79. J. Walker, *Scientific American* **237** (4), 142–150 (1977).

80. Y. Yen, *ASME J. Heat Transfer* **102**, 550–556 (1980).

81. E. M. Sparrow, J. W. Ramsey, and R. G. Kemink, *ASME J. Heat Transfer* **101**, 578–584 (1979).

82. D. W. Van De Pol and J. K. Tierney, *ASME J. Heat Transfer* **95**, 542–543 (1973).

83. A. Bar-Cohen, *ASME J. Heat Transfer* **101**, 564–566 (1979).

84. C. D. Jones and L. F. Smith, *ASME J. Heat Transfer* **92**, 6–10 (1970).

85. W. Elenbaas, *Physica* **IX** (1), 2–28 (1942).

86. P. Cheng, in S. Kakac, W. Aung, and R. Viskanta, Eds., *Natural Convection*, Hemisphere, 1985, pp. 475–513.

87. A. Bejan, in S. Kakac, R. Shah, and W. Aung, Eds., *Handbook of Single-Phase Convective Heat Transfer*, Wiley-Interscience, 1987, pp. 16-1–16-34.

88. M. Kaviany, *Principles of Heat Transfer in Porous Media*, Springer-Verlag, 1991.

89. D. A. Nield and A. Bejan, *Convection in Porous Media*, Springer-Verlag, 1992.

90. P. Cheng and W. J. Minkowycz, *J. Geophys. Res.* **82**, 2040–2044 (1977).

91. C. S. Yih, *Dynamics of Nonhomogeneous Fluids*, Macmillan, 1965.

92. H. S. Carslaw and J. C. Jaeger, *Conduction of Heat in Solids*, 2nd ed., Clarendon, 1959.

93. P. Cheng, *Adv. Heat Transfer* **14**, 1–105 (1979).

94. N. Kladias and V. Prasad, *J. Thermophys. Heat Transfer* **5**, 560–576 (1991).

95. A. Hadim and L. C. Burmeister, *J. Thermophys. Heat Transfer* **2**, 343–351 (1988).

96. M. Yovanovich and K. Jafarpur, "Bounds on Laminar Natural Convection from Isothermal Disks and Finite Plates of Arbitrary Shape for all Orientations and Prandtl Numbers," Session on Fundamentals of Natural Convection, 1993 ASME Winter Annual Meeting, Nov. 28–Dec. 3, 1993, New Orleans, LA.

11

BOILING

The large change of density that accompanies a change of phase can give rise to vigorous natural convection. For example, heating or cooling air at atmospheric pressure through the temperature range 90–100°C induces a 2.7% density change, whereas heating water under the same conditions induces a much larger density change of 99.9%. The large difference in density between the liquid and the vapor state is accompanied by two other effects: (1) the vapor is much less viscous than the liquid; and (2) there is usually a well-defined interface between a vapor-filled region and a liquid-filled region.

The formation of vapor from a liquid is called *boiling* when bubbles are formed; formation of liquid from a vapor is called *condensation*. Because of their importance to technology, a considerable body of literature has been developed concerning these two phenomena.

11.1 POOL BOILING

Pool boiling, in which a surface heated above a liquid saturation temperature is immersed in a large pool of the liquid, is the simplest boiling situation. When the liquid is at its saturation temperature, the process is termed *saturated boiling*. If the liquid is subcooled below its saturation temperature, the process is termed *subcooled boiling*. The sequence of events is illustrated in Fig. 11-1a as the temperature of the submerged surface is increased for pool boiling from an electrically heated 0.04-in. horizontal wire in saturated water at 1 atm of pressure. Referring to Fig. 11-1b, at the small temperature differences in region *A–B* there is no vapor formation since the liquid is capable of sustaining a small amount of superheat, and natural convection without phase change occurs. At higher temperature differences in region B–DNB small vapor bubbles form at a few points on the hot surface but condense after they detach from the surface and move into the liquid. At higher temperature differences in region DNB vapor bubbles form at many points and rise to the top of the pool without condensing. The boiling in region B–C is termed *nucleate boiling*. In region C–D bubble formation occurs so rapidly and at so many points on the heated surface that the surface is partially covered by an intermittent vapor film. The peak heat flux occurs at point C and is called the *burnout heat flux*; hence point C is often referred to as the *boiling crises*, or *burnout*. In region D the increased temperature difference results in the surface being increasingly blanketed with an insulating vapor film. The heat flux decreases, as a result, until the surface is entirely covered by a stable vapor film; a further increase in temperature difference above the value at point D then again gives increased heat flux. This regime is called *stable film boiling*. At the elevated temperatures in region E–F thermal radiation contributes appreciably to heat flow across the vapor film.

A peculiarity of boiling is that the boiling curve typified by Fig. 11-1 has multiple values at heat fluxes between the values of point C and point D. Above or below these values, a specified heat flux corresponds to a single temperature difference. A specified heat flux intermediate to these values can be provided by as many as three distinct temperature differences, however; such a situation tends to be unstable since the surface temperature can oscillate between the three allowed values if subjected to external disturbances. Most importantly, the highest of the three allowed temperatures, in region E–F, is sufficiently high for many materials to fail. Troubles of this sort are not encountered with a surface heated by a condensing vapor since its temperature is specified—Fig. 11-1 is single valued if the abscissa is specified—but they are important to the integrity of nuclear reactors and electrically heated surfaces.

11.2   NUCLEATE BOILING

The formation of a vapor bubble on a solid surface occurs at discrete sites, called *nucleation sites*, which are usually gas- or vapor-filled cavities in the heated surface as sketched in Fig. 11-2. Alternatively, a nucleation site could be an impurity or a boundary between crystals. The general idea for a cavity is that heat addition causes a vapor pocket to grow by evaporation from the liquid interface. Eventually the bubble grows large enough that buoyancy forces cause it to detach, leaving some vapor behind for a later repetition of these events. If the liquid is subcooled, the bubble may collapse without detaching, although the chain of events is otherwise unchanged. Nucleation usually occurs at a solid surface although it can occur in a homogeneous

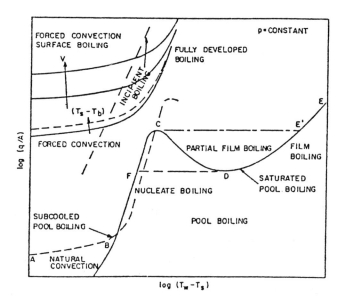

(a)

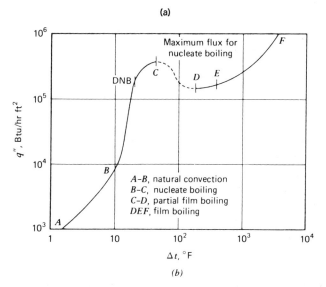

(b)

Figure 11-1 Pool boiling fundamentals. (*a*) Physical interpretation of boiling curve [By permission from W. M. Rohsenow, *Ann. Rev. Fluid Mech.* **3**, 211–236 1971 [1]. ©1971 by Annual Reviews, Inc.]. (*b*) Boiling of water at 212°F on an electrically heated platinum wire measured by Nukiyama [2]; DNB signifies departure from nucleate boiling. [By permission from W. H. McAdams, *Heat Transmission*, 3rd ed., McGraw-Hill, 1954.]

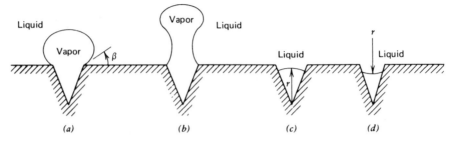

Figure 11-2   Nucleation sites.

liquid as discussed by Leppert and Pitts [3] and can cause explosive vapor formation in a superheated liquid, called "bumping," as discussed by
Westwater [4], which can be prevented by the common practice of adding
"boiling stones." Cole [5] provides additional information on the rate of
homogeneous nucleation and on the limits of superheat a liquid can endure;
the vapor explosions that can occur when a molten material suddenly
contacts a cooler liquid are also surveyed.

In Fig. 11-2c, the curvature of the liquid–vapor interface in a cavity is such
that surface tension requires the pressure in the vapor to exceed the pressure
in the liquid. As the surface cools below the liquid saturation temperature
following a bubble formation and detachment, the vapor all condenses and
the cavity completely fills with liquid, rendering it unable to form a new
bubble later. If, on the other hand, the liquid–vapor interface in the cavity is
as shown in Fig. 11-2d, the pressure in the vapor is less than the pressure in
the liquid. As the cavity cools, the interface recedes into the cavity, decreasing the radius of curvature and reducing the pressure in the trapped vapor.
Since p_{vapor} decreases, its saturation temperature also decreases, and vapor
can exist in equilibrium with the liquid at the lowered temperature; this
cavity always contains some vapor and can participate repetitively in cycles of
bubble formation. If an inert gas is trapped in a cavity, the cavity can be an
active nucleation site even without any vapor initially present. Inert gases can
be gradually absorbed into the liquid after a number of bubble-producing
cycles, leading to a hysteresis effect [3]; after exposure to an inert atmosphere, initial conditions are restored.

The effects of radius of curvature of the liquid–vapor interface and
inert-gas presence can be illustrated by imagining a vapor sphere in a liquid.
The surface tension requires that the internal pressure be greater than the
external pressure at equilibrium. Hence

$$(p_v + p_g) - p_L = \frac{2\sigma}{r}$$

where p_v and p_g are the partial pressures of vapor and inert gas, respectively, p_L is the exterior pressure in the liquid, σ is surface tension, and r is the bubble radius. The Clausius–Clapeyron relationship with the perfect gas approximation relates saturation temperature T_v and saturation pressure p_v as

$$\frac{dp_v}{dT_v} \approx \frac{h_{fg}\rho_v}{T_v} \approx \frac{h_{fg}p_v}{R_v T_v^2}$$

where R_v is the vapor gas constant and h_{fg} is the latent heat of vaporization. Since $p_v - p_L \approx (T_v - T_{sat}) \, dp/dT$, these two relations can be combined to obtain, with T_{sat} the saturation temperature at p_L and T_L the liquid temperature,

$$T_v - T_{sat} \approx \frac{R_v T_{sat}^2}{h_{fg} p_L} \left(\frac{2\sigma}{r} - p_g \right) \tag{11-1}$$

This result suggests that if $T_L - T_{sat} > T_v - T_{sat}$ from this equation, a bubble will grow; if it is smaller, the bubble will collapse. Evidently, some superheat is required in the liquid for bubble growth. Also, inert gases will affect the process of bubble growth and collapse. Equation (11-1) was experimentally confirmed by Griffith and Wallis [6].

Vapor generation for bubble growth in nucleate boiling has been found to occur by evaporation, Koffman and Plesset [7] reported as much as 50% of the total heat flow for subcooled nucleate boiling, from a very thin liquid layer, the microlayer, between the bubble and the hot surface. This physical situation, first reported by Moore and Mesler [8], is illustrated in Fig. 11-3. Plesset and Sadhal [9] give the average microlayer thickness δ_0 in terms of liquid kinematic viscosity ν and bubble lifetime t_0 as $\delta_0 = \frac{8}{7}(3\nu t_0)^{1/2}$. For water with $t_0 \approx 0.04$ s, $\delta_0 = 0.2$ mm, which is roughly the thickness of four sheets of paper.

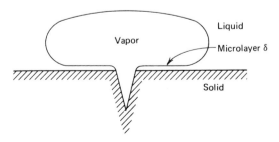

Figure 11-3   Liquid microlayer under a vapor bubble at a nucleation site.

Extensive data on nucleate pool boiling in regions 13-DNB and DNB has been correlated within 20% by Rohsenow [10] for clean surfaces as

$$\frac{q}{A} = \mu_L h_{fg} \left[\frac{g(\rho_L - \rho_v)}{g_c \sigma} \right]^{1/2} \left[\frac{C_{p_L}(T_w - T_{sat})}{h_{fg} P_{r_L}^{1.7} C_{sf}} \right]^3 \qquad (11\text{-}2)$$

which predicts heat flux to increase as ΔT^3 and accounts for the effects of pressure. In Eq. (11-2):

C_{p_L}	saturated liquid specific heat
C_{sf}	surface–fluid constant
g	gravitational acceleration
g_c	constant, 32.17 lb_m ft/lb_f sec^2 or 1 kg m/N s^2
h_{fg}	heat of vaporization
P_{r_L}	saturated liquid Prandtl number
q/A	heat flux
$T_w - T_{sat}$	temperature difference
μ_L	liquid dynamic viscosity
σ	surface tension
ρ_L	saturated liquid density
ρ_v	saturated vapor density

For dirty surfaces, the Prandtl number exponent varies from 0.8 to 2.0. C_{sf} values for various surface–liquid combinations are given in Table 11-1. There, as well as in Fig. 11-5, a substantial influence of surface condition is apparent. More detailed correlations are available both without [15] and with [16] forced convection. The surface tension of water is approximately given in units of lb_f/ft (1 lb_f/ft = 14.594 N/m) by $\sigma = 58 \times 10^{-4}(1 - 0.00142T)$ with T in Fahrenheit, a linear fit to data between 212°F and the critical-point temperature of 705°F. The condition of the solid surface can substantially affect nucleate boiling heat flux as was shown by Berenson's [17] measurements and the study of Vachon et al. [12]. Nucleate pool boiling is insensitive to geometry, so Eq. (11-2) applies to horizontal and vertical surfaces as well as to wires whose diameter exceeds 0.01 cm. Subcooling of the liquid pool mostly increases the $T_w - T_{sat}$ value at which nucleation begins and thereafter has little effect [3]. Surfactants reduce surface tension and can nearly double nucleate boiling heat flux [18].

When the liquid pool is so thin as to be a film, nucleate boiling heat fluxes are underpredicted by Eq. (11-2). Nishikawa et al. [19] found that nucleate boiling heat flux from a horizontal surface to water increases when the water film thickness decreases below 3 mm; Fakeeha [20] found the increase to be a factor of 2 or greater, depending on the system pressure, at a film thickness of 0.44 mm on a vertical surface. Mesler and Mailen [21] found this increase

TABLE 11-1   Values of C_{sf} for Various Liquid–Surface Combinations[a]

Liquid–Surface Combination	C_{sf}
Water–copper	0.013 [11]
Water–scored copper	0.0068 [12]
Water–emery-polished copper	0.0128 [12]
Water–emery-polished, paraffin-treated copper	0.0147 [12]
Water–chemically etched stainless steel	0.0133 [12]
Water–mechanically polished stainless steel	0.0132 [12]
Water–ground and polished stainless steel	0.008 [12]
Water–Teflon pitted stainless steel	0.058 [12]
Water–platinum	0.013 [13]
Water–brass, water–nickel	0.006 [10, 11]
Ethyl alcohol–chromium	0.027 [14]
Benzene–chromium	0.01 [14]
Carbon tetrachloride–copper	0.013 [11]
Carbon tetrachloride–emery-polished copper	0.007 [12]
n-Pentane–chromium	0.015 [10]
n-Pentane–emery-polished copper	0.0154 [12]
n-Pentane–emery-rubbed copper	0.0074 [12]
n-Pentane–lapped copper	0.0049 [12]
n-Pentane–emery-polished nickel	0.0127 [12]
Isopropyl alcohol–copper	0.0025 [10]
n-Butyl alcohol–copper	0.003 [10]
35% K_2CO_3–copper	0.0054 [10]
50% K_2CO_3–copper	0.0027 [10]

[a] Numbers in brackets denote references at end of chapter.

to be due to fragments from bubbles bursting at the free surface acting as secondary nucleation sites.

The burnout heat flux at point C in Fig. 11-1 has been correlated from hydrodynamic considerations by Zuber [22] for a saturated liquid as

$$\frac{q_{max}}{A} = K\rho_v h_{fg} \left[\frac{\sigma(\rho_L - \rho_v)gg_c}{\rho_v^2} \right]^{1/4} \left(\frac{\rho_L + \rho_v}{\rho_L} \right)^{1/2} \tag{11-3}$$

where K can be taken as 0.18 for water; Kutateladze [23] achieved the general form of Eq. (11-3) earlier by a different approach. Borishanskii [24] included the effects of liquid viscosity in a dimensional analysis and found by comparison with data that

$$K = 0.13 + 4 \left\{ \frac{\rho_L(g_c\sigma)^{3/2}/\mu_L^2}{[g(\rho_L - \rho_v)]^{1/2}} \right\}^{-0.4}$$

Appendix E can be consulted for details of the stability analysis that leads to Eq. (11-3). A maximum value occurs at about one-third the critical pressure [2].

The burnout heat flux increases with subcooling of the liquid according to Zuber et al. [25] (also see Elkassabgi and Lienhard [26]) as

$$\frac{q_{max}}{q_{max,\,sat}} = 1 + \frac{5.3(T_{sat} - T_L)}{\rho_v h_{fg}}$$

$$\times (k_L \rho_L C_{p,\,L})^{1/2} \left[\frac{\sigma(\rho_L - \rho_v)gg_c}{\rho_v^2} \right]^{-1/8} \left[\frac{g(\rho_L - \rho_v)}{g_c \sigma} \right]^{1/4}$$

and with cross flow at velocity V over cylinders of diameter D, according to Lienhard and Eichhorn [27] (also see [28, 29]), as

$$\frac{\pi q_{max}/A}{V \rho_v h_{fg}} = \frac{1}{169} \left(\frac{\rho_L}{\rho_v} \right)^{3/4} + \frac{1}{19.2} \left(\frac{\rho_L}{\rho_v} \right)^{1/2} \mathrm{We}^{-1/3}$$

when the left-hand side exceeds $1 + 0.275(\rho_L/\rho_v)^{1/2}$, the Weber number We being $\mathrm{We} = D\rho_v V^2/\sigma$. See Lienhard [29] for more discussion and refined correlations.

Information on peak boiling heat flux for horizontal plates, cylinders, and spheres has been given by Lienhard et al. [30, 31]; Eq. (11-3) gives close results, but there are interesting differences.

Nucleate boiling can produce characteristic sounds that are sometimes called "boiling songs." Aoki and Welty [32] measured the acoustic spectra from a horizontal plate; the emitted sound pressure level increased with increasing nucleate heat flux, diminished in the peak heat-flux region as a result of attenuation by bubble coalescence above the boiling surface, and rose again when the film boiling conditions were attained. The reverse effect, the influence of acoustic vibrations on nucleate boiling, is [33] noticeable under certain conditions but is seldom large. Low-frequency, mechanical, vertical vibrations of a horizontal cylinder do not increase burnout heat flux more than about 8%, according to Bergles [34], although in the nucleate boiling regime substantial heat-flux increase can be obtained as a result of induced motion of the liquid pool. The ability of a vertically vibrating liquid column to render bubbles motionless can influence boiling in a manner different from simple mechanical vibrations as was found by Fuls and Geiger [35], who report a heat-flux increase of about 10% in nucleate boiling in some cases.

Electric fields influence the boiling curve [36]. Though Cooper [37] reported an order of magnitude increase, Markels and Durfee [38] found that nucleate pool boiling from a horizontal tube was not increased much, but that burnout heat flux could be increased by a factor of 5 or more and that film boiling could be delayed in its onset for dc (direct-current) voltages of the order of 10,000 V—similar results were found for ac (alternating-current) voltages at 60 Hz. Boiling with forced convection in the presence of 60-Hz ac electric fields had like results [39]. The physical mechanisms involved have been discussed by Jones [40] with reference to both boiling and condensation and by Crowley [41]. A magnetic field reduces mercury nucleate boiling heat fluxes and encourages film boiling [42], a phenomenon important to heat removal from fusion reactors with magnetic confinement systems.

Forced convection about the boiling surface affects the heat flux. Rohsenow [43] suggested

$$q = q_{conv} + q_{boil} \tag{11-4}$$

Here q_{conv} is the heat flow associated with pure forced convection, and q_{boil} is the heat flow due to boiling in the absence of forced convection. Although Eq. (11-4) correlates data with some success [44], the alternative suggested by Kutateladze [45] is to relate the heat-transfer coefficients according to

$$\frac{h}{h_{conv}} = \left[1 + \left(\frac{h_{boil}}{h_{conv}} \right)^n \right]^{1/n} \tag{11-5}$$

He found that $n = 2$ correlates data for flow inside tubes, whereas others [46] found that 5.5 correlates nucleate boiling data for forced convection perpendicular to cylinders. Consult Fand et al. [46] for calculation of h_{conv} and h_{boil} and Kandlibar [47] for a survey of correlations for forced convection boiling.

For water flowing in a pipe of diameter D and length L at velocity V, a correlation for burnout heat flux due to Lowdermilk et al. [48] is

$$\frac{q_{max}}{A} = \frac{270(\rho V D/L)^{0.85}}{D^{0.2}}, \qquad 1 < \rho V \left(\frac{D}{L} \right)^2 < 150$$

$$= \frac{1400(\rho V)^{0.5}(D/L)^{0.15}}{D^{0.2}}, \qquad 150 < \rho V \left(\frac{D}{L} \right)^2 < 10^4$$

with q/A in Btu/hr ft^2, ρ the liquid density in lb$_m$/ft^3, V in ft/sec, and L and D in in. Here 14.7 psia $< p < 100$ psia, 0.1 ft/sec $< V < 98$ ft/sec, $25 < L/D < 250$, 0.051 in. $< D < 0.188$ in. This result applies to inlet sub-cooling from 0 to 140°F and the full range of quality. A more complete correlation based on the work of Tippets is reviewed by Leppert and Pitts [3].

The general trend of events of the two-phase flow that occurs when a fluid boils while flowing through a tube is illustrated in Fig. 11-4 for a vertical-tube evaporator. There a subcooled liquid enters the bottom and contacts the hot wall where local nucleate boiling is soon experienced; the flow is bubbly when there is less than about 10% vapor. Here the local heat-transfer coefficient is above its value for single-phase flow of the liquid. As the quality increases, the flow becomes annular with a thin liquid layer on the wall and a core of

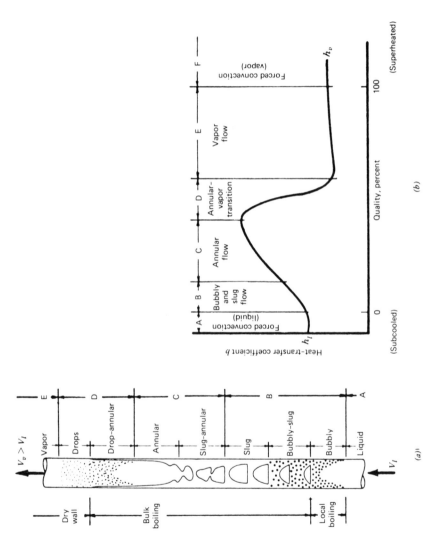

Figure 11-4 Two-phase flow regimes of a liquid boiling during flow through a tube.

vapor of velocity greater than the liquid's. Although some bubbles form at the wall, most evaporation occurs at the liquid–vapor interface. At even higher qualities the heat-transfer coefficient drops as a result of a transition to vapor (sometimes called *mist* or *fog*) flow. Burnout sometimes occurs at this point because low-conductivity vapor contacts the wall and cannot remove heat as rapidly. Vapor flow continues until the quality achieves 100%; then the heat-transfer coefficient can be estimated by the relationships for single-phase forced convection in a duct. For horizontal orientations, vapor can congregate in the upper portion of the tube. The survey by Kalinin et al. [49] can be consulted for additional discussion. Griffith [50] in a survey of two-phase flow discusses procedures for estimating pressure drop (Martinelli's [51, 52] is best at pressures near 1 atm, but Thom's [52] is better at higher pressures and choked flow) as does Fraas [53].

11.3   TRANSITION BOILING

At point C in Fig. 11-1 the liquid intermittently touches the solid surface and is separated from the surface by a vapor film. Measurements in this region are difficult, particularly if electrically heated test sections are used; credit is given to Nukiyama [2] for the first studies.

Because the temperature difference at point C produces vapor at a greater rate than can be removed by bubble formation during the contact phase, a vapor film forms. However, the temperature difference at point C is not large enough to produce vapor at the rate required to maintain a stable film, and so the film collapses. As the temperature difference is increased in this transition region, the liquid contacts the surface for smaller fractions of the cycle.

Heat flux versus temperature difference curves are shown in Fig. 11-5. The dependence of the boiling curve in the transition region on surface roughness suggests that the liquid contacts the surface.

Forced convection influences heat flux in the transition region [41, 55]. The effects of subcooling, pressure, surface properties, coatings, and cavity distribution are predicted by the formulas of Pan and Lin [56]. Additives can increase transition heat flux (more than for nucleate boiling) but can also cause a decrease [57, 58].

The conservative predictive procedure usually adopted is to assume that stable film boiling starts immediately after the nucleate boiling regime.

11.4   FILM BOILING MINIMUM HEAT FLUX

At the surface temperature corresponding to point D in Fig. 11-1 the liquid pool is separated from the heated surface by a continuous vapor film.

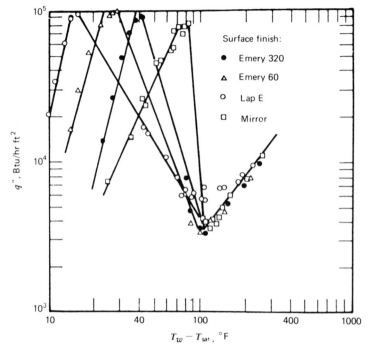

Figure 11-5 Pool boiling of *n*-pentane from an upward-directed copper surface. [By permission from P. J. Berenson, *Int. J. Heat Mass Transfer* **5**, 985–999 (1962) [54].]

Because the vapor generation rate just equals the rate at which vapor leaves the film in detaching bubbles, the film is steadily present. The heat flux at point *D* is termed the *film boiling minimum heat flux*.

To give insight into the physical mechanisms, the horizontal plane is considered in detail, largely following Jordan's [59] discussion of the original work by Zuber [22] and Chang [60].

The heat flux from a horizontal surface facing up in a pool of saturated liquid is determined by the rate at which vapor bubbles detach from the vapor film covering the surface. The unstable condition of dense liquid overlaying the less dense vapor is responsible for the bubble detachment and is called *Taylor instability* (see Appendix E for details). The minimum film boiling heat flux can be computed as

$$\frac{q}{A} = \frac{\text{energy transport}}{\text{bubble}} \times \frac{\text{bubble}}{\text{area-time}} \qquad (11\text{-}6)$$

As discussed in Appendix E the liquid–vapor interface is wavy, with the nodes and antinodes arranged in a nearly square pattern that has a characteristic wavelength λ_D. Bubbles are presumed to be spherical at detachment (photographs show this to be reasonable) of diameter proportional to λ_D, usually $\lambda_D/2$. Then, if liquid and vapor are both saturated,

$$\frac{\text{energy transport}}{\text{bubble}} = \rho_v h_{fg} \left(\frac{\lambda_D}{4}\right)^3 \frac{4\pi}{3} \tag{11-7}$$

where ρ_v is vapor density and h_{fg} is heat of vaporization. Each node above the interface nominal equilibrium surface grows until a bubble detaches. Then what was an antinode (below the equilibrium interface position) becomes a node and a bubble detaches from there. Thus, in one cycle of interface oscillation, two bubbles are detached from a nearly square area λ_D^2 of side λ_D and so

$$\frac{\text{bubbles}}{\text{area-oscillation}} = \frac{2}{\lambda_D^2} \tag{11-8}$$

The time required for a node to grow to a height equal to the diameter of a detaching bubble can be obtained from the representation of the interface displacement η from its equilibrium position

$$\eta = \eta_0 e^{b^* t} e^{\pm i(m_1 x + m_2 y)} \tag{E-21b}$$

At a node the second exponential term is unity so that there

$$\eta = \eta_0 e^{b^* t}$$

The time t required for the interface to achieve a displacement η_0 is

$$t = (b^*)^{-1} \ln\left(\frac{\eta}{\eta_0}\right)$$

Lewis [Eq. (E-6)] in experimental confirmation of Taylor's stability analysis found that bubbles detached when $\eta = 0.4\lambda_D$. Since this is near the assumed detaching-bubble diameter D_b of $\lambda_D/2$, it is plausible that at detachment the interface displacement equals D_b—since the initial disturbance η_0 is unknown but can be assumed to be proportional to D_b, now merely let $C = \ln(D_b/\eta_0)$. Then, by use of Eq. (E-27) to relate b^* to fluid properties, the frequency $1/t$ of bubble emission is

$$\frac{[(4\pi/3)(\rho_L - \rho_v)g/(\rho_L + \rho_v)]^{1/2}}{C\lambda_D^{1/2}} \tag{11-9}$$

where the "most dangerous" wavelength λ_D is

$$\lambda_D = 2\pi\left[\frac{3\sigma(g_c/g)}{(\rho_L - \rho_v)}\right]^{1/2}$$ (E-25)

Introduction of Eqs. (11-7)–(11-9) and (E-24) into Eq. (11-6) then gives the minimum film boiling heat flux as

$$\frac{q}{A} = \frac{\pi^2}{2^{3/2}3^{5/4}C}\rho_v h_{fg}\left[\frac{\sigma(\rho_L - \rho_v)(gg_c)}{(\rho_L + \rho_v)^2}\right]^{1/4}$$

Berenson [61] empirically found that the coefficient of this equation is 0.09, from which it can be found that $C = 10$, to finally give [62]

$$\frac{q}{A} = 0.09\rho_v h_{fg}\left[\frac{g(\rho_L - \rho_v)}{\rho_L + \rho_v}\right]^{1/2}\left[\frac{\sigma(g_c/g)}{\rho_L - \rho_v}\right]^{1/4}$$ (11-10)

for the minimum film boiling heat flux from a horizontal surface.

The temperature difference that accompanies the minimum heat flux of Eq. (11-10) was ascertained by Berenson from the physical model shown in Fig. 11-6. At any radial position in the film between bubbles the inward vapor mass flow rate $\dot{m}$ is

$$\dot{m} = \rho_v V 2\pi r a$$ (11-11)

With the assumption that heat flows by conduction across the vapor film of thickness a and that all this heat is used in vaporization (the liquid is

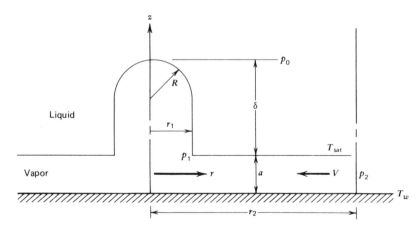

Figure 11-6   Physical model for minimum film boiling from a horizontal plate.

saturated), conservation of energy requires

$$\dot{m} h_{fg} = \frac{k_v \pi (r_2^2 - r^2)(T_w - T_s)}{a} \tag{11-12}$$

The surface area that generates vapor for a single bubble is $\lambda_D^2/2$ (see Appendix E for details) so that $\pi r_2^2 = \lambda_D^2/2$. Then Eqs. (11-11) and (11-12) yield

$$V = \frac{k_v(T_w - T_s)}{\rho_v h_{fg} a^2} \frac{\lambda_D^2/2 - \pi r^2}{2\pi r} \tag{11-13}$$

The pressure drop $p_2 - p_1$ between the interbubble position r_2 and the bubble position r_1 can be expressed in two ways. First, a Poiseuille flow assumption gives (see Appendix C), neglecting inertial terms,

$$0 = -\frac{\partial p}{\partial r} + \mu \frac{\partial^2 V_r}{\partial z^2} \tag{11-14}$$

Solution for V_r gives

$$V_r = \frac{1}{2\mu} \frac{\partial p}{\partial r} z^2 + C_1 z + C_2$$

For one extreme case of a stationary liquid $V_r(z = a) = 0$,

$$V_r = \frac{1}{2\mu} \frac{\partial p}{\partial r} (z^2 - az)$$

from which $\tau_w = \mu \, \partial V_r(z = 0)/\partial z = -(a/2)(\partial p/\partial r)$, which equals the shear stress τ_a at the interface by symmetry. Thus the average velocity V

$$V = \frac{(\tau_w + \tau_a)a}{12\mu} \tag{11-15}$$

For another extreme case of zero shear stress at the interface, $\tau_a = \mu \, \partial V_r(z = a)/\partial z = 0$,

$$V_r = \frac{a}{2\mu} \frac{\partial p}{\partial r} \left(\frac{z^2}{a} - 2z \right)$$

from which $\tau_w = \mu \, \partial V_r(z = 0)/\partial z = -(a/\mu)(\partial p/\partial r)$. Thus the average velocity is

$$V = (\tau_w + \tau_a) \frac{a}{3\mu} \tag{11-16}$$

Integration of Eq. (11-14) between $z = 0$ and $z = a$ gives

$$\frac{\partial p}{\partial r} = \tau_w + \tau_a \tag{11-17}$$

Now Eqs. (11-15) and (11-16) suggest that

$$\tau_w + \tau_a = \frac{\beta\mu V}{a}$$

where $3 < \beta < 12$, which, when used in Eq. (11-17), yields

$$\frac{\partial p}{\partial r} = \frac{\beta\mu_v V}{a^2}$$

Insertion of the average radial vapor velocity from Eq. (11-13) into this relationship gives

$$\frac{\partial p}{\partial r} = \frac{\beta\mu_v k_v (T_s - T_s)}{a^4 \rho_v h_{fg}} \frac{\lambda_D^2/2 - \pi r^2}{2\pi r}$$

On integration from $r = r_1$ to $r = r_2$, the pressure difference $p_2 - p_1$ can be found. Berenson's measurements showed that

$$R = r_1 = 2.35 \left(\frac{\sigma g_c/g}{\rho_L - \rho_v} \right)^{1/2} \tag{11-18}$$

Having assumed $\pi r_2^2 = \lambda_D^2/2$ with λ_D from Eq. (E-25),

$$r_2 = \left(\frac{6\pi\sigma g_c/g}{\rho_L - \rho_v} \right)^{1/2}$$

Integration of the radial pressure gradient from $r = r_1$ to $r = r_2$ then gives

$$p_2 - p_1 = \frac{8\beta}{\pi} \frac{\mu_v k_v (T_w - T_s)}{a^4 \rho_v h_{fg}} \frac{\sigma g_c/g}{\rho_L - \rho_v} \tag{11-19}$$

Second, appeal can be made to hydrostatic considerations for additional information. Referring to Fig. (11-6) one can see that

$$p_2 = p_0 + \rho_L \delta \frac{g}{g_c} \quad \text{and} \quad p_1 = p_0 + \rho_v \frac{\delta g}{g_c} + \frac{2\sigma}{r}$$

Thus

$$p_2 - p_1 = (\rho_L - \rho_v)\delta \frac{g}{g_c} - \frac{2\sigma}{R} \tag{11-20}$$

Equation (11-18) gives R, and the measurements of Borishanskii [24] provide

$$\delta = 0.68 D_{\text{bubble}} = 1.36 R = 3.2 \left(\sigma \frac{g_c/g}{\rho_L - \rho_v} \right)^{1/2} \tag{11-21a}$$

Combination of Eqs. (11-19)–(11.21a) allows the vapor film thickness between bubbles to be obtained as

$$a = 1.68 \left[\frac{\mu_v k_v (T_w - T_s)(g_c/g)}{\rho_v h_{\text{fg}}(\rho_L - \rho_v)} \left(\frac{\sigma g_c/g}{\rho_L - \rho_v} \right)^{1/2} \right]^{1/4} \tag{11-21b}$$

in which comparison with experiment shows that $(1.09\beta)^{1/4} = 1.68$—the liquid is intermediate between a rigid and a no-shear surface.

The temperature difference $T_w - T_s$ can now be obtained by equating the heat conducted across the vapor film between bubbles to the total heat flow given by Eq. (11-10) since

$$\frac{q}{A} = k_v \left(\frac{A_{\text{film}}}{A_{\text{total}}} \right) \frac{T_w - T_s}{a}$$

The surface area not covered by a bubble is $A_{\text{film}} = \lambda_D^2/2 - \pi r_1^2$ and $A_{\text{total}} = \lambda_D^2/2$ so that $A_{\text{film}}/A_{\text{total}} = 0.707$. Appeal to Eq. (11-21a) for a gives

$$T_w - T_s = C \frac{\rho_v h_{\text{fg}}}{k_v} \left[\frac{g(\rho_L - \rho_v)}{\rho_L + \rho_v} \right]^{2/3} \left[\frac{\sigma(g_c/g)}{\rho_L - \rho_v} \right]^{1/2} \left[\frac{\mu_v(g_c/g)}{\rho_L - \rho_v} \right]^{1/3} \tag{11-22}$$

with $C = 0.127$ for the minimum temperature difference for film boiling from a horizontal surface to a saturated pool with a reported [60, 61] accuracy of $\pm 10\%$. All vapor properties should be evaluated at $T_f = (T_w + T_s)/2$.

The heat-transfer coefficient is obtained by dividing the heat flux given in Eq. (11-10) by the temperature difference of Eq. (11-22) to yield

$$h = 0.425 \left[\frac{\rho_v h_{fg}(\rho_L - \rho_v) k_v^3 (g/g_c)}{\mu_v (T_w - T_s) [\sigma(g_c/g)/(\rho_L - \rho_v)]^{1/2}} \right]^{1/4} \quad (11\text{-}23)$$

with $\pm 10\%$ accuracy. To account for vapor superheat, $h_{fg} = (h_{vapor} - h_{liq}) + 0.34 C_{p_v}(T_w + T_{sat})$. The assumed laminar vapor flow might not hold at very low values of g.

Phase-change processes in which a dense fluid overlays a lighter fluid are not confined to boiling. Dhir et al. [63, 64] performed experiments for a horizontal slab of dry ice (CO_2) sublimating while under liquid water. They found that the heat-transfer coefficient is predicted by Eq. (11-23) if the coefficient is reduced to 0.36, provided the water–CO_2 vapor interfacial temperature exceeds the freezing point. Taghavi-Tafreshi et al. [65] found that for melting of a horizontal slab of frozen olive oil under liquid water the heat-transfer coefficient is predicted by Eq. (11-23) if the coefficient is reduced to 0.18—a photograph of the two-dimensional interfacial wave pattern is presented.

The minimum temperature difference for film boiling is not necessarily determined by a Taylor instability phenomenon as given by Eq. (11-22). It is possible for it to be limited by spontaneous nucleation on contact of the collapsing film under a detached bubble with the heated surface as explained by Yao and Henry [66]. The limiting mechanism is one that is stable at the lowest wall temperature. Cryogenic fluids, which have a comparatively small temperature difference between their normal boiling and critical temperatures, are usually limited by rates of spontaneous nucleation on contact. Liquid metals, on the other hand, are usually limited by hydrodynamic instability. If hydrodynamic instability is the limiting mechanism, nucleation sites at the solid surface must be sufficiently numerous to generate sufficient vapor during liquid–solid contact to prevent the liquid from completely covering the wall—such a case is not possible in poorly wetting liquid metal–solid or liquid–liquid systems, such as water over mercury, since insufficient surface nucleation sites exist and in such a case the limiting mechanism might be spontaneous nucleation on contact, which can occur near saturation temperatures. Spontaneous nucleation can be on contact (bubbles form by density fluctuations at a liquid–solid interface as a result of imperfect wetting), or it can be homogeneous (bubbles form by density fluctuations entirely within the liquid bulk). Only homogeneous spontaneous nucleation is easily calculable (but serves as an upper bound and illustrates typical spontaneous nucleation behavior) by the expression

$$J = \frac{nKT}{h} \exp(-4\pi r^2 \sigma / 3KT)$$

in which J is homogeneous nucleation/cm^3 s, n is molecule/cm^3, K is Boltzmann's constant, h is Planck's constant, and T is absolute temperature. Typically, J becomes appreciable only when T exceeds a threshold value of approximately 195°C for ethanol and 305°C for water. Application to vapor explosions, perhaps caused by Landau–Darrieus instability [67], was made by Henry and Fauske [68]. Surveys of homogeneous nucleation by Springer [69] and boiling liquid superheat in general by Afgan [70] can be consulted for greater detail. The homogeneous nucleation temperature T_{hn} is given by Lienhard et al. [71] as

$$\frac{T_{hn}}{T_{crit}} = 0.923 + 0.077\left(\frac{T_{sat}}{T_{crit}}\right)^9$$

The minimum temperature for stable film boiling is often referred to as the *Leidenfrost temperature*, after J. G. Leidenfrost who in 1756 investigated the evaporation rate of water droplets from a hot spoon [72]. He found that at sufficiently high spoon temperatures, the droplet was spherical in shape and did not touch the spoon, resting instead on a vapor film. This is called the *Leidenfrost phenomenon* or, because of the droplet shape, the spheroidal state—the latter term includes flat disk droplets and large, bubbly liquid masses that are two forms possible in addition to a sphere [73]. The maximum metastable superheat T_{MAX} the liquid can sustain has been calculated for fluids with a van der Waals equation of state by Lienhard [74] who imposed a Maxwell criterion and included modifications of an empirical nature to find

$$\frac{T_{MAX}}{T_c} = 1 - 0.095\left[1 - \left(\frac{T_{sat}}{T_c}\right)^8\right]$$

where T_c and T_{sat} are the absolute critical and saturation temperatures, respectively. For liquid metals consult Gunnerson and Cronenberg [75]. Generally speaking, the Leidenfrost temperature lies between T_{sat} and T_{MAX} as

$$T_{sat} < T_{Leid} \lesssim T_{MAX}$$

For intermittent liquid–surface contact, surface properties are important.

The minimum film boiling heat flux from horizontal cylinders, investigated by Lienhard and Wong [76], differs from the horizontal-plate case only in the need to account for surface tension in the circumferential direction. If the cylinder radius is R, the effect of surface tension is to require an additional pressure difference P_{eff} across the interface of

$$P_{eff} = \frac{\sigma}{radius} \approx \frac{\sigma}{R + \eta}$$

At the peak of an interfacial wave $P_{eff} = \dot{\sigma}/(R + \eta_0)$, whereas at the valley $P_{eff} = \sigma/R$. The average value of P_{eff} is approximately $\sigma/(R + \eta_0/2)$, and the departure from this average is approximately

$$P_{eff, amplitude} \approx \frac{\sigma \eta}{2R^2}$$

Addition of this term to the right-hand side of Eq. (E-6) ultimately leads to the most dangerous wavelength λ_D being given by

$$\lambda_D = 2\pi \frac{[3\sigma(g_c/g)/(\rho_L - \rho_v)]^{1/2}}{[1 + \sigma(g_c/g)/2R^2(\rho_L - \rho_v)]^{1/2}} \tag{11-24}$$

The growth rate parameter b^* is then given by

$$b^* = \left(\frac{4\pi}{3\lambda_D}\right)^{1/2} \left[\frac{g(\rho_L - \rho_v)}{\rho_L + \rho_v}\right]^{1/2} \left[1 + \frac{\sigma(g_c/g)}{2R^2(\rho_L - \rho_v)}\right]^{1/2} \tag{11-25}$$

Proceeding as for the horizontal plate, the frequency of bubble emission is

$$\left[\frac{4\pi}{3}\frac{(\rho_L - \rho_v)g}{\rho_L + \rho_v}\right]^{1/2} \frac{[1 + \sigma(g_c/g)2R^2(\rho_L - \rho_v)]^{1/2}}{C'\lambda_D^{1/2}} \tag{11-26}$$

whereas

$$\frac{bubbles}{area\text{-}oscillation} = \frac{2}{\pi D \lambda_D} \tag{11-27}$$

where interfacial waviness is presumed to exist only along the length of the cylinder. Introduction of Eqs. (11-24)–(11-27) into Eq. (11-6) then results in the minimum heat flux being given by [62]

$$\frac{q}{A} = 0.06 \left(\frac{\rho_v h_{fg}}{D}\right) \left[\frac{\sigma(g_c/g)}{\rho_L - \rho_v}\right]^{1/2} \left[\frac{gg_c\sigma(\rho_L - \rho_v)}{(\rho_L + \rho_v)^2}\right]^{1/4}$$

$$\times \left[1 + \frac{2\sigma(g_c/g)}{D^2(\rho_L - \rho_v)}\right]^{-1/4} \tag{11-28}$$

and $C' = 20$. Equation (11-28) is intended for use with cylinders for which the vapor film thickness is much smaller than the cylinder diameter and for which the cylinder diameter is less than the critical wavelength given by Eq. (E-24). As pointed out by Lienhard and Sun [77], the cylinder end supports make the minimum heat flux for a short cylinder greater than predicted by Eq. (11-28).

An overview of hydrodynamic boiling, elements of which have been presented here, is in the survey of Lienhard and Witte [78]. So-called vapor explosions are also discussed.

11.5   FILM BOILING FROM A HORIZONTAL CYLINDER

Film boiling from a horizontal cylinder immersed in a stagnant pool of saturated liquid was analyzed by Bromley [79]. The physical situation is shown in Fig. 11-7. For small to moderate cylinder diameters, heat conduction across the vapor film is the controlling physical mechanism; for large cylinder diameters, vapor removal and bubble formation are the controlling physical mechanisms. The following analysis focuses on heat conduction through the vapor film.

The momentum principle applied to the control volume shown in Fig. 11-6b states that

$$\textbf{force} + \textbf{momentum}_{\text{in}} - \textbf{momentum}_{\text{out}} = \textbf{momentum}_{\text{stored}} \quad (2\text{-}17)$$

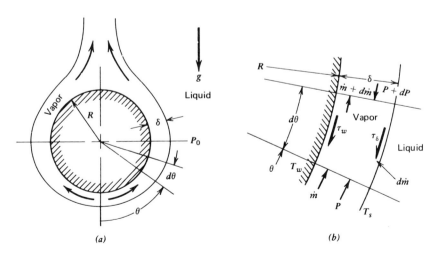

Figure 11-7 Film boiling from a horizontal cylinder with (a) the coordinate system and (b) the control volume to which conservation principles are applied.

In steady state, the right-hand side vanishes. Additionally, it is assumed that vapor motion is sufficiently slow that momentum fluxes are negligible compared to forces—an assumption found to be reasonable in Chapter 10 for natural convection which this film boiling is. Taking the θ-direction component of all forces acting on the control volume then gives

$$[P - (P + dP)]\delta - (\tau_w + \tau_\delta)R\,d\theta - (g\sin\theta)\rho_v\frac{\delta R\,d\theta}{g_c} = 0 \quad (11\text{-}29)$$

if it is assumed that $\delta/R \ll 1$ and that δ varies slowly with θ. Here τ_w and τ_δ are shear stresses at the cylinder and interface, respectively. As in natural convection, the hydrostatic pressure variation outside the vapor film gives the imposed pressure variation as

$$P = P_0 + \rho_L gR\frac{\cos\theta}{g_c}$$

Insertion of this result into Eq. (11-29) gives

$$(\rho_L - \rho_v)g\delta\frac{\sin\theta}{g_c} = (\tau_w + \tau_L) \quad (11\text{-}30)$$

Assumption of Poiseuille flow of the vapor between the solid and the interface gives the shears at the two surfaces and average vapor velocity V to be related as

$$\tau_w + \tau_\delta = \frac{\beta\mu_v V}{\delta} \quad (11\text{-}17)$$

in which $\beta = 3$ if the liquid exerts no shear and $\beta = 12$ if the liquid is stationary. Introduction of Eq. (11-17) into Eq. (11-30) results in

$$(\rho_L - \rho_v)g\delta\frac{\sin\theta}{g_c} = \frac{\beta\mu_v V}{\delta}$$

Recognition that the mass flow rate of the vapor $\dot{m}$ is related to V by

$$\dot{m} = \rho_v\,\delta V$$

allows the previous relationship to be restated as

$$(\rho_L - \rho_v)g\frac{\sin\theta}{g_c} = \frac{\beta\mu_v\dot{m}}{\rho_v\delta^3} \quad (11\text{-}31)$$

The conservation of energy principle for the control volume of Fig. 11-6b gives

$$\dot{Q}_{in} - \dot{W}_{out} = \text{internal}^{\cdot}\ \text{energy}_{stored} - \text{internal}^{\cdot}\ \text{energy}_{in} + \text{internal}^{\cdot}\ \text{energy}_{out}$$

(2-32)

With the control volume outer surface on the liquid side of the liquid–vapor interface, Eq. (2-32) becomes

$$\dot{Q}_{in} - \left[-\frac{P_1 \dot{m}}{\rho_{v_1}} + \frac{P_2 \dot{m}_2}{\rho_{v_2}} - \frac{P(\dot{m}_2 - \dot{m}_1)}{\rho_L} \right]$$

$$= -e_{v_1}\dot{m}_1 - e_{L,\,sat}(\dot{m}_2 - \dot{m}_1) + e_{v_2}\dot{m}_2$$

Subscripts 1 and 2 refer to vapor conditions at the bottom and top of the control volume, respectively; e is internal energy. Rearrangement results in

$$\dot{Q}_{in} = \left[e_{v_2} + \frac{P_2}{\rho_{v_2}} - \left(e_{L,\,sat} + \frac{P}{\rho_L} \right) \right]\dot{m}_2 - \left[e_{v_1} + \frac{P_1}{\rho_{v_1}} - \left(e_{L,\,sat} + \frac{P}{\rho_L} \right) \right]\dot{m}_1$$

$$= h_{fg_2}\dot{m}_2 - h_{fg_1}\dot{m}_1 = h_{fg}\,d\dot{m} + \dot{m}\,dh_{fg}$$

where h_{fg} is latent heat of vaporization and $d\dot{m} = \dot{m}_2 - \dot{m}_1$. The latent heat of vaporization h_{fg} requires accounting for a slight amount of superheating [80] as

$$h_{fg} = (h_{v,\,sat} - h_{L,\,sat}) + 0.34 C_{p,\,v}(T_w - T_{sat}) \qquad (11\text{-}32)$$

$\dot{Q}_{in}$, entering the control volume, is conducted across the vapor film and is

$$\dot{Q}_{in} = k_v(T_w - T_{sat})R\frac{d\theta}{\delta} \qquad (11\text{-}33)$$

It is expended in vaporization at the interface, the liquid being uniformly at the saturation temperature, and the energy required for superheating vapor is accounted for by Eq. (11-32). Thus

$$\dot{Q}_{in} = h_{fg}\,d\dot{m} \qquad (11\text{-}34)$$

Equations (11-33) and (11-34) substituted into Eq. (11-31) allow δ to be eliminated and yield

$$\dot{m}^{1/3}\, d\dot{m} = \left[(\rho_L - \rho_v) \frac{g}{g_c} \frac{\rho_v}{\mu_v \beta} \right]^{1/3} \left[k_v (T_w - T_{sat}) \frac{R}{h_{fg}} \right] \sin^{1/3} \theta \, d\theta$$

Integration* of this differential equation from $\theta = 0$ to $\theta = \pi$, with account for the fact that such a procedure gives only half the vapor production, results in

$$\dot{m} = 2 \left[\frac{(3.45)^{3/4}}{\beta^{1/4}} \right] \left[\frac{(\rho_L - \rho_v)(g/g_c)\rho_v k_v^3 (T_w - T_{sat})^3 R^3}{(\mu_v h_{fg}^3)} \right]^{1/4}$$

An average heat-transfer coefficient is defined by

$$q = \dot{m} h_{fg} = \bar{h}_c 2\pi R (T_w - T_{sat})$$

which gives

$$\overline{Nu}_c = \bar{h}_c \frac{D}{k_v} = C \left[Ra \frac{h_{fg}}{C_{\rho v}(T_w - T_{sat})} \right]^{1/4} \tag{11-35}$$

where the Rayleigh number Ra is

$$Ra = \left[g \frac{\rho_L - \rho_v}{\rho_v} \frac{D^3}{\nu_v^2} \right] Pr_v$$

and $C = 0.958/\beta^{1/4}$. Comparison with experiment shows that $C = 0.62$ within 15%, in good agreement with the expected variation (for $3 < \beta < 12$) of $0.515 \leqslant C \leqslant 0.728$. As expected, a natural-convection-like correlation results. Problem 11-18 contains additional discussion. Pitschmann and Grigull [81] extended the Bromley model to obtain, verified by Hesse et al. [82],

$$\overline{Nu}_c = \bar{h}_c \frac{D}{k} = 0.9(Ra^*)^{0.08} + 0.8(Ra^*)^{0.2} + 0.02(Ra^*)^{0.4}$$

where $Ra^* = Ra\, Pr^*$ and $Pr^* = Pr[\frac{1}{3} + h_{fg}/C_p(T_w - T_s)]$.

*Note that, with $\Gamma(m)$ being the gamma function,

$$\int_0^{\pi/2} \sin^n x \, dx = \frac{\pi^{1/2}}{2} \frac{\Gamma(n/2 + 1/2)}{\Gamma(1 + n/2)} \qquad \text{for} \quad n > -1$$

For $n = \frac{1}{3}$, this definite integral equals 1.2946.

The effect of thermal radiation must be included at the elevated temperatures that often accompany film boiling. Order of magnitude arguments led Bromley to, rigorous derivation was by Lubin [83],

$$\frac{\bar{h}}{\bar{h}_c} = \left(\frac{\bar{h}}{\bar{h}_c}\right)^{-1/3} + \frac{h_r}{\bar{h}_c} \approx 1 + \frac{3}{4}\frac{h_r}{\bar{h}_c} \tag{11-36}$$

Here $\bar{h}_c$ is the heat-transfer coefficient for conduction across the vapor acting alone as given by Eq. (11-35), h_r is the radiative coefficient for the assumed parallel plate geometry, and $\bar{h}$ is the net heat-transfer coefficient. With the liquid taken to be a perfect absorber ($\epsilon_L = 1$),

$$h_r = \frac{\sigma'}{1/\epsilon_w + 1/\epsilon_L - 1} \frac{T_w^4 - T_{sat}^4}{T_w - T_{sat}}$$

where $\sigma' = 0.1714 \times 10^{-8}$ Btu/hr ft^2 R^4 is the Stefan–Boltzmann constant, ϵ_w is the wall thermal emittance, and T is absolute temperature. Sakurai et al. [84] treated the subcooled case.

The assumptions underlying Eq. (11-35) restrict its applicability to cylinder diameters above 0.1 cm. At large diameters waves at the liquid–vapor interface make the prediction of Eq. (11-35) too low. To account for this latter effect, Breen and Westwater [84] correlated measurements for diameters of 2.2×10^{-4}–1.895 in., by

$$\frac{\bar{h}_c \lambda_c^{1/4}}{F} = 0.59 + 0.069 \frac{\lambda_c}{D}, \qquad 0.15 \leqslant \frac{\lambda_c}{D} \leqslant 300 \tag{11-37}$$

in which the critical wavelength is $\lambda_c = 2\pi[\sigma(g_c/g)/(\rho_L - \rho_v)]^{1/2}$ and $F^4 = (g/g_c)(\rho_L - \rho_v)\rho_v k_v^3 h_{fg}/[\mu_v(T_w - T_{sat})]$. Bromley's Eq. (11-35) is accurate for $0.8 \leqslant \lambda_c/D \leqslant 8$. Pomerantz [86] verified Breen and Westwater's correlation under increased gravitational fields. Thermal radiation is still accounted for by Eq. (11-36).

The effect of upward forced convection at velocity V on film boiling from a horizontal cylinder to a saturated pool was studied by Bromley et al. [87]. They found that Eq. (11-35) applies when $V/(gD)^{1/2} < 1$ while for $V/(gD)^{1/2} > 2$,

$$\overline{Nu}_c = \bar{h}_c \frac{D}{k} = 2.7 \left[Re_{D_v} Pr_v \frac{h_{fg}}{C_p(T_w - T_{sat})} \right]^{1/2} \tag{11-38}$$

in which $Re_{D_v} = VD/\nu_v$ and $Pr_v = \mu_v C_{p_v}/k_v$. The effect of thermal radiation is accounted for in an approximate way by

$$\bar{h} = \bar{h}_c + \frac{7h_r}{8}$$

The effect of subcooling for upward flow was examined by Motte and Bromley [88]. A comprehensive survey of the pertinent literature was made by Kalinin et al. [49]. Witte [89] recommends that the lead coefficient of Eq. (11-38) be changed to 2.98 for flow around a sphere. Measurements [44] confirm Eq. (11-38) but suggest a better correlation.

Film boiling from a horizontal plate has been correlated within 25% by Klimenko [90] for laminar conditions with

$$\text{Nu} = 0.19\left[\text{Ga}\left(\frac{\rho_L}{\rho_v} - 1\right)\right]^{1/3} \text{Pr}_v^{1/3} f_1, \qquad \text{Ga}\left(\frac{\rho_L}{\rho_v} - 1\right) < 10^8$$

where

$$f_1 = \begin{cases} 1, & \dfrac{h_{fg}}{C_{p,v}(T_w - T_{\text{sat}})} \leqslant 1.4 \\[3mm] 0.89\left[\dfrac{h_{fg}}{C_{p,v}(T_w - T_{\text{sat}})}\right]^{1/3} & \text{otherwise} \end{cases}$$

and for turbulent conditions with

$$\text{Nu} = 0.0086\left[\text{Ga}\left(\frac{\rho_L}{\rho_v} - 1\right)\right]^{1/2} \text{Pr}_v^{1/3} f_2, \qquad \text{Ga}\left(\frac{\rho_L}{\rho_v} - 1\right) > 10^8$$

where

$$f_2 = \begin{cases} 1, & \dfrac{h_{fg}}{C_{p,v}(T_w - T_{\text{sat}})} < 2 \\[3mm] 0.71\left[\dfrac{h_{fg}}{C_{p,v}}(T_w - T_{\text{sat}})\right]^{1/2} & \text{otherwise} \end{cases}$$

where $\text{Ga} = g\lambda_c^3/2\nu_v^2$ is the Galileo number, $\text{Nu} = h\lambda_c/k$ is the Nusselt number, $\lambda_c = 2\pi[\sigma(g_c/g)/(\rho_L - \rho_v)]^{1/2}$ is the critical wavelength, and all vapor properties are evaluated at $(T_w + T_{\text{sat}})/2$. This correlation is tested for $7 \times 10^4 < \text{Ga}(\rho_L/\rho_v - 1) < 3 \times 10^8$, $0.69 < \text{Pr} < 3.45$, $0.031 < h_{fg}/C_{p,v}$

$(T_w - T_{sat}) < 7.3$, $0.0045 < P/P_{critical} < 0.98$, and $1 < g/g_{earth\ normal} < 21.7$. The effect of plate size is accounted for, L is the minimum dimension of the plate, by

$$
\frac{h}{h_\infty} = \begin{cases} 1, & \dfrac{L}{\lambda_c} > 5 \\[2ex] 2.9\left(\dfrac{\lambda_c}{L}\right)^{0.67}, & \dfrac{L}{\lambda_c} < 5 \end{cases}
$$

11.6   FILM BOILING FROM A VERTICAL SURFACE

Film boiling from a vertical surface immersed in a stagnant pool is a boundary-layer problem if vapor flows up in a film between the solid and the liquid as shown in Fig. 11-8. If the liquid is subcooled, only part of the heat is used to form vapor; otherwise, all the heat entering the liquid forms vapor. The vapor formed enters the vapor film, "blowing" toward the solid surface. Near the bottom, flow is laminar, with eventual transition to turbulent flow if the solid is sufficiently high.

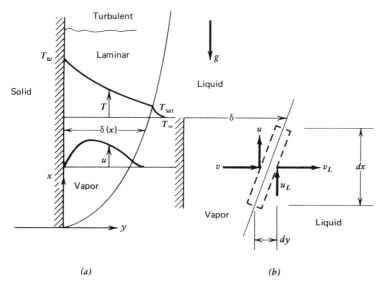

(a)   (b)

Figure 11-8 Film boiling from a vertical plate with (a) the coordinate system and (b) a control volume fixed in the liquid–vapor interface used to obtain boundary conditions.

Assuming steady state and constant properties, the laminar boundary-layer equations are

$$\frac{\partial u}{\partial x} + \frac{\partial v}{\partial y} = 0 \tag{11-39a}$$

$$u\frac{\partial u}{\partial x} + v\frac{\partial u}{\partial y} = (\rho_L - \rho)\frac{g}{\rho} + \nu\frac{\partial^2 u}{\partial y^2} \tag{11-39b}$$

$$u\frac{\partial T}{\partial x} + v\frac{\partial T}{\partial y} = \alpha\frac{\partial^2 T}{\partial y^2} \tag{11-39c}$$

The liquid's stagnant condition far from the plate (near the plate it is slightly dragged along by the vapor) gives $\partial P/\partial x = -\rho_L g/g_c$, where the subscript L denotes a liquid quantity. The body force B_x is $-\rho g/g_c$. The assumption that there is no pressure change across the interface is accurate but inexact, as was discussed in Problem 2-13—the fluid is accelerated because of its density change, and this requires a pressure difference.

Boundary conditions at the interface are obtained by applying a conservation principle to a control volume straddling the interface as shown in Fig. 11-8b. For mass

$$\rho v\, dx - \rho u\, dy = \rho_L v_L\, dx - \rho_L u_L\, dy$$

With $y = \delta$ at the interface, this relationship is rearranged as

$$\rho\left(v - u\frac{d\delta}{dx}\right) = \rho_L\left(v_L - u_L\frac{d\delta}{dx}\right) \tag{11-40a}$$

For energy, with e representing internal energy per unit mass and KE kinetic energy per unit mass,

$$\left[-k\frac{\partial T}{\partial y} + (e + \text{KE})\rho v + Pv\right] dx$$

$$-\left[-k\frac{\partial T}{\partial x} + (e + \text{KE})\rho u + Pu\right] dy + q_r\, dx$$

$$= \left[-k_L\frac{\partial T_L}{\partial y} + (e_L + \text{KE}_L)\rho_L v_L + P_L v_L\right] dx$$

$$-\left[-k_L\frac{\partial T_L}{\partial x} + (e_L + \text{KE}_L)\rho_L u_L + P_L u_L\right] dy$$

where q_r is the net radiative heat flux between the plate and the interface. Incorporating Eq. (11-40a), neglecting changes in kinetic energy, defining the latent heat of vaporization as $h_{fg} = (e + P/\rho) - (e_L + P_L/\rho_L)$, and recognizing $\partial T/\partial x \ll \partial T/\partial y$ and $d\delta/dx \ll 1$ in the boundary layer assumptions at the interface, one obtains

$$q_r - k\frac{\partial T}{\partial y} = -k_L\frac{\partial T_L}{\partial y} - h_{fg}\rho\left(v - u\frac{d\delta}{dx}\right) \qquad (11\text{-}40b)$$

at $y = \delta$. The momentum principle applied to the interfacial control volume leads to two equations, one for the perpendicular direction and one for the parallel direction. The one in the perpendicular direction is of little utility since it has already been assumed that there is no pressure difference across the interface. Convective momentum transport is negligible at the interface as a result of the low velocities, and the interface curvature is also negligible (except near the origin where the boundary-layer equations are inaccurate anyway); therefore, the parallel direction yields only the requirement of equal tangential shears on the liquid and the vapor sides

$$\mu\frac{\partial u}{\partial y} = \mu_L\frac{\partial u_L}{\partial y} \qquad (11\text{-}40c)$$

at $y = \delta$. With no interfacial velocity slip or temperature jump, and an impermeable plate:
At $y = \delta$

$$u = u_L \qquad (11\text{-}40d) \qquad \text{and} \qquad T = T_{sat} \qquad (11\text{-}40e)$$

At $y = 0$

$$u = 0 \quad (11\text{-}40f), \qquad v = 0 \quad (11\text{-}40g), \qquad \text{and} \qquad T = T_w \quad (11\text{-}40h)$$

An integral method of solution of Eqs. (11-39) and (11-40) will first be essayed. It gives reasonable results and indicates the functional forms that must appear in the exact solution. First, the continuity equation [Eq. (11-39a)] is integrated across the vapor film to obtain

$$v_\delta = \int_0^\delta \frac{\partial u}{\partial x}\,dy = -\frac{\partial\left(\int_0^\delta u\,dy\right)}{\partial x} + u_\delta\frac{d\delta}{dx} \qquad (11\text{-}41)$$

in which Eq. (11-40g) has been used. Second, the x-motion equation [Eq. (11-39b)] in conservation form

$$\frac{\partial(uu)}{\partial x} + \frac{\partial(vu)}{\partial y} - u\left(\frac{\partial u}{\partial x} + \frac{\partial v}{\partial y}\right) = (\rho_L - \rho)\frac{g}{\rho} + v\frac{\partial^2 u}{\partial y^2}$$

is integrated across the vapor film to obtain

$$\frac{d\left(\int_0^\delta u^2 \, dy\right)}{dx} - u_\delta^2 \frac{d\delta}{dx} + v_\delta u_\delta = \frac{(\rho_L - \rho) g \delta}{\rho} + v\left[\frac{\partial u(\delta)}{\partial y} - \frac{\partial u(0)}{\partial y}\right]$$

in which Eq. (11-40f) has been used. Use of Eq. (11-41) yields, with $\eta = y/\delta$,

$$\frac{d\left(\delta \int_0^1 u^2 \, d\eta\right)}{dx} - u_\delta \frac{d\left(\delta \int_0^1 u \, d\eta\right)}{dx} = (\rho_L - \rho)\frac{g\delta}{\rho} + \frac{v}{\delta}\left[\frac{du(1)}{d\eta} - \frac{du(0)}{d\eta}\right]$$

$$(11\text{-}42)$$

Third, the energy equation [Eq. (11-39c)] in conservation form

$$\frac{\partial(uT)}{\partial x} + \frac{\partial(vT)}{\partial y} - T\left(\overset{0 \text{ by continuity}}{\cancel{\frac{\partial u}{\partial x}} + \frac{\partial v}{\partial y}}\right) = \alpha\frac{\partial^2 T}{\partial y^2}$$

is integrated across the vapor film to obtain

$$\frac{d\left(\int_0^\delta uT \, dy\right)}{dx} - u_\delta T_\delta \frac{d\delta}{dx} + v_\delta T_\delta = \alpha\left[\frac{\partial T(\delta)}{\partial y} - \frac{\partial T(0)}{\partial y}\right]$$

in which Eq. (11-40g) has been used. Use of Eq. (11-41) yields, with $\eta = y/\delta$,

$$\frac{d\left(\delta \int_0^1 uT \, d\eta\right)}{dx} - T_0\frac{d\left(\delta \int_0^1 u \, d\eta\right)}{dx} = \frac{\alpha[dT(1)/d\eta - dT(0)/d\eta]}{\delta}$$

This can be rearranged, with the help of Eq. (11-40e), into

$$\frac{d\left[\delta \int_0^1 u(T - T_s) \, d\eta\right]}{dx} = \frac{\alpha}{\delta}\left[\frac{dT(1)}{d\eta} - \frac{dT(0)}{d\eta}\right]$$

$$(11\text{-}43)$$

A velocity distribution is assumed of the form

$$u = a_0 + a_1\eta + a_2\eta^2 + a_3\eta^3 + \cdots$$

which can be subjected to the conditions:

At $\eta = 0$

$$0 = (\rho_L - \rho)\frac{g}{\rho} + v\frac{\partial^2 u}{\partial y^2} \qquad (11\text{-}39a) \quad \text{and} \qquad u = 0 \qquad (11\text{-}40f)$$

At $\eta = 1$

$$\left[u_\delta\frac{d\delta}{dx} - \frac{d\left(\delta \int_0^1 u \, d\eta\right)}{dx}\right]\frac{\partial u}{\partial y} = (\rho_L - \rho)\frac{g}{\rho} + v\frac{\partial^2 u}{\partial y^2} \qquad (11\text{-}39a)$$

$$\mu\frac{\partial u}{\partial y} = \mu_L\frac{\partial u_L}{\partial y} \qquad (11\text{-}40c) \qquad \text{and} \qquad u = u_L \qquad (11\text{-}40d)$$

Meeting the first, second, and fifth of the preceding conditions, with the simplifications that the liquid is saturated ($\partial T_L / \partial y = 0$) and is so viscous compared to the vapor as to be immovable, and thermal radiation is unimportant ($q_r = 0$), gives

$$u = \frac{(\rho_L - \rho) g \delta^2}{2 \rho \nu} (\eta - \eta^2) \qquad (11\text{-}44)$$

The "blowing" of stagnant vapor toward the plate from the interface, neglected in Eq. (11-44), causes the velocity profile to be slightly displaced toward the plate from the parabolic distribution that is its essential characteristic. A temperature distribution is next assumed of the form

$$T - T_s = b_0 + b_1 \eta + b_2 \eta^2 + b_3 \eta^3 + b_4 \eta^4 + \cdots$$

which can be subjected to the conditions:

At $\eta = 0$ $\qquad \dfrac{d^2 T}{d\eta^2} = 0 \qquad (11\text{-}39c) \qquad$ and $\qquad T = T_w \ (11\text{-}40h)$

At $\eta = 1$ $\qquad \dfrac{1}{\delta}\left[u_\delta \dfrac{d\delta}{dx} - \dfrac{d\left(\delta \int_0^1 u \, d\eta\right)}{dx} \right] \dfrac{dT}{d\eta} = \dfrac{\alpha}{\delta^2} \dfrac{d^2 T}{d\eta^2} \qquad (11\text{-}39c)$

$$T = T_s \qquad (11\text{-}40e)$$

$$q_r - \frac{k}{\delta} \frac{dT}{d\eta} = -k_L \frac{\partial T_L}{\partial y} + \rho h_{\mathrm{fg}} \frac{d\left(\delta \int_0^1 u \, d\eta\right)}{dx} \qquad (11\text{-}40b)$$

Compliance with the first, second, fourth, and fifth of these possible five conditions gives, with $q_r = 0$ (see Sparrow [129] for $q_r \neq 0$) and $\partial T_L / \partial y = 0$,

$$\frac{T - T_s}{T_w - T_s} = 1 - \left(1 + \frac{1 - C}{2}\right)\eta + \frac{1 - C}{2}\eta^3 \qquad (11\text{-}45)$$

where C is the constant defined by Eq. (11-44a). Substitution of Eqs. (11-44) and (11-45) into the integrated energy Eq. (11-43) gives

$$\frac{1}{3} \frac{C_p(T_w - T_s)}{h_{\mathrm{fg}}}\left[1 - (1 - C)\frac{3}{10}\right]C = 1 - C$$

Note that for water, $C_p(T_w - T_s)/h_{\mathrm{fg}} \sim \frac{1}{10}$, as is the subject of Problem 11-20. Then $C \approx 1$ and the corrected value is

$$1 - C \approx \frac{1}{3} \frac{C_p(T_w - T_s)}{h_{\mathrm{fg}}} \qquad (11\text{-}46)$$

From the definition of C one can next obtain the vapor film thickness δ as

$$\delta = 2\left[1 - \frac{1}{3}\frac{C_p(T_w - T_s)}{h_{fg}}\right]^{1/4}\left[\frac{x(T_w - T_s)\nu k}{gh_{fg}(\rho_L - \rho)}\right]^{1/4} \tag{11-47}$$

which shows, as expected, that $\delta \sim x^{1/4}$. The heat flux q_w at the wall is

$$q_w = -k\frac{\partial T(0)}{\partial y} = \frac{k}{\delta}(T_w - T_s)\left(1 + \frac{1 - C}{2}\right)$$

The heat-transfer coefficient follows, with the use of Eqs. (11-46) and (11-47), as

$$h = \frac{q_w}{T_w - T_s} = \frac{1}{2}k\left[\frac{g(\rho_L - \rho)\,\mathrm{Pr}^*}{\rho\nu^2 x}\right]^{1/4} K$$

where $\mathrm{Pr}^* = \mu[C_p + h_{fg}/(T_w - T_s)]/k$.

Interfacial "blowing" was only approximately determined here. Problem 11-27 illustrates that a better correction could easily be made, leading to a $\frac{3}{10}$ coefficient for C_p in Pr^*; a refined estimate of the coefficient is $0.968 - 0.163/\mathrm{Pr}$ for $0.5 \leq \mathrm{Pr} < \infty$ as provided by Sadasivan and Lienhard [91].

The average heat-transfer coefficient $\bar{h}$ is related to the local coefficient by $\bar{h} = 4h/3$. Hence

$$\overline{\mathrm{Nu}} = \frac{\bar{h}x}{k} = \frac{2}{3}\mathrm{Ra}^{*1/4} \tag{11-48}$$

where the Rayleigh number is $\mathrm{Ra}^* = g(\rho_L - \rho)\mathrm{Pr}^*x^3/\rho\nu^2$. Bromley [79] suggested that the lead coefficient should range between $\frac{2}{3}$ for an immovable liquid and 0.943 for a liquid that exerts no shear; Ellion [92] correlated measurements with the lead coefficient equal to 0.714.

A detailed integral analysis of laminar film boiling from a vertical surface to a saturated liquid has been accomplished by Frederking [93].

If x is sufficiently large, the vapor will undergo transition to turbulent flow and Eq. (11-48) will not give accurate predictions. Hsu and Westwater [59, 94] found that transition begins at a Reynolds number of

$$\mathrm{Re} = u_{max}\,\delta/\nu \approx 100$$

Their accurate predictors are cumbersome; refer to references for details. For turbulent film boiling from a vertical surface, it is recommended [49] that the correlation by Labuntsov [95]

$$\overline{Nu} = \frac{\overline{h}x}{k} = 0.25\,Ra_x^{*1/3}$$

be used for $Ra^* \geqslant 2 \times 10^7$.

Exact solutions to the describing boundary-layer Eqs. (11-39) and associated boundary conditions of Eq. (11-40) are in good agreement with those of the foregoing integral method. Koh [96] treated the saturated-liquid case and accounted for the liquid boundary-layer motion that is described by the parameter of $(\rho\mu/\rho_L\mu_L)^{1/2}$, accounting for vapor Prandtl numbers of 0.5 and 1.0. The subcooled liquid case was solved by Sparrow and Cess [97], simultaneously solving the boundary-layer equations in the liquid and vapor films. The calculations are shown in Fig. 11-9 to demonstrate the change in local Nusselt number for laminar film boiling of water with $T_s = 467°F$ at a vertical surface with $T_w = 767°F$. Subcooling can have a substantial effect—the pure free convection limit at large liquid subcooling represents the fact that the liquid seems to be in laminar free convection at an impermeable vertical plate of temperature T_s. Experimental support for these trends of liquid subcooling is exhibited in Fig. 11-10. An integral analysis of the same problem was performed by Frederking and Hopenfeld [99]. The

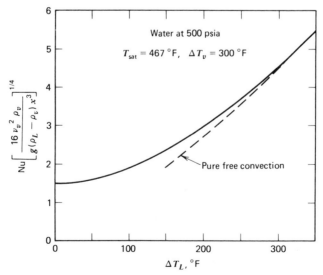

Figure 11-9 Illustration of the effect of subcooling on Nusselt number for film boiling from a vertical plate. [From E. M. Sparrow and R. D. Cess, *ASME J. Heat Transfer* **84**, 149–156 (1962) [97].]

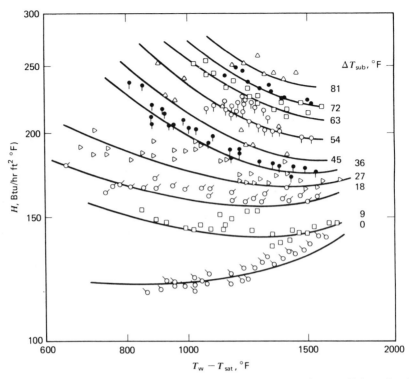

Figure 11-10 Effect of liquid subcooling on the heat-transfer coefficient for film boiling of distilled water from a 0.0195-in.-diameter wire. [From F. Tachibana and S. Fukui, *International Developments in Heat Transfer*, ASME, 1961, p. 219 [98].]

effect of intermediate subcooling can be taken into account by the superposition

$$q_w^4 = q_{w_{\text{sat}}}^4 + q_{w_{\text{large subcool}}}^4$$

which leads to

$$\overline{\text{Nu}} = \frac{hx}{k} = \frac{2}{3}\left\{\text{Ra}_{x,v}\left[\frac{h_{\text{fg}}}{C_p(T_w - T_s)}\right] + \frac{k_L(T_s - T_\infty)}{k_v(T_w - T_s)}\frac{\text{Ra}_{x,L}\,\text{Pr}_L}{(1 + \text{Pr}_L)}\right\}^{1/4}$$

Parallel forced convection laminar film boiling from a constant temperature plate to a saturated liquid was treated by Cess and Sparrow [100]. They solved the laminar boundary-layer equations in the vapor and adjacent liquid. Their results are represented by

$$\text{Nu Re}^{-1/2}\frac{\mu}{\mu_L}\left(1 + \pi^{1/2}\,\text{Nu Re}^{-1/2}\frac{\mu}{\mu_L}\right)^{1/2} = \frac{1}{2}K^{1/2} \quad (11\text{-}49a)$$

for $10^{-3} \leqslant K = (\rho\mu/\rho_L\mu_L) \Pr h_{fg}/C_p(T_w - T_s) \leqslant 1$ where $\mathrm{Nu} = hx/k$, $\mathrm{Re} = U_\infty x/\nu_L$, x is the distance from the plate leading edge, and U_∞ is the parallel approach velocity of the liquid far from the plate. This result cannot be taken to the limit of $U_\infty \to 0$ since no body force term was retained in the x-motion equations. The local wall shear τ is represented by

$$\frac{\left(\tau/\rho_L U_\infty^2\right) \mathrm{Re}^{1/2}}{1 - \pi^{1/2} \mathrm{Re}^{1/2}\left(\tau/\rho_L U_\infty^2\right)} = 0.5K^{1/2} \qquad (11\text{-}49b)$$

and is useful in assessing the benefit of film boiling in drag reduction of waterborne vehicles [101, 102] (see Problems 13-24 and 13-25). Subcooled forced convection film boiling from a constant-temperature plate was studied by Cess and Sparrow [103] and Nakayama [104], who found liquid subcooling to increase both heat transfer and plate shear.

11.7 OTHER GEOMETRIES AND CONDITIONS

Other geometries and conditions have been studied. Pilling and Lynch [105] investigated the quenching of hot objects of various shapes. For film boiling of cryogenic fluids from spheres, the data due to Merte and Clark [106] give the same heat flux versus $T_w - T_s$ as predicted by Eq. (11-35) for normal Earth-gravitation and near-zero accelerations. Frederking et al. [107] give similar data for liquid nitrogen and liquid helium at normal Earth-gravitation acceleration. Frederking and Clark [108] utilized an integral analysis, generalized to all shapes by Nakayama and Koyama [109], for the lower part of the sphere where no bubbles are released; Frederking and Daniels [110] studied the frequency of bubble release from the upper part of the sphere and bubble diameter. Film boiling from fully submerged spheres was analyzed by Hendricks and Baumeister [111]. Marschall and Farrar [112] analyzed half-submerged spheres and applied the results to film boiling in a scaling liquid. The presence of scale on the sphere suggests that liquid–solid contact occurs, perhaps by liquid droplets ejected into the vapor film from the wavy interface.

The effects of 11–26 Hz externally imposed pressure fluctuations of 14.7–90 psi amplitude above atmospheric pressure on film boiling from a wire were measured by DiCicco and Schoenhals [114]; film boiling heat flux was nearly double that at the steady peak pressure. An analytical study in the framework of the unsteady boundary-layer equations by Burmeister and Schoenhals [115] for laminar film boiling from a vertical surface suggests that 10% departure of the time-averaged local heat flux from steady-state results can be expected if the imposed pressure sinusoidal amplitude and frequency ω combine as

$$\frac{p_{\max} - p_{\min}}{p_{\min}}\left[\frac{2\omega\rho_v}{g(\rho_L - \rho_v)}\frac{T_w + T_\infty}{T_w - T_\infty}\right]^{1/2} = 0.02$$

Measurements for film boiling from a sphere of diameter D oscillating sinusoidally in saturated Freon-11 and liquid nitrogen are correlated by Schmidt and Witte [116] by

$$\overline{\mathrm{Nu}} = \frac{\bar{h}D}{k} = 0.14 \left[\mathrm{Ra} \frac{h_{\mathrm{fg}}^*}{C_p(T_w - T_s)}(1 + \mathrm{Fr}) \right]^{1/3}$$

for the turbulent case of

$$\mathrm{Ra} \frac{h_{\mathrm{fg}}^*}{C_p(T_w - T_s)} \geqslant 6 \times 10^7$$

Here $\mathrm{Ra} = g(\rho_L - \rho)\Pr D^3/\rho \nu^2$, $h_{\mathrm{fg}}^* = h_{\mathrm{fg}} + C_p(T_w - T_s)/2$, $\mathrm{Fr} = X^2 f^2/1.5gD$ is a Froude number, $Xf/(1.5)^{1/2}$ is the rms average velocity, X is displacement amplitude, and f is frequency. The effect of diameter on film boiling from a sphere is the same as for cylinders [49]. For laminar film boiling from a sphere [108],

$$\overline{\mathrm{Nu}} = \frac{\bar{h}D}{k} = 0.586 \left[\mathrm{Ra} \frac{h_{\mathrm{fg}}}{C_p(T_w - T_s)} \right]^{1/4}$$

The critical heat flux for a liquid jet impinging on a hot surface is increased by a factor of 2 (and the single-phase heat-transfer coefficient is increased by a factor of 100) by applying about 20 kV of electrical potential difference between the surface and a ring through which the jet issues, according to Yabe and Maki [118].

11.8 SCULPTURED SURFACES

Sculptured surfaces are often used to increase heat-transfer rates to fluids whose low heat-transfer coefficients are the controlling thermal resistances. This can be the case even in a boiling situation (see Problem 13-28).

The variable heat-transfer coefficients on a fin whose temperature varies from values at the root that cause film boiling (with moderate coefficients) to midvalues that cause nucleate boiling (with high coefficients) to tip values that cause natural convection without boiling (with low coefficients) lead to unusual optimal fin shapes. Information on fin analytical and experimental considerations is available [119–122].

The evolution of enhanced surface geometries for nucleate boiling is surveyed by Webb [123], Bergles [124] and Thome [125]. The effects of surface roughness formed by mechanical working or chemical etching of the base surface, porous metallic and nonmetallic coatings applied to the base surface, and promoters such as pierced cover sheets and wires or screens are reported. In some cases the nucleate boiling heat flux is nearly six times that from a smooth surface at the same temperature difference—see C_{sf} values in Table 11-1 to see some effects. Nucleate boiling heat-transfer coefficients of enhanced surfaces in a tube bundle approximately equal those measured in single-tube tests. Liquid superheats of as much as 8°C are sometimes required for initiation of nucleate boiling on enhanced surfaces.

As explained by Ünal [126], tubes with longitudinal straight fins of rectangular profile provide substantial enhancement of nucleate boiling heat-transfer coefficients while still being simple enough to be reliable and of low manufacturing cost. In contrast, more exotic sculptured surfaces are more easily degraded in performance (oil in the boiling liquid can clog the pores of a porous surface, for example, whose cavities then cease to be nucleation sites) and are more costly. Four types of exotic sculptured surfaces are the High-Flux, Thermoexcel-E, Gewa-T, and ECR40 surfaces. The High-Flux surface consists of a 50%–65% porosity, 0.25–0.50-mm-thick metallic matrix bonded to a metallic substrate. The Thermoexcel-E surface has continuous tunnels (0.2 mm wide, 0.4 mm deep, 0.55 mm pitch) atop which are isolated pores (0.1 mm diameter, 260 pore/cm^2); the tunnels are formed by bending the ridges of small fins. The Gewa-T surface is formed by flattening the tips of fins in an integral fin tube; the ECR40 surface is like that of the Gewa-T except that porous plates are atop the flattened fin tips. Such exotic sculptured surfaces have been used in evaporators of refrigeration machines due to the coefficient of performance improvement that results.

For boiling inside tubes with forced convection, Bergles [124] observed that the foregoing "minor" surface treatments have less effect. However, "major" surface treatments (spiral internal fins and twisted tape inserts, spiral indentations, etc.) that impart swirling motion to the flow can as much as double the heat flux over that of a smooth tube at the same temperature difference and flow rate. Burnout heat fluxes are increased by a similar amount. A method for predicting forced convection boiling heat-transfer coefficients inside augmented and smooth tubes was devised by Kandlikar [127].

PROBLEMS

11-1 Use Eq. (11-2) to predict the heat flux and the coefficient of heat transfer for nucleate pool boiling of saturated water at 212°F on a 242°F copper heater surface. *Answer*: $q/A = 6 \times 10^4$ Btu/hr ft^2, $h = 2 \times 10^3$ Btu/hr ft^2 °F.

11-2 Compare the results of Problem 11-1 with those of Fig. 11-1.

11-3 Use Eq. (11-3) to predict the maximum heat flux for nucleate pool boiling of saturated water at 212°F. *Answer*: $q_{max}/A = 4 \times 10^5$ Btu/hr ft^2.

11-4 Equate Eqs. (11-2) and (11-3) to derive an expression for the maximum temperature difference $T_w - T_{sat}$ in nucleate pool boiling of a saturated liquid. For saturated water at 212°F and a copper heater surface, show that the heater temperature at which the peak heat flux occurs is $T_w = 270$°F. Use this result and that of Problem 11-3 to show that $h = 1.5 \times 10^4$ Btu/hr ft² °F.

11-5 Predict the maximum heat flux for nucleate pool boiling of saturated water at 380 psia and compare the result with that of Problem 11-3 at 14.7 psia. *Answer*: $q_{max}/A = 10^6$ Btu/hr ft².

11-6 Estimate the maximum heat flux for nucleate pool boiling of water at 14.7 psia that is subcooled 20°F and compare with the result for saturated water. *Answer*: $q = 6 \times 10^5$ Btu/hr ft² °F.

11-7 A frying pan is electrically heated, dissipating 100 W across a surface of 520 cm². The frying pan is filled with saturated water. Estimate the temperature of the frying pan. *Answer*: 230°F.

11-8 A stainless-steel kettle whose 0.3-m-diameter bottom is maintained at 250°F contains saturated water at 212°F. Estimate the time required to boil away a 20-cm water layer. *Answer*: 23 min.

11-9 A long brass rod of 1 cm diameter is quenched in saturated water at 212°F. At a certain instant the rod surface temperature is 225°F.

    (a) Estimate the temperature gradient at the rod surface. *Answer*: 70°F/ft.

    (b) Estimate the amount of water subcooling required to increase the temperature gradient by 10%.

11-10 Repeat Problem 11-3 for a gravitational acceleration one-sixth that at the Earth's surface.

11-11 An immersion heater for a coffee mug is to be made of stainless steel. It must heat water to its saturation temperature in 5 min and, to avoid possible early material failure, the heater must operate in the nucleate boiling regime. Estimate the heater area and the wattage of the heater *Answer*: $A = 5$ in.², $q = 400$ W.

11-12 A horizontal plate is exposed to a stagnant pool of saturated water at 212°F. Determine the numerical values of:

    (a) The minimum film boiling heat flux.

    (b) The temperature difference between plate and liquid associated with the minimum film boiling heat flux.

    (c) The heat-transfer coefficient for this situation.

    (d) Compare the results for parts a–c with those at the point of burnout boiling heat flux. *Answers*: (a) $q/A = 6000$ Btu/hr ft²; (b) 160°F; (c) $h = 40$ Btu/hr ft² °F.

(e) Compare the results for parts a and c for natural convection to liquid water.

11-13 For the conditions stated in Problem 11-12, determine the numerical values of:

(a) The vapor film thickness between bubbles.

(b) The "most dangerous" wavelength λ_D.

(c) The distance between detaching bubbles.

(d) The average vapor velocity in the film between bubbles at an average distance from a bubble center of $\lambda_D/3$.

(e) The Reynolds number Va/ν_v in the vapor film.

(f) The diameter of a detaching bubble.

(g) The pressure difference $p_2 - p_1$ in the vapor film between the bubble position and a position between bubbles. *Answers*: (a) 0.05 mm; (b) 27 mm; (f) 10 mm.

11-14 A horizontal cylinder of 0.1 in. diameter is in a stagnant pool of saturated water. Minimum film boiling heat flux occurs.

(a) Is the cylinder diameter less than the critical wavelength?

(b) Determine the minimum film boiling heat flux. *Answer*: $q = 3000$ Btu/hr ft^2.

11-15 Estimate the Leidenfrost temperature for water at atmospheric pressure. Comment on the ease of obtaining the Leidenfrost temperature on a kitchen stove or a skillet. *Answer*: 190°C.

11-16 Show that the minimum film boiling heat flux varies as $q_{min}/A = \lambda_D b^*$ for a horizontal plate, according to Eq. (11-10), and as $q_{min}/A = \lambda_D^2 b^*$ for a horizontal cylinder, according to Eq. (11-28). Here λ_D is the most dangerous wavelength and b^* is the maximum growth rate. Review the formulation for bubbles/area–oscillation given by Eqs. (11-8) and (11-27) to trace the origin of the difference in q_{min}/A formulas.

11-17 A horizontal stainless-steel cylinder of $\frac{1}{4}$ in. diameter is at 600°F while immersed in a pool of stagnant water saturated at 212°F.

(a) Determine the numerical value of the film boiling heat-transfer coefficient $\bar{h}$, neglecting radiation.

(b) Determine the numerical value of the coefficient, taking thermal radiation into account. What is the percentage influence of thermal radiation?

(c) Determine the heat flux and the vapor production rate.

(d) Estimate the thickness of the vapor film and if $\delta/R \ll 1$.

11-18 (a) Compare the correlation of Eq. (11-35) for film boiling from a horizontal cylinder with the correlation for laminar natural convection from a horizontal cylinder. Note that if an equivalent specific heat were defined as $C_{p_{eff}} = h_{fg}/(T_w - T_s)$, Eq. (11-35) could be expressed as $\overline{Nu}_c = 0.62 \; Ra^{*1/4}$. Are the coefficients and exponents of Rayleigh number nearly equal?

(b) Show that the heat-transfer coefficient for minimum film boiling for a horizontal plate given by Eq. (11-23) can be expressed as

$$
Nu = \frac{hL}{k_v} = 0.425 \left[Ra \frac{h_{fg}}{C_{p_v}(T_w - T_{sat})} \right]^{1/4}
$$

where the Rayleigh number is $Ra = g[(\rho_L - \rho_v)/\rho_v]L^3 \, Pr_v/\nu_v^2$ and $L = [\sigma(g_c/g)/(\rho_L - \rho_v)]^{1/2}$, which is related to the most dangerous wavelength λ_D by $L = \lambda_D/[2\pi(3^{1/2})]$. Show that an alternative expression is

$$
\frac{h\lambda_c^{1/4}}{F} = 0.673 \pm 10\%
$$

where

$$
\lambda_c = 2\pi \left[\frac{\sigma(g_c/g)}{\rho_L - \rho_v} \right]^{1/2} \quad \text{and} \quad F^4 = \frac{\rho_v h_{fg} g(\rho_L - \rho_v)k_v^3}{\mu_v(T_w - T_{sat})}
$$

Compare this with Eq. (11-37a) for large horizontal cylinders.

11-19 Determine the film boiling heat-transfer coefficient for saturated water at atmospheric pressure on a 0.08-in.-diameter horizontal tube at a wall temperature of 1200°F. Also determine the energy supply rate required for a 6-in. length of this tube. *Answer*: $q = 35$ W.

11-20 Show that for film boiling of water, the typical value of $C_p(T_w - T_s)/h_{fg} \sim \frac{1}{10}$. The results of Problem 11-12 can be used to estimate typical $T_w - T_s$ values.

11-21 In film boiling of a saturated liquid at a vertical surface show that:

(a) "Blowing" of vapor into the film toward the plate slightly increases the temperature gradient at the wall—plot $(T - T_s)/(T_w - T_s)$ against y/δ for $C_p(T_w - T_s)/h_{fg} = \frac{1}{10}$.

(b) The maximum velocity in the vapor film is

$$
u_{max} = 2^{-1} \left[1 + \frac{C_p(T_w - T_s)}{3h_{fg}} \right]^{-1/2} \left[\frac{x(T_w - T_s)(\rho_L - \rho)gk}{\rho^2 \nu h_{fg}} \right]^{1/2}
$$

which occurs at $y/\delta = \frac{1}{2}$—plot u/u_{max} against y/δ for

$C_p(T_w - T_s)/h_{fg} = \frac{1}{10}$. Calculate a typical numerical value of u_{max} for saturated water. Calculate the maximum value of x for which laminar vapor flow can be expected for saturated water. *Answer*: (b) $u_{max} = 4$ m/s; (c) $x = 0.04$ m.

11-22 To explore the effect of liquid motion on film boiling of a saturated liquid at a vertical plate in a simple way:

(a) Neglect convective terms in the boundary-layer x-motion equation [Eq. (11-39b)] assuming no interfacial shear [$du(y = \delta)/dy = 0$], to show that the velocity profile is (with $\eta = y/\delta$)

$$u = \frac{(\rho_L - \rho)g\delta^2}{\rho\nu}\left(\eta - \frac{\eta^2}{2}\right)$$

(b) Neglect convective terms in the boundary-layer energy equation [Eq. (11-39c)] to show that the temperature profile is

$$\frac{T - T_s}{T_w - T_s} = 1 - \eta$$

(c) Substitute the results of part b into the interfacial energy boundary condition of Eq. (11-40b) to find

$$\delta = 2^{1/2}\left[\frac{x(T_w - T_s)\nu k}{gh_{fg}(\rho_L - \rho)}\right]^{1/4}$$

(d) Use the result of part c to obtain the average heat-transfer coefficient $\bar{h}$ as

$$\overline{Nu} = \frac{\bar{h}x}{k} = \frac{2^{3/2}}{3}\left[Ra\frac{h_{fg}}{C_p(T_w - T_{sat})}\right]^{1/4}$$

where $Ra = g(\rho_L - \rho)Pr\, x^3/\rho\nu^2$. Compare this coefficient with that of Ellion and that of Eq. (11-48).

11-23 Compare the expression for $\overline{Nu}$ for film boiling of a saturated liquid at a vertical surface with that for laminar free convection from a vertical surface without phase change.

11-24 Compare the laminar drag of a vehicle submerged in water with and without an enveloping vapor film.

11-25 Consider parallel forced convection laminar film boiling from a constant-heat-flux plate to a saturated liquid.

(a) Assuming the velocity at the liquid–vapor interface to be the same as the free-stream velocity U_∞, show that the vapor velocity profile is $u = U_\infty y/\delta$, where δ is the vapor film thickness.

(b) Substitute the result of part a into the integral form of the interface energy boundary condition, with $\eta = y/\delta$,

$$-k\frac{\partial T(y = \delta)}{\partial y} = \rho h_{fg}\frac{d(\delta\int_0^1 u\,d\eta)}{dx} \qquad (11\text{-}40b)$$

and show that, with a linear vapor temperature profile, this leads to

$$\delta = \frac{2q_w x}{\rho U_\infty h_{fg}}$$

(c) Since $q_w \approx k(T_w - T_s)/\delta$, show that

$$T_w - T_s = \frac{2q_w^2 x}{k\rho U_\infty h_{fg}}$$

(d) Since $h \approx k(T_w - T_s)/\delta$, show that the local Nusselt number is given by

$$\text{Nu} = \frac{hx}{k} = \frac{\rho U_\infty h_{fg}}{2q_w} = 0.707\left[\frac{x\rho U_\infty h_{fg}}{k(T_w - T_s)}\right]^{1/2}$$

(e) Compare the result of part d with Eq. (11-49a) for constant plate temperature and show that h is 41% higher for constant heat flux than for constant temperature.

11-26 Consider laminar film boiling from a vertical plate to a saturated liquid in which the liquid is immovable (as a result of its large viscosity relative to that of the vapor). A similarity solution to the describing boundary-layer equations [Eqs. (11-39)] and boundary conditions of Eqs. (11-40) for the vapor film is to be attempted.

(a) Show that if a stream function ψ such that $u = \partial\psi/\partial y$ and $v = -\partial\psi/\partial x$ is chosen to be

$$\psi = 4\nu Cx^{3/4}f$$

where $f = f(\eta)$ and $\eta = Cy/x^{1/4}$ with $C^4 = g(\rho_L - \rho)/4\nu^2\rho$, then $u = 4C^2\nu x^{1/2}f'$ and $v = C\nu x^{-1/4}(\eta f' - 3f)$. Here a prime denotes a derivative with respect to η.

(b) On the basis of the results of part a, show that the vapor x-motion and energy boundary-layer equations become

$$f''' + 3ff'' - 2(f')^2 + 1 = 0, \qquad 0 \leqslant \eta \leqslant \eta_\delta$$

$$\theta'' + 3\,\text{Pr}\,f\theta' = 0, \qquad 0 \leqslant \eta \leqslant \eta_\delta$$

where the dimensionless temperature is $\theta = (T - T_s)/(T_w - T_s)$.

(c) Show that the boundary conditions for the results of part b are

At $\eta = 0$ $\qquad f = 0 = f'$ $\qquad$ and $\qquad \theta = 1$

At $\eta = \eta_\delta$ $\qquad f' = 0 = \theta$ $\qquad$ and $\qquad \left[\text{Pr}\dfrac{C_p(T_w - T_s)}{h_{\text{fg}}}\right]\theta' = 3f$

where $\eta_\delta = C\delta/x^{1/4}$. Solution of the equations of part b requires that the unknown values of $f''(0)$, $\theta'(0)$, and η_δ be determined simultaneously in such a manner as to meet the conditions at the interface.

(d) Show that, if the unknown location of the interface is set at unity by employing the transformation $Z = \eta/\eta_\delta$, the results of parts b and c are obtained in the more convenient form of

$$\frac{d^3f}{dZ^3} + 3\eta_\delta f\frac{d^2f}{dZ^2} - 2\eta_\delta\left(\frac{df}{dZ}\right)^2 + \eta_\delta^3 = 0, \qquad 0 \leqslant Z \leqslant 1$$

$$\frac{d^2\theta}{dZ^2} + 3\,\text{Pr}\,\eta_\delta f\frac{d\theta}{dZ} = 0, \qquad 0 \leqslant Z \leqslant 1$$

$$f(0) = 0 = \frac{df(0)}{dZ} \qquad \text{and} \qquad \theta(0) = 1$$

$$\frac{df(1)}{dZ} = 0 = \theta(1) \quad \text{and} \quad \left[\text{Pr}\frac{C_p(T_w - T_s)}{h_{\text{fg}}}\right]\frac{d\theta(1)}{dZ} = 3\eta_\delta f(1)$$

in which $d^2f(0)/dZ^2$, $d\theta(0)/dZ$, and η_δ are to be determined. Even though liquid motion is still neglected, the influences of the two parameters Pr and $C_p(T_w - T_s)/h_{\text{fg}}$ are elucidated. And the exact solution is useful to ascertain the integral method accuracy. Consult Koh [96] for a treatment of drag-induced liquid motion and Nakayama and Koyama [109] for generalization to curved surfaces.

(e) Show that the local heat-transfer coefficient is given by

$$\text{Nu} = \frac{hx}{k} = -\left(\frac{Cx^{3/4}}{\eta_\delta}\right)\frac{d\theta(0)}{dZ}$$

11-27 The latent heat of vaporization h_{fg} used in Problem 11-22 needs correction to account for the fact that the flowing vapor film is superheated. The upward enthalpy flow in the vapor film at a vertical plate can be investigate to ascertain the required correction since

$$\int_0^\delta u\left[h_{\text{fg}} + C_p(T - T_s)\right] dy = \delta\bar{u}\left[h_{\text{fg}} + aC_p(T_w - T_s)\right]$$

in which $\bar{u}$ is the average vapor velocity and a is the fraction of the wall superheat that constitutes the correction. Show that the corrected latent heat of vaporization is $h_{\text{fg}} + C_p(T_w - T_s)/2$.

11-28 Consider a very hot surface of constant temperature that is immersed in a liquid. Discuss the manner in which the heat flux from the hot surface varies as the thickness of a poorly conducting material bonded onto the surface is increased. Note that if film boiling originally ensues, the intermediate insulating material can have a surface temperature sufficiently low to allow nucleate boiling to occur with a resultant increase in heat flux.

11-29 Briefly comment of the use of water–fuel mixtures to enhance breakup of droplets injected into an engine or combustor (see Cho et al. [128]).

REFERENCES

1. W. M. Rohsenow, *Ann. Rev. Fluid Mech.* **3**, 211–236 (1971).
2. S. Nukiyama, *J. Soc. Mech. Eng.* (*Japan*) **37**, 367–374 (1934); English translation: *Proc. ASME/JSME Thermal Engineering Joint Conference* **1**, 441–452 (1991) (see also pp. 437–440 of the same volume); English translation: *Int. J. Heat Mass Transfer* **27**, 959–970 (1984).
3. G. Leppert and C. C. Pitts, *Adv. Heat Transfer* **1**, 185–266 (1964).
4. J. W. Westwater, *Adv. Chem. Eng.* **1**, 1–31 (1958).
5. R. Cole, *Adv. Heat Transfer* **10**, 85–166 (1974).
6. P. Griffith and J. D. Wallis, *Chem. Eng. Prog. Symp.*, Ser. 30 **56**, 49–63 (1960).
7. L. D. Koffman and M. S. Plesset, *ASME J. Heat Transfer* **105**, 625–632 (1983).
8. F. D. Moore and R. B. Mesler, *AIChE J.* **7**, 620–624 (1961); *AIChE J.* **10**, 656–660 (1964); A. Kovacs and R. B. Mesler, *Rev. Sci. Instr.* **35**, 485–488 (1964); R. R. Olander and R. G. Watts, *ASME J. Heat Transfer* **91**, 178–180 (1969).

9. M. S. Plesset and S. S. Sadhal, *ASME J. Heat Transfer* **101**, 180–182 (1979).

10. W. M. Rohsenow, *Trans. ASME* **74**, 969–976 (1952).

11. E. L. Piret and H. S. Isbin, *Chem. Eng. Prog.* **50**, 305–311 (1954).

12. R. I. Vachon, G. H. Nix, and G. E. Tanger, *ASME J. Heat Transfer* **90**, 239–247 (1968).

13. J. N. Addoms, Heat transfer at high rates to water boiling outside cylinders, D.Sc. Thesis, Department of Chemical Engineering, Massachusetts Institute of Technology, Cambridge, MA, 1948.

14. M. T. Chichelli and C. F. Bonilla, *Trans. AIChE* **41**, 755–787 (1945).

15. K. Stephan and M. Abdelsalam, *Int. J. Heat Mass Transfer* **23**, 73–87 (1980).

16. K. Stephan and H. Auracher, *Int. J. Heat Mass Transfer* **24**, 99–107 (1981).

17. P. Berenson, *ASME J. Heat Transfer* **83**, 351–358 (1961).

18. Y. L. Tzan and Y. M. Yang, *ASME J. Heat Transfer* **112**, 207–212 (1990).

19. N. Nishikawa, H. Kusuda, K. Yamasaki, and K. Tanaka, *Bull. JSME* **10**, 328–338 (1967).

20. A. H. Fakeeha, Heat transfer on a vertical surface, M.S. Thesis, Chemical and Petroleum Engineering Department, University of Kansas, Lawrence, KS, 1982.

21. R. B. Mesler and G. Mailen, *AIChE J.* **23**, 954–957 (1977).

22. N. Zuber, *Trans. ASME* **80**, 711–720 (1958).

23. S. S. Kutateladze, *Zh. Tekh. Fiz.* **20**, 1389 (1950).

24. V. M. Borishanskii, *Zh. Tekh. Fiz.* **26**, 452 (1956) (see also AEC-tr-3405 translation from *Problems of Heat Transfer during a Change of State*, by S. S. Kutateladze, 1959).

25. N. Zuber, M. Tribus, and J. W. Westwater, *Int. Developments in Heat Transfer*, ASME, No. 27, 1961, pp. 230–236.

26. Y. Elkassabgi and J. H. Lienhard, *ASME J. Heat Transfer* **110**, 479–486 (1988).

27. J. H. Lienhard and R. Eichhorn, *Int. J. Heat Mass Transfer* **19**, 1135–1142 (1976).

28. A. Sharan, J. H. Lienhard, and R. Kaul, *ASME J. Heat Transfer* **107**, 392–397 (1985).

29. J. Lienhard, *ASME J. Heat Transfer* **110**, 1271–1286 (1988).

30. J. H. Lienhard and M. M. Hasan, *ASME J. Heat Transfer* **10**, 276–279 (1979).

31. J. H. Lienhard, V. K. Dkir, and D. M. Riherd, *ASME J. Heat Transfer* **95**, 477–482 (1973).

32. A. Aoki and J. R. Welty, *ASME J. Heat Transfer* **92**, 542–544 (1970).

33. F. W. Schmidt, D. F. Torok, and G. E. Robinson, *ASME J. Heat Transfer* **89**, 289–294 (1967).

34. A. E. Bergles, *ASME J. Heat Transfer* **91**, 152–154 (1969).

35. G. M. Fuls and G. E. Geiger, *ASME J. Heat Transfer* **92**, 635–640 (1970).

36. J. Berghmans, *Int. J. Heat Mass Transfer* **19**, 791–797 (1976).

37. P. Cooper, *ASME J. Heat Transfer* **112**, 458–464 (1990).

38. M. Markels and R. L. Durfee, *AIChE J.* **10**, 106–110 (1964).

39. M. Markels and R. L. Durfee, *AIChE J.* **11**, 716–723 (1965).

40. T. B. Jones, ASME Paper 76-WA/HT-48 (1976); *Adv. Heat Transfer* **14**, 107–148 (1978); T. B. Jones and R. C. Schaeffer, *AIAA J.* **14**, 1759–1765 (1976).

41. J. M. Crowley, *AIAA J.* **15**, 734–736 (1977).

42. L. Y. Wagner and P. S. Lykoudis, *Int. J. Heat Mass Transfer* **24**, 635–643 (1981).

43. W. M. Rohsenow, in *Heat Transfer*, University of Michigan Press, 1953.

44. S. Yilmaz and J. W. Westwater, *ASME J. Heat Transfer* **102**, 26–31 (1980).

45. S. S. Kutateladze, *Int. J. Heat Mass Transfer* **4**, 31–45 (1961).

46. R. M. Fand, K. K. Keswani, M. M. Jotwani, and R. C. C. Ho, *ASME J. Heat Transfer* **98**, 395–400 (1976).

47. S. G. Kandlikar, *ASME J. Heat Transfer* **112**, 219–228 (1990).

48. W. H. Lowdermilk, C. D. Lanzo, and B. L. Siegel, Investigation of boiling burnout and flow stability for water flowing in tubes, NACA TN 4382, 1958.

49. E. K. Kalinin, I. I. Berlin, and V. V. Kostyuk, *Adv. Heat Transfer* **11**, 51–197 (1975).

50. P. Griffith, in W. Rohsenow, J. Hartnett, and E. Ganic, Eds., *Handbook of Heat Transfer Fundamentals*, 2nd ed., McGraw-Hill, 1985, pp. 13-1–13-41.

51. R. W. Lockhart and R. C. Martinelli, *Chem. Eng. Prog.* **45**, 39–48 (1947); A. Rouet, *Int. J. Heat Mass Transfer* **26**, 145–146 (1982).

52. J. R. S. Thom, *Int. J. Heat Mass Transfer* **7**, 709–724 (1964).

53. A. P. Fraas, *Heat Exchanger Design*, 2nd ed., Wiley-Interscience, 1989, pp. 87–126.

54. P. J. Berenson, *Int. J. Heat Mass Transfer* **5**, 985–999 (1962).

55. F. S. Pramuk and J. W. Westwater, *Chem. Eng. Prog. Symp.*, *Ser. 18* **52**, 79–83 (1956).

56. C. Pan and T. L. Lin, *Int. J. Heat Mass Transfer* **35**, 1355–1370 (1991).

57. A. J. Lowery and J. W. Westwater, *Ind. Eng. Chem.* **49**, 1445–1448 (1957).

58. T. Dunskus and J. W. Westwater, *Chem. Eng. Prog. Symp.*, *Ser. 32* **57**, 173–181 (1961).

59. D. P. Jordan, *Adv. Heat Transfer* **5**, 55–128 (1968).

60. Y. P. Chang, *ASME J. Heat Transfer* **81**, 1–12 (1959).

61. P. J. Berenson, *ASME J. Heat Transfer* **83**, 351–358 (1961).

62. J. H. Lienhard and V. K. Dhir, *ASME J. Heat Transfer* **102**, 457–460 (1980).

63. V. K. Dhir, J. N. Castle, and I. Catton, *ASME J. Heat Transfer* **99**, 411–418 (1977).

64. V. K. Dhir, *ASME J. Heat Transfer* **102**, 380–382 (1980).

65. K. Taghavi-Tafreshi, V. K. Dhir, and I. Catton, *ASME J. Heat Transfer* **101**, 318–325 (1979).

66. S. Yao and R. E. Henry *ASME J. Heat Transfer* **100**, 260–267 (1978).

67. D. Frost and B. Sturtevant, *ASME J. Heat Transfer* **108**, 418–424 (1986).

68. R. E. Henry and H. K. Fauske, *ASME J. Heat Transfer* **101**, 280–287 (1979).

69. G. S. Springer, *Adv. Heat Transfer* **14**, 281–346 (1978).

70. N. H. Afgan, *Adv. Heat Transfer* **11**, 1–49 (1975).

71. J. H. Lienhard, N. Shamsundar, and P. O. Biney, *Nucl. Eng. Design* **95**, 297–314 (1986).

72. J. G. Leidenfrost, De aquae comminis nonnullis qualitatibus tractatus, Duisburg, 1756; J. G. Leidenfrost (translated by C. Wares), *Int. J. Heat Mass Transfer* **9**, 1153–1166 (1966).

73. K. J. Baumeister, T. D. Hamill, F. L. Schwartz, and G. J. Schoessow, *Chem. Eng. Prog. Symp., Ser. 64* **62**, 52–61 (1966).

74. J. H. Lienhard, *Chem. Eng. Sci.* **31**, 847–949 (1976).

75. F. S. Gunnerson and A. W. Cronenberg, *ASME J. Heat Transfer* **100**, 734–737 (1978).

76. J. H. Lienhard and P. T. Y. Wong, *ASME J. Heat Transfer* **86**, 220–226 (1964).

77. J. H. Lienhard and K. Sun, *ASME J. Heat Transfer* **92**, 292–298 (1970).

78. J. H. Lienhard and L. C. Witte, *Rev. Chem. Eng.* **3**, 187–280 (1985).

79. L. A. Bromley, *Chem. Eng. Prog.* **46**, 221–227 (1950).

80. W. H. Rohsenow, *Trans. ASME* **78**, 1645–1648 (1956).

81. P. Pitschmann and U. Grigull, *Wärme-und Stoffübertragung* **3**, 75–84 (1970).

82. G. Hesse, E. M. Sparrow, and R. J. Goldstein, *ASME J. Heat Transfer* **98**, 166–172 (1976).

83. B. T. Lubin, *ASME J. Heat Transfer* **91**, 452–453 (1969).

84. A. Sakurai, M. Shiotsu, and K. Hatta, *ASME J. Heat Transfer* **112**, 430–450 (1990).

85. B. P. Breen and J. W. Westwater, *Chem. Eng. Prog.* **58** (1), 67–72 (1962).

86. M. L. Pomerantz, *ASME J. Heat Transfer* **86**, 213–219 (1964).

87. L. A. Bromley, N. R. LeRoy, and J. A. Robbers, *Ind. Eng. Chem.* **45**, 2639–2646 (1953).

88. F. I. Motte and L. A. Bromley, *Ind. Eng. Chem.* **49**, 1921–1928 (1957).

89. L. C. Witte, *Ind. Eng. Chem., Fundamentals* **7**, 517–518 (1968).

90. V. V. Klimenko, *Int. J. Heat Mass Transfer* **24**, 69–79 (1981).

91. P. Sadasivan and J. H. Lienhard, *ASME J. Heat Transfer* **109**, 545–547 (1987).

92. M. E. Ellion, A study of the mechanism of boiling heat transfer, Memo 20-88, Jet Propulsion Laboratory, Pasadena, CA, 1954.

93. T. H. K. Frederking, *Zeitschrift für angewandte Mathematik und Physik* **14**, 207–218 (1963).

94. Y. Y. Hsu and J. W. Westwater, *Chem. Eng. Prog. Symp., Ser. 30* **56**, 15–24 (1960).

95. D. A. Labuntsov, *Teploenergetika* **10**, 60 (1963).

96. J. C. Y. Koh, *ASME J. Heat Transfer* **84**, 55–62 (1962).

97. E. M. Sparrow and R. D. Cess, *ASME J. Heat Transfer* **84**, 149–156 (1962).

98. F. Tachibana and S. Fukui, *International Developments in Heat Transfer*, ASME, 1961, p. 219.

99. T. H. K. Frederking and J. Hopenfeld, *Zeitschrift für angewandte Mathematik und Physik* **15**, 388–399 (1964).

100. R. D. Cess and E. M. Sparrow, *ASME J. Heat Transfer* **83**, 370–376 (1961).

101. W. S. Bradfield, R. O. Barkdoll, and J. T. Byrne, *Int. J. Heat Mass Transfer* **5**, 615–622 (1962).

102. R. D. Cess, *ASME J. Heat Transfer* **84**, 395 (1962).

103. R. D. Cess and E. M. Sparrow, *ASME J. Heat Transfer* **87**, 377–379 (1961).

104. A. Nakayama, *AIAA J.* **24**, 230–236 (1986).

105. N. B. Pilling and T. D. Lynch, *Trans. Amer. Inst. Mining Met. Eng.* **62**, 665–688 (1920).

106. H. Merte and J. A. Clark, *Adv. Cryog. Eng.* **7**, 546–550 (1961).

107. T. H. K. Frederking, R. C. Chapman, and S. Wang, in K. D. Timmerhaus, Ed., *International Advances in Cryogenic Engineering*, Plenum, 1965, Paper T-3.

108. T. H. K. Frederking and J. A. Clark, *Adv. Cryog. Eng.* **8**, 501–506 (1963).

109. A. Nakayama and H. Koyama, *ASME J. Heat Transfer* **108**, 490–493 (1986).

110. T. H. K. Frederking and D. J. Daniels, *ASME J. Heat Transfer* **88**, 87–93 (1966).

111. R. C. Hendricks and K. J. Baumeister, Film boiling from submerged spheres, NASA TN D-5124, 1969.

112. E. Marschall and L. C. Farrar, *Int. J. Heat Mass Transfer* **18**, 875–878 (1975).

113. L. C. Farrar and E. Marschall, *ASME J. Heat Transfer* **98**, 173–177 (1976).

114. D. A. DiCicco and R. J. Schoenhals, *ASME J. Heat Transfer* **86**, 457–461 (1964).

115. L. C. Burmeister and R. J. Schoenhals, *Prog. Heat Mass Transfer* **2**, 371–394 (1969).

116. W. E. Schmidt and L. C. Witte, *ASME J. Heat Transfer* **94**, 491–493 (1972).

117. F. S. Gunnerson and A. W. Cronenberg, *ASME J. Heat Transfer* **102**, 335–341 (1980).

118. A. Yabe and H. Maki, *Int. J. Heat Mass Transfer* **31**, 407–417 (1988).

119. D. R. Cash, G. J. Klein, and J. W. Westwater, *ASME J. Heat Transfer* **93**, 19–24 (1971).

120. D. L. Bondurant and J. W. Westwater, *Chem. Eng. Prog. Symp.*, Ser. *113* **67**, 30–37 (1971).

121. G. R. Rubin, L. I. Royzen, and I. N. Dul'kin, *Heat Transfer Soc. Res.* **3**, 130–134 (1971); from *Inzehenerno-Fisicheskii Zhurnal* **20**, 26–30 (1971).

122. C. C. Shih and J. W. Westwater, *Int. J. Heat Mass Transfer* **15**, 1965–1968 (1972).

123. R. L. Webb, *Heat Transfer Eng.* **2**, 46–69 (1981).

124. A. E. Bergles, *ASME J. Heat Transfer* **10**, 1082–1096 (1988).

125. J. R. Thome, *Enhanced Boiling Heat Transfer*, Hemisphere, 1990.

126. H. C. Ünal, *Int. J. Heat Mass Transfer* **29**, 640–644 (1986).

127. S. G. Kandlikar, *ASME J. Heat Transfer* **113**, 966–972 (1991).

128. P. Cho, C. K. Law, and M. Mizomoto, *AMSE J. Heat Transfer* **113**, 272–274 (1991).

129. E. M. Sparrow, *Int. J. Heat Mass Transfer* **7**, 229–238 (1964).

12

CONDENSATION

In many respects condensation on a surface resembles natural convection in a single-phase fluid, just as is the case for boiling. But, as remarked in Chapter 11 for boiling, the change in density that accompanies a change from the vapor to the liquid phase gives larger body forces per unit volume. Hence more vigorous motion than for a single-phase fluid can occur by natural means.

Because condensation generally still is characterized by low velocities, forces other than gravity can play an important role. Centrifugal forces and shear stresses due to externally forced vapor motion can induce motion of a condensed liquid film, for example, as can electromagnetic fields. Surface tension also can generate liquid motion when the liquid—vapor interface has curvature or when surface tension varies over the interface due to gradients in temperature or chemical composition.

When condensation occurs on a cooled solid surface, the condensed liquid either forms a continuous film on the surface or coalesces into droplets. The former behavior, called *film condensation*, is the more common; the latter, called *dropwise condensation*, is accompanied by higher condensation rates but is difficult to achieve continuously. Examples of equipment in which condensation occurs on a cooled solid surface are condensers for Rankine power generation cycles and vapor compression refrigeration cycles, dehumidifiers for air conditioning, and heat pipes. Condensation sometimes changes a vapor into a solid as in vacuum deposition of metals and in cryopumping. High heat-transfer coefficients typically accompany boiling and condensation.

Condensation can occur away from solid boundaries by homogeneous bulk nucleation in a pure fluid or by heterogeneous nucleation on small foreign particles. Expansion in a steam turbine nozzle accompanied by spontaneous formation of liquid nuclei that grow rapidly (leading to a condensation shock), the events in a Wilson cloud chamber, and expansion of a vapor in space are examples of the former. Water droplets that form around dust, salt particles, or AgI particles in fogs, clouds, and rocket exhaust plumes [1] are examples of the latter.

Surveys by Merte [2] and Fajii [3] can be consulted for details and references beyond those in the following treatment.

12.1 SURFACE TENSION

Liquids differ from gases by having larger cohesive forces between molecules that tend to restrict separation of molecules. One consequence of this difference between liquids and gases is that a body of liquid tends to contract to the configuration of smallest surface area, or surface energy, as sketched in Fig. 12-1. At the surface a molecule is subjected to tangential forces F_s that cancel and a larger inward attractive force F_i to the many nearby liquid molecules than the small outward attractive force F_0 to the few distant gas molecules. This net inward force causes inward movement of surface molecules until the maximum number are in the interior, leading to a surface of minimum area. In water, these forces have their origin in the fact that hydrogen atoms in one molecule attract oxygen atoms of neighboring molecules.

Increase of the surface area above this minimum requires exertion of external force. As this external force moves through the required displacement, it does work to bring additional molecules to the surface from the interior. The liquid surface, as a result, possesses a surface energy that could do work in the process of moving back to the minimum-surface-area configu-

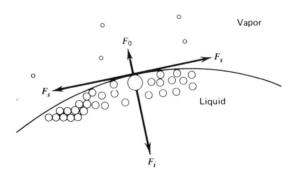

Figure 12-1 Origin of surface tension in strong inward attractive force between many nearby liquid molecules and a surface liquid molecule greatly exceeding weak outward attractive force between few distant exterior vapor molecules.

ration. The surface free energy per unit area, commonly called the *surface tension*, has dimensions of energy/area, which is equivalent to force-length/length^2 or force/length.

It can be appreciated that a flat liquid surface can be due only to the predominant effect of some force, such as gravitational attraction, other than surface tension forces. The Bond number Bo characterizes the relative magnitude of gravitational and body forces [4] according to

$$Bo = \frac{F_{gravity}}{F_{surface\ tension}} = \frac{(\rho_L - \rho_v)L^3 g/g_c}{\sigma L} = \frac{(\rho_L - \rho_v)L^2 g}{\sigma g_c}$$

where L is a characteristic length of the system considered. If a liquid surface is curved, a pressure difference exists across it. The magnitude of this pressure difference can be ascertained by consideration of the situation sketched in Fig. 12-2 where a small interfacial area is shown. The normal force exerted by the pressure difference acting on the projected area of the interface is $(P_1 - P_2)2R_1 \sin(\Delta\theta)2R_2 \sin(\Delta\phi)$. The normal force due to surface tension pulling tangentially is $\sigma R_1 \Delta\theta \sin \Delta\phi + \sigma R_2 \Delta\phi \sin \Delta\theta$. Equation of these two normal forces and realization that small interfacial areas require that $\Delta\phi, \Delta\theta \to 0$ leads to

$$P_1 - P_2 = \sigma\left(\frac{1}{R_1} + \frac{1}{R_2}\right) \tag{12-1}$$

A more precise derivation of Eq. (12-1) [5] would have allowed the interface to move a small distance and then equated the work done by the pressure forces to the change in surface free energy. In such a derivation the question of the variation of surface tension σ with interface radius occurs. Tolman, as well as Kirkwood and Buff, proposed [6], on the basis of statistical mechanics considerations, that $\sigma = \sigma_\infty(1 + \delta/r)$ where σ_∞ is the surface tension at a flat surface, r is the interface curvature, and δ is a length between 0.25 and 0.6 of the liquid molecular radius. Surface tension magnitude depends on whether the liquid is in contact with its own vapor or a second fluid.

For most substances, the surface tension decreases as temperature increases. The derivation of Eq. (12-1) suggests that a temperature variation along an interface would result in surface tension variation and a consequent force imbalance that could induce liquid motion. The Marangoni number Ma [129] is the ratio of the imbalance in surface tension forces to liquid tangential viscous forces according to

$$Ma = \frac{\Delta F_{surface\ tension}}{F_{tang\ viscous}} = \frac{(L\,d\sigma/dT)(L\nabla T)}{\mu_L VL^2/L}$$

where L is a characteristic system dimension (which could be a boundary-layer

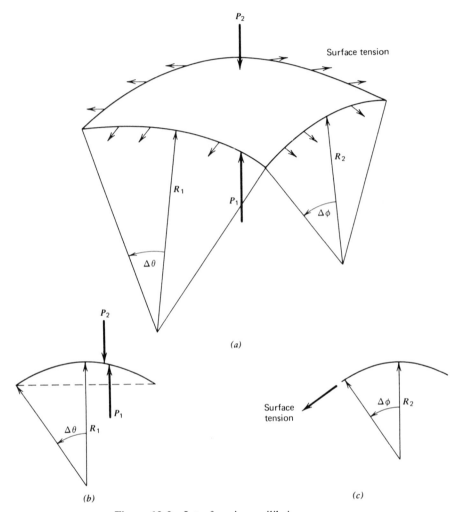

Figure 12-2   Interface in equilibrium.

thickness) and ∇T is a temperature gradient on the interface. Reviews on surface tension effects are available [7, 8]. Inasmuch as the velocity V is not specified, it is reasonable to relate it to the stabilizing effect of the liquid thermal conductivity, which tends to equalize surface temperatures. The distances δ a step change in temperature propagates by diffusion through a medium was found in Chapter 3 to be given by

$$\delta \sim (\alpha_L t)^{1/2} \tag{3-53}$$

If the derivative is taken with respect to time and allowed to be the characteristic velocity, then $\dot{\delta} \sim \alpha_L/\delta$, which suggests that the characteristic velocity is represented by $V \sim \alpha_L/L$. Hence the Marangoni number is [4]

$$\text{Ma} = \frac{L^2 \nabla T (d\sigma/dT)}{\mu_L \alpha_L}$$

Fluid flows driven by surface tension effects at low-gravity conditions would be encountered in the processing of materials in outer space. There bubble migration through a liquid could be influenced by temperature gradients [9] that could cause liquid motion whenever a free surface exists [10, 11] even in a horizontal liquid layer cooled from below [12].

Surface tension σ is related to latent heat of vaporization h_{fg}. To demonstrate this, surface tension is interpreted as a surface energy per unit area. Then

$$\sigma = u(z - z')n'$$

where u is the mutual potential energy between two molecules of mean spacing r, z is the number of nearest neighbors within the bulk liquid, z' is the number of nearest neighbors at the interface $z' \sim z/2$, and n' is the number of molecules per unit area at the interface $(n' \sim 1/r^2)$. Also, evaporation requires that the potential energy between the surface molecules and those beneath must be overcome. This energy is the latent heat of vaporization and can be expressed as

$$\rho h_{fg} = \frac{nzu}{2}$$

where ρ is the mass density of the liquid, n is the number of molecules per unit volume $(n \sim 1/r^3)$, and the factor of 2 accounts for the attractive forces only acting from below. Division of this expression for h_{fg} into that for σ gives

$$\frac{\sigma}{\rho h_{fg}} \approx r \sim 10^{-10} \text{ m} \qquad (12\text{-}2)$$

since the intermolecular spacing r of most liquids is of the order of 10^{-10} m.

The tensile stress (negative pressures can be sustained in a liquid [13]) required to rupture a liquid can be estimated by considering surface tension σ to be surface energy per unit area. Formation of surface area A consequently requires energy $2\sigma A$. If rupture occurs when liquid molecules are separated by a distance that is of the order of the molecular spacing, the work $P_{rupt} A r$ done by the rupturing force must equal that stored on the newly created two surfaces. Hence

$$P_{rupt} = \frac{2\sigma}{r} \qquad (12\text{-}3)$$

The fact that cavitation occurs, usually at a solid surface, at lower values than the upper bound of Eq. (12-3) is due to the effects of dissolved gases, oil films, and differing intermolecular forces between liquid and solid.

12.2   BULK CONDENSATION

Surface tension plays an important role in bulk condensation. In particular, it is involved in the determination of the critical liquid droplet size, which is in equilibrium with the surrounding vapor. To see this, consider a liquid droplet of radius r at pressure P_L and temperature T in equilibrium with a vapor at pressure P_v and temperature T, which is also in equilibrium with a pool of flat-surface liquid as sketched in Fig. 12-3. From Eq. (12-1) the pressure in the droplet exceeds that in the vapor, according to

$$P_L = P_v + \frac{2\sigma}{r}$$

in which $2\sigma/r$ can be considered to be a pressure differential ΔP.

The total free energy G of the system is [14]

$$G = n_v g_v(P_v, T) + n_L g_L(P_v, T) + 4\pi r^2 \sigma \qquad (12\text{-}4)$$

in which n_v and n_L are the number of moles of vapor and liquid whereas g_v and g_L are Gibbs free energy per mole of vapor and liquid, respectively. The last term is the surface free energy, representing the work required to form a surface area $4\pi r^2$ with constant surface tension. At thermodynamic equilibrium the system free energy is at a minimum. Hence

$$dG = n_v\, dg_v + g_v\, dn_v + n_L\, dg_L + g_L\, dn_L + 4\pi\sigma\, dr^2 = 0$$

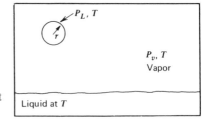

Figure 12-3   Vapor in equilibrium with flat liquid pool and a liquid droplet.

Since the number of moles is constant, $dn_v = -dn_L$. Also, $n_L = 4\pi r^3/3v_L$. At constant temperature and pressure neither g_v nor g_L varies. As a result of these three observations, the expression for dG shows that at equilibrium

$$g_v(P_v, T) = g_L(P_v, T) + \frac{2v_L\sigma}{r} \tag{12-5}$$

From Eq. (12-5), the value of P_v required to maintain the drop in equilibrium at the specified temperature T depends on the drop radius r. To ascertain this dependence, Eq. (12-5) is differentiated with T constant to obtain

$$dg_v - dg_L = 2v_L\sigma\, d\!\left(\frac{1}{r}\right)$$

The thermodynamic relationship that $dg = v\,dP$ is then introduced to obtain

$$(v_v - v_L)\, dP_v = 2v_L\sigma\, d\!\left(\frac{1}{r}\right)$$

Recognizing that $v_v \gg v_L$ and taking the vapor to be a perfect gas $v_v = RT/P_v$, where R is the universal gas constant, one can restate this expression as

$$\frac{1}{P_v}\, dP_v = \frac{2v_L\sigma}{RT}\, d\!\left(\frac{1}{r}\right)$$

Integration between the limits of $r = \infty$, where $P_v = P_{sat}$, the saturation pressure normally associated with T, and the equilibrium radius of $r = r^*$ gives

$$\frac{P_v}{P_{sat}} = \exp\!\left(\frac{2v_L\sigma}{r^*RT}\right) \tag{12-6}$$

Equation (12-6) shows that a drop is in equilibrium at a specified temperature only if the pressure of the surrounding vapor exceeds what is normally considered to be its saturation pressure P_{sat}. If the surrounding vapor pressure is lower or greater than that required by Eq. (12-6), the drop will either shrink by evaporation or grow by condensation, respectively.

Equation (14-6) gives the supersaturation pressure ratio for a drop of size r^* to be in equilibrium. If the supersaturation pressure ratio is specified, the equilibrium drop size is obtained from a rearrangement of Eq. (12-6) as

$$r^* = \frac{2v_L\sigma/RT}{\ln(P_v/P_{sat})} \tag{12-7}$$

It is apparent that when a liquid–vapor interface is curved, equilibrium requires that the vapor be supersaturated. The amount of supersaturation is illustrated for water at 68°F in Table 12-1. If condensation is to occur with negligible supersaturation of the vapor, foreign nuclei must be available since it is unlikely that the number of molecules required (about 2.7×10^{11} for a 10^{-4}-in. drop) would come together spontaneously. On the other hand, when the vapor is extremely supersaturated only a few (approximately eight) liquid molecules comprise a drop of the corresponding equilibrium size and could come together spontaneously. The minimum drop size cannot be smaller than the average distance between liquid molecules; such a limitation allows an estimate of the maximum supersaturation ratio to be $P_v/P_{sat} \approx 10^2$ as is the subject of Problem 12-4. An equilibrium of a drop with its vapor is unstable since the drop will grow if its diameter suddenly experiences either an infinitesimal size increase as a result of condensation or a sudden infinitesimal size decrease due to evaporation.

The formation of vapor cavities in a bulk liquid (a boiling phenomenon) is similar to the formation of liquid drops in a bulk vapor (a condensation phenomenon). This similarity is explored in Problem 12-7.

Bulk condensation is important in the flow of a vapor through a turbine nozzle since droplets can seriously erode turbine blades. The fluids involved are often water but are occasionally liquid metals. Hill et al. [15] applied classical liquid drop nucleation theory to the calculation of the rate per unit volume J at which molecules are added to a drop of equilibrium size. They found, a refined form accounting for difference liquid and vapor temperatures was used by Skillings and Jackson [16] to study nucleating and condensing steam flows,

$$ J = \left(\frac{P_v}{KT} \right)^2 \frac{M}{N_A \rho_L} \left(\frac{2\sigma}{\pi m} \right)^{1/2} \exp\left(\frac{-4\pi\sigma r^{*2}}{3KT} \right) \tag{12-8} $$

where K is the Boltzmann constant, M is molecular weight, N_A is Avogadro's number, and r^* is given by Eq. (12-7). The rate per unit volume at which molecules are added to a drop for water at 35°F versus the supersaturation pressure ratio P_v/P_{sat} is shown in Fig. 12-4. There it is seen that the curve is so steep at low values of J that a critical supersaturation pressure ratio can

TABLE 12-1   Water Drop Diameter and Supersaturation Pressure Ratio at 68°F

Diameter, in.	Number of Molecules	Supersaturation Ratio
10^{-2}	2.7×10^{17}	1.000009
10^{-4}	2.6×10^{11}	1.00086
10^{-6}	2.7×10^5	1.0905
10^{-7}	2.7×10^2	2.38

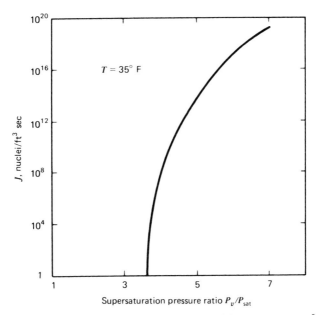

Figure 12-4 Rate of formation of condensation nuclei in water vapor. [From P. G. Hill, H. Witting, and E. P. Demetri, *ASME J. Heat Transfer* **85**, 303–317 (1963) [15].]

be defined above which condensation occurs spontaneously, a ratio that equals 4.2 for water at 35°F. This corresponds to a critical drop radius of 6.5×10^{-10} m, having about 40 molecules of liquid, that is substantially less than a mean free path 10^{-8} m in the vapor. The data due to Volmer and Flood [17] in Table 12-2 verify the prediction of Eq. (12-8). More recent data on condensation in nozzles [18] show that although water and many other fluids are accurately treated by Eq. (12-8), the Lothe–Pound theory [19] which accounts for surface tension variation with drop size and includes

TABLE 12-2 Supersaturation Pressure Ratio [17] P_v / P_{sat}

Vapor	Temperature, K	Measured	Calculated
Water	275.2	4.2 ± 0.1	4.2
Water	261.0	5.0	5.0
Methanol	270.0	3.0	1.8
Ethanol	273.0	2.3	2.3
I-Propanol	270.0	3.0	3.2
Isopropyl alcohol	265.0	2.8	2.9
n-Butyl alcohol	270.0	4.6	4.5
Nitromethane	252.0	6.0	6.2
Ethyl acetate	242.0	8.6–12.3	10.4

quantum-statistical effects must be used for NH_3. The condensation of water from still, moist air is not accurately predicted by these methods [15]. A review of nucleation theory [20] can be consulted for additional details.

12.3 DROPWISE CONDENSATION ON SOLID SURFACES

A vapor condensing on a solid surface can either form widely separated liquid drops, a mode termed *dropwise condensation*, or a uniform liquid film, a mode termed *film condensation*. In this respect there are similarities between condensation and boiling at solid surfaces. In both nucleate boiling and dropwise condensation, nucleation occurs at preferred nucleation sites, the number of which increases with increasing temperature difference—as does the heat flux, as illustrated in Fig. 12-5; the drops on which condensation begins originate in recesses in the surface where liquid remains after a drop leaves, much as for nucleate boiling. In both film boiling and film condensation, a layer of fluid acts as an insulator between the solid surface and the pool of fluid of unchanged phase. One difference is that a peak condensation heat flux, similar to the burnout heat flux in boiling, as temperature difference increases has yet to be observed. A second difference is that small vapor bubbles in nucleate boiling are readily detached from their surface of formation since buoyant forces are larger than surface forces

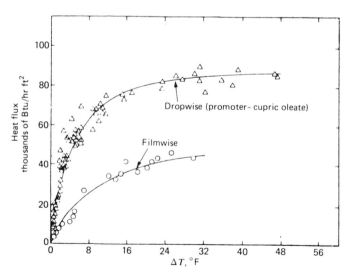

Figure 12-5 Comparison of drop and film condensation of steam at atmospheric pressure on a vertical copper surface. [By permission of the American Society of Mechanical Engineers and J. F. Welch and J. W. Westwater, *Proc. International Heat Transfer Conference, University of Colorado*, Part II, 1961, pp. 302–309 [21].]

holding the bubble to the surface, whereas small liquid drops in dropwise condensation often find gravity forces for possible removal smaller than the surface forces holding them to the surface. It has been speculated [2] that with a suitable liquid drop removal mechanism such as centrifugal force, a peak condensation heat flux could be achieved, beyond which individual drops will merge into continuous liquid film just as nucleate boiling merges into film boiling as the imposed temperature difference exceeds burnout conditions. A third difference is that film condensation rather than dropwise condensation might occur at low imposed temperature differences, in contrast to the usual case for boiling.

Generally, condensation on surfaces of industrial interest is film condensation. For this reason designers of heat exchange equipment often conservatively assume film condensation. As illustrated in Fig. 12-5, film condensation heat fluxes are lower than for dropwise condensation. Efforts consistently to have dropwise condensation in industrial equipment for lengthy periods have not been successful [22–24].

Prediction of the occurrence of dropwise condensation requires a determination of the ability of the liquid to wet the solid surface. Dropwise condensation is likely when the liquid cannot wet the surface, whereas film condensation is likely when the liquid can wet the surface. The judgment of wetting ability proceeds as follows. For the drop resting on a solid and surrounded by vapor as illustrated by Fig. 12-6, if gravity is negligible there are three forces, due to three surface tensions, in equilibrium: (1) σ_{Lv}, which acts in the liquid–vapor interface that has a contact angle θ with the solid; (2) σ_{Ls}, which is determined by the properties of the solid surface and the liquid and acts along the solid surface on the liquid side; and (3) σ_{vs}, which is determined by the properties of the solid surface and the vapor and acts along the solid surface on the vapor side. Horizontal equilibrium of these three forces requires that

$$\sigma_{vs} = \sigma_{Ls} + \sigma_{Lv} \cos \theta$$

which suggests that the surface properties of the solid are important to the determination of θ. A liquid is considered not to wet a surface if $\theta > 90°$ and

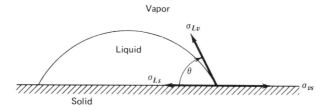

Figure 12-6 Equilibrium of surface tension forces for a liquid drop resting on a solid surface in a vapor.

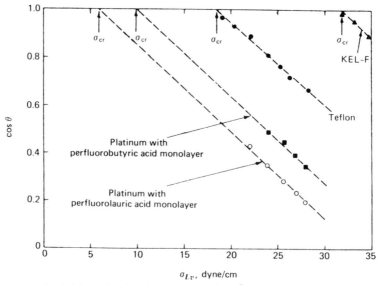

Figure 12-7 Definition of critical surface tension. [Reprinted with permission from E. G. Shafrin and W. A. Zisman, *J. Phys. Chem.* **64**, 519–524 (1960) [25]; © 1960, American Chemical Society.]

to wet a surface if $\theta < 90°$. Measurement of the contact angle for a series of homologous liquids on different surfaces results in an approximately linear relationship between $\cos(\theta)$ and σ_{Lv} as illustrated in Fig. 12-7. Extrapolation to $\theta = 0$, where the liquid can be considered to completely wet the surface, yields a critical surface tension σ_{cr} that is a characteristic of the surface alone. Dropwise condensation occurs when $\sigma_{Lv} > \sigma_{cr}$. Critical surface tensions for several solids are listed in Table 12-3. The fact that film condensation occurs with water on most metals suggests that most metals have

TABLE 12-3   Critical Surface Tensions of Some Solids [25]

Solid	σ_{cr}, dyne/cm
TFE (Teflon)	18
Polyethylene	31
Kel-F	31
Polystyrene	33
Polyvinyl chloride	39
Nylon	46
Platinum with perfluorobutyric acid monolayer	10
Platinum with perfluorolauric acid monolayer	6

$\sigma_{\rm cr} > \sigma_{Lv}$, although so-called promoters have been found to be effective on metals in promoting dropwise condensation. Probably the promoters form a layer that reduces $\sigma_{\rm cr}$ below σ_{Lv}.

As suggested in Table 12-3, it is possible to coat a metal surface on which film condensation normally occurs with a substance on which dropwise condensation naturally occurs. Tests by Brown and Thomas [26] showed that the change from film to dropwise condensation allowed by a 0.0001-in. film of polytetrafluorethylene on a horizontal 0.75-in. brass tube increased the overall heat-transfer coefficient for steam condensation by a factor of nearly 3. This procedure is similar to that of coating a very hot surface on which film boiling occurs with a poorly conducting material in order to achieve nucleate boiling to achieve a net increase in heat flux as is the subject of Problem 11-28. Noble metal coatings have also been tried [27] but are usually rejected on economic grounds. Organic coatings have increased condensing steam heat-transfer coefficients by factors of 8 [28].

To alter the properties of the solid surface by a promoter, the entire system must be saturated with the promoter (not just the condensing surface) since the vapor becomes contaminated and the promoter concentration must be maintained. Promoters tried include oleic, stearic, or linoleic acids, benzyl mercaptan, sulfur or selenium compounds, and long-chain hydrocarbons [29]. These promoters fail by surface fouling or oxidation, although operation for up to 1 year has been achieved [21]. It is important that noncondensable gases be removed for dropwise condensation heat fluxes to achieve their possible large values [30].

The measurements of dropwise condensation heat-transfer coefficients have been correlated by a number of workers [2]. For dropwise condensation of steam and ethylene glycol, Peterson and Westwater [31] recommend, as illustrated in Fig. 12-8,

$$\mathrm{Nu} = 1.46 \times 10^{-6} \, \mathrm{Re}^{-1.63} \, \mathrm{Pr}^{0.5} \, \pi_k^{1.16} \qquad (12\text{-}9)$$

for $1.65 \leqslant \mathrm{Pr} \leqslant 23.6$ and $7.8 \times 10^{-4} \leqslant \pi_k \leqslant 2.65 \times 10^{-2}$ where

$$\mathrm{Nu} = \frac{2\sigma T_v h}{h_{\rm fg} \rho_L k_L (T_w - T_v)} \qquad \mathrm{Pr} = \mu_L \frac{C_{p_L}}{k_L}$$

$$\mathrm{Re} = \frac{k_L (T_w - T_v)}{\mu_L h_{\rm fg}} \qquad \pi_k = \frac{2\sigma (d\sigma/dT) T_v}{\mu_L^2 h_{\rm fg}}$$

where $h_{\rm fg}$ is the latent heat of vaporization, the subscript L refers to liquid properties, and all properties are evaluated at the saturation temperature.

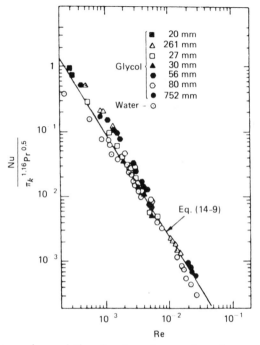

Figure 12-8 Improved correlation for dropwise condensation on vertical surfaces. [By permission from A. C. Peterson and J. W. Westwater, *Chem. Eng. Prog. Symp. Ser.* **62**, 135–142 (1966) [31].]

12.4   LAMINAR FILM CONDENSATION ON VERTICAL PLATES

As mentioned in Section 12-3, it is difficult to cause dropwise condensation to occur over long periods of time. Most commonly a solid condensing surface is covered by a liquid film that separates the solid surface from the vapor.

Initial discussion of film condensation begins with a vertical plate as illustrated in Fig. 12-9. Similarity to the film boiling discussion in Section 11.6, and illustration in Fig. 11-8 is apparent—the vapor and the liquid have merely exchanged positions. Indeed, the analysis of laminar film condensation accomplished by Nusselt [32] in 1916 was later applied to film boiling.

The constant-property boundary-layer equations for the condensate film are

$$\frac{\partial u_L}{\partial x} + \frac{\partial v_L}{\partial y} = 0 \tag{12-10a}$$

$$u_L \frac{\partial u_l}{\partial x} + v_L \frac{\partial u_L}{\partial y} = \frac{(\rho_L - \rho)g}{\rho_L} + \nu_L \frac{\partial^2 u_L}{\partial y^2} \tag{12-10b}$$

$$u_L \frac{\partial T_L}{\partial x} + v_L \frac{\partial T_L}{\partial y} = \alpha_L \frac{\partial^2 T_L}{\partial y^2} \tag{12-10c}$$

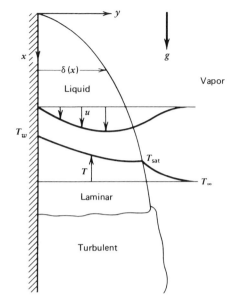

Figure 12-9 Boundary layer of condensate on a vertical flat plate.

Unsubscripted variables refer to vapor quantities. The boundary conditions are

At $y = \delta$

$$\rho_L\left(v_L - u_L\frac{d\delta}{dx}\right) = \rho\left(v - u\frac{d\delta}{dx}\right) \tag{12-11a}$$

$$-k_L\left(\frac{\partial T_L}{\partial y} - \frac{\partial T_L}{\partial x}\frac{\partial\delta}{\partial x}\right) = -k\left(\frac{\partial T}{\partial y} - \frac{\partial T}{\partial x}\frac{\partial\delta}{dx}\right) - h_{fg}\rho_L\left(v_L - u_L\frac{d\delta}{dx}\right) \tag{12-11b}$$

$$\mu_L\frac{\partial u_L}{\partial y} = \mu\frac{\partial u}{\partial y} \quad (12\text{-}11c), \quad u_L = u \quad (12\text{-}11d), \quad \text{and} \quad T_L = T_s \quad (12\text{-}11e)$$

At $y = 0$

$$u_L = 0 \quad (12\text{-}11f), \quad v_L = 0 \quad (12\text{-}11g), \quad \text{and} \quad T_L = T_w \tag{12-11h}$$

It is assumed for simplicity that the vapor is saturated and exerts no shear on the liquid so that there is no need to determine the temperature and velocity distributions in the vapor.

Solutions of high accuracy can be obtained from Eqs. (12-10) and (12-11) by employing similarity transformations to reduce them to ordinary differential equations that are then solved by numerical methods. Here, however, insight is first attained from an integral method, a simple version of which is the subject of Problem 12-25. As for film boiling, the integral form of the boundary-layer equations is, with $\eta = y/\delta$,

$$\frac{d\left(\delta \int_0^1 u_L^2 \, d\eta\right)}{dx} - u_{L,\delta} \frac{d\left(\delta \int_0^1 u_L \, d\eta\right)}{dx}$$
$$= \frac{(\rho_L - \rho)g\delta}{\rho} + \frac{\nu_L}{\delta}\left[\frac{du_L(1)}{d\eta} - \frac{du_L(0)}{d\eta}\right] \qquad (12\text{-}12)$$

which is the x-motion equation and

$$\frac{d\left[\delta \int_0^1 u_L(T - T_s)\, d\eta\right]}{dx} = \frac{\alpha_L}{\delta}\left[\frac{dT(1)}{d\eta} - \frac{dT(0)}{d\eta}\right] \qquad (12\text{-}13)$$

which is the energy equation. The integral form of the continuity equation is

$$v_{L,\delta} = -\frac{d\left(\delta \int_0^1 u_L \, d\eta\right)}{dx} + u_{L,\delta}\frac{d\delta}{dx} \qquad (12\text{-}14)$$

A liquid velocity profile is obtained by requiring the assumed polynomial

$$u_L = a_0 + a_1\eta + a_2\eta^2 + a_3\eta^3 + \cdots$$

to meet the conditions of no slip at the wall [Eq. (12-11f)], no shear at the liquid–vapor interface [Eq. (12-11c)], and to satisfy the x-motion equation [Eq. (12-10b)] at the wall. The result is

$$u_L = \frac{(\rho_L - \rho)g\delta^2}{\mu_L}\left[\eta - \frac{\eta^2}{2}\right] \qquad (12\text{-}15a)$$

An approximate polynomial temperature distribution is assumed as

$$T_L - T_s = b_0 + b_1\eta + b_2\eta^2 + b_3\eta^3 + b_4\eta^4 + \cdots$$

which is required to meet the conditions of no thermal jump at the wall [Eq. (12-11h)], equilibrium at the liquid–vapor interface [Eq. (12-11e)], conservation of energy at the liquid–vapor interface [Eq. (12-11b)], and to satisfy the energy equation [Eq. (12-10c)] at the wall. The result is

$$\frac{T_L - T_s}{T_w - T_s} = 1 - \left(1 + \frac{1 - C}{2}\right)\eta + \left(\frac{1 - C}{2}\right)\eta^3 \qquad (12\text{-}15b)$$

Introduction of the temperature and velocity profiles of Eq. (12-15) into the integral energy equation [Eq. (12-13)] gives

$$\frac{(\rho_L - \rho)g}{8\rho_L\nu_L}\left[1 - \frac{11}{30}(1 - C)\right]\delta\frac{d\delta^3}{dx} = \frac{3\alpha_L(1 - C)}{2}$$

This relation is rearranged into

$$\frac{1}{4}\frac{C_p(T_s - T_w)}{h_{fg}}\left[1 - \frac{11}{30}(1 - C)\right]C = 1 - C$$

Seeing that $C \approx 1$ for small $C_{p_L}(T_s - T_w)/h_{fg}$, the most commonly encountered case, leads to the approximation that $1 - C \approx C_p(T_s - T_w)/4h_{fg}$. The effect of condensation at the outer edge of the laminar boundary layer is small, causing the profiles to be "blown" toward the plate from the essentially linear form for temperature and the parabolic form for velocity as stagnant fluid is added.

The condensate film thickness follows from the definition of C as

$$\delta = 2^{1/2}\left[1 - \frac{1}{4}\frac{C_{p_L}(T_s - T_w)}{h_{fg}}\right]^{1/4}\left[\frac{x(T_s - T_w)\nu_L k_L}{gh_{fg}(\rho_L - \rho)}\right]^{1/4} \quad (12\text{-}16)$$

The local heat-transfer coefficient found from

$$q_w = -k\frac{\partial T(0)}{\partial y} = \frac{k}{\delta}(T_w - T_s)\left[1 + \frac{(1 - C)}{2}\right]$$

and Eq. (12-16) to be

$$h = \frac{q_w}{T_w - T_s} = \frac{1}{2^{1/2}}k\left[\frac{g(\rho_L - \rho)\,\text{Pr}}{x\rho_L\nu_L^2}\frac{h_{fg}}{C_{p_L}(T_s - T_w)}\right]^{1/4}K \quad (12\text{-}17a)$$

Here $K \approx [1 + \frac{3}{4}C_{p_L}(T_s - T_w)/h_{fg}]^{1/4}$; Koh [33] and Rohsenow [34] found a coefficient of 0.68 instead of $\frac{3}{4}$, whereas Sadasivan and Lienhard [35] refined the estimate to $0.683 - 0.228/\text{Pr}$ for $0.5 \le \text{Pr} < \infty$. A general integral treatment was provided by Churchill [36]. The average heat-transfer coefficient $\bar{h}$ is related to the local value by $\bar{h} = 4h/3$. Hence

$$\overline{\text{Nu}} = \frac{\bar{h}x}{k} = 0.943\,\text{Ra}^{*1/4} \quad (12\text{-}17b)$$

whose coefficient best fits measurements if set to 1.13 as discussed later and where the Rayleigh number is $\text{Ra}^* = g(\rho_L - \rho)\text{Pr}^* x^3/\rho_L\nu_L^2$ and $\text{Pr}^* = \mu[3C_p/4 + h_{fg}/(T_s - T_w)]/k$. Liquid property variations with temperature are taken into account by evaluating liquid properties at $T_{\text{ref}} = T_w + (T_s - T_w)/4$, according to Poots and Miles [37] and Lott and Parker [38].

The temperature difference on which the transfer coefficient depends in Eq. (12-16) might not be specified. In such cases a form that employs the condensate flow rate, often specified when temperature difference is not, is convenient. This form, the derivation of which is the subject of Problem 12-12, is

$$\frac{\overline{h}}{k}\left(\frac{\nu_L^2}{g}\right)^{1/3} = 1.47\left(\frac{\rho_L - \rho}{\rho_L}\right)^{1/3} \text{Re}^{-1/3} \tag{12-18}$$

where $\text{Re} = 4\dot{m}/\mu_L$ is a Reynolds number and $\dot{m}$ is the condensate mass flow rate per unit width of the vertical plate. Best agreement with data is found with a coefficient of 1.88, as discussed later.

Equations (12-17) and (12-18) can be applied to film condensation from the upper surface of plates inclined at an angle θ from the horizontal if the gravitational component $g \sin \theta$ parallel to the plate is used. Gerstmann and Griffith [39] treated film condensation on the lower surface of inclined plates, whereas Leppert and Nimmo [40] gave results for the extreme case of condensation atop a horizontal strip; surface tension has an effect in both cases.

The accuracy of Eqs. (12-17) and (12-18) can be judged from the measurements for vertical plates shown in Fig. 12-10. Heat-transfer coefficients can be as much as 50% above or below the predicted value. It is believed that rough surfaces often retain the condensate, thus thickening the boundary

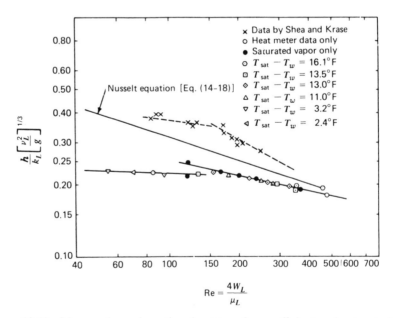

Figure 12-10 Measured condensation heat-transfer coefficients of saturated and superheated steam on a vertical surface. [By permission from D. L. Spencer and W. E. Ibele, *Proc. Third International Heat Transfer Conference*, Vol. 2, 1966, pp. 337–347 [41].]

layer and reducing the coefficient of heat transfer below the predicted value. Heat-transfer coefficients above the predicted value may be due to surface waves, an instability that not only increases surface area and intermittently thins the liquid film but also promotes mixing. These surface waves are formed when $Re = 4\dot{m}/\mu_L \geqslant 33$ and are influenced by surface tension, as Kapitsa [42] found. For vertical tubes, for which Eqs. (12-17) and (12-18) should apply for either exterior or interior condensation as long as the condensate film is thin relative to the tube diameter. Selin [43] recommends that the coefficients of Eqs. (12-17) and (12-18) be changed to 1.13 and 1.88, respectively, suggesting that the effects of surface waves are usually predominant.

Assumption of negligible vapor drag can cause over overestimation of condensation heat-transfer coefficients in some cases. Koh et al. [44] accounted for interfacial vapor drag during laminar film condensation of a stagnant saturated vapor on a vertical plate in a similarity solution of the boundary-layer equations. Problem 12-13 deals with this topic. Vapor drag has little effect for large liquid Prandtl numbers, as shown in Fig. 12-11, reducing local transfer coefficients by less than 10% for $Pr = 1$ and having a diminishing effect for increasing Pr. For small liquid Prandtl numbers, such as are characteristic of liquid metals (see Fig. 12-12), vapor drag is influential. Chen [45] employed a perturbed integral technique to solve the describing equations, finding that the effect of all factors on the average heat-transfer

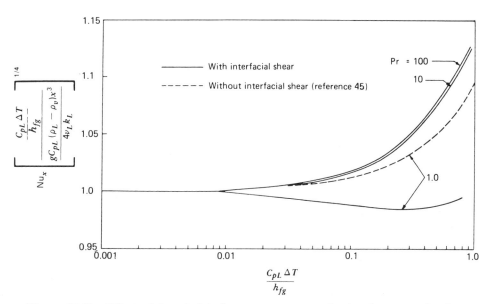

Figure 12-11 Effect of interfacial shear stress on condensing heat transfer for $Pr > 1$. [By permission from J. C. Y. Koh, E. M. Sparrow, and J. P. Hartnett, *Int. J. Heat Mass Transfer* **2**, 69–82 (1961) [44].]

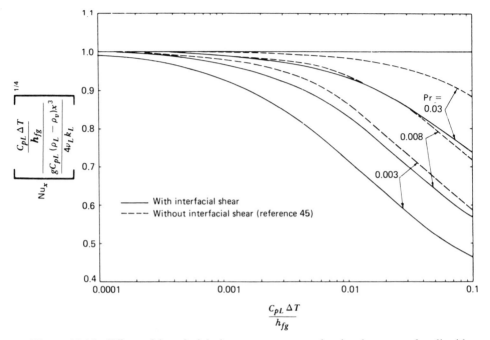

Figure 12-12 Effect of interfacial shear stress on condensing heat transfer, liquid metal range of Pr. [By permission from J. C. Y. Koh, E. M. Sparrow, and J. P. Hartnett, *Int. J. Heat Mass Transfer* **2**, 69–89 (1961) [44].]

coefficient is closely predicted by

$$\left(\frac{\bar{h}}{\bar{h}_{\text{Nusselt}}}\right)^4 = \frac{1 + (0.68 + 0.02/\text{Pr})A}{1 + (0.85 - 0.15A)A/\text{Pr}} \qquad (12\text{-}19)$$

subject to the restrictions that $A \leqslant 2$ for $\text{Pr} \geqslant 1$ and $A \leqslant 20\,\text{Pr}$ for $\text{Pr} \leqslant 0.05$ —the cases of $0.05 < \text{Pr} < 1$ are excluded—in which $A = C_p(T_s - T_w)/h_{\text{fg}}$; the $(\rho\mu)_L/(\rho\mu)_v$ parameter that arises in an exact analysis has only a second-order affect on the heat transfer [44, 45]. Chen found that measurements for liquid metals were from 0.3 to 0.9 of his predictions, a discrepancy seemingly not explained by the hypothesis that the fraction of vapor molecules striking a surface and actually condensing is less than unity; see Rohsenow [46] for discussion of the depression of the interface temperature below the saturation temperature due to kinetic-theory-of-gases effects, usually negligible except for liquid metals, that give an apparent interfacial resistance. The data compiled by Sukhatme and Rohsenow [47] are shown in Fig. 12-13.

The time t_s required to achieve steady condensation after a step change in wall temperature from T_s to T_w was found by Sparrow and Siegel [48] for

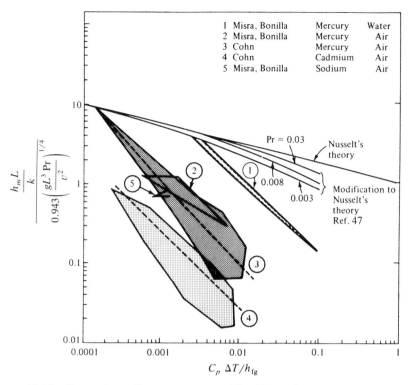

Figure 12-13   Comparison of measurements of liquid metal condensation rates with Nusselt's theory and modifications of it. [From S. P. Sukhatme and W. M. Rohsenow, *ASME J. Heat Transfer* **88**, 19–28 (1966) [47].]

vertical plates to be, with $h'_{\text{fg}} = h_{\text{fg}} + 3C_p(T_s - T_w)/8$,

$$t_s = \left[\frac{h_{\text{fg}} \rho \mu x}{kg(T_s - T_w)(\rho_L - \rho)} \right]^{1/2} \left[1 + \frac{C_p(T_s - T_w)}{8h_{\text{fg}}} \right] \qquad (12\text{-}20)$$

During this transition period the condensate film thickness δ and the local heat-transfer coefficient h vary as

$$\delta = \begin{cases} \delta_0 \left(\dfrac{t}{t_s} \right)^{1/2} \\ \delta_0 \end{cases} \qquad h = \begin{cases} h_0 \left(\dfrac{t}{t_s} \right)^{-1/2}, & t \leqslant t_s \\ h_0, & t \geqslant t_s \end{cases}$$

where δ_0 and h_0 are given by Eqs. (12-16) and (12-17), respectively. An integral method and the method of characteristics, the subject of Problem 12-15, were employed. Later Wilson [49] employed the method of characteristics to achieve solutions for transient condensation on horizontal cylinders, steady condensation and transient evaporation from inclined cylinders, and steady condensation on upward-pointing cones. For a horizontal cylinder, it is found that the instantaneous film thickness atop the cylinder approaches its steady-state value δ_0 (see Problem 12-26) asymptotically as

$$\frac{\delta^2}{\delta_0^2} = \tanh\frac{t}{2t_s}$$

where $t_s = 3\nu_L D/8g\delta_0^2$. See Flick and Tien and Reed et al. [50] for more detail.

Analytical studies for a vertical plate [51, 52] confirm that vapor superheat generally increases condensation rates by less than 10%, an effect that is appreciable only if the wall and saturation temperatures differ by less than about 3°C. To account for the effect of vapor superheat, an effective heat of vaporization can be employed as

$$h_{fg_{eff}} = h_{fg} + C_{p_v}(T_v - T_s)$$

and used in relationships for saturated vapor.

Admission of vapor into a condenser is often accompanied by vapor velocities high enough, because of the disparity between vapor and liquid densities, for shear at the interface to appreciably either aid or oppose condensate motion. Rohsenow et al. [53] showed the effect of vapor shear in the following way. The conditions are as shown in Fig. 12-9 with the addition of a vapor shear τ_v at the outer edge of the condensate film. The liquid x-motion equation is

$$\frac{d^2u}{dy^2} = -\frac{(\rho_L - \rho)g}{\rho_L \nu_L}$$

subject to $u(0) = 0$ and $du(\delta)/dy = \tau_v/\mu_L$, neglecting convective effects. The energy equation, also neglecting convective effects, is

$$\frac{d^2T_L}{dy^2} = 0$$

subject to $T_L(0) = T_w$ and $T_L(\delta) = T_s$. The approximate velocity and temperature distributions are then

$$u = \frac{(\rho_L - \rho)g\delta^2}{\rho_L \nu_L}\left(\eta - \frac{\eta^2}{2}\right) + \frac{\tau_0}{\mu_L \delta}\eta$$

and

$$\frac{T - T_s}{T_w - T_s} = 1 - \eta$$

An energy balance at the liquid–vapor interface requires

$$-k_L \frac{\partial T_L(\delta)}{\partial y} = -h_{\text{fg}}\rho_L \frac{d\left(\delta \int_0^1 u\, d\eta\right)}{dx}$$

Introduction of the velocity and temperature profiles yields

$$\delta_0^4 = \left[1 + \frac{4\tau_v}{3g(\rho_L - \rho)\delta}\right]\delta^4 \qquad (12\text{-}20a)$$

where δ_0 is given by Eq. (12-16). For $\tau_v/g(\rho_L - \rho)\delta \ll 1$, the simplification that

$$\frac{\delta}{\delta_0} \approx \left[1 + \frac{4\tau_v}{3g(\rho_L - \rho)\delta_0}\right]^{-1/4}$$

follows. On the basis of this result, the local heat flux at the wall is

$$q_w = -k\frac{\partial T(0)}{\partial y} = q_{w0}\left[1 + \frac{4\tau_v}{3g(\rho_L - \rho)\delta_0}\right]^{1/4}$$

where q_{w0} is given by Eq. (12-17). Although rough, this estimate suggests the vapor drag in the direction of the body force increases condensation rates, reducing them if its direction is opposite that of the body force. The results for cocurrent vapor drag and body force are shown in Fig. 12-14 where

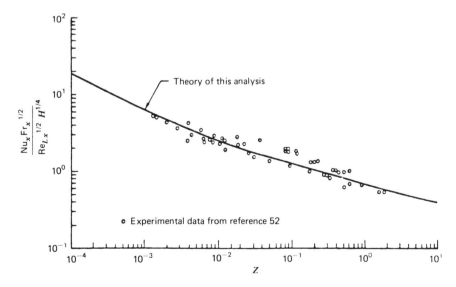

Figure 12-14 Comparison of theory and measurement for combined body force and forced convection for Freon-113. [By permission from H. R. Jacobs, *Int. J. Heat Transfer* **9**, 637–648 (1966) [54].]

$Re_{Lx} = U_\infty x / \nu_L$, $Fr_x = U_\infty^2 / gx$, $Nu_x = hx / k_L$, $H = C_{p_L}(T_s - T_w) / Pr_L h_{fg}$, $Z = 1/Fr_x$, and the ordinate is $Nu_x Fr_x^{1/2} H^{1/4} Re_x^{-1/2}$. A similarity solution, including the effect of noncondensable gases, was executed by Sparrow et al. [55]. The extreme case of vapor drag with negligible body force was solved by Cess [56] and Koh [57].

Laminar film condensation on a vertical plate fin was analyzed by Patankar and Sparrow [58]. They found that condensation rates on such a fin are a fraction of that which would occur on a vertical plate of uniform temperature. Problem 12-36 treats a variation of this situation.

12.5   TURBULENT FILM CONDENSATION VERTICAL PLATES

The information in Section 12.4 pertains to a laminar condensate film, the usual case. Occasionally, however, the condensate film is turbulent. This circumstance can occur on tall vertical surfaces or on banks of horizontal tubes as discussed in a later section.

The criterion for the transition from laminar to turbulent flow is based on the Reynolds number $Re_x = 4\dot{m}/\mu_L$, with terms as defined following Eq. (12-18). Since the heat flow into the wall must have come from the released heat of vaporization,

$$\dot{m} h_{fg} = \bar{h} x (T_s - T_w) \qquad (12\text{-}21)$$

giving the Reynolds number as $Re = 4\bar{h}x(T_s - T_w)/\mu_L h_{fg}$. For a laminar film, Eq. (12-17) substituted into the latter Re expression gives

$$\frac{Re}{4} = 0.9426 \left[\frac{g^{1/3} \rho_L^{2/3} k (T_s - T_w) x}{h_{fg} \mu_L^{5/3}} \right]^{3/4}$$

In Fig. 12-15 this laminar relationship is shown to fail at $Re \geq 1400$, which is the sought-after transition criterion in the absence of vapor drag due to forced convection. On the basis of turbulent flow principles, the form of the turbulent correlation was deduced; the suggested fit to the data is

$$\frac{Re}{4} = 0.003 \left[\frac{g^{1/3} \rho_L^{2/3} k (T_s - T_w) x}{h_{fg} \mu_L^{5/3}} \right]^{3/2} \qquad (12\text{-}22)$$

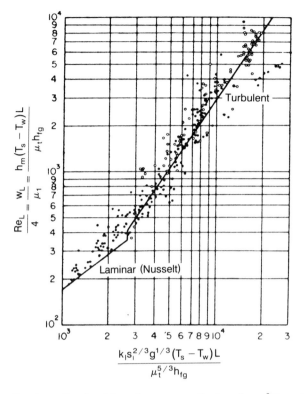

Figure 12-15 Condensation heat transfer on a vertical surface. [From H. Grober, S. Erk, and U. Grigull, *Fundamentals of Heat Transfer*, McGraw-Hill, 1961 [59].]

Solution of Eq. (12-22) for the temperature difference and insertion of that result into Eq. (12-21) gives the average turbulent heat-transfer coefficient as

$$\frac{\overline{h}\left(\nu_L^2/g\right)^{1/3}}{k} = 0.0131\,\mathrm{Re}^{1/3}, \qquad \mathrm{Re} \geqslant 1400 \qquad (12\text{-}23)$$

or

$$\frac{\overline{h}x}{k} = 0.003\left[\frac{gk(T_s - T_w)x^3\rho_L^2}{h_{\mathrm{fg}}\mu_L^3}\right]^{1/2}, \qquad \mathrm{Re} \geqslant 1400$$

Equation (12-23) is fitted to data whose Prandtl number ranges from 1 to 5.

The condensing heat-transfer coefficient for turbulent flow of a low Prandtl number condensate such as a liquid metal is less satisfactorily correlated. Dukler [60] applied the eddy diffusivity described by Deissler and the von Karman mixing length (both discussed in Section 7.2) to such a

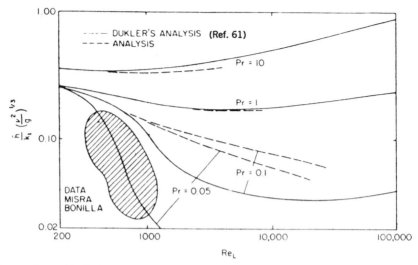

Figure 12-16 Turbulent condensation on vertical surfaces for various Prandtl numbers. [By permission from J. Lee, *AIChE J.* **10**, 540–544 (1964) [61].]

calculation and obtained good agreement with measurements, despite neglecting molecular conductivity relative to eddy conductivity and having some inconsistencies in the velocity profile. Lee [61] removed these assumptions and repeated the calculations to find less satisfactory agreement with measurements, as shown in Fig. 12-16, possibly due to either imprecise measurements or the presence of noncondensable gases.

Condensing heat-transfer coefficients are often roughly the same for laminar and turbulent cases, as shown in Fig. 12-17. Chan and Kim [63] suggest the correlation

$$\overline{Nu} = 1.33/Re^{1/3} + 9.56 \times 10^{-6} \, Re^{0.89} \, Pr^{0.94} + 0.0822$$

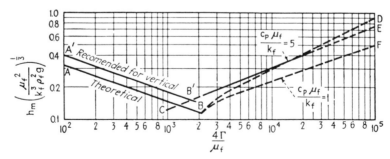

Figure 12-17 McAdams' recommended curves $A'B'$ and CE for film condensation of single vapors on vertical tubes or plates. [By permission from W. H. McAdams, *Heat Transmission*, 3rd ed., McGraw-Hill, 1954 [62].]

12.6 LAMINAR FILM CONDENSATION ON TUBES

The relationships presented for condensation on vertical plates are applicable to condensation on the exterior of vertical tubes when the condensate film is thin compared to the tube radius, a condition that is usually met, as illustrated in Problem 12-17. Condensation on the interior surface of a vertical tube can be complicated by the need to account for the vapor draft along the tube.

For a horizontal tube, modification of vertical plate results is necessary. This modification, first done by Nusselt [32], proceeds as outlined in Section 11.5 for film boiling from a horizontal cylinder—historically, Nusselt's condensation analysis preceded the film boiling analysis. The geometric basis for the developments that follow is shown in Fig. 12-18. An energy balance on the condensate film between θ and $\theta + d\theta$ gives

$$h_{\text{fg}} \frac{dm}{dx} = \frac{k_L(T_s - T_w)}{\delta} \qquad (12\text{-}24)$$

Note that $\theta = x/R$. The component of gravity acting tangentially to the tube surface is $g \sin \theta$; thus, with the parabolic velocity distribution used for a vertical flat plate, the condensate mass flow rate is given by

$$\dot{m} = \rho_L \int_0^\delta u \, dy = \rho_L(\rho_L - \rho) \delta^3 g \frac{\sin \theta}{3\mu_L}$$

Insertion of this relationship into Eq. (12-24) gives

$$\dot{m}^{1/3} \, d\dot{m} = \frac{Rk(T_w - T_s)}{h_{\text{fg}}} \left[\frac{g(\rho_L - \rho)}{3\nu_L} \right]^{1/3} \sin^{1/3} \theta \, d\theta \qquad (12\text{-}24a)$$

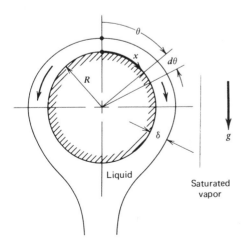

Figure 12-18 Film condensation on a horizontal tube.

Integration from $\theta = 0$ to $\theta = \pi$ gives the condensate production from one side as, using the footnote on p. 478,

$$\dot{m} = 1.924 \left[\frac{R^3 k^3 (T_s - T_w)^3 g (\rho_L - \rho)}{h_{fg}^3 \nu_L} \right]^{1/4} \qquad (12\text{-}24\text{b})$$

Accounting for the fact that this relation gives only half of the condensate flow from the tube, an energy balance on the entire tube yields

$$2 h_{fg} \dot{m} = \bar{h} 2 \pi R (T_s - T_w)$$

Solution for $\bar{h}$ gives the overall Nusselt number as

$$\overline{\text{Nu}} = 0.728 \left[\text{Ra} \frac{h_{fg}}{C_{p_L} (T_s - T_w)} \right]^{1/4} \qquad (12\text{-}25)$$

where $\overline{\text{Nu}} = \bar{h} D / k$ and $\text{Ra} = g(\rho_L - \rho) \Pr D^3 / \rho_L \nu_L^2$. A fuller discussion of the analysis leading to Eq. (12-25) is provided by Jakob [64] and by Chen [65], who points out that this problem, when treated in detail, does not possess similarity. Nusselt numbers for arbitrary bodies were obtained by Nakayama and Koyama [66] by integral treatment.

To have a correlation in terms of condensate flow rate rather than temperature difference, a rearrangement of Eq. (12-25) is possible. First, a Reynolds number is defined as for a vertical plate as $\text{Re} = 4 \dot{M} / \mu_L$, where $\dot{M}$ is the rate of condensate flow per unit length of tube. Expressing the temperature difference in terms of other variables by use of Eq. (12-24b), substituting that result into Eq. (12-25), and appealing to the definition of the Reynolds number, one obtains

$$\frac{\bar{h}}{k} \left[\frac{\nu_L^2 \rho_L}{g (\rho_L - \rho)} \right]^{1/3} = \frac{1.52}{\text{Re}^{1/3}}, \qquad \text{Re} \leqslant 2800 \qquad (12\text{-}26)$$

for laminar flow of condensate. Viewing flow around each half of the tube as the equivalent of flow down a single vertical plate for which laminar flow occurs when $\text{Re} = 4 \dot{m} / \mu_L \leqslant 1400$, it is estimated that laminar flow would exist on the tube when $\text{Re} = 4 \dot{M} / \mu_L \leqslant 2800$. Because of its small vertical dimension a horizontal tube rarely has turbulent condensate flow; use Eq. (12-23) according to McAdams [62] if turbulence occurs.

The similarity between the Reynolds number relationships for heat-transfer coefficients for vertical plates and horizontal cylinders is revealed by comparison of Eqs. (12-26) and (12-18). It is seen in Fig. 12-19 that better predictions can be made if the coefficients in Eqs. (12-25) and (12-26) are

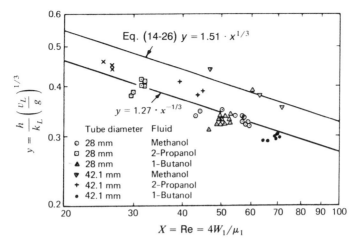

Figure 12-19 Condensation heat-transfer measurements for horizontal tubes. [By permission of the American Society of Mechanical Engineers and G. Selin, *Proc. International Heat Transfer Conference, University of Colorado*, Part II, 1961, pp. 279–289 [43].]

changed to 0.61 and 1.27, respectively, weakening the initially perceived similarity. The similarity is yet strong enough, however, to enable use of Fig. 12-12 to ascertain the effect of low Prandtl numbers such as are characteristic of liquid metals. As with vertical plates, measurements for liquid metals are 0.3–0.9 of predicted values.

The effect of surface tension on the manner in which condensate leaves the bottom of the tube has been studied by Henderson and Marchello [67].

For condensation on the outside of tubes inclined at an angle θ from the horizontal, Selin [43] correlated measurements within 15% for $0° \leqslant \theta \leqslant 60°$ by taking the component $g \cos \theta$ of gravity perpendicular to the tube for use in Eq. (12-25) in place of g. The success of this simple stratagem may be due to the fact that most of the condensation occurs atop the tube, where the condensate film is thin and surface tension effects are small; the cross flow that occurs down the top of the inclined tube has negligible effect compared to the flow around the circumference of the tube. The bottom of the tube, where the condensate film is thick and surface tension effects are large, experiences a minor part of the total condensation. Hence condensate drainage down the bottom of the inclined tube has a small effect on the overall condensation rate. Sheynkman and Linetskiy [68] present a refined treatment of condensation on inclined tubes, treating the top as covered by circumferential flow without surface tension effects and separately treating the thick film along the bottom where surface tension effects are considered.

In some condensers, banks of horizontal tubes are in vertical rows as illustrated in Fig. 12-20. In this situation, first treated by Nusselt [32], the

Figure 12-20   Film condensation on a bank of horizontal tubes.

condensate from the topmost tube drains onto the second tube, and so on. The thicker condensate film on lower tubes renders them less effective condensing surfaces than the upper tubes. For each tube, the analysis leading to Eq. (12-24a) applies. Hence for each tube

$$\dot{m}_{\text{bottom},n}^{4/3} = \dot{m}_{\text{top},n}^{4/3} + B$$

where $B^{3/4} = 1.924[R^3k^3(T_w - T_s)^3g(\rho_L - \rho)/h_{fg}^3\nu_L]^{1/4}$. For tube 1,

$$\dot{m}_{b,1}^{4/3} = B$$

For tube 2,

$$\dot{m}_{b,2}^{4/3} = \dot{m}_{b,1}^{4/3} + B = 2B$$

Generalization to the nth tube results in

$$\dot{m}_{b,n}^{4/3} = nB$$

As before, the overall heat-transfer coefficient is found from

$$2h_{fg}\dot{m}_{b,n} = 2\pi Rn\bar{h}(T_s - T_w)$$

to be

$$\frac{\bar{h}(nD)}{k} = \frac{2h_{fg}}{\pi(T_s - T_w)}(nB)^{3/4}$$

This result shows that Eq. (12-25) for a single horizontal tube can be taken over directly if nD is used in place of D. For n tubes in a vertical row,

$$\frac{\bar{h}(nD)}{k} = 0.728 \left[\frac{g(\rho_L - \rho)(nD)^3 h_{fg}}{k_L \nu_L (T_s - T_w)} \right]^{1/4} \tag{12-27}$$

a result that suggests that the condensation rate will vary as $n^{3/4}$. The effects of splashing, uneven bowing of tubes, external vibrations, surface ripples, and so forth make it unnecessary to modify the coefficient of Eq. (12-27) to correlate measurements. A minor modification to Eq. (12-27) developed by Chen [65], accounting for additional condensation on the subcooled condensate stream between tubes, consists of multiplying it by

$$\frac{[1 + 0.2A(n - 1)][1 + (0.68 + 0.02A)A/\text{Pr}]}{1 + (0.95 - 0.15A)A/\text{Pr}}$$

Here $A = C_{p_L}(T_s - T_w)/h_{fg}$, and restrictions are that $A(n - 1) \leqslant 2$, $\text{Pr} \leqslant 0.05$ or $\text{Pr} \geqslant 1$, and $A/\text{Pr} \leqslant 20$ for $n = 1$ whereas $A/\text{Pr} \leqslant 0.1$ for $n > 1$.

Laminar film condensation of a vapor flowing perpendicular to a horizontal cylinder has been treated by Denny and Mills [69] by an integral method with a lengthy result. For water with $T_{sat} = 110°F$, $T_{sat} - T_{wall} = 40°F$, and $D_{tube} = 0.25$ in., vapor approach velocities of 4 and 25 ft/sec increase the average condensing heat-transfer coefficients by about 8% and 40%, respectively.

12.7   CONDENSATION INSIDE TUBES

When a vapor is condensed on the inside of a tube, shear stress on the condensate film that separates the tube wall from the vapor can be important. The general trend, suggested in Section 12.4 by a simple treatment for flat-plate geometry, is that vapor-induced shear increases the condensation rate if the vapor motion aids the local body force in causing condensate motion. The condensate film may either form an annulus of uniform thickness or it may stratify with a thicker liquid film along the bottom of the tube if gravity acts perpendicular to the tube axis. Because the high density ratios commonly encountered produce a low liquid fraction and the vapor velocity is usually high, an annular condensate film of uniform thickness is most common.

Nusselt [32] extended his vertical-plate analysis to laminar film condensation inside a vertical tube with forced vapor flow. He assumed that the shear at the liquid–vapor interface (the condensate formed an annular film on the tube wall) could be predicted from pipe flow relations as $\tau_v = f\rho_v U_v^2/8g_c$, where f is the friction factor for head loss. Equation (12-20a) resulted and was solved for vapor flow both aiding and opposing the body force. His results, discussed by Jakob [64], yield low predictions for heat-transfer coefficients at either high vapor velocities or when the condensate flow is turbulent as a result of the inaccurate manner of shear calculation and the assumption of laminar condensate flow. This sort of analysis was extended by Rohsenow et al. [53] to include turbulent condensate films.

Over the years a substantial amount of data has been collected [2]. The correlations most generally applicable were developed by Soliman et al. [70] who applied the correlation by Lockhart and Martinelli [71] for friction in two-phase adiabatic flow to the basic ideas set forth by Carpenter and Colburn [72]. Their resulting correlation, predicting measurements within 40%, is

$$\frac{h\mu_L}{k_L \rho_L^{1/2}} = 0.036 \, \mathrm{Pr}_L^{0.65} \, F_0^{1/2} \tag{12-28}$$

where F_0 is an effective stress given by the sum of three parts as

$$F_0 = F_f + F_m \pm F_a$$

and the positive sign is used for vapor flow assisting gravity, whereas the negative sign is used for vapor flow opposing gravity. The friction term F_f dominates at high to intermediate quality. The momentum term F_m arises from the momentum given to the slow liquid by the fast vapor upon condensation. The gravity term F_a dominates at low quality. Here

$$F_f \frac{\pi^2 \rho_v D^4}{8W^2} = 0.045 \, \mathrm{Re}^{-0.2}\left[X^{1.8} + 5.7\left(\frac{\mu_L}{\mu_v}\right)^{0.0523}(1-X)^{0.47} X^{1.33}\left(\frac{\rho_v}{\rho_L}\right)^{0.261} \right.$$

$$\left. + 8.11\left(\frac{\mu_L}{\mu_v}\right)^{0.105}(1-X)^{0.94} X^{0.86}\left(\frac{\rho_v}{\rho_L}\right)^{0.522} \right]$$

$$F_m \frac{\pi^2 \rho_v D^4}{8W^2} = 0.5\left(D\frac{dX}{dz}\right)\left[2(1-X)\left(\frac{\rho_v}{\rho_L}\right)^{2/3} + \left(2X - 3 + \frac{1}{X}\right)\left(\frac{\rho_v}{\rho_L}\right)^{4/3} \right.$$

$$+ (2X - 1 - \beta X)\left(\frac{\rho_v}{\rho_L}\right)^{2/3} + \left(2\beta - \beta X - \frac{\beta}{X}\right)\left(\frac{\rho_v}{\rho_L}\right)^{5/3}$$

$$\left. + 2(1 - X - \beta + \beta X)\frac{\rho_v}{\rho_L} \right]$$

$$F_a \frac{\pi^2 \rho_v D^4}{8W^2} = \frac{0.5}{\mathrm{Fr}}\left\{ 1 - \left[1 + \frac{(1-X)(\rho_v/\rho_L)^{2/3}}{X} \right]^{-1} \right\}$$

where $Fr = 16W^2/[\pi^2 D^5 a(\rho_L - \rho_v)\rho_v]$ is a Froude number, a is the gravity component along the tube axis, W is the total mass flow rate of liquid and vapor, D is tube diameter, $Re = 4W/\pi D\mu_v$ is a Reynolds number, X is local quality, β is 1.25 for a turbulent film and 2.0 for a laminar film, and (assuming uniform heat removal along the tube with unity entering quality) $dX/dz = -1/L_0$ if L_0 is the total length of the tube.

For vapor flow against gravity, slugging and plugging can occur as a result of folding and runback of the condensate film when vapor-induced friction and momentum exchanges are insufficient to drag the condensate against gravity. The criterion for the onset of such a condition is that $F_0 = F_f + F_m - F_a = 0$; from the foregoing relationships, the quality at which this condition occurs can be ascertained.

At high quality ($X \approx 1$) Eq. (12-28) reduces to

$$\mathrm{Nu} = 0.0054 \, \mathrm{Pr}_L^{0.65} \, \mathrm{Re}_v^{0.9} \, \frac{\mu_v}{\mu_L} \left(\frac{\rho_L}{\rho_v}\right)^{1/2} \tag{12-29}$$

with $\mathrm{Nu} = hD/k_L$, $\mathrm{Re}_v = 4W_v/\pi D\mu_v$, where W is the mass flow rate of vapor. Equation (12-29) resembles the correlation for Nusselt number in single-phase turbulent flow, as comparison with Eq. (9-16e) shows. Shah [73] gives a refined correlation. Tandon et al. [74] give a flow-regimes map for the spray, annular and semiannular, wavy, slug, and plug flows that are observed for condensation with through flow in a horizontal tube.

12.8 NONGRAVITATIONAL CONDENSATE REMOVAL

Actual condensation rates are only about 1% of the maximum rate that would result if all vapor molecules incident on a cooled surface were condensed. This realization, the subject of Problem 12-32, has led many to seek practical nongravitational means of causing condensate flow. When gravity is absent, nongravitational condensate removal is required for steady operation.

Condensation on a horizontal rotating disc was studied by Sparrow and Gregg [75], who achieved a similarity solution of the laminar boundary-layer equations. The heat-transfer coefficient, neglecting vapor drag that was studied later [76], is constant as

$$\frac{h(\nu/\omega)^{1/2}}{k} = 0.904 \left[\frac{\mathrm{Pr}_L \, h_{\mathrm{fg}}}{C_{p_L}(T_s - T_w)}\right]^{1/4}$$

for $Pr \approx 1$ and $C_{p_L}(T_s - T_w)/h_{\mathrm{fg}} < 0.1$; see the original work for other conditions. Laminar flow of condensate is expected for $Re = r^2\omega/\nu \leq 3 \times 10^5$. Also, h_{rot} for a rotating disk is related to h_{vert} for a vertical plate under the sole influence of gravity as

$$\frac{h_{\text{rot}}}{h_{\text{vert}}} = \left(\frac{8x\omega^2}{3g}\right)^{1/4}$$

The measurements by Nadapurkar and Beatty [77] are about 30% less than predicted. Condensation on a rotating cone [78] with its apex uppermost was found to have nearly the same coefficient as for a rotating disk since

$$\frac{h_{\text{cone}}}{h_{\text{disk}}} = (\sin\phi)^{1/2}$$

where ϕ is the half-angle of the cone. A rotating cone was studied by Bromley [79]. For condensation on the inside of rotating cones, consult the study by Marto [80].

A vertical tube spinning about its own axis was experimentally studied by Nicol and Gacesa [81]. At low angular velocity ω the overall heat-transfer coefficient measurements were correlated by, for $L/D = 10$,

$$\frac{\overline{\text{Nu}}}{\left[gL^3 h_{\text{fg}}\rho_L/\nu_L k_L(T_s - T_w)\right]^{1/4}} = \begin{cases} 0.0943, & \text{We} \leqslant 250 \\ 0.00923\,\text{We}^{0.39}, & \text{We} > 250 \end{cases}$$

where $\text{We} = \rho_L\omega^2 D^3/4\sigma$ is a Weber number and $\overline{\text{Nu}} = \bar{h}D/k$. This correlation holds until the Nusselt number is nearly trebled. At higher angular velocities, the effect of gravity is unimportant and the correlation is

$$\overline{\text{Nu}} = 12.26\,\text{We}^{0.496}$$

These measurements for a vertical cylinder are compared with those of others for a rotating horizontal tube and additional references are cited.

Condensation for a vertical tube rotating about a vertical axis displaced from the tube axis was studied by Mochizuki and Shiratori [82] for internal condensation and Suryanarayana [83] for external condensation. They found increases of a factor of as much as 4 in heat-transfer rate over that for the stationary case.

The effect of a nonuniform electric field was noted in Chapter 11 in connection with boiling. A similar large effect can be found for condensation —Jones [84] discusses the physical mechanisms involved for both boiling and condensation as well as for natural convection, fluidized beds, solidification, and heat pipes. Velkoff and Miller [85] found that the condensation rate of Freon-113 on a plate could be increased by a factor of as much as 3 by placing electrodes with dc voltage parallel to the plate. Choi [86] doubled the condensation rate of Freon-113 in an electrically grounded vertical tube with a central electrode to which 30,000 V dc was applied. The efficacy of an ac electrical field was established by Holmes and Chapman [87], who applied up to 50,000 V to electrodes parallel to a plate to increase condensation rates by a factor of 10. Seth and Lee [88] found a more modest increase in condensation rate of Freon-113 on a horizontal cylinder as dc voltage was applied when air was present as a noncondensable gas. In all cases a voltage threshold must be exceeded before the effect of electrical fields is manifested.

Suction of condensate through the porous surface on which it condenses has been considered by Jain and Bankoff [89]. As much as a 50% increase in local heat-transfer coefficient could be achieved, but only after a large fraction of the condensate layer is sucked away. These results were confirmed by Yang [90] in a more rigorous manner and by Lienhard and Dhir [91] with an integral method. The same general conclusions apply to horizontal tubes [92].

The effect of transverse mechanical vibrations on a vertical tube was explored by Dent [93], for steam condensation on a vertical tube. Heat-transfer coefficients were increased up to 60%, not by throwing off condensate, but rather by causing condensate to slosh from side to side. His correlation is, a_0 is vibration amplitude and ω is vibration frequency,

$$\frac{h_x}{h_{\text{no vibr}}} = \left[1 + \left(\frac{a_0 \omega}{\pi} \right)^2 \frac{\mu_L h_{\text{fg}}}{g k_L x (T_s - T_w)} \right]^{1/4}$$

Capillary forces are used to move condensate in heat pipes [94, 95].

12.9 EFFECT OF NONCONDENSABLE GAS AND MULTICOMPONENT VAPORS

The presence of a noncondensable gas in a vapor can significantly lower the condensation rate below that for a pure vapor. Only a few parts by volume of air could reduce condensation heat-transfer coefficients by up to 50%. Such effects are important in power plant condensers where buildup of such noncondensable gases as air is unavoidable despite provisions to remove noncondensable gases. Similarly, separation of ammonia from air by condensation requires that the presence of air be tolerated. In such cases design procedures must allow for the presence of noncondensables.

To acquire understanding of the physical mechanisms at work, consider the results for Stefan's diffusion problem presented in Section 3.3 and illustrated in Figs. 3-6 and 3-7. As the vapor flows to the condensate–vapor

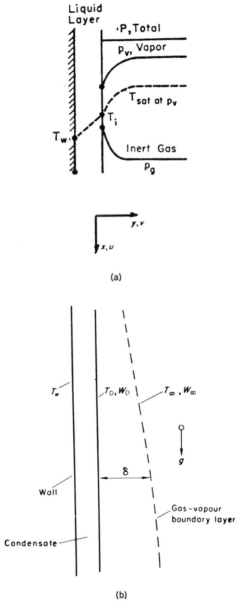

(a)

(b)

Figure 12-21 (*a*) Film condensation in the presence of noncondensable gases. (*b*) Coordinate system for integral analysis of film condensation in the presence of noncondensable gases. [By permission from J. W. Rose, *Int. J. Heat Mass Transfer* **12**, 233–237 (1960) [97].]

interface, the noncondensable gas is swept along with it. Since the interface is impermeable to the noncondensable gas, its concentration and partial pressure there are increased above their values at more distant locations. Correspondingly, the partial pressure of the vapor at the interface is reduced below its value at more distant locations. Because of this reduced partial vapor pressure, the equilibrium temperature at the condensate–vapor interface is reduced and the driving temperature difference for heat transfer is reduced, causing reduction in the condensation rate. This situation is illustrated in Fig. 12-21. The equilibrium level of the noncondensable gas at the interface requires that noncondensable gas be removed by (1) diffusion, (2) drag from the falling condensate, (3) natural convection caused by density variations as a result of varying composition, or (4) forced convection. Approximate procedures to account for the effect of a noncondensable gas on condensation rates by Kern [96] have been succeeded by boundary-layer approximations. In general, the effects of noncondensable gases are greater in ducts than in a chamber since the noncondensables' concentrations increase down the duct as condensation occurs [98].

The boundary-layer description of this problem considers the condensate film characteristics to be as given in Section 12.4 for condensation on a vertical plate. The condensate film thickness δ is

$$\delta = \left[\frac{C_{p_L}(T_s - T_w)}{h_{fg} \Pr_L} \right]^{1/4} \left(\frac{4v_L^2 x}{g} \right)^{1/4} \tag{12-16}$$

the condensate interface velocity u_0 is

$$u_0 = 2\left[\frac{C_{p_L}(T_s - T_w)}{h_{fg} \Pr_L} \right]^{1/2} \left(\frac{gx}{4} \right)^{1/2} \tag{12-15a}$$

and the local heat flux q is

$$q = \left[\frac{C_{p_L}(T_s - T_w)}{h_{fg} \Pr_L} \right]^{3/4} \left(\frac{gh_{fg}^4 \mu_L^4}{4v_L^2 x} \right)^{1/4} \tag{12-17a}$$

The challenge is to calculate the saturation temperature T_s since then all would follow inasmuch as the heat flux is available from Eq. (12-17a).

The situation in the vapor–noncondensable mixture adjacent to the condensate film is illustrated in Fig. 12-21b, where a boundary layer in the gaseous region is shown. Its behavior is described by the boundary-layer equations:

Continuity
$$\frac{\partial u}{\partial x} + \frac{\partial v}{\partial y} = 0$$

x Motion
$$u\frac{\partial u}{\partial x} + v\frac{\partial u}{\partial y} = g\left(1 - \frac{\rho_\infty}{\rho}\right) + \nu\frac{\partial^2 u}{\partial y^2}$$

Diffusion
$$u\frac{\partial \omega}{\partial x} + v\frac{\partial \omega}{\partial y} = D\frac{\partial^2 \omega}{\partial y^2}$$

The temperature variation is known so the energy equation is unnecessary. Exact solutions were achieved by Sparrow and Lin [99] and Minkowycz and Sparrow [52] for some common cases. A more general result was obtained by Rose [97] by the integral method. The boundary-layer equations in integral form are

$$\frac{d\left(\delta_v \int_0^1 u^2 \, d\eta\right)}{dx} - v_0 u_0 + \frac{\nu}{\delta_v}\frac{\partial u(0)}{d\eta} - gX\delta_v \int_0^1 W \, d\eta = 0$$

and

$$\frac{d\left(\delta_v \int_0^1 uW \, d\eta\right)}{dx} + \frac{D}{\delta_v}\frac{\omega_\infty}{\omega_0}\frac{\partial W(0)}{\partial \eta} = 0$$

where $\eta = y/\delta_v$, $W = \omega - \omega_\infty$, $X = (M_g - M_v)/[M_g - \omega_\infty(M_g - M_v)]$, $W = \omega - \omega_\infty$, and $\omega = \rho_{gas}/(\rho_{vapor} + \rho_{gas})$ is the mass fraction of the noncondensable gas. The solution of these two equations, with use of the assumed profiles of

$$u = u_0(1 - \eta)^2 + \bar{u}\eta(1 - \eta)^2 \quad \text{and} \quad \frac{\omega - \omega_\infty}{\omega_0 - \omega_\infty} = (1 - \eta)^2$$

gives

$$10F \, \text{Sc} \frac{\mu_L \rho_L}{\mu \rho}\left(\frac{\omega_\infty}{W_0}\right)^2\left(\frac{20}{21} + \text{Sc}\frac{\omega_0}{\omega_\infty}\right) + \frac{8}{F^2 \, \text{Sc}}\frac{\mu \rho}{\mu_L \rho_L}\left(\frac{W_0}{\omega_0}\right)^2\left(\frac{5F}{28} - X\frac{W_0}{3}\right)$$

$$= \frac{100\omega_\infty}{21\omega_0} - \frac{2W_0}{\omega_0} + 8 \, \text{Sc} \tag{12-30}$$

The accuracy of Eq. (12-30), for which $F = C_{p_L}(T_s - T_w)/h_{fg} \, \text{Pr}_L$ and $W_0 =$

$\omega_0 - \omega_\infty$ and Sc = ν/D is the Schmidt number for the gas mixture, was confirmed by the measurements of Al-Diwany and Rose [100]. Best agreement resulted when the noncondensable gas was heavier than the vapor. Possible fogging in the boundary layer due to cooling is not taken into account by Eq. (12-30) but will not always occur since some supersaturation is tolerable. Such fogging is taken into account in the integral analysis by Mori and Hijikata [101].

Suppose now that the wall temperature T_w, total pressure p, and mass fraction of noncondensable gas in the bulk mixture ω_∞ are specified and the heat-transfer rate is to be computed. One begins by guessing an equilibrium interface temperature T_s from which a value of F is computed for use in Eq. (12-30). Equation (12-30) is solved for ω_0, the noncondensable gas mass fraction at the interface. From this result, the mass fraction $1 - \omega_0$ of the vapor at the interface is known and is substituted into Eq. (3-32) to obtain the partial pressure of the vapor at the interface. Since the partial pressure of the vapor is known, the corresponding saturation temperature T_s can be ascertained from steam tables. If this value of T_s differs appreciably from that initially assumed, the procedure is repeated until no appreciable difference results. Once T_s has been found, the heat flux is obtained by using T_s in Eq. (12-17a).

Removal of noncondensable gases from the interface by forced convection will be beneficial as concluded by Citakoglu and Rose [102], and Rose [103, 104]. Liquid metals were studied by Turner et al. [105].

Condensation of two condensable vapors whose condensates are miscible was studied by Sparrow and Marschall [106]. Marschall and Hickman [107] treated condensation of two condensable vapors whose condensates are immiscible.

Sage and Estrin [108] and Taifel and Tamir [109] analyzed vapor condensation on a vertical plate in the presence of two noncondensable gases. Such a model is applicable to fission product removal from contaminated vapors and some industrial processes. Exact treatments of multicomponent condensation are complicated by the fact that the mass diffusivities involved are dependent on concentration (see Bird et al. [110] for an introductory discussion and Sparrow and Niethammer [111] for a recent application). A linearization involves use of an effective binary diffusivity as explained by Toor [112, 113].

12.10 SCULPTURED SURFACES

The rate of heat transfer in film condensation is limited by conduction through the condensate film. Consequently, augmentation of film-condensation heat-transfer coefficients depends on finding some way to thin the

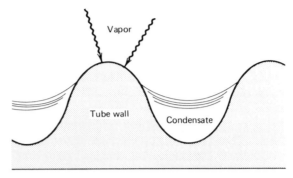

Figure 12-22 Section through a vertical tube with axial flutes, showing the effects of surface tension on the configuration of the condensate film. [From E. Lusterader, R. Richter, and R. N. Neugebauer, *ASME J. Heat Transfer* **81**, 297–307 (1959) [115].]

condensate film. The sculptured surfaces that result from such an attempt are analogous to the augmented surfaces discussed in Chapter 13 to improve boiling heat flow rates.

In 1953 Gregorig [114] noted that surface tension could cause pressure gradients in a condensate film whose curvature varied with position. The wavy surface shown in Fig. 12-22 in a top view is vertical. At the crest, the convex curvature requires the pressure in the condensate to exceed the vapor pressure; at the trough, the concave curvature requires the pressure in the condensate to be less than the vapor pressure. Hence the condensate is pushed from the crest down to the trough by this pressure gradient. The thin condensate film at the crest is accompanied by high heat-transfer coefficients that can, with proper design, more than compensate for the diminished heat-transfer coefficients in the trough resulting from condensate accumulation in the trough. Tests by Lusterader et al. [115] revealed a heat-transfer coefficient about four times larger than could be achieved on a smooth surface. Webb [116] developed designs for practical Gregorig surfaces.

For downward-facing surfaces, a three-dimensional (or doubly rippled) surface was studied by Markowitz et al. [117] for electronic equipment cooling. Their measured twofold improvement was considered to be adversely affected by the presence of noncondensable gases that accumulate near the high-heat-flux crests.

Twisted-tape inserts and internal fins can increase average condensing heat-transfer coefficients by as much as a factor of 2 over values in smooth horizontal tubes [24]. Sculptured exterior surfaces can increase condensing heat-transfer coefficients by a factor of 2 to 6 over values for smooth horizontal tubes, according to the measurements by Briggs et al. [118] and others in the survey by Bergles [119] of recent developments in condensation augmentation. Fraas [120] discusses application to condenser design.

PROBLEMS

12-1 Briefly estimate the annual national cost saving that would result from a 10% increase in the condensing heat-transfer coefficient in steam power plant condensers.

12-2 (a) Confirm Eq. (12-2) for water by substituting property values into the equation and estimating r from $r^3 = M/\rho N_A$, where M is molecular weight and N_A is Avogadro's number.

   (b) Use the result of part a to show that the tensile stress that would rupture water is of the order of 10^4 times atmospheric pressure.

12-3 Show that the pressure inside a liquid drop of radius r^* in equilibrium with a supersaturated vapor is

$$P_L = \frac{2\sigma}{r^*} + P_{\text{sat}} \exp\frac{2v_L\sigma}{r^*RT}$$

Evaluate P_L/P_{sat} for water at 68°F when $2r^* = 10^{-6}$ in. and compare it with the corresponding P_v/P_{sat} from Table 12-1. *Answer*: $P_L/P_{\text{sat}} = 4600$.

12-4 To estimate the limiting vapor supersaturation (or subcooling) that can exist without condensation, introduce the drop radius, essentially the distance between liquid molecules, that corresponds to about eight liquid molecules into Eq. (12-7) for water at 68°F. *Answer*: $P_v/P_{\text{sat}} = 19$, obtained in a nozzle [121]; only $P_v/P_{\text{sat}} \approx 8$ was obtained in a nonflow device [122].

12-5 The change in free energy ΔG of a liquid–vapor system as shown in Fig. 12-3 is to be ascertained.

   (a) Note that when the system is all vapor, the total free energy is given by Eq. (12-4) as $G_0 = (n_v + n_L)g_v$.

   (b) Subtract the result of part a from the total free energy of the system when part of the system is a liquid drop to obtain the change in free energy from the all-vapor case as

$$\Delta G = G - G_0 = n_L(g_L - g_v) + 4\pi\sigma r^2$$

   (c) Note from Eq. (12-5) that $g_L - g_v = -2v_L\sigma r^*$ and that $n_L v_L = 4\pi r^3/3$, and substitute these results into the result of part b to obtain the free energy of formation of a drop of radius r

$$\Delta G = 4\pi\sigma\left(r^2 - \frac{2r^3}{3r^*}\right)$$

(d) Show that the maximum value of G, the free energy of formation of a drop of radius r^*, is

$$\Delta G_{max} = \frac{4\pi\sigma r^{*2}}{3}$$

12-6 A supersaturated vapor at temperature T and pressure P_v that is in equilibrium with a liquid drop of radius r^* can be considered to be subcooled below the saturation temperature T_{sat} [similar to the manner in which P_v exceeds P_{sat}, shown by Eq. (12-6)]. The subcooling $T_v - T_{sat}$ is to be estimated. To do this:

(a) Utilize the Clausius–Clapeyron equation, relating saturation pressure and temperature,

$$T\frac{dP}{dT} = \frac{h_{fg}}{v_v - v_L}$$

to obtain, assuming the vapor to be a perfect gas and $v_v \gg v_L$,

$$\ln\left(\frac{P_v}{P_{sat}}\right) = h_{fg}\frac{1/T_v - 1/T_{sat}}{R}$$

since $P = P_v$ when $T = T_{sat}$ and $P = P_{sat}$ when $T = T_v$. Here R is the universal gas constant and h_{fg} is the molar heat of vaporization per mole.

(b) Combine the result of part a with Eq. (12-6) to find

$$r^* = \frac{2v_L\sigma}{h_{fg}}\frac{T_{sat}}{T_{sat} - T_v}$$

(c) The result of part b is the smallest equilibrium drop and also the smallest drop that could be expected in dropwise condensation. Show that for water at 100°C and atmospheric pressure,

$$r^* = 2 \times 10^{-6} \text{ cm } °C/(T_{sat} - T_0)$$

which gives $r^* = 10^{-6}$ cm for 2°C subcooled vapor.

12-7 The formation of vapor cavities in a bulk liquid is similar to the formation of liquid drops in a bulk vapor. To see this:

(a) Show by application of Eq. (12-3), where $r \approx 10^{-10}$ m is the distance between liquid molecules, that formation of a vapor cavity in a liquid requires

$$P_\infty - P_{\text{rupture}} \approx 10^4 \text{ atmospheric pressure}$$

which is a negative pressure and a superheating (just as subcooling is required for condensation).

(b) Apply the Clausius–Clapeyron equation, relating saturation pressure and temperature, with $dP \approx 2\sigma/r$ and $dT \approx T_v - T_{\text{sat}}$,

$$T\frac{dP}{dT} = \frac{h_{\text{fg}}}{v_v - v_L}$$

to show that the liquid superheat $T_v - T_{\text{sat}}$ required for a vapor cavity of radius r^* to be in equilibrium with the liquid is

$$T_v - T_{\text{sat}} = \frac{2\sigma v_v}{r^* h_{\text{fg}}} T_{\text{sat}}$$

Superheat limits for liquid mixtures are available [123].

(c) Noting that the result of part b determines the minimum size of vapor bubble whose preexistence is required for bulk nucleation to start, show that for water at 100°C and atmospheric pressure

$$r^* = 4 \times 10^{-3} \text{ cm °C}/(T_v - T_{\text{sat}})$$

which gives $r^* = 2 \times 10^{-3}$ cm for 2°C superheated liquid. Compare these results with the corresponding ones for condensation from Problem 12-6.

12-8 Compare the heat-flux–temperature-difference variation for dropwise condensation shown in Fig. 12-5 with that predicted by Eq. (12-9) for steam condensing at atmospheric pressure.

12-9 Compare the heat-flux variation with imposed temperature difference for dropwise condensation against that for nucleate boiling.

12-10 Compare the magnitude of the heat-transfer coefficient for dropwise condensation with that for nucleate boiling for water at 1 atm of pressure.

12-11 The maximum condensation heat flux, also the peaking boiling heat flux, is to be estimated on the basis that all vapor molecules arriving at a surface condense and yield their latent heat of vaporization h_{fg} with none reemitted by evaporation. From the kinetic theory of gases, show that the estimate is

$$q_{cond,max}/A = h_{fg}P(M/2\pi RT)^{1/2}$$

where P is the vapor pressure, T is the vapor temperature, M is the vapor molecule weight, and R is the universal gas constant. Related discussion is given by Sukhatme and Rohsenow [47], Gambill and Lienhard [124], and Hickman [125].

12-12 The local heat-transfer coefficient for laminar condensation on a vertical plate given in terms of temperature difference by Eq. (12-17) is to be cast into the form of Eq. (12-18), involving condensate flow rate.

(a) Show that the Reynolds number

$$Re = \frac{\text{average velocity} \times \text{hydraulic diameter}}{\text{kinematic viscosity}}$$

is $Re = 4u_{av}\delta/\nu_L = 4\dot{m}/\mu_L$, where $\dot{m}$ is the condensate mass flow rate per unit width.

(b) Use the fact that $\dot{m}h_{fg} = \bar{h}x(T_s - T_w)$, since the heat of condensation was absorbed by the plate, and eliminate $T_s - T_w$ by use of Eq. (12-17) to obtain Eq. (12-18).

12-13 A similarity transformation for laminar film condensation of a stagnant saturated vapor on a vertical plate is to be obtained. The boundary-layer equations for the liquid are given in Eq. (12-10), whereas for the vapor only the continuity and x-motion equations are needed as $\partial u/\partial x + \partial v/\partial y = 0$ and $u\,\partial u/\partial y + v\,\partial u/\partial y = \nu\,\partial^2 u/\partial y^2$ subject to the boundary conditions in Eq. (12-11) and $u_{vapor}(y \to \infty) \to 0$.

(a) Show that the similarity variable $\eta_L = C_L y/x^{1/4}$ with $C_L^4 = g(\rho_L - \rho_v)/4\nu_L^2\rho_L$ and the stream function $\psi_L = 4\nu_L C_L x^{3/4}F(\eta_L)$ give the liquid velocity as $u_L = \partial\psi_L/\partial y = 4\nu_L C_L^2 x^{1/2}F'$ and $v_L = -\partial\psi_L/\partial x = \nu_L C_L x^{-1/4}(\eta_L F' - 3F)$, where $F' = dF/d\eta_L$.

(b) Show that the similarity variable $\eta_v = C_v y/x^{1/4}$ with $C_v^4 = g/4\nu_v^2$ and the stream function $\psi_v = 4\nu_v C_v x^{3/4}f(\eta_v)$ give the vapor velocity as $u_v = \partial\psi_v/\partial y = 4\nu_v C_v^2 x^{1/2}f'$ and $v_v = -\partial\psi_v/\partial x = \nu_v C_v x^{-1/4}(\eta_v f' - 3f)$.

(c) Show that the boundary-layer equations are

$$F''' + 3FF'' - 2(F')^2 + 1 = 0 \qquad \text{liquid } x \text{ motion}$$

$$\theta'' + 3\,Pr\,F\theta' = 0 \qquad \text{liquid energy}$$

$$f''' + 3ff'' - 2(f')^2 = 0 \qquad \text{vapor } x \text{ motion}$$

with $\theta = (T - T_s)/(T_w - T_s)$ and subject to

$$F(0) = 0 = F'(0), \qquad \theta(0) = 1$$

$$f(0) = \left[\frac{(\rho\mu)_L}{(\rho\mu)_v}\right]^{1/2}\left(\frac{\rho_L - \rho_v}{\rho_L}\right)^{1/4} F(\eta_{L\delta})$$

$$f'(0) = \left(\frac{\rho_L - \rho_v}{\rho_L}\right)^{1/2} F'(\eta_\delta)$$

$$f''(0) = \left[\frac{(\rho\mu)_L}{(\rho\mu)_v}\right]^{1/2}\left(\frac{\rho_L - \rho_v}{\rho_L}\right)^{3/4} F''(\eta_{L\delta})$$

$$\theta(\eta_{L\delta}) = 0, \qquad f'(\eta_v \to \infty) \to 0$$

where the unknown quantities to be determined are $F''(0)$, $\theta'(0)$, $f'(0)$, and $\eta_{L\delta} = C_L\delta/x^{1/4}$.

(d) Show that the local Nusselt number is

$$\mathrm{Nu}_x\left[\frac{C_{p_L}(T_s - T_w)/h_{\mathrm{fg}}}{gC_p(\rho_L - \rho_v)x^3/4\nu_L k_L}\right]^{1/4} = \frac{C_{p_L}(T_s - T_w)}{\mathrm{Pr}_L\, h_{\mathrm{fg}}}[-\theta'(0)]$$

with parameters Pr_L, $(\rho\mu)_L/(\rho\mu)_v$, and $C_{p_L}(T_s - T_w)/h_{\mathrm{fg}}$.

(e) Comment on the agreement between the functional form for $\eta_{L\delta}$ of part a and the result of a simpler integral analysis [Eq. (12-16)].

12-14 A vertical plate is exposed to stagnant saturated water vapor at 100°C. Suddenly the plate temperature, initially at 100°C, is permanently changed to 38°C. Show that steady condensate rates are achieved after 0.7 s at 15 cm from the top of the plate and after 1 s at 30 cm from the top of the plate. Show that the average heat-transfer coefficient over the entire plate $\bar{h}$ varies with time as

$$\frac{\bar{h}}{\bar{h}_0} = \begin{cases} \left[(t/t_s)^{3/2} + 3(t_s/t)^{1/2}\right]/4, & t < t_s \\ 1 + \dfrac{3t_s}{5t}, & t \geqslant t_s \end{cases}$$

where $\bar{h}_0$ is the steady-state average coefficient for the entire plate and t_s is the transient period for the bottom of the plate.

12-15   Problem 12-14 is to be solved by an integral technique.

(a) Show that the energy equation in integral form is

$$\frac{\partial\left(\delta\int_0^1 \theta\, d\eta\right)}{\partial t} + \frac{\partial\left(\delta\int_0^1 u\theta\, d\eta\right)}{\partial x} = \frac{h_{fg}}{C_p(T_w - T_s)}\frac{\partial(M/\rho)}{\partial x} - \frac{\alpha}{\delta}\frac{\partial\theta(0)}{\partial\eta}$$

where M is the rate of condensation per unit width of plate and $\theta = (T - T_s)/(T_w - T_s)$, whereas the conservation of mass gives

$$\frac{\partial\left[\delta\int_0^1 u\, d\eta\right]}{\partial x} + \frac{\partial\delta}{\partial t} = \frac{\partial(M/\rho)}{\partial x}$$

(b) Assume the velocity and temperature profiles to be the steady-state relations

$$u = \frac{(\rho_L - \rho)g\delta^2}{\mu}\left(\eta - \frac{\eta^2}{2}\right) \qquad \text{and} \qquad \theta = 1 - \eta$$

(c) Use the profiles of part b to find, from the energy equation of part a, that

$$\frac{\partial(M/\rho)}{\partial x} = \frac{C_p(T_s - T_w)}{h_{fg}}\left[\frac{\alpha}{\delta} - \frac{3(\rho_L - \rho_v)g}{8\mu}\delta^2\frac{\partial\delta}{\partial x} - \frac{1}{2}\frac{\partial\delta}{\partial t}\right]$$

(d) Use the velocity profile of part b in the continuity equation of part a to obtain an expression for $\partial(M/\rho)/\partial x$. Combine this result with that of part c to find

$$P\frac{\partial\delta}{\partial x} + Q\frac{\partial\delta}{\partial t} = R$$

where $P = [\rho_L(\rho_L - \rho_v)gh'_{fg}/k\mu(T_s - T_w)]\delta^3$, $Q = [\rho h'_{fg}(1 + C_p(T_s - T_w)/8h'_{fg}/k(T_s - T_w)]\delta$, $R = 1$, and $h'_{fg} = h_{fg} + 3C_p(T_s - T_w)/8$.

(e) The method of characteristics* requires

$$\underbrace{\frac{dx}{P}}_{1} = \underbrace{\frac{dt}{Q}}_{2} = \underbrace{\frac{d\delta}{R}}_{3}$$

subject to $\delta(x = 0, t) = 0 = \delta(x, t = 0)$. This result follows from

---

*F. B. Hildebrand, *Advanced Calculus for Engineers*. Prentice-Hall, 1949; F. H. Miller, *Partial Differential Equations*, Wiley, 1958.

a geometric interpretation of the partial differential equation of part d. Since $\partial\delta/\partial x$, $\partial\delta/\partial t$, and -1 are direction numbers of the normal N to the surface $\delta(x, t)$, the equation of part d states that N is perpendicular to a line L through the same surface and with direction numbers P, Q, and R. Now let the plane containing N and L cut the surface in the curve C that has direction numbers dx, dt, and $d\delta$. Since C and L have the same direction, the two sets of direction numbers are proportional as stated previously. If groups 1 and 2 are equated, the characteristic line separating the steady-state and transient regions in the x–t plane is

$$\frac{dt}{dx} = \frac{Q}{P} = \frac{\mu\left(1 + C_p(T_s - T_w)/8h'_{fg}\right)}{g(\rho_L - \rho_v)\delta^2}$$

(f) On each characteristic, if groups 1 and 3 are equated,

$$\delta^3 \, d\delta = \frac{k\mu(T_s - T_w)}{\rho_L(\rho_L - \rho_v)gh'_{fg}} \, dx$$

and if groups 2 and 3 are equated,

$$\delta \, d\delta = \frac{k(T_s - T_w)}{\rho h'_{fg}\left(1 + C_p(T_s - T_w)/8h'_{fg}\right)} \, dt$$

Combine these relationships with the result of part e to obtain Eq. (12-20) and relationships for $\delta(x, t)$ and $h(x, t)$ (see Fig. 12P-15).

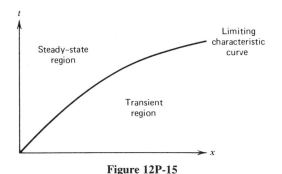

Figure 12P-15

12-16 Determine the time required to achieve steady condensation on a horizontal tube of 4 cm diameter and 50°C cooler than the saturated steam to which it is suddenly exposed. *Answer*: $t \approx 0.5$ s.

12-17 Saturated steam at 54°C condenses on the outside surface of a 3.7-m-long vertical tube of 2.5 cm o.d. whose outer wall temperature is held constant at 43°C. The total condensation rate is to be determined.

(a) Assuming a laminar condensate film, show that the heat-transfer coefficient for the entire tube is $\bar{h} = 4600$ W/m² K.

(b) Show that the condensate flow at the bottom of the tube is $\dot{m} = \pi DL\bar{h}(T_s - T_w)/h_{fg} = 0.006$ kg/s.

(c) Show that Re $= 4\dot{m}/\mu_L = 530 < 1400$ to confirm the original laminar flow assumption.

(d) Show that the condensate film thickness at the bottom of the tube is $\delta = 0.019$ cm (appeal to Eq. (12-16) if necessary, or realize $h = k/\delta$). Comment on the applicability of flat-plate relations to vertical tubes at this value of δ/D.

(e) On the basis of the fact that $\rho\bar{u}\delta = \dot{m}$, show that the average velocity is $\bar{u} = 0.1$ m/s.

12-18 Saturated steam at 54°C condenses on the outside surface of a vertical tube of 2.5 cm o.d. whose outer wall temperature is held at 43°C. For film condensation the tube length required to have a condensate flow of 0.0032 kg/s from the bottom of the tube is to be determined.

(a) First, show that laminar conditions exist since Re $= 4\dot{m}/\mu_L = 280 < 1400$.

(b) Show that, with the use of Eq. (12-18) with a coefficient of 1.88, the average heat-transfer coefficient is $h = 5600$ W/m² K.

(c) Realizing that $\dot{m} = \pi DL\bar{h}(T_s - T_w)/h_{fg}$, show that the required tube length is $L = 1.6$ m.

(d) If Re < 33, would it have been necessary to modify the coefficient of Eq. (12-18) to 1.88 from 1.47?

12-19 A square vertical plate 1 ft on a side and at 208°F contacts saturated steam at 212°F.

(a) Show from Eq. (12-17) with a coefficient of 1.13 that the average heat-transfer coefficient is $\bar{h} = 2700$ Btu/hr ft² °F if condensate flow is laminar.

(b) Noting that the condensate flow rate $\dot{m}$ from the bottom of the plate is related to the heat-transfer coefficient by $\dot{m} = \bar{h}A(T_s - T_w)/h_{fg}$, show that $\dot{m} = 11$ lb$_m$/hr.

(c) Show that the flow in the condensate film is laminar since Re $= 4\dot{m}/\mu_L = 64 < 1400$.

12-20 Consider laminar film condensation of a saturated and stagnant vapor on the inside of a vertical circular tube of uniform wall temperature, neglecting vapor drag at the interface. Estimate the distance from the top of the tube at which liquid fills the tube.

12-21 A shallow pan is exposed to stagnant saturated steam at 100°C. The sides of the pan are insulated and the bottom is kept at 90°C. Estimate the rate of condensate accumulation in the pan. *Answer*: $\delta = 7.9 \times 10^{-5}\ \mathrm{m}(t/s)^{-1/2}$.

12-22 A saturated stagnant vapor condenses on the outside of a constant-temperature cone, with its apex up. Derive the differential equation relating local condensate thickness to temperature difference, fluid properties, and distance from cone apex. Solve the differential equation if possible [46].

12-23 Stagnant saturated steam at 100°C condenses on a 65°C vertical tube. At what tube length would turbulent condensate flow occur? *Answer*: 2 m.

12-24 Saturated steam at 54°C condenses on the outside of a 37-m-long vertical tube of 2.5 cm o.d. whose outer wall temperature is held at 43°C. For film condensation, the total condensation rate on the tube is to be determined.

(a) Show from Eq. (12-22) that the Reynolds number for the condensate is $\mathrm{Re} = 4\dot{m}/\mu_L = 5300 > 1400$ and thus turbulent conditions exist after about 17 m from the top of the tube.

(b) From the result of part a show that $\dot{m} = 0.06\ \mathrm{kg/s}$.

(c) On the basis of Eq. (12-23), show that the average turbulent heat-transfer coefficient is $\bar{h} = 4500\ \mathrm{W/m^2\ K}$.

(d) Are the $\bar{h}$ values of parts a and c of Problem 12-17 for laminar condensation approximately equal?

12-25 A simple determination of the heat-transfer coefficient for laminar film condensation of a stagnant vapor on a vertical plate is to be made. Referring to Fig. 12-9, an energy balance over the film gives

$$\frac{d\left\{ \int_0^\delta \rho_L u \left[C_{p_L}(T - T_s) + h_{fg} \right] dy \right\}}{dx} = \frac{k_L(T_s - T_w)}{\delta}$$

(a) Use linear temperature and parabolic velocity distributions for the liquid [Eqs. (12-15a) and (12-15b)] to find

$$\delta^4 = \frac{4\nu_L k_L(T_w - T_s)x/gh_{fg}(\rho_L - \rho)}{1 + 3C_{p_L}(T_w - T_s)/8h_{fg}}$$

(b) Show that the local heat-transfer coefficients as

$$h = \frac{k}{\delta} = 2^{-1/2}k \left[\frac{g(\rho_L - \rho)\,\mathrm{Pr}}{x\rho_L \nu_L^2} \frac{h_{fg}}{C_{p_L}(T_w - T_s)} \right]^{1/4} K$$

where $K = [1 + 3C_{p_L}(T_w - T_s)/8h_{fg}]^{1/4}$, in close agreement with Eq. (12-17a).

12-26 The condensate film thickness is to be estimated at the top of a horizontal tube on which stagnant saturated vapor condenses.

(a) Show that near the top of the tube ($\theta \approx 0$), $\dot{m} = C^{3/4}\theta$, where

$$C = \frac{Rk(T_w - T_s)}{h_{fg}} \left[\frac{g(\rho_L - \rho)}{3\nu_L} \right]^{1/3}$$

(b) Utilize the result of part a and Eq. (12-24) to finally obtain

$$\frac{\delta(\theta = 0)}{D} = \frac{1.11}{\left\{ \text{Ra}[h_{fg}/C_p(T_s - T_w)] \right\}^{1/4}}$$

12-27 Briefly explain how to estimate the condensation rate on the underside of a cool horizontal plate exposed to a stagnant saturated vapor. *Hint*: Utilize relations for film boiling from the topside of a hot horizontal plate (references 39 and 40 may also be helpful).

12-28 For horizontal condenser tubes arranged vertically as sketched in Fig. 12-20, the lower tubes have a condensation rate that is substantially diminished below that of the upper tubes. A staggered scheme to partially circumvent this diminution has all the condensate from an upper tube passing onto the left side of the lower tube, leaving the right side bare of condensate for rapid condensation. Show that in the limit as the number of tubes becomes large and condensation on the left side makes a negligible contribution, the average heat-transfer coefficient is half that of a single tube.

12-29 A horizontal 2-in.-o.d. tube at 90°F is surrounded by saturated steam at 2 psia. Show that the average heat-transfer coefficient is $\bar{h} = 1000$ Btu/hr ft^2 °F and that the condensate flow is laminar.

12-30 What length/diameter ratio will provide the same condensation rate in both horizontal and vertical orientations for a tube assuming that condensate flow is laminar?

12-31 Compare the average heat-transfer coefficient for condensation of a saturated stagnant vapor on a 0.61-m-high vertical surface with that for condensation on twenty-four 2.5-cm-o.d. horizontal tubes arranged in a vertical tier. *Answer*: $\bar{h}_{tube}/\bar{h}_{plate} = 0.65$.

12-32 Compare the laminar film condensation rate on a horizontal tube with the maximum condensation rate that would result if all vapor molecules incident on a surface were condensed. Take the conditions to be those of Problem 12-29. *Answer*: $\dot{m}_{max}/\dot{m}_{actual} \approx 280$.

12-33 Estimate the angular velocity at which the horizontal tube in Problem 12-29 must spin about its own axis in order to double the condensation rate (see [77]). *Answer*: 1100 rpm.

12-34 Estimate the angular velocity at which the vertical tube in Problem 12-24 must spin about its own axis to double the condensation rate. *Answer*: 1900 rpm.

12-35 Estimate the condensation rate for saturated steam at 54°C condensing on the outside of a 3.7-m-long vertical tube of 2.5 cm o.d. whose outer wall temperature is 43°C and that spins about an axis perpendicular to the tube length and located at one end of the tube. The angular velocity is 120 rpm. *Answer*: 0.011 kg/s.

12-36 Compare the condensation rate of stagnant saturated steam at 20°C on a steel plate fin of 0.64 cm thickness, 2.5 cm length, and 14°C root temperature with the condensation rate if the fin were uniformly maintained at the root temperature (see [126–128]. Estimate the accuracy of neglecting temperature variation across the fin width by evaluating the Biot number $tk_L/\delta k_f$ at the fin tip and root. *Answer*: $\eta = 0.3$.

12-37 Use the integral method to solve the problem of laminar film condensation on a rotating disk (see [72]).

12-38 Use the integral method to solve the problem of laminar film condensation on a plate with a linearly varying body force (see [81]).

12-39 Determine the condensation rate for the conditions stated in Problem 12-19 if air is present in the bulk mixture as a mass fraction of $\omega_\infty = 0.02$. Express it as a fraction of the condensation rate in the absence of a noncondensable gas. *Answer*: $\dot{m}(\omega_\infty = 0.02) = 0.28\dot{m}(\omega_\infty = 0)$.

REFERENCES

1. J. W. Meyer, *AIAA J.* **17**, 135–144 (1979).
2. H. Merte, *Adv. Heat Transfer* **9**, 181–272 (1973).
3. T. Fujii, *Theory of Laminar Film Condensation*, Springer-Verlag, 1991.
4. H. J. Palmer, *J. Fluid Mech.* **75**, 487–511 (1976).
5. E. A. Guggenheim, *Trans. Faraday Soc.* **36**, 397–412 (1940).
6. R. C. Tolman, *J. Chem. Phys.* **17**, 333–337 (1949); J. G. Kirkwood and F. P. Buff, *J. Chem. Phys.* **17**, 338–343 (1949).
7. L. E. Scriven and C. V. Sternling, *Nature* **187**, 186–188 (1960).

8. D. R. B. Kenning, *Appl. Mech. Rev.* **21**, 1101–1111 (1968).

9. J. M. Papazian, *AIAA J.* **17**, 1111–1117 (1979); J. M. Papazian and R. L. Kosson, *AIAA J.* **17**, 1279–1280 (1979).

10. S. Ostrach and A. Pradhan, *AIAA J.* **16**, 419–424 (1978).

11. J. R. A. Pearson, *J. Fluid Mech.* **4**, 489–500 (1958).

12. F. G. Collins and B. N. Antar, *AIAA J.* **20**, 1464–1466 (1982).

13. A. T. J. Hayward, *American Scientist* **59**, 434–443 (July–August 1971).

14. J. Frenkel, *Kinetic Theory of Liquids*, Clarendon, 1946, pp. 366–374.

15. P. G. Hill, H. Witting, and E. P. Demetri, *ASME J. Heat Transfer* **85**, 303–317 (1963).

16. S. A. Skillings and R. Jackson, *Int. J. Heat Fluid Flow* **8**, 139–144 (1987).

17. M. Volmer and H. Flood, *Zeitschrift für Physikalische Chemie, Abteilung A* **170**, 273–285 (1934).

18. H. L. Jaeger, E. D. Wilson, P. G. Hill, and K. C. Russell, *J. Chem. Phys.* **51**, 5380–5388 (1969).

19. J. Lothe and G. M. Pound, *J. Chem. Phys.* **45**, 630–634 (1966).

20. J. Feder, K. C. Russell, J. Lothe, and G. M. Pound, *Adv. Phys.* **15**, 111–178 (1966).

21. J. F. Welch and J. W. Westwater, Microscopic study of dropwise condensation, *International Developments in Heat Transfer* (*Proc. International Heat Transfer Conference, University of Colorado, 1961*), Part II, 1961, pp. 302–309.

22. H. Tanaka, *ASME J. Heat Transfer* **101**, 603–611 (1979).

23. C. Bonacina, S. Del Giudice, and G. Comini, *ASME J. Heat Transfer* **101**, 441–452 (1979).

24. J. H. Royal and A. E. Bergles, *ASME J. Heat Transfer* **100**, 17–24 (1978).

25. E. G. Shafrin and W. A. Zisman, *J. Phys. Chem.* **64**, 519–524 (1960).

26. A. R. Brown and M. A. Thomas, Filmwise and dropwise condensation of steam at low pressures, *Proc. of Third International Heat Transfer Conference, Chicago*, Vol. 2, 1966, pp. 300–305.

27. R. A. Erb and E. Thelen, *Ind. Eng. Chem.* **57**, 49–52 (1965).

28. K. M. Holden, A. S. Wanniarachchi, P. J. Marto, D. H. Boone, and J. W. Rose, *ASME J. Heat Transfer* **109**, 768–774 (1987).

29. H. Hampson and N. Ozisik, *Proc. Inst. Mech. Eng., London* **1B**, 282–294 (1952); L. C. F. Blackman and M. J. S. Dewar, Parts I–IV, *J. Chem. Soc.*, 162–176 (January–March 1957); L. C. F. Blackman, M. J. S. Dewar, and H. Hampson, *J. Appl. Chem.* **7**, 160–171 (1957); B. D. J. Osment, D. Tudor, R. M. M. Speirs, and W. Rugman, *Trans. Inst. Chem. Eng.* **40**, 152–160 (1962); E. G. Lefevre and J. Rose, *Int. J. Heat Mass Transfer* **8**, 1117–1133 (1965); J. L. McCormick and J. W. Westwater, *Chem. Eng. Sci.* **20**, 1021–1036 (1965).

30. E. Citakoglu and J. W. Rose, *Int. J. Heat Mass Transfer* **11**, 523–537 (1968).

31. A. C. Peterson and J. W. Westwater, *Chem. Eng. Prog. Symp. Ser.* **62**, 135–142 (1966).

32. W. Nusselt, *Zeitschrift des Vereins deutscher Inginuere* **60**, 541–575 (1916).

33. J. C. Y. Koh, *ASME J. Heat Transfer* **83**, 359–362 (1961).

34. W. M. Rohsenow, *Trans. ASME* **78**, 1645–1648 (1956).

35. P. Sadasivan and J. H. Lienhard, *ASME J. Heat Transfer* **109**, 545–547 (1987).

36. W. Churchill, *Int. J. Heat Mass Transfer* **29**, 1219–1226 (1986).

37. C. Poots and R. G. Miles, *Int. J. Heat Mass Transfer* **10**, 1677–1692 (1967).

38. R. L. Lott and J. D. Parker, *ASME J. Heat Transfer* **95**, 267–268 (1973).

39. J. Gerstmann and P. Griffith, *Int. J. Heat Mass Transfer* **10**, 567–580 (1967).

40. G. Leppert and B. Nimmo, *ASME J. Heat Transfer* **90**, 178–179 (1968).

41. D. L. Spencer and W. E. Ibele, *Proc. Third International Heat Transfer Conference, Chicago*, Vol. 2, 1966, pp. 337–347.

42. P. L. Kapitsa, *Zh. Eksp. Teoret. Fiz.* **18**, 1 (1948).

43. G. Selin, *International Developments in Heat Transfer* (*Proc. International Heat Transfer Conference, University of Colorado, 1961*), Part II, 1961, pp. 279–289.

44. J. C. Y. Koh, E. M. Sparrow, and J. P. Hartnett, *Int. J. Heat Mass Transfer* **2**, 69–82 (1961).

45. M. M. Chen, *ASME J. Heat Transfer* **83**, 48–54 (1961).

46. W. M. Rohsenow, in W. Rohsenow, J. Hartnett, and E. Ganic, Eds., *Handbook of Heat Transfer Fundamentals*, McGraw-Hill, 1985, pp. 11-1–11-36.

47. S. P. Sukhatme and W. M. Rohsenow, *ASME J. Heat Transfer* **88**, 19–28 (1966).

48. E. M. Sparrow and R. Siegel, *ASME J. Appl. Mech.* **26**, 120–121 (1959).

49. S. D. R. Wilson, *ASME J. Heat Transfer* **98**, 313–315 (1976).

50. M. I. Flik and C. L. Tien, *ASME J. Heat Transfer* **111**, 511–517 (1989); J. G. Reed, F. M. Gerner, and C. L. Tien, *J. Thermophys. Heat Transfer* **2**, 257–263 (1988).

51. W. J. Minkowycz and E. M. Sparrow, *Int. J. Heat Transfer* **12**, 147–154 (1969).

52. W. J. Minkowycz and E. M. Sparrow, *Int. J. Heat Mass Transfer* **9**, 1125–1144 (1966).

53. W. M. Rohsenow, J. H. Webber, and A. T. Ling, *Trans. ASME* **78**, 1637–1644 (1956).

54. H. R. Jacobs, *Int. J. Heat Mass Transfer* **9**, 637–648 (1966).

55. E. M. Sparrow, W. J. Minkowycz, and M. Saddy, *Int. J. Heat Mass Transfer* **10**, 1829–1845 (1967).

56. R. D. Cess, *Zeitschrift für angewandte Mathematik und Physik* **11**, 426–433 (1960).

57. J. C. Y. Koh, *Int. J. Heat Mass Transfer* **5**, 941–954 (1962).

58. S. V. Patankar and E. M. Sparrow, *ASME J. Heat Transfer* **101**, 434–440 (1979); J. E. Wilkins, *ASME J. Heat Transfer* **102**, 186–187 (1980).

59. H. Grober, S. Erk, and U. Grigull, *Fundamentals of Heat Transfer*, McGraw-Hill, 1961.

60. A. E. Dukler, *Chem. Eng. Prog. Symp. Ser.* **56** (30), 1–10 (1960).

61. J. Lee, *AIChE J.* **10**, 540–544 (1964).

62. W. H. McAdams, *Heat Transmission*, 3rd ed., McGraw-Hill, 1954.

63. M.-H. Chun and K.-T. Kim, *Proc. 1991 ASME/JSME Thermal Engineering Joint Conference*, ASME, 1991, pp. 459–464.

64. M. Jakob, *Heat Transfer*, Vol. 1, Wiley, 1949.

65. M. M. Chen, *ASME J. Heat Transfer* **83**, 55–60 (1961).

66. A. Nakayama and H. Koyama, *ASME J. Heat Transfer* **107**, 417–423 (1985).

67. C. L. Henderson and J. M. Marchello, *AIChE J.* **13**, 613–614 (1967).

68. A. G. Sheynkman and V. N. Linetskiy, *Heat Transfer—Sov. Res.* **1**, 90–97 (1969).

69. V. E. Denny and A. F. Mills, *ASME J. Heat Transfer* **91**, 495–510 (1969).

70. M. Soliman, J. R. Schuster, and P. J. Berenson, *ASME J. Heat Transfer* **90**, 267–276 (1968).

71. R. W. Lockhart and R. C. Martinelli, *Chem. Eng. Prog.* **45**, 39–48 (1949).

72. F. G. Carpenter and A. P. Colburn, *Proc. General Discussion of Heat Transfer, The Institute of Mechanical Engineers and the ASME*, July 1951, pp. 20–26.

73. M. M. Shah, *Int. J. Heat Mass Transfer* **22**, 547–556 (1979).

74. T. N. Tandon, H. K. Varma, and C. P. Gupta, *ASME J. Heat Transfer* **104**, 763–768 (1983).

75. E. M. Sparrow and J. L. Gregg, *Trans. ASME* **81**, 113–120 (1959).

76. E. M. Sparrow and J. L. Gregg, *Trans. ASME* **82**, 71–72 (1960).

77. S. S. Nadapurkar and K. O. Beatty, *Chem. Eng. Prog. Symp. Ser.* **56**, 129–137 (1960).

78. E. M. Sparrow and J. P. Hartnett, *ASME J. Heat Transfer* **83**, 101–102 (1961).

79. L. A. Bromley, *Ind. Eng. Chem.* **50**, 233–236 (1958).

80. P. J. Marto, *ASME J. Heat Transfer* **95**, 270–272 (1973).

81. A. A. Nicol and M. Gacesa, *ASME J. Heat Transfer* **92**, 144–152 (1970).

82. S. Mochizuki and T. Shiratori, *ASME J. Heat Transfer* **102**, 158–162 (1980).

83. N. V. Suryanarayana, *Proc. Fifth International Heat Transfer Conference*, Vol. II, 1974, pp. 279–285.

84. T. B. Jones, *Adv. Heat Transfer* **14**, 107–148 (1978).

85. H. R. Velkoff and J. H. Miller, *ASME J. Heat Transfer* **87**, 197–201 (1965).

86. H. Y. Choi, *ASME J. Heat Transfer* **90**, 98–102 (1968).

87. R. E. Holmes and A. J. Chapman, *ASME J. Heat Transfer* **92**, 616–620 (1970).

88. A. K. Seth and L. Lee, *ASME J. Heat Transfer* **94**, 237–260 (1974).

89. K. C. Jain and S. G. Bankoff, *ASME J. Heat Transfer* **86**, 481–489 (1964).

90. J. W. Yang, *ASME J. Heat Transfer* **92**, 252–256 (1970).

91. J. Lienhard and V. Dhir, *ASME J. Heat Transfer* **94**, 334–336 (1972).

92. N. A. Frankel and S. G. Bankoff, *ASME J. Heat Transfer* **87**, 95–102 (1965).

93. J. C. Dent, *Int. J. Heat Mass Transfer* **12**, 991–996 (1969).

94. P. Dunn and D. A. Reay, *Heat Pipes*, 2nd ed., Pergamon, 1978.

95. S. W. Chi, *Heat Pipe Theory and Practice*, McGraw-Hill, 1976.

96. D. Q. Kern, *Process Heat Transfer*, McGraw-Hill, 1950.

97. J. W. Rose, *Int. J. Heat Mass Transfer* **12**, 233–237 (1969).

98. C.-Y. Wang and C.-J. Tu, *Int. J. Heat Mass Transfer* **31**, 2339–2345 (1988).

99. E. M. Sparrow and S. H. Lin, *ASME J. Heat Transfer* **86**, 430–436 (1964).

100. H. K. Al-Diwany and J. Rose, *Int. J. Heat Mass Transfer* **16**, 1359–1369 (1973).

101. Y. Mori and K. Kijikata, *Int. J. Heat Mass Transfer* **16**, 2229–2240 (1973).

102. E. Citakoglu and J. W. Rose, *Int. J. Heat Mass Transfer* **11**, 523–537 (1968).

103. W. C. Lee and J. W. Rose, *Int. J. Heat Mass Transfer* **27**, 519–528 (1984).

104. J. W. Rose, *Int. J. Heat Mass Transfer* **23**, 539–546 (1980).

105. R. H. Turner, A. F. Mills, and V. E. Denny, *ASME J. Heat Transfer* **95**, 6–11 (1973).

106. E. M. Sparrow and E. Marschall, *ASME J. Heat Transfer* **91**, 205–211 (1969).

107. E. Marschall and R. S. Hickman, *ASME J. Heat Transfer* **95**, 1–5 (1973).

108. F. E. Sage and J. Estrin, *Int. J. Heat Mass Transfer* **19**, 323–333 (1976).

109. Y. Taitel and A. Tamir, *Int. J. Multiphase Flow* **1**, 697 (1974).

110. R. B. Bird, W. E. Stewart, and E. N. Lightfoot, *Transport Phenomena*, Wiley, 1960.

111. E. M. Sparrow and J. E. Niethammer, *ASME J. Heat Transfer* **101**, 404–410 (1979).

112. H. L. Toor, *AIChE J.* **3**, 198–207 (1957).

113. H. L. Toor, *AIChE J.* **10**, 448–455, 460–465 (1964).

114. R. Gregorig, *Zeitschrift für angewandte Mathematik and Physik* **5**, 36–49 (1954).

115. E. L. Lusterader, R. Richter, and R. N. Neugebauer, *ASME J. Heat Transfer* **81**, 297–307 (1959).

116. M. A. Kedzierski and R. L. Webb, *ASME J. Heat Transfer* **112**, 479–485 (1990).

117. A. Markowitz, B. B. Mikic, and A. E. Bergles, *ASME J. Heat Transfer* **94**, 315–320 (1972).

118. A. Briggs, X.-L. Wen, and J. W. Rose, *ASME J. Heat Transfer* **114**, 719–726 (1992).

119. A. E. Bergles, *Heat Exchanger Design Handbook*, Hemisphere, 1986, pp. 2.6.6-1–2.6.6-4.

120. A. P. Fraas, *Heat Exchanger Design*, 2nd ed., Wiley-Interscience, 1989, pp. 87–126.

121. G. L. Goglia and G. J. Van Wylen, *ASME J. Heat Transfer* **83**, 27–32 (1961).

122. C. T. R. Wilson, *Proc. Roy. Soc. London* **61**, 240–243 (1897).

123. E. L. Pinnes and W. K. Mueller, *ASME J. Heat Transfer* **101**, 617–621 (1979).

124. W. R. Gambill and J. H. Lienhard, *Proc. 1987 ASME/JSME Thermal Engineering Joint Conference*, Vol. 3, 1987, pp. 621–626.

125. K. C. D. Hickman, *Ind. Eng. Chem.* **46**, 1442–1446 (1954).

126. J. H. Lichard and V. K. Dhir, *ASME J. Heat Transfer* **96**, 197–203 (1974).

127. L. C. Burmeister, *ASME J. Heat Transfer* **104**, 391–393 (1982).

128. W. K. Nader, Extended surface heat transfer with condensation, *Proc. Sixth International Heat Transfer Conference, Toronto*, August 1978, Paper CS-5, pp. 407–412.

129. C. Marangoni, *Ann. Phys. Chem.* **143**, 337 (1871).

Appendix A

HEAT AND MASS DIFFUSION

The steady diffusive fluxes of heat and mass through a homogeneous material are related to gradients of temperature and mass fraction by

$$\mathbf{q} = -k\,\nabla T \qquad \text{and} \qquad \dot{\mathbf{m}}_1 = -\rho D_{12}\,\nabla\omega_1$$

where the thermal conductivity k and the density–mass diffusivity product ρD_{12} are material properties that are independent of direction.

There occasionally arise circumstances in which there is such a large departure from either the steady state or the homogeneous material case, that the diffusive fluxes of heat and mass are not accurately predicted by the preceding two relations. Also, under extreme conditions the diffusive heat and mass fluxes influence each other to a significant extent. To account for these effects, more complicated predictive relations are needed.

STEADY HEAT AND MASS DIFFUSION IN ANISOTROPIC MATERIALS

Materials whose properties do not depend on direction are *isotropic*, and those whose properties do depend on direction are *anisotropic*. For example, the thermal conductivity of an oak log [1] is 0.12 Btu/hr ft °F perpendicular to the grain and 0.23 Btu/hr ft °F parallel to the grain, a twofold variation; the permeability of a porous medium, often layered, shows similar directional variation [2]. Other materials, notably pyrolytic graphite with a 200-fold thermal conductivity ratio in its two principal directions and individual crystals, also exhibit substantial anisotropy. Anisotropic behavior is observed for mass diffusion as well as for heat conduction. Almost all fluids can be considered to be isotropic, although the preferential orientation of fluids

558

such as liquid crystals, blood, lubricants, and suspensions can exhibit anisotropy in both viscosity and thermal conductivity [3, 4]—a thermal conductivity ratio in the two principal directions of about 30% has been measured for some lubricants, for example.

Because heat and mass diffusion are similarly affected by anisotropy, only heat conduction is discussed in further detail. In general, the component of the heat flux vector $\mathbf{q}$ in a specified direction depends not only on the temperature gradient in that direction, but also on the temperature gradients in the other two coordinate directions. Thus for steady state in rectangular coordinates

$$-q_x = k_{xx}\frac{\partial T}{\partial x} + k_{xy}\frac{\partial T}{\partial y} + k_{xz}\frac{\partial T}{\partial z}$$

$$-q_y = k_{yx}\frac{\partial T}{\partial x} + k_{yy}\frac{\partial T}{\partial y} + k_{yz}\frac{\partial T}{\partial z} \qquad \text{(A-1)}$$

$$-q_z = k_{zx}\frac{\partial T}{\partial x} + k_{zy}\frac{\partial T}{\partial y} + k_{zz}\frac{\partial T}{\partial z}$$

in which $k_{xy} = k_{yx}$, $k_{xz} = k_{zx}$, and $k_{yz} = k_{zy}$ by symmetry considerations. Or, more compactly written in tensor notation with Einstein's convention for summation of repeated indices as

$$\mathbf{q}_i = k_{ij}\partial T/\partial x_j$$

In an isotropic material only k_{xx}, k_{yy}, and k_{zz} are nonzero. Although these three values of thermal conductivity need not necessarily be equal in an isotropic material, they usually are.

By a coordinate transformation it is always possible to find a new set of coordinate axes, called the *principal axes* and denoted by ζ_1, ζ_2, and ζ_3 for which the heat flux components are more simply given by

$$q_{\zeta_1} = -k_{\zeta_1}\frac{\partial T}{\partial_{\zeta_1}}, \qquad q_{\zeta_2} = -k_{\zeta_2}\frac{\partial T}{\partial_{\zeta_2}}, \qquad \text{and} \qquad q_{\zeta_3} = -k_{\zeta_3}\frac{\partial T}{\partial_3} \quad \text{(A-2)}$$

where k_{ζ_1}, k_{ζ_2}, and k_{ζ_3} are the *principal conductivities*. The principal axes are dictated by the material structure; for instance, they are specified for crystals. Alternatively, two-dimensional laminated materials such as plywood and transformer cores would have one principal axis parallel to the laminae and one principal axis perpendicular to them. When the principal axes coincide with the shape of the object, Eqs. (A-2) can be employed; otherwise, the more complicated Eqs. (A-1) must be used. Additional information on anisotropic diffusion is available [5–7].

Fortunately, the random orientation of anisotropic crystals causes most metal objects of appreciable size to behave in an isotropic manner. Similarly, whatever crystalline nature a liquid might possess that would tend toward anisotropy is masked by random orientations for liquid bodies of appreciable size with consequent effective isotropic behavior.

STEADY HEAT AND MASS DIFFUSION IN BINARY FLUIDS

For a pure fluid in the steady state, the rates at which heat and mass diffuse relative to the mass-averaged velocity are proportional to the temperature and mass fraction gradients, respectively. If gradients become extremely large, the linear relationships begin to lose accuracy. Likewise, if the fluid simultaneously experiences both heat and mass diffusion, the two fluxes influence each other in a manner predictable by a linear combination of the driving gradients of temperature and mass fraction. On physical grounds, this interdependence is plausible since the motion of the particles that transfers mass also transfers energy and vice versa. The Onsager relations of thermodynamics give information regarding the linearized coupling between effects that can be expected.

The general expressions for a multicomponent fluid are complex [8–10]. For a two-component fluid, they are simpler and display the essential interplay of effects more comprehensibly. In the binary fluid case [11] the diffusive heat and mass fluxes are given by

$$\dot{\mathbf{m}}_1 = \underbrace{\left[-\rho D_{12} \nabla \omega_1 \right]}_{\text{Fick's diffusion}} + \underbrace{\left[\frac{M_1 M_2 D_{12} \omega_1}{p} \nabla p \right]}_{\text{pressure diffusion}} + \underbrace{\left[\frac{M_1 M_2 \omega_1 \omega_2 D_{12}}{RT} (\mathbf{B}_2 - \mathbf{B}_1) \right]}_{\text{body-force diffusion}}$$

$$+ \underbrace{\left[\frac{-\rho D_{12} \omega_1 \omega_2 \alpha}{T} \nabla T \right]}_{\substack{\text{thermal diffusion} \\ \text{(Soret effect)}}} \tag{A-3}$$

$$\mathbf{q} = \underbrace{\left[-k \nabla T \right]}_{\substack{\text{Fourier} \\ \text{conduction}}} + \underbrace{\left[(H_1 - H_2) \dot{\mathbf{m}}_1 \right]}_{\substack{\text{interdiffusional} \\ \text{convection}}} + \underbrace{\left[\frac{-RTM^2 \alpha}{M_1 M_2} \dot{\mathbf{m}}_1 \right]}_{\substack{\text{diffusion thermo} \\ \text{(Dufour effect)}}} \tag{A-4}$$

where α is the thermal diffusion factor. Applications of such expressions for fluxes as Eq. (A-4) to boundary-layer problems have been made by Baron [12] and Hoshizaki [13]. The viscous stresses in a multicomponent fluid are affected by diffusive mass fluxes because the velocities used in the relationships for a Newtonian fluid employ mass-averaged velocities whereas viscous stresses are actually connected with momentum fluxes. In rectangular coordinates, the surface stresses are given by

$$\tau_{a_{ij}} = \tau_{ij} + \underbrace{\left[-\frac{\rho}{g_c} \left(\omega_1 \frac{\dot{m}_{1i}}{\rho_1} \frac{\dot{m}_{1j}}{\rho_1} + \omega_2 \frac{\dot{m}_{2i}}{\rho_2} \frac{\dot{m}_{2j}}{\rho_2} \right) \right]}_{\text{diffusional momentum transport}} \qquad \text{(A-5)}$$

where $\tau_{a_{ij}}$ is the apparent surface stress, τ_{ij} is the surface stress on the ith face and acting in the jth direction computed for a Newtonian fluid with the use of mass-averaged velocities, and m_{1i} is the diffusive flux of species in the ith direction relative to the mass-averaged velocity in the ith direction.

The pressure diffusion term in Eq. (A-3) indicates that a net movement of species 1 can occur if a pressure gradient is imposed. Although this is normally a negligible effect, it can be important in swirling flows where tremendous pressure gradients occasionally exist, as in some centrifuges. The body-force diffusion term is nonzero only when different body forces act on the two components. This occurs in plasma technology where the fluid interacts with electric and magnetic forces and in ionic systems, generally. If gravity is the only body force, then $\mathbf{B}_1 = \mathbf{B}_2$ and the body-force diffusion term vanishes. The thermal diffusion term describes the tendency of a species of mass to diffuse in an imposed temperature gradient and is negligible unless very large gradients are encountered. Thermal diffusion is often called the *Soret effect* and was independently predicted by Chapman and Enskog from a refined kinetic theory analysis. The thermal diffusion effect has been utilized for isotope separation in the Clusius–Dickel column [10, 14], which combines convection to achieve continuous separation. A gas mixture is placed in a vertical column whose walls are cool and with a heated wire in the center. Thermal diffusion enriches the fraction of one species in the hot central region and enriches the fraction of the other species in the cool wall region. Natural convection moves the cool species to the bottom and the hot species to the top, thereby achieving continuous species separation.

The diffusion thermo term in Eq. (A-4) is often called the *Dufour effect* in recognition of his discovery of this influence in 1873. It indicates that a diffusive mass flux gives rise to an energy flux through the thermal diffusion factor α. Although normally negligible, the diffusion thermo effect can be appreciable when, for example, helium is blown through a porous surface into a hot gas stream to protect the surface from the hot gas [15, 16]. The interdiffusional convection term of Eq. (A-4), usually negligible, indicates that diffusional mass transfer leads to a net energy flux, even when the net mass diffusion is zero if the different species of mass particles carry different amounts of energy at the same temperature. As seen in Eq. (A-4), this effect, like the other coupling effects, is of second order inasmuch as it involves the product of two small terms (the difference in species enthalpy times the mass diffusion rate relative to the mass-averaged velocity).

The diffusional momentum transport term in Eq. (A-5) is usually negligible. As discussed previously in a related case, it is a second-order term and is usually negligible since it involves the square of diffusional mass fluxes.

As pointed out by Eckert [15], the definition of properties in a multicomponent fluid requires a careful statement of conventions. To illustrate this, consider Eqs. (A-3) and (A-4), which when taken together give the heat flux, neglecting pressure diffusion and body-force diffusion, as

$$
\mathbf{q} = - \left[k + \frac{\rho D_{12} \omega_1 \omega_2 \alpha}{T} \left(H_1 - H_2 - \frac{RTM^2\alpha}{M_1 M_2} \right) \right] \nabla T
$$
$$
- \rho D_{12} \left[H_1 - H_2 - \frac{RTM^{12}\alpha}{M_1 M_2} \right] \nabla \omega_1 \tag{A-6}
$$

For a binary mixture, therefore, the thermal conductivity which describes diffusional heat flux under the condition of zero net diffusional mass flux is given in Eq. (A-4) as k, the conventionally understood property for a pure fluid. The thermal conductivity that describes diffusional heat flux under the condition of zero concentration gradient is different, as it is the coefficient of ∇T in Eq. (A-6). Only if the thermal diffusion and diffusion thermo effects are negligible are the two thermal conductivities really equivalent.

TRANSIENT HEAT, MASS, AND MOMENTUM DIFFUSION

The partial differential equations normally used to describe heat, mass, and momentum transfer provide predictions that are in excellent agreement with experiment in most cases. It is usually true that gradients in space and time are not large, and so there is no sacrifice in accuracy in letting the time interval and control volume size become infinitesimally small in the derivation of the equations of continuity, motion, energy, and mass diffusion. Rate equations (e.g., Fourier's law) that are accurate only for small gradients in the steady state are then substituted into the equations resulting from this step. Most often, departure from local equilibrium is so slight that excellent accuracy is still achieved, but the limiting behavior of the resulting equations for short times in the face of sudden disturbances (e.g., as might occur in shock waves, detonations, and pulsed later heating of a solid) is incorrect.

As pointed out by Weymann [17], whose discussion guides the following remarks, the essential difficulty can be understood by consideration of the parabolic equation that describes unsteady heat, mass, and viscous shear diffusion. This equation has the one-dimensional form,

$$
\frac{\partial \psi}{\partial t} = D \frac{\partial^2 \psi}{\partial x^2} \tag{A-7}
$$

and the fundamental solution of

$$\psi(x,t) = (4\pi Dt)^{-1/2} e^{-x^2/4Dt}$$

At $t \equiv 0$, this solution gives $\psi \equiv 0$ everywhere except at the origin, where it is large. For $t > 0$, ψ differs from zero at all x regardless of how nearly t approaches zero. Hence a disturbance is apparently propagated with an infinitely large speed, and this is physically unreasonable. Of course, this unreasonable result is usually of no interest since the predictions at "large" times are the ones of real interest and are accurate.

On physical grounds, it is apparent that the "grainy" structure of matter gives a lower limit on space resolutions. This lower limit λ is the mean free path in gases and the intermolecular spacing in liquids and solids. Similarly, the time needed for a particle of mass to adjust to changed conditions provides a lower limit on time resolutions. This lower limit τ is the inverse of the collision frequency for a simple billiard ball gas, the relaxation time needed to distribute energy in an equilibrium fashion among the internal storage modes of a complex molecule, or the time required for a disturbance to move the least space interval at the characteristic velocity (the mean thermal velocity in a gas and the speed of sound in a liquid or a solid). Although the precise value of these lower limits cannot be found, these thoughts are useful in achieving a criterion by which the applicability of the usually used describing equations [of the form of Eq. (A-7)] can be judged. In their derivation, either explicitly or implicitly, the approximation is made that

$$\psi(x_0 \pm \lambda, t) = \psi(x_0, t) \pm \left(\frac{\partial \psi}{\partial x} \bigg|_{x_0, t} \right) \lambda$$

An approximation of the same form is made for time. For these approximations and the solutions to the usual describing equations to be accurate, it is necessary that

$$\left| \frac{1}{\psi} \frac{\partial \psi}{\partial x} \right| \ll \frac{1}{\lambda} \quad \text{and} \quad \left| \frac{1}{\psi} \frac{\partial \psi}{\partial t} \right| \ll \frac{1}{\tau} \tag{A-8}$$

Solutions from Eq. (A-7) can be judged for satisfaction of physical assumptions by insertion into the criteria of Eq. (A-8).

The form of the equation more suited than Eq. (A-7) to a description of physical events at short times is derivable from a one-dimensional random walk. A walker moves along the x axis in discrete steps of size λ, each step taking a time interval τ. At the end of each step, a new one is started immediately. The probability of a step in the $+x$ direction is p, and the probability of a step in the $-x$ direction is q. The probabilities must sum to unity, $p + q = 1$. The probability that a walker is at the location $x = k\lambda$

after n steps is denoted by $u_{n,k}$. To reach $x = k\lambda$ after the $(n + 1)$th step, the walker must have been either at $x = (k - 1)\lambda$ or $x = (k + 1)\lambda$ after the nth step. In symbolic form, this is expressed as

$$u_{n+1,k} = pu_{n,k-1} + qu_{n,k+1} \tag{A-9}$$

For sufficiently large n, the discrete function $u_{n,k}$ can be approximated by a continuous function as

$$u_{n,k} \approx u(t = n\tau, x = k\lambda) \tag{A-10}$$

Insertion of Eq. (A-10) in Eq. (A-9) gives

$$u(t + \tau, x) = pu(t, x - \lambda) + qu(t, x + \lambda)$$

Introduction of a Taylor series expansion about t, x into this relationship gives

$$
\cdots + \frac{\partial^3 u}{\partial t^3} \frac{\tau^3}{3!} + \frac{\partial^2 u}{\partial t^2} \frac{\tau^2}{2} + \frac{\partial u}{\partial t} \tau + u(t, x)
$$
$$
= p\left[u(t, x) - \frac{\partial u}{\partial x}\lambda + \frac{\partial^2 u}{\partial x^2} \frac{\lambda^2}{2!} - \frac{\partial^3 u}{\partial x^3} \frac{\lambda^3}{3!} + \frac{\partial^4 u}{\partial x^4} \frac{\lambda^4}{4!} + \cdots \right]
$$
$$
+ q\left[u(t, x) + \frac{\partial u}{\partial x}\lambda + \frac{\partial^2 u}{\partial x^2} \frac{\lambda^2}{2!} + \frac{\partial^3 u}{\partial x^3} \frac{\lambda^3}{3!} + \frac{\partial^4 u}{\partial x^4} \frac{\lambda^4}{4!} + \cdots \right]
$$

which can be slightly simplified into

$$
\cdots + \frac{\partial^3 u}{\partial t^3} \frac{\tau^3}{3!} + \frac{\partial^2 u}{\partial t^2} \frac{\tau^2}{2!} + \frac{\partial u}{\partial t} \tau = (q - p)\left[\lambda \frac{\partial u}{\partial x} + \frac{\partial^3 u}{\partial x^3} \frac{\lambda^3}{3!} + \cdots \right]
$$
$$
+ \frac{\partial^2 u}{\partial x^2} \lambda^2 + \frac{\partial^4 u}{\partial x^4} \frac{2\lambda^4}{4!} + \cdots
$$

It is seen that any desired accuracy can be achieved by retaining more terms in the series expansion. Restriction of attention to the accuracy improvement over Eq. (A-7) offered by a single additional term in the time series of this expression results in, for the additional reasonable condition that $q = p$,

$$\frac{\partial^2 u}{\partial t^2} + \frac{2}{\tau} \frac{\partial u}{\partial t} = 2\frac{\lambda^2}{\tau^2} \frac{\partial^2 u}{\partial x^2} \tag{A-11}$$

Equation (A-11) is a modified wave equation that has the property that the speed of disturbance propagation is less than the average velocity $c = \lambda/\tau$ of the walker. It can be shown from kinetic theory that $\lambda^2/\tau \approx 2D$; thus Eq. (A-11) can also be written as

$$\frac{D}{c^2}\frac{\partial^2 u}{\partial t^2} + \frac{\partial u}{\partial t} = D\frac{\partial^2 u}{\partial x^2}$$

in which form it is seen that the diffusion equation form of Eq. (A-7) is accurate if the characteristic speed c is large.

An alternative derivation can be made in which, taking heat conduction in a solid as an example, it is first assumed that conductive heat flux is given by

$$\tau\frac{\partial \mathbf{q}}{\partial t} + \mathbf{q} = -k\,\nabla T \qquad (A\text{-}12)$$

Now, the result of applying the energy conservation principle to an infinitesimal control volume is

$$\rho C_p \frac{\partial T}{\partial t} = -\nabla \cdot \mathbf{q} \qquad (A\text{-}13)$$

Taking the time derivative of Eq. (A-13) in combination with the result of a $\nabla \cdot$ operation applied to Eq. (A-12), one obtains

$$\tau\frac{\partial^2 T}{\partial t^2} + \frac{\partial T}{\partial t} = \alpha\,\nabla^2 T$$

The discussion by Eckert and Drake [7, pp. 23–27] gives historical details, pointing out that unsteady (non-Fourier) effects might be important at lower temperatures in liquid helium. Among the numerous findings is that impositing a step change in heat flux q_s on the surface of a solid gives an instantaneous surface temperature jump of

$$\lim_{t \to 0} \Delta T_s = q_s\left(\frac{\tau}{\rho C_p k}\right)^{1/2}$$

for a relaxation time of τ—only 1% departure from Fourier model results is achieved after $t \sim 50\tau$ [18]. Baumeister and Hamill [19] point out that the maximum conductive heat flux that can be achieved by a step change of a surface temperature is

$$q_{max} = q(t = \tau) = \frac{\rho C_p}{3}\Delta T_s c$$

where c is the velocity of propagation. From this result it is found that the maximum convective heat-transfer coefficient is

$$h_{max} = \frac{q_{max}}{\Delta T_s} = \frac{pC_p}{3}c$$

which for water at room temperature ($c \approx c_{sound} = 1600$ m/s and $\tau = 2D/c^2 \approx 1.5 \times 10^{-13}$ s) amounts to

$$h_{max} = 3.4 \times 10^8 \text{ Btu/hr ft}^2 \text{ °F} = 19.3 \times 10^8 \text{ W/m}^2 \text{ °C}$$

whereas for air at room temperature ($c \approx c_{sound} = 1100$ ft/sec and $\tau \approx 0.5 \times 10^{-9}$ sec), one finds

$$h_{max} = 7.5 \times 10^4 \text{ Btu/hr ft}^2 \text{ °F} = 42.5 \times 10^4 \text{ W/m}^2 \text{ °C}$$

The possibility of resonant amplification that is presented by a describing equation of the wave-equation form was analytically explored by Lumsdaine [20]. It was found that thermal energy could be amplified to a higher temperature in a solid by imposing on one face a sinusoidally oscillating surface temperature whose frequency nearly equals the resonant frequency $c/2L$, where L is the characteristic size of the solid. The nonlinear behavior of temperature waves in a solid was analytically studied by Lindsay and Straughan [21] and Glass et al. [22].

REFERENCES

1. W. M. Rohsennow and H. Y. Choi, *Heat, Mass, and Momentum Transfer*, Prentice-Hall, 1961, p. 517.

2. M. Muskat, *The Flow of Homogeneous Fluids Through Porous Media*, McGraw-Hill, 1937, pp. 102–112.

3. L. N. Novichenok, G. V. Gnilitsky, and L. N. Khokhlenkov, *Progress in Heat and Mass Transfer*, Vol. 4, Pergamon, 1971, pp. 159–164.

4. M. S. Khader and R. I. Vachon, *ASME J. Heat Transfer* **99**, 684–687 (1977).

5. H. S. Carslaw and J. C. Jaeger, *Conduction of Heat in Solids*, Clarendon, 1959, pp. 38–49.

6. M. N. Ozisik, *Boundary Value Problems of Heat Conduction*, International Textbook Company, 1968, pp. 455–479.

7. E. R. G. Eckert and R. M. Drake, *Analysis of Heat and Mass Transfer*, McGraw-Hill, 1972, pp. 12–17.

8. J. O. Hirshfelder, C. F. Curtis, and R. B. Bird, *Molecular Theory of Gases and Liquids*, Wiley, 1954, pp. 514–523, 584–585.

9. S. Chapman and T. G. Cowling, *The Mathematical Theory of Non-Uniform Gases*, Cambridge University Press, 1958, pp. 244–258, 399–409.

10. R. B. Bird, W. E. Stewart, and E. N. Lightfoot, *Transport Phenomena*, Wiley, 1960, pp. 563–580.

11. E. R. G. Eckert and R. M. Drake, *Analysis of Heat and Mass Transfer*, McGraw-Hill, 1972, p. 721.

12. J. R. Baron, *ARS J.* **32**, 1053–1059 (1962).

13. H. Hoshizaki, *ARS J.* **32**, 1544–1552 (1962).

14. R. D. Present, *Kinetic Theory of Gases*, McGraw-Hill, 1958, pp. 117–127.

15. E. M. Sparrow, W. J. Minkowycz, E. R. G. Eckert, and W. E. Ibele, *ASME J. Heat Transfer* **86**, 311–319 (1964).

16. E. R. Eckert, Diffusion thermo effects in mass transfer cooling, *Proc. Fifth U.S. National Congress on Applied Mechanics*, 1966, p. 639.

17. H. D. Weymann, *Amer. J. Phys.* **35**, 448–496 (1967).

18. M. J. Maurer and H. A. Thompson, *ASME J. Heat Transfer* **95**, 284–286 (1973).

19. K. J. Baumeister and T. D. Hamill, *ASME J. Heat Transfer* **93**, 126–127 (1977).

20. E. Lumsdaine, *Mech. Eng. News* **9**, 34–37 (1972).

21. K. A. Lindsay and B. Straughan, *Zeitschrift für angewandte Mathematik und Physik* **27**, 653–662 (1976).

22. D. E. Glass, K. K. Tamma, and S. B. Railkar, *J. Thermophys. Heat Transfer* **5**, 110–116 (1991).

Appendix B

NEWTONIAN AND NON-NEWTONIAN FLUIDS

The linear relationships between velocities and the stresses acting on a fluid element which characterize a Newtonian fluid can either be determined on fundamental grounds from a detailed consideration of molecular behavior or by analogy with the relationships that describe the behavior of solids. The latter procedure is more readily comprehended and thus is pursued here.

First, consider an initially rectangular element of material that is subjected to shear as illustrated in Fig. B-1. For a solid, stresses are proportional to strains, and it is observed that

$$\tau_{yx} = G\gamma_{yx} \tag{B-1}$$

for small strains, where G is the shear modulus of the solid material and γ_{yx} is the angle in the y–x plane through which the vertical side of the material element has displaced as a consequence of the imposed shear stress τ_{yx}. For a fluid, stresses are proportional to the rate of strain, and it is observed that

$$\tau_{yx} = \mu \frac{du}{dy} \tag{B-2}$$

That these two relationships are analogous can be demonstrated by realizing that, for the fluid-flow case, the change in angular displacement that occurs in a time interval Δt is

$$\Delta\gamma_{yx} = \arctan\left\{[u(y + \Delta y) - u(y)]\frac{\Delta t}{\Delta y}\right\}$$

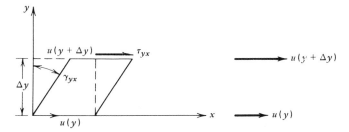

Figure B-1 Deformation rate of an initially rectangular fluid element subjected to shear.

In the limit as all differential quantities become small, this result yields

$$\dot{\gamma}_{yx} = \frac{du}{dy} \tag{B-3}$$

Insertion of this result into Eq. (B-2) shows that for a fluid

$$\tau_{yx} = \mu \dot{\gamma}_{yx} \tag{B-4}$$

which is analogous to the relationship for a solid [Eq. (B-1)], with viscosity replacing shear modulus and angular strain *rate* replacing angular strain.

In general, a fluid element experiences rotation of both the initially vertical and the initially horizontal sides. As shown in Fig. B-2, the included angle at the lower left corner of the fluid element undergoes an angular

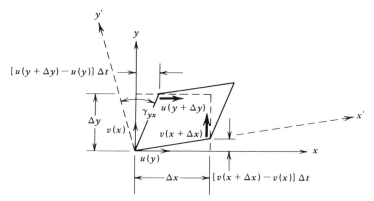

Figure B-2 Rotation of adjoining sides of an initially rectangular fluid element subjected to shear.

displacement given by

$$
\frac{\pi}{2} - \arctan\left\{ [u(y + \Delta y) - u(y)] \frac{\Delta t}{\Delta y} \right\}
$$
$$
- \arctan\left\{ [v(x + \Delta x) - v(x)] \frac{\Delta t}{\Delta x} \right\} - \frac{\pi}{2}
$$

Referring this angular displacement to the rotated $x'-y'$ axes and taking the limit as all differential quantities become small, one finds that

$$
\dot{\gamma}_{yx} = \frac{\partial u}{\partial y} + \frac{\partial v}{\partial x}
$$

In similar fashion it is found that the rates of angular strain viewed in the $y-z$ and $x-z$ planes are given by

$$
\dot{\gamma}_{yz} = \frac{\partial v}{\partial z} + \frac{\partial w}{\partial y} \quad \text{and} \quad \dot{\gamma}_{xz} = \frac{\partial w}{\partial x} + \frac{\partial u}{\partial z}
$$

As a consequence of the analogy previously developed, it then follows that

$$
\tau_{yx} = \mu\left(\frac{\partial u}{\partial y} + \frac{\partial v}{\partial x} \right) = \tau_{xy} \qquad \text{(B-5a)}
$$

$$
\tau_{yz} = \mu\left(\frac{\partial v}{\partial z} + \frac{\partial w}{\partial y} \right) = \tau_{zy} \qquad \text{(B-5b)}
$$

$$
\tau_{xz} = \mu\left(\frac{\partial w}{\partial x} + \frac{\partial u}{\partial z} \right) = \tau_{zx} \qquad \text{(B-5c)}
$$

The need for equality of τ_{ij} and τ_{ji} can be demonstrated by summing the moments shown in the free-body diagram of Fig. B-3 about the z axis. Such a

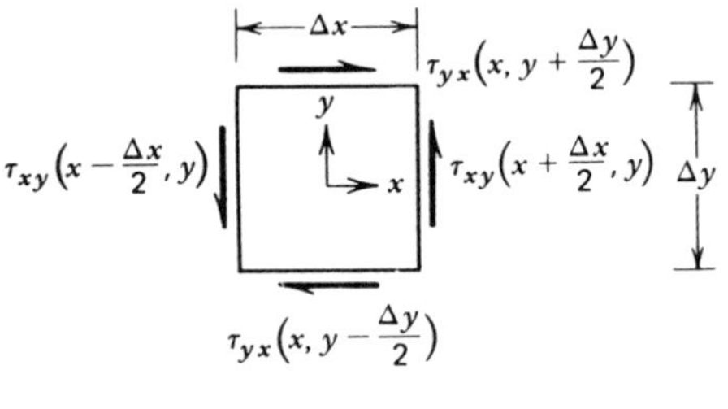

Figure B-3 Free-body diagram to demonstrate that $\tau_{ij} = \tau_{ji}$.

procedure gives

$$\sum \text{torque} = \text{moment of inertia about } z \text{ axis} \times \text{angular acceleration}$$

$$\left[\tau_{yx}\left(x, y + \frac{\Delta y}{2}\right) + \tau_{yx}\left(x, y - \frac{\Delta y}{2}\right) \right]\frac{\Delta x\,\Delta y}{2}$$

$$- \left[\tau_{xy}\left(x + \frac{\Delta x}{2}, y\right) + \tau_{xy}\left(x - \frac{\Delta x}{2}, y\right) \right]\frac{\Delta x\,\Delta y}{2}$$

$$= \left(\rho\,\Delta x\,\Delta y\,\frac{\Delta x^2 + \Delta y^2}{12} \right)\frac{\text{angular acceleration}}{g_c}$$

In the limit as all differential quantities become small, this relation becomes

$$\tau_{yx} - \tau_{xy} = \rho(\Delta x^2 + \Delta y^2)\frac{\text{angular acceleration}}{12 g_c}$$

Since it is not observed that a small fluid element spins rapidly, it must be that

$$\tau_{yx} = \tau_{xy}$$

or, in general, $\tau_{ij} = \tau_{ji}$. The existence of couple stresses has been postulated, however, that can invalidate this equality [1].

The relationship between the normal stresses and strain is more complicated than for the shear cases of Eqs. (B-1)–(B-5), although an analogy is still developed here. For a solid, imposition of a balanced normal stress in only the x direction τ_{xx} will cause a strain ε_x in the x direction whose magnitude is

$$\varepsilon_x = \frac{\tau_{xx}}{E}$$

where E is the modulus of elasticity. In addition, the imposed normal stress in the x direction will cause strains in the other perpendicular directions, ε_y and ε_z, which are proportional to ε_x so that

$$\varepsilon_y = -\frac{\varepsilon_x}{m} = -\frac{\tau_{xx}}{mE} \quad \text{and} \quad \varepsilon_z = -\frac{\varepsilon_x}{m} = -\frac{\tau_{xx}}{mE}$$

where m denotes Poisson's ratio, which is about $\frac{3}{10}$ for many solids. The net strain in the x direction for normal stresses acting in all three coordinated directions is then

$$\varepsilon_x = \frac{\tau_{xx}}{E} - \frac{\tau_{yy}}{mE} - \frac{\tau_{zz}}{mE} \tag{B-6a}$$

Similarly,

$$\varepsilon_y = \frac{\tau_{yy}}{E} - \frac{\tau_{xx}}{mE} - \frac{\tau_{zz}}{mE} \tag{B-6b}$$

and

$$\varepsilon_z = \frac{\tau_{zz}}{E} - \frac{\tau_{xx}}{mE} - \frac{\tau_{yy}}{mE} \tag{B-6c}$$

On addition of Eqs. (B-6), the volume dilation $dV/V = e = \varepsilon_x + \varepsilon_y + \varepsilon_z$ is found to be

$$e = \frac{m-2}{mE}(\tau_{xx} + \tau_{yy} + \tau_{zz}) \tag{B-7}$$

Rearrangement of Eq. (B-6a) into the form

$$E\varepsilon_x = \tau_{xx}\left(1 + \frac{1}{m}\right) - \frac{\tau_{xx} + \tau_{yy} + \tau_{zz}}{m}$$

with a following substitution from Eq. (B-7) gives

$$E\varepsilon_x = \tau_{xx}\frac{m+1}{m} - e\frac{E}{m-2}$$

In this way the normal stresses are expressed in terms of the strains as

$$\tau_{xx} = 2G\left(\varepsilon_x + \frac{e}{m-2}\right), \tag{B-8a}$$

$$\tau_{yy} = 2G\left(\varepsilon_y + \frac{e}{m-2}\right) \tag{B-8b}$$

$$\tau_{zz} = 2G\left(\varepsilon_z + \frac{e}{m-2}\right) \tag{B-8c}$$

where $G = (mE/2)/(m+1)$ relates the shear modulus G to the modulus of elasticity E and Poisson's ratio m.

To develop the analogy further, recall that the volume dilatation is

$$\varepsilon = \varepsilon_x + \varepsilon_y + \varepsilon_z$$

and that the strains in the x, y, and z directions are given by

$$\varepsilon_x = \frac{\partial \xi}{\partial x}, \qquad \varepsilon_y = \frac{\partial \eta}{\partial y}, \qquad \varepsilon_z = \frac{\partial \zeta}{\partial z}$$

in which ξ, η, and ζ are displacements of a point located at x, y, z before deformation. With this understanding, it is seen that

$$\varepsilon = \text{div}(\mathbf{s}) \tag{B-9}$$

where $\mathbf{s} = \xi\hat{\mathbf{i}} + \eta\hat{\mathbf{j}} + \zeta\hat{\mathbf{k}}$ is the vector displacement described previously.

It was demonstrated earlier that the viscous shear stress relationships for a fluid can be obtained from those for a solid by replacing G with μ and the angular strain with the rate of angular strain. In the same vein, the analogy is extended to obtain the normal stress relationships for a fluid from those for a solid by replacing G with μ (as before) and the strain by the rate of strain. Substitution of this idea into Eqs. (B-8) and (B-9) gives

$$\tau_{xx} = -p + 2\mu\frac{\partial u}{\partial x} - \mu_2\,\text{div}(\mathbf{V}) \tag{B-10a}$$

$$\tau_{yy} = -p + 2\mu\frac{\partial v}{\partial y} - \mu_2\,\text{div}(\mathbf{V}) \tag{B-10b}$$

$$\tau_{zz} = -p + 2\mu\frac{\partial w}{\partial x} - \mu_2\,\text{div}(\mathbf{V}) \tag{B-10c}$$

Here μ_2 is the second coefficient of viscosity that cannot be evaluated by analogy to the behavior of a solid. The pressure p has been added* to Eqs. (B-10) because the normal stresses must be equal to the pressure in a static situation, with the analogy to a solid's behavior giving only the viscous stresses.

Evaluation of the second coefficient of viscosity μ_2 can proceed by summing Eqs. (B-10) to obtain

$$p = -\frac{\tau_{xx} + \tau_{yy} + \tau_{zz}}{3} + \left(\frac{2\mu}{3} - \mu_2\right)\text{div}(\mathbf{V})$$

in which the coefficient $2\mu/3 - \mu_2$ is the *bulk viscosity*. The average of the normal stresses in a static situation $(\mathbf{V} = 0)$ is defined as the pressure. Pursuit of this convention into the unsteady or compressible situation then requires that there be a zero-valued coefficient of the nonzero div($\mathbf{V}$) term. This

---

*Addition of Eqs. (B-8) yields

$$\bar{\tau} = \frac{\tau_{xx} + \tau_{yy} + \tau_{xx}}{3} = \frac{2G(m+1)e}{3(m-2)}$$

Addition and substraction of $\bar{\tau}$ to and from Eq. (B-8a), for example, then yields

$$\tau_{xx} = \bar{\tau} + 2G\left(\varepsilon_x - \frac{e}{3}\right)$$

requires that the bulk viscosity, $2\mu/3 - \mu_2$, be zero; or

$$\mu_2 = \frac{2\mu}{3} \tag{B-11}$$

which is referred to as *Stoke's hypothesis*. The accuracy of Eq. (B-11) is confirmed by statistical analysis for gases of low density; in dense gases these methods show that bulk viscosity is quite small. However, the developments of this appendix all presume steady-state conditions; if markedly unsteady conditions are encountered, the bulk viscosity can differ from zero (see Marcy [2] for values for air). Processes with a relatively slow transfer of energy between translational, vibrational, and rotational degrees of freedom (e.g., as occur in rapid chemical reactions or ultrasonic vibrations) would be characterized by a nonzero bulk viscosity, in which case p cannot be the average normal stress when density variations are very large. Experiment shows that it is possible for μ_2 to be negative and of a magnitude that is 200 times greater than μ.

A discussion of the second coefficient of viscosity is given by Rosenhead [3], and the effect of bubbles in a liquid on the second coefficient of viscosity is discussed by Taylor [4] and Davies [5]. A parallel derivation of Stokes' viscosity relation is given by Schlichting [6]. The tensor calculus is used (more rigorously than the method of analogy of this appendix, but also less readably) by Schlichting [7] as well as by Deissler [8], who points out that the validity of assuming a linear relationship between stresses and velocities is as important a question as the value of bulk viscosity to be used in a linear model. By analogy for a fluid then

$$\tau_{xx} = \bar{\tau} + 2\mu\frac{\partial\mu}{\partial x} - (2/3)\mu \operatorname{div} \mathbf{V}$$

with similar results for τ_{yy} and τ_{zz}. Thus, if there is no wish to deal with thermodynamic pressure (which is really well defined only for equilibrium conditions), use of the average normal stress $\bar{\tau}$ makes consideration of the second coefficient of viscosity unnecessary. The departure of μ_2 from $2\mu/3$ is, accordingly, a measure of the lack of equilibrium in the considered process —μ_2 is then seen to not be entirely a property of the fluid.

Externally imposed electrical fields can influence the apparent viscosity of a fluid [9], although such effects are usually small.

Not all fluids possess a linear relationship between stresses and velocities. Those possessing the linear relationship are called *Newtonian fluids* and those without, *non-Newtonian fluids*. In general, mathematical models can be characterized by the number of adjustable parameters in the relationship between shear and velocity gradient. Thus a Newtonian fluid is a one-parameter model. Worthwhile qualitative discussions are given by Walker [10] and Collyer [11].

An ideal Bingham plastic fluid is one for which the shear rate is proportional to the shear stress only after a yield stress τ_0 has been surpassed. A two-parameter representation for one-dimensional Couette flow is

$$\tau_{yx} \pm \tau_0 = \mu \frac{du}{dy}, \qquad |\tau_{yx}| > \tau_0$$

$$0 = \frac{du}{dy}, \qquad |\tau_{yx}| < \tau_0$$

and the proper sign is to be affixed to τ_0 in keeping with the direction of flow. Examples are cements (typically $\mu \sim 2.4$ N s/m^2 and $\tau_0 \sim 48$ N/m^2), toothpaste, drilling mud, sewage sludge, aqueous nuclear fuel slurries, and slurpers [12] (water-saturated with cornstarch as one constituent). A plastic fluid may be better characterized as having a sharply curved early portion to its stress–strain rate curve that passes through the origin. An explanation of plastic fluid behavior is that such fluids have a three-dimensional structure that withstands shear stress below the field stress but that breaks down above this value. Many plastic fluids have a nonstraight stress–strain rate curve after surpassing the yield stress.

Shear-thinning (pseudoplastic) fluids have a shear stress–strain rate curve whose slope decreases with increasing strain rate. The apparent viscosity decreases, and hence the fluid is "thinned" by the shear. A logarithmic plot of the curve for such a fluid is generally linear, giving rise to the two-parameter power-law mathematical model for Couette flow of

$$\tau_{yx} = \mu \left| \frac{du}{dy} \right|^{n-1} \frac{du}{dy}$$

For shear-thinning fluids, $n < 1$. When $n = 1$, the fluid is Newtonian. Examples of shear-thinning fluids are dilute solutions of high polymers (e.g., polyethylene oxide or polyacrylamide), most printing inks, paper pulp, and napalm. In one explanation of this behavior, asymmetric particles are progressively aligned with streamlines, an alignment that responds nearly instantaneously to changes in imposed shear; after complete alignment at high shear, the apparent viscosity becomes constant and the curve is straight. In another explanation, a shear-thinning fluid has a structure when undisturbed that is broken down as shear stresses are increased. Shear-thinning fluids seem to be "lumpy" when poured.

Shear-thickening (dilatant) fluids are less common than shear-thinning fluids and behave conversely. Their apparent viscosity increases with increasing strain rate with $n > 1$ in the power-law model written above. Examples of shear-thickening fluids are close-packed suspensions such as some paints, quicksand, wet sand, and starch in water. A mixture of 66.7 g of wheat starch in 100 ml of water provides a good demonstration; slow stirring with a rod meets little resistance, but vigorous or rapid motion meets great resistance. Shear-thickening fluids were apparently first investigated by Osborne Reynolds in 1885. He called them dilatant fluids because his explanation of their behavior required dilation (volume increase) on shearing. In this explanation it is helpful to visualize wet sand. At low shear rates, water lubricates the movement of sand particles past one another and forces are small. At high shear rates, the close packing of sand grains is disrupted because they must lift over one another to move past and the water lubrication is insufficient and, thus, forces are large. Such a physical model requires that all suspensions of solids in liquids be dilatant at high solid content. Although this expectation is confirmed by observation, it has not been shown that *all* shear-thickening fluids undergo a volume increase when sheared. Hence the term "dilatant" is less descriptive than "shear-thickening." One alternative physical explanation is that shearing forces fluid particles together so that bonds and structure can be formed—rather than disrupted, as would be the case for a shear-thinning fluid. Hence increase of shear leads to a thicker fluid. Alternatively, it has been suggested that fluid particles rubbing against one another can acquire an electrical charge; the resultant electrical attraction leads to an increase in apparent viscosity. Walker [13] discussed the physical phenomena involved and pointed out that the effects of electrical charges explain the viscous properties of sand, muds, and powders. Yet another physical explanation is advanced for shear-thickening fluids that are suspensions or solutions of long-chain molecules. When the long-chain molecules are at rest, they are coiled. When sheared, the molecules are stretched and the apparent viscosity increases. When completely uncoiled, further shear merely orients the molecules along a streamline and shear thinning begins.

Thixotropic fluids are shear-thinning fluids whose apparent viscosity depends on both the shearing rate and the length of time the shearing has been applied. When shear is removed, a substantial amount of time is required to regain initial conditions. Margarine, some paints, shaving cream, and catsup are examples of a thixotropic fluid. One physical explanation for this effect in the case of asymmetric particles is that, as for shear-thinning, the structure possessed by an undisturbed fluid is broken down as shear stresses are increased, only now the bonds would take a longer time to be ruptured. For suspensions of clay, however, an alternative explanation seems more likely. The initial structure might be a gel; under shearing action, the gel is transformed into a less solid colloidal fluid. Third, a long-chain polymer may be aligned along streamlines with accompanying uncoiling, disentangling, and stretching to decrease viscosity in a noninstantaneous manner. Negative thixotropic fluids (rheopectic) increase their apparent viscosity and the direction of the preceding physical explanations is reversed.

Viscoelastic fluids are elastic to an important extent as well as being viscous. In general, the elasticity is due to the coiling of long-chain polymers; shearing either compresses or extends these coils, which then behave somewhat like springs. For solutions of polyethylene oxide in water, hydrogen bonding between water molecules and the oxygen atoms of the coiled polyethylene chain gives a structural order that is the origin of the elasticity of the solution. An emulsion of one Newtonian fluid in another is also viscoelastic; distortion of the shape of the dispersed droplets by shearing causes their surface energy to increase, an increase that is released when shear stress is removed. Some soap solutions are observed to be viscoelastic, as well. Silicone putty, STP Oil Treatment, some condensed soups, the thick part of egg white, Slime (a toy manufactured by the Mattel Corporation), some shampoo, some condensed milk, and gelatin in water also exhibit viscoelastic effects. A viscoelastic fluid, when strained a small amount, will spring back to its original condition when it is unstressed. A related swelling after issuing from the orifice of a die is called the *die-swell*, or *Barus*, *effect*. Such fluids have the capability of being drawn out into long fine threads, a phenomenon called *spinnability*. Self-siphoning occurs when, for example, a stream is poured from a container—after the container is again placed level, the flow continues until the container is emptied. In the Weissenberg effect a viscoelastic fluid climbs up a spinning rod in the center of a container, a behavior that is contrary to that of a Newtonian fluid (which would be depressed in the vicinity of the central rod into a parabola). The Weissenberg effect is a consequence of the shear stress created by the spinning rod leading in turn to an inward-directed radial force that pushes the fluid inward and then up the rod. The *Kaye effect* refers to the fact that when a viscoelastic fluid is poured onto a surface, small streamers of fluid occasionally emerge from the heap of fluid formed at the point of impact. The Kaye effect is believed to be due to the fact that the poured stream initially falls slowly and, because of this low shearing rate, has a high apparent viscosity and thus a heap forms at the impact point; after the heap is formed, later portions of the impacting stream can experience higher shearing rates and low apparent viscosity, enabling it to bounce elastically from the relatively rigid surface of the heap.

Old and commonly used mathematical models of these non-Newtonian fluids are, as explained in the survey by Armstrong [14], called generalized Newtonian fluid models with the viscous stress tensor τ dependent on the strain rate $\dot{\gamma}$ as

$$\tau = \eta \dot{\gamma}$$

in which the effective viscosity η is nonlinearly dependent on the magnitude $|\dot{\gamma}|$ of the strain rate as illustrated in Fig. B-4. As for the anisotropic thermal conductivity in Eq. (A-1), the strain rate $\dot{\gamma}$, sometimes expressed in vector notation as $\dot{\gamma} = \nabla V + (\nabla V)^T$ where the superscript T denotes transpose, is a

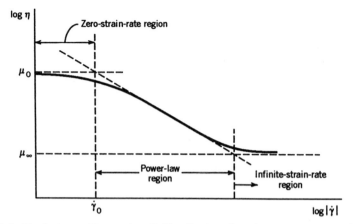

Figure B-4 Typical non-Newtonian fluid effective viscosity η versus magnitude of strain rate $|\dot{\gamma}|$.

symmetric $(\dot{\gamma}_{ij} = \dot{\gamma}_{ji})$ tensor

$$
\dot{\gamma} = \begin{pmatrix} \dot{\gamma}_{xx} & \dot{\gamma}_{xy} & \dot{\gamma}_{xz} \\ \dot{\gamma}_{yx} & \dot{\gamma}_{yy} & \dot{\gamma}_{yz} \\ \dot{\gamma}_{zx} & \dot{\gamma}_{zy} & \dot{\gamma}_{zz} \end{pmatrix}
$$

whose components, given in rectangular coordinates following Eq. (B-4) and as the coefficient of μ in Eq. (B-10), are generally expressed in tensor notation as

$$
\dot{\gamma}_{ij} = \frac{\partial v_i}{\partial x_j} + \frac{\partial v_j}{\partial x_i}
$$

The magnitude of the strain rate generally is (see Bird [15, pp. 103, 729])

$$
|\dot{\gamma}| = \left(\frac{1}{2} \dot{\gamma} : \dot{\gamma} \right)^{1/2} = \left(\frac{1}{2} \sum_i \sum_j \dot{\gamma}_{ij} \dot{\gamma}_{ji} \right)^{1/2}
$$

or, specifically, as the square root of the viscous dissipation function Φ taken from Appendix C without the $\frac{2}{3}(\nabla \cdot \mathbf{V})^2$ term. Mathematical forms of common generalized Newtonian fluid models are

Bingham

$$
\tau = \left(\mu_0 + \frac{\tau_0}{|\dot{\gamma}|} \right) \dot{\gamma}, \qquad \frac{1}{2} \tau : \tau > \tau_0^2
$$

$$
0 = \dot{\gamma}, \qquad \frac{1}{2} \tau : \tau \leqslant \tau_0^2
$$

Power law

$$\tau = \mu |\dot{\gamma}|^{n-1}\dot{\gamma}$$

Truncated power law

$$\tau = \mu_0\dot{\gamma}, \qquad\qquad |\dot{\gamma}| < \dot{\gamma}_0$$

$$\tau = \mu_0\big(|\dot{\gamma}|/\dot{\gamma}_0\big)^{n-1}\dot{\gamma}, \qquad |\dot{\gamma}| \geq \dot{\gamma}_0$$

Modified power law [16, 17]

$$\tau = \Big\{\eta_0/\big[1 + \eta_0|\dot{\gamma}|^{1-n}/K_1\big]\Big\}\dot{\gamma}$$

or $\qquad \tau = \Big\{\mu_0/\big[1 + \eta_0^{1/(1-n)}|\dot{\gamma}|/K_2\big]^{1-n}\Big\}\dot{\gamma}$

Carreau

$$\tau = \Big\{\mu_\infty + (\mu_0 - \mu_\infty)\big[1 + \big(\lambda|\dot{\gamma}|^2\big)\big]^{(n-1)/2}\Big\}\dot{\gamma}$$

The truncated power-law model avoids the flaw of too high effective viscosity as $|\dot{\gamma}| \to 0$ in the power-law model; μ_0 is the zero-strain-rate viscosity, and $\dot{\gamma}_0$ is the strain rate at which the power-law and zero-strain-rate effective viscosities are equal. The Carreau model provides continuous transition from the zero-strain-rate constant viscosity μ_0 region, through the power-law region, and to the infinite-strain-rate constant viscosity μ_∞ region. The modified power-law model only provides smooth transition between the zero-strain-rate and power-law regions, having K_1 as a constant.

Newer models contain the ideas of both viscosity and elasticity to enable representation of elastic and unsteady phenomena, especially in manufacturing processes, as surveyed by Armstrong [14]. The convected Maxwell model relating viscous stress and strain rate is

$$\tau + \lambda_0\frac{dd(\tau)}{dt} = \eta\dot{\gamma}$$

with η the effective viscosity, λ_0 a time constant (sometimes λ_0 is set equal to η/G where G is a constant unique to the fluid), and $dd(\ )/dt$ is a "convected derivative"

$$\frac{dd\tau}{dt} = \frac{D\tau}{Dt} - \big[(\nabla\mathbf{V})^T \cdot \tau + \tau \cdot \nabla\mathbf{V}\big]$$

Here $D(\ )/Dt$ is the substantial derivative while the last term $[(\nabla\mathbf{V})^T$ is the transpose of $\nabla\mathbf{V}]$ accounts for the deformation of the fluid element for which τ is calculated. Oldroyd [18] suggested that since the product of velocity gradient and stress appears in the convected Maxwell model, all possible products of velocity gradient with stress and with itself would appear in a generalization called the Oldroyd eight-constant model. Many common models can be obtained by choices for the eight constants. See Armstrong [14] for amplified discussion that includes the primary, $\tau_{xx} - \tau_{yy}$, and secondary, $\tau_{yy} - \tau_{zz}$, stress differences that are zero for a Newtonian fluid but nonzero for a non-Newtonian fluid. A survey by Bird [19] includes discussion of memory-integral mathematical models.

REFERENCES

1. V. K. Stokes, *ASME J. Heat Transfer* **91**, 182–184 (1969).

2. S. J. Marcy, *AIAA J.* **28**, 171–173 (1990).

3. L. Rosenhead, *Proc. Roy. Soc. London, Ser. A* **226**, 1–6 (1954).

4. G. I. Taylor, *Proc. Roy. Soc. London, Ser. A* **226**, 34–39 (1954).

5. R. O. Davies, *Proc. Roy. Soc. London, Ser. A* **226**, 39 (1954).

6. H. Schlichting, *Boundary Layer Theory*, 4th ed., McGraw-Hill, 1960, pp. 46–51.

7. H. Schlichting, *Boundary Layer Theory*, 6th ed., McGraw-Hill, 1968, pp. 49–60.

8. R. G. Deissler, *Amer. J. Phys.* **44**, 127–132 (1971).

9. A. A. Kim, *Prog. Heat Mass Transfer* (*Int. J. Heat Mass Transfer*) **4**, 127–132 (1971).

10. J. Walker, *Scientific American* **239**, 186–196 (1978).

11. A. A. Collyer, *Phys. Ed.* **8**, 333–338 (1973); **8**, 111–116 (1973); **9**, 313–321 (1974).

12. R. A. Hardin and L. C. Burmeister, *Fundamentals of Heat Transfer in Non-Newtonian Fluids* (Proc. 28th National Heat Transfer Conference, Minneapolis, Minnesota, July 28–31, 1991), ASME HTD-Vol. 174, 1991, pp. 29–39.

13. J. Walker, *Scientific American* **246** (1), 174–179 (1982).

14. R. C. Armstrong, *Heat Exchanger Design Handbook*, Hemisphere, 1986, pp. 2.2.8-1–2.2.8-13.

15. R. B. Bird, W. E. Stewart, and E. N. Lightfoot, *Transport Phenomena*, Wiley, 1965.

16. R. A. Brewster and T. F. Irvine, *Wärme-und Stoffübertragung* **21**, 83–86 (1987).

17. J. E. Donleavy and S. Middleman, *Trans. Soc. Rheol.* **10**, 151–168 (1966).

18. J. G. Oldroyd, *Proc. Roy. Soc. London, Ser. A* **245**, 278–297 (1958).

19. R. B. Bird, *Ann. Rev. Fluid Mech.* **8**, 13–34 (1976).

20. A. Peterline, *Nature* **227**, 598–599 (1970).

21. R. J. Gordon, *Nature* **227**, 599–600 (1970).

22. Y. Dimant and M. Porch, *Adv. Heat Transfer* **12**, 77–113 (1976).

23. H. K. Yoon and A. J. Ghajar, *ASME J. Heat Transfer* **106**, 898–900 (1984).

24. D. D. Kale and A. B. Metzner, *AIChE J.* **22**, 669–674 (1976).

25. A. B. Metzner, *Adv. Heat Transfer* **2**, 357–397 (1965).

26. Y. I. Cho and J. P. Hartnett, *Adv. Heat Transfer* **15**, 59–141 (1982).

Appendix C

CONTINUITY, MOTION, ENERGY, AND DIFFUSION EQUATIONS

Equations of continuity, motion, energy, and diffusion in several coordinate systems are presented in this appendix.

CONTINUITY

Rectangular coordinates (x, y, z)

$$\frac{\partial \rho}{\partial t} + \frac{\partial}{\partial x}(\rho v_x) + \frac{\partial}{\partial y}(\rho v_y) + \frac{\partial}{\partial z}(\rho v_z) = 0$$

Cylindrical coordinates (r, θ, z)

$$\frac{\partial \rho}{\partial t} + \frac{1}{r}\frac{\partial}{\partial r}(\rho r v_r) + \frac{1}{r}\frac{\partial}{\partial \theta}(\rho v_\theta) + \frac{\partial}{\partial z}(\rho v_z) = 0$$

Spherical coordinates (r, θ, ϕ)

$$\frac{\partial \rho}{\partial t} + \frac{1}{r^2}\frac{\partial}{\partial r}(\rho r^2 v_r) + \frac{1}{r \sin \theta}\frac{\partial}{\partial \theta}(\rho v_\theta \sin \theta) + \frac{1}{r \sin \theta}\frac{\partial}{\partial \phi}(\rho v_\phi) = 0$$

MOTION

Motion in Rectangular Coordinates

In terms of viscous stresses

x Component $\quad \rho\left(\dfrac{\partial v_x}{\partial t} + v_x\dfrac{\partial v_x}{\partial x} + v_y\dfrac{\partial v_x}{\partial y} + v_z\dfrac{\partial v_x}{\partial z}\right)$

$$= -\frac{\partial p}{\partial x} + \left(\frac{\partial \tau_{xx}}{\partial x} + \frac{\partial \tau_{yx}}{\partial y} + \frac{\partial \tau_{zx}}{\partial z}\right) + \rho g_x$$

y Component $\quad \rho\left(\dfrac{\partial v_y}{\partial t} + v_x\dfrac{\partial v_y}{\partial x} + v_y\dfrac{\partial v_y}{\partial y} + v_z\dfrac{\partial v_y}{\partial z}\right)$

$$= -\frac{\partial p}{\partial y} + \left(\frac{\partial \tau_{xy}}{\partial x} + \frac{\partial \tau_{yy}}{\partial_y} + \frac{\partial \tau_{zy}}{\partial z}\right) + \rho g_y$$

z Component $\quad \rho\left(\dfrac{\partial v_z}{\partial t} + v_x\dfrac{\partial v_z}{\partial x} + v_y\dfrac{\partial v_z}{\partial y} + v_z\dfrac{\partial v_z}{\partial z}\right)$

$$= -\frac{\partial p}{\partial z} + \left(\frac{\partial \tau_{xz}}{\partial x} + \frac{\partial \tau_{yz}}{\partial y} + \frac{\partial \tau_{zz}}{\partial z}\right) + \rho g_z$$

In terms of velocity gradients for a Newtonian fluid with constant ρ and μ

x Component $\quad \rho\left(\dfrac{\partial v_x}{\partial t} + v_x\dfrac{\partial v_x}{\partial x} + v_y\dfrac{\partial v_x}{\partial y} + v_z\dfrac{\partial v_x}{\partial z}\right)$

$$= -\frac{\partial p}{\partial x} + \mu\left(\frac{\partial^2 v_x}{\partial x^2} + \frac{\partial^2 v_x}{\partial y^2} + \frac{\partial^2 v_x}{\partial z^2}\right) + \rho g_x$$

y Component $\quad \rho\left(\dfrac{\partial v_y}{\partial t} + v_x\dfrac{\partial v_y}{\partial x} + v_y\dfrac{\partial v_y}{\partial y} + v_z\dfrac{\partial v_y}{\partial z}\right)$

$$= -\frac{\partial p}{\partial y} + \mu\left(\frac{\partial^2 v_y}{\partial x^2} + \frac{\partial^2 v_y}{\partial y^2} + \frac{\partial^2 v_y}{\partial z^2}\right) + \rho g_y$$

z Component $\quad \rho\left(\dfrac{\partial v_z}{\partial t} + v_x\dfrac{\partial v_z}{\partial x} + v_y\dfrac{\partial v_z}{\partial y} + v_z\dfrac{\partial v_z}{\partial z}\right)$

$$= -\frac{\partial p}{\partial z} + \mu\left(\frac{\partial^2 v_z}{\partial x^2} + \frac{\partial^2 v_z}{\partial y^2} + \frac{\partial^2 v_z}{\partial z^2}\right) + \rho g_z$$

Motion in Cylindrical Coordinates

In terms of viscous stresses

r Component
$$\rho\left(\frac{\partial v_r}{\partial t} + v_r\frac{\partial v_r}{\partial r} + \frac{v_\theta}{r}\frac{\partial v_r}{\partial \theta} - \frac{v_\theta^2}{r} + v_z\frac{\partial v_r}{\partial z}\right)$$

$$= -\frac{\partial p}{\partial r} + \left(\frac{1}{r}\frac{\partial}{\partial r}(r\tau_{rr}) + \frac{1}{r}\frac{\partial \tau_{r\theta}}{\partial \theta} - \frac{\tau_{\theta\theta}}{r} + \frac{\partial \tau_{rz}}{\partial z}\right) + \rho g_r$$

θ Component
$$\rho\left(\frac{\partial v_\theta}{\partial t} + v_r\frac{\partial v_\theta}{\partial r} + \frac{v_\theta}{r}\frac{\partial v_\theta}{\partial \theta} + \frac{v_r v_\theta}{r} + v_z\frac{\partial v_\theta}{\partial z}\right)$$

$$= -\frac{1}{r}\frac{\partial p}{\partial \theta} + \left(\frac{1}{r^2}\frac{\partial}{\partial r}(r^2\tau_{r\theta}) + \frac{1}{r}\frac{\partial \tau_{\theta\theta}}{\partial \theta} + \frac{\partial \tau_{\theta z}}{\partial z}\right) + \rho g_\theta$$

z Component
$$\rho\left(\frac{\partial v_z}{\partial t} + v_r\frac{\partial v_z}{\partial r} + \frac{v_\theta}{r}\frac{\partial v_z}{\partial \theta} + v_z\frac{\partial v_z}{\partial z}\right)$$

$$= -\frac{\partial p}{\partial z} + \left(\frac{1}{r}\frac{\partial}{\partial r}(r\tau_{rz}) + \frac{1}{r}\frac{\partial \tau_{\theta z}}{\partial \theta} + \frac{\partial \tau_{zz}}{\partial z}\right) + \rho g_z$$

In terms of velocity gradients for a Newtonian fluid with constant ρ and μ

r Component
$$\rho\left(\frac{\partial v_r}{\partial t} + v_r\frac{\partial v_r}{\partial r} + \frac{v_\theta}{r}\frac{\partial v_r}{\partial \theta} - \frac{v_\theta^2}{r} + v_z\frac{\partial v_r}{\partial z}\right)$$

$$= -\frac{\partial p}{\partial r} + \mu\left[\frac{\partial}{\partial r}\left(\frac{1}{r}\frac{\partial}{\partial r}(rv_r)\right)\right.$$

$$\left. + \frac{1}{r^2}\frac{\partial^2 v_r}{\partial \theta^2} - \frac{2}{r^2}\frac{\partial v_\theta}{\partial \theta} + \frac{\partial^2 v_r}{\partial z^2}\right] + \rho g_r$$

θ Component
$$\rho\left(\frac{\partial v_\theta}{\partial t} + v_r\frac{\partial v_\theta}{\partial r} + \frac{v_\theta}{r}\frac{\partial v_\theta}{\partial \theta} + \frac{v_r v_\theta}{r} + v_z\frac{\partial v_\theta}{\partial z}\right)$$

$$= -\frac{1}{r}\frac{\partial p}{\partial \theta} + \mu\left[\frac{\partial}{\partial r}\left(\frac{1}{r}\frac{\partial}{\partial r}(rv_\theta)\right)\right.$$

$$\left. + \frac{1}{r^2}\frac{\partial^2 v_\theta}{\partial \theta^2} + \frac{2}{r^2}\frac{\partial v_r}{\partial \theta} + \frac{\partial^2 v_\theta}{\partial z^2}\right] + \rho g_\theta$$

z Component
$$\rho\left(\frac{\partial v_z}{\partial t} + v_r\frac{\partial v_z}{\partial r} + \frac{v_\theta}{r}\frac{\partial v_z}{\partial \theta} + v_z\frac{\partial v_z}{\partial z}\right)$$

$$= -\frac{\partial p}{\partial z} + \mu\left[\frac{1}{r}\frac{\partial}{\partial r}\left(r\frac{\partial v_z}{\partial r}\right) + \frac{1}{r^2}\frac{\partial^2 v_z}{\partial \theta^2} + \frac{\partial^2 v_z}{\partial z^2}\right] + \rho g_z$$

Motion in Spherical Coordinates

In terms of viscous stresses

r Component $\rho\left(\dfrac{\partial v_r}{\partial t} + v_r\dfrac{\partial v_r}{\partial r} + \dfrac{v_\theta}{r}\dfrac{\partial v_r}{\partial \theta} + \dfrac{v_\phi}{r\sin\theta}\dfrac{\partial v_r}{\partial \phi} - \dfrac{v_\theta^2 + v_\phi^2}{r}\right)$

$$= -\frac{\partial p}{\partial r} + \left(\frac{1}{r^2}\frac{\partial}{\partial r}\left(r^2\tau_{rr}\right) + \frac{1}{r\sin\theta}\frac{\partial}{\partial\theta}\left(\tau_{r\theta}\sin\theta\right)\right.$$

$$\left. + \frac{1}{r\sin\theta}\frac{\partial\tau_{r\phi}}{\partial\phi} - \frac{\tau_{\theta\theta} + \tau_{\phi\phi}}{r}\right) + \rho g_r$$

θ Component $\rho\left(\dfrac{\partial v_\theta}{\partial t} + v_r\dfrac{\partial v_\theta}{\partial r} + \dfrac{v_\theta}{r}\dfrac{\partial v_\theta}{\partial \theta} + \dfrac{v_\phi}{r\sin\theta}\dfrac{\partial v_\theta}{\partial \phi} + \dfrac{v_r v_\theta}{r} - \dfrac{v_\phi^2\cot\theta}{r}\right)$

$$= -\frac{1}{r}\frac{\partial p}{\partial\theta} + \left(\frac{1}{r^2}\frac{\partial}{\partial r}\left(r^2\tau_{r\theta}\right) + \frac{1}{r\sin\theta}\frac{\partial}{\partial\theta}\left(\tau_{\theta\theta}\sin\theta\right)\right.$$

$$\left. + \frac{1}{r\sin\theta}\frac{\partial\tau_{\theta\phi}}{\partial\phi} + \frac{\tau_{r\theta}}{r} - \frac{\cot\theta}{r}\tau_{\phi\phi}\right) + \rho g_\theta$$

ϕ Component $\rho\left(\dfrac{\partial v_\phi}{\partial t} + v_r\dfrac{\partial v_\phi}{\partial r} + \dfrac{v_\theta}{r}\dfrac{\partial v_\phi}{\partial \theta} + \dfrac{v_\phi}{r\sin\theta}\dfrac{\partial v_\phi}{\partial \phi} + \dfrac{v_\phi v_r}{r} + \dfrac{v_\theta v_\phi}{r}\cot\theta\right)$

$$= -\frac{1}{r\sin\theta}\frac{\partial p}{\partial\phi} + \left(\frac{1}{r^2}\frac{\partial}{\partial r}\left(r^2\tau_{r\phi}\right) + \frac{1}{r}\frac{\partial\tau_{\theta\phi}}{\partial\theta} + \frac{1}{r\sin\theta}\frac{\partial\tau_{\phi\phi}}{\partial\phi}\right.$$

$$\left. + \frac{\tau_{r\phi}}{r} + \frac{2\cot\theta}{r}\tau_{\theta\phi}\right) + \rho g_\phi$$

In terms of velocity gradients for a Newtonian Fluid with constant ρ and μ

r Component $\rho\left(\dfrac{\partial v_r}{\partial t} + v_r\dfrac{\partial v_r}{\partial r} + \dfrac{v_\theta}{r}\dfrac{\partial v_r}{\partial \theta} + \dfrac{v_\phi}{r\sin\theta}\dfrac{\partial v_r}{\partial \phi} - \dfrac{v_\theta^2 + v_\phi^2}{r}\right)$

$$= -\frac{\partial p}{\partial r} + \mu\left(\nabla^2 v_r - \frac{2}{r^2}v_r - \frac{2}{r^2}\frac{\partial v_\theta}{\partial\theta} - \frac{2}{r^2}v_\theta\cot\theta\right.$$

$$\left. - \frac{2}{r^2\sin\theta}\frac{\partial v_\phi}{\partial\phi}\right) + \rho g_r$$

θ Component $\rho\left(\dfrac{\partial v_\theta}{\partial t} + v_r\dfrac{\partial v_\theta}{\partial r} + \dfrac{v_\theta}{r}\dfrac{\partial v_\theta}{\partial \theta} + \dfrac{v_\phi}{r\sin\theta}\dfrac{\partial v_\theta}{\partial \phi} + \dfrac{v_r v_\theta}{r} - \dfrac{v_\phi^2\cot\theta}{r}\right)$

$$= -\frac{1}{r}\frac{\partial p}{\partial\theta} + \mu\left(\nabla^2 v_\theta + \frac{2}{r^2}\frac{\partial v_r}{\partial\theta} - \frac{v_\theta}{r^2\sin^2\theta}\right.$$

$$\left. - \frac{2\cos\theta}{r^2\sin^2\theta}\frac{\partial v_\phi}{\partial\phi}\right) + \rho g_\theta$$

ϕ Component $\quad \rho \left(\dfrac{\partial v_\phi}{\partial t} + v_r \dfrac{\partial v_\phi}{\partial r} + \dfrac{v_\theta}{r} \dfrac{\partial v_\phi}{\partial \theta} + \dfrac{v_\phi}{r \sin \theta} \dfrac{\partial v_\phi}{\partial \phi} + \dfrac{v_\phi v_r}{r} + \dfrac{v_\theta v_\phi}{r} \cot \theta \right)$

$$= -\dfrac{1}{r \sin \theta} \dfrac{\partial p}{\partial \phi} + \mu \left(\nabla^2 v_\phi - \dfrac{v_\phi}{r^2 \sin^2 \theta} + \dfrac{2}{r^2 \sin \theta} \dfrac{\partial v_r}{\partial \phi} \right.$$

$$\left. + \dfrac{2 \cos \theta}{r^2 \sin^2 \theta} \dfrac{\partial v_\theta}{\partial \phi} \right) + \rho g_\phi$$

where

$$\nabla^2 = \dfrac{1}{r^2} \dfrac{\partial}{\partial r} \left(r^2 \dfrac{\partial}{\partial r} \right) + \dfrac{1}{r^2 \sin \theta} \dfrac{\partial}{\partial \theta} \left(\sin \theta \dfrac{\partial}{\partial \theta} \right) + \dfrac{1}{r^2 \sin^2 \theta} \left(\dfrac{\partial^2}{\partial \phi^2} \right)$$

VISCOUS STRESS COMPONENTS FOR NEWTONIAN FLUIDS

Rectangular Coordinates (x, y, z)

$$\tau_{xx} = \mu \left[2 \dfrac{\partial v_x}{\partial x} - \dfrac{2}{3} (\nabla \cdot \mathbf{v}) \right], \qquad \tau_{yy} = \mu \left[2 \dfrac{\partial v_y}{\partial y} - \dfrac{2}{3} (\nabla \cdot \mathbf{v}) \right]$$

$$\tau_{zz} = \mu \left[2 \dfrac{\partial v_z}{\partial z} - \dfrac{2}{3} (\nabla \cdot \mathbf{v}) \right]$$

$$\tau_{xy} = \tau_{yx} = \mu \left[\dfrac{\partial v_x}{\partial y} + \dfrac{\partial v_y}{\partial x} \right], \qquad \tau_{yz} = \tau_{zy} = \mu \left[\dfrac{\partial v_y}{\partial z} + \dfrac{\partial v_z}{\partial y} \right]$$

$$\tau_{zx} = \tau_{xz} = \mu \left[\dfrac{\partial v_z}{\partial x} + \dfrac{\partial v_x}{\partial z} \right]$$

Cylindrical Coordinates (r, θ, z)

$$\tau_{rr} = \mu \left[2 \dfrac{\partial v_r}{\partial r} - \dfrac{2}{3} (\nabla \cdot \mathbf{v}) \right], \qquad \tau_{\theta\theta} = \mu \left[2 \left(\dfrac{1}{r} \dfrac{\partial v_\theta}{\partial \theta} + \dfrac{v_r}{r} \right) - \dfrac{2}{3} (\nabla \cdot \mathbf{v}) \right]$$

$$\tau_{zz} = \mu \left[2 \dfrac{\partial v_z}{\partial z} - \dfrac{2}{3} (\nabla \cdot \mathbf{v}) \right]$$

$$\tau_{r\theta} = \tau_{\theta r} = \mu \left[r \dfrac{\partial}{\partial r} \left(\dfrac{v_\theta}{r} \right) + \dfrac{1}{r} \dfrac{\partial v_r}{\partial \theta} \right], \qquad \tau_{\theta z} = \tau_{z\theta} = \mu \left[\dfrac{\partial v_\theta}{\partial z} + \dfrac{1}{r} \dfrac{\partial v_z}{\partial \theta} \right]$$

$$\tau_{zr} = \tau_{rz} = \mu \left[\dfrac{\partial v_z}{\partial r} + \dfrac{\partial v_r}{\partial z} \right]$$

Spherical Coordinates (r, θ, ϕ)

$$\tau_{rr} = \mu\left[2\frac{\partial v_r}{\partial r} - \frac{2}{3}(\nabla \cdot \mathbf{v})\right], \qquad \tau_{\theta\theta} = \mu\left[2\left(\frac{1}{r}\frac{\partial v_\theta}{\partial \theta} + \frac{v_r}{r}\right) - \frac{2}{3}(\nabla \cdot \mathbf{v})\right]$$

$$\tau_{\phi\phi} = \mu\left[2\left(\frac{1}{r \sin \theta}\frac{\partial v_\phi}{\partial \phi} + \frac{v_r}{r} + \frac{v_\theta \cot \theta}{r}\right) - \frac{2}{3}(\nabla \cdot \mathbf{v})\right]$$

$$\tau_{r\theta} = \tau_{\theta r} = \mu\left[r\frac{\partial}{\partial r}\left(\frac{v_\theta}{r}\right) + \frac{1}{r}\frac{\partial v_r}{\partial \theta}\right], \qquad \tau_{\phi r} = \tau_{r\phi} = \mu\left[\frac{1}{r \sin \theta}\frac{\partial v_r}{\partial \phi} + r\frac{\partial}{\partial r}\left(\frac{v_\phi}{r}\right)\right]$$

$$\tau_{\theta\phi} = \tau_{\phi\theta} = \mu\left[\frac{\sin \theta}{r}\frac{\partial}{\partial \theta}\left(\frac{v_\phi}{\sin \theta}\right) + \frac{1}{r \sin \theta}\frac{\partial v_\theta}{\partial \phi}\right]$$

ENERGY

Energy in Rectangular Coordinates

In terms of viscous stresses and heat fluxes

$$\rho C_p \left(\frac{\partial T}{\partial t} + u\frac{\partial T}{\partial x} + v\frac{\partial T}{\partial y} + w\frac{\partial T}{\partial x}\right)$$

$$= -\left(\frac{\partial q_x}{\partial x} + \frac{\partial q_y}{\partial y} + \frac{\partial q_z}{\partial z}\right) + q'''$$

$$+ \beta T\left(\frac{\partial p}{\partial t} + u\frac{\partial p}{\partial x} + v\frac{\partial p}{\partial y} + w\frac{\partial p}{\partial z}\right) + p\left(\frac{\partial u}{\partial x} + \frac{\partial v}{\partial y} + \frac{\partial w}{\partial z}\right)$$

$$+ \left[\left(\tau_{xx}\frac{\partial u}{\partial x} + \tau_{yx}\frac{\partial u}{\partial y} + \tau_{zx}\frac{\partial u}{\partial z}\right) + \left(\tau_{xy}\frac{\partial v}{\partial x} + \tau_{yy}\frac{\partial v}{\partial y} + \tau_{zy}\frac{\partial v}{\partial z}\right)\right.$$

$$\left. + \left(\tau_{xz}\frac{\partial w}{\partial x} + \tau_{yz}\frac{\partial w}{\partial y} + \tau_{zz}\frac{\partial w}{\partial z}\right)\right]$$

In terms of velocity and temperature gradients for $\rho, k = $ const

$$\rho C_p\left(\frac{\partial T}{\partial t} + u\frac{\partial T}{\partial x} + v\frac{\partial T}{\partial y} + w\frac{\partial T}{\partial z}\right)$$

$$= k\left(\frac{\partial^2 T}{\partial x^2} + \frac{\partial^2 T}{\partial y^2} + \frac{\partial^2 T}{\partial z^2}\right) + q'''$$

$$+ \beta T\left(\frac{\partial p}{\partial t} + u\frac{\partial p}{\partial x} + v\frac{\partial p}{\partial y} + w\frac{\partial p}{\partial z}\right) + \mu\Phi$$

Energy in Cylindrical Coordinates

In terms of viscous stresses and heat fluxes

$$\rho C_p \left(\frac{\partial T}{\partial t} + v_r \frac{\partial T}{\partial r} + \frac{v_\theta}{r} \frac{\partial T}{\partial \theta} + v_z \frac{\partial T}{\partial z} \right)$$

$$= -\left[\frac{1}{r} \frac{\partial (r q_r)}{\partial r} + \frac{1}{r} \frac{\partial q_\theta}{\partial \theta} + \frac{\partial q_z}{\partial z} \right]$$

$$+ q''' + \beta T \left(\frac{\partial p}{\partial t} + v_r \frac{\partial p}{\partial r} + \frac{v_\theta}{r} \frac{\partial p}{\partial \theta} + v_z \frac{\partial p}{\partial z} \right)$$

$$- p \left[\frac{1}{r} \frac{\partial (r v_r)}{\partial r} + \frac{1}{r} \frac{\partial v_\theta}{\partial \theta} + \frac{\partial v_z}{\partial z} \right]$$

$$+ \left\{ \tau_{rr} \frac{\partial v_r}{\partial r} + \tau_{\theta\theta} \frac{1}{r} \left(\frac{\partial v_\theta}{\partial \theta} + v_r \right) + \tau_{zz} \frac{\partial v_z}{\partial z} \right.$$

$$\left. + \tau_{r\theta} \left[r \frac{\partial (v_\theta/r)}{\partial r} + \frac{1}{r} \frac{\partial v_r}{\partial \theta} \right] + \tau_{rz} \left(\frac{\partial v_z}{\partial r} + \frac{\partial v_r}{\partial z} \right) \right\}$$

In terms of velocity and temperature gradients for $\rho, k = \text{const}$

$$\rho C_p \left(\frac{\partial T}{\partial t} + v_r \frac{\partial T}{\partial r} + \frac{v_\theta}{r} \frac{\partial T}{\partial \theta} + v_z \frac{\partial T}{\partial z} \right)$$

$$= k \left[\frac{1}{r} \frac{\partial (r \, \partial T/\partial r)}{\partial r} + \frac{1}{r^2} \frac{\partial^2 T}{\partial \theta^2} + \frac{\partial^2 T}{\partial z^2} \right]$$

$$+ q''' + \beta T \left(\frac{\partial p}{\partial t} + v_r \frac{\partial p}{\partial r} + \frac{v_\theta}{r} \frac{\partial p}{\partial \theta} + v_z \frac{\partial p}{\partial z} \right) + \mu \Phi$$

Energy in Spherical Coordinates

In terms of viscous stresses and heat fluxes

$$\rho C_p \left(\frac{\partial T}{\partial t} + v_r \frac{\partial T}{\partial r} + \frac{v_\theta}{r} \frac{\partial T}{\partial \theta} + \frac{v_\phi}{r \sin \theta} \frac{\partial T}{\partial \phi} \right)$$

$$= -\left[\frac{1}{r^2} \frac{\partial (r^2 q_r)}{\partial r} + \frac{1}{r \sin \theta} \frac{\partial (q_\theta \sin \theta)}{\partial \theta} + \frac{1}{r \sin \theta} \frac{\partial q_\phi}{\partial \phi} \right]$$

$$+ q''' + \beta T \left(\frac{\partial p}{\partial t} + v_r \frac{\partial T}{\partial r} + \frac{v_\theta}{r} \frac{\partial T}{\partial \theta} + \frac{v_\phi}{r \sin \theta} \frac{\partial T}{\partial \phi} \right)$$

$$+ p \left[\frac{1}{r^2} \frac{\partial (r^2 v_r)}{\partial r} + \frac{1}{r \sin \theta} \frac{\partial (v_\theta \sin \theta)}{\partial \theta} + \frac{1}{r \sin \theta} \frac{\partial v_\phi}{\partial \phi} \right]$$

$$+ \left\{ \tau_{rr} \frac{\partial v_r}{\partial r} + \tau_{\theta\theta} \left(\frac{1}{r} \frac{\partial v_\theta}{\partial \theta} + \frac{v_r}{r} \right) + \tau_{\phi\phi} \left(\frac{1}{r \sin \theta} \frac{\partial v_\phi}{\partial \phi} + \frac{v_r}{r} + \frac{v_\theta \cot \theta}{r} \right) \right.$$

$$+ \tau_{r\theta} \left(\frac{\partial v_\theta}{\partial r} + \frac{1}{r} \frac{\partial v_r}{\partial \theta} - \frac{v_\theta}{r} \right) + \tau_{r\phi} \left(\frac{\partial v_\phi}{\partial r} + \frac{1}{r \sin \theta} \frac{\partial v_r}{\partial \phi} - \frac{v_\phi}{r} \right)$$

$$+ \left. \tau_{\theta\phi} \left(\frac{1}{r} \frac{\partial v_\phi}{\partial \theta} + \frac{1}{r \sin \theta} \frac{\partial v_\theta}{\partial \phi} - \frac{\cot \theta}{r} v_\phi \right) \right\} $$

In terms of velocity and temperature gradients for $\rho, k = \mathrm{const}$

$$\rho C_p \left(\frac{\partial T}{\partial t} + v_r \frac{\partial T}{\partial r} + \frac{v_\theta}{r} \frac{\partial T}{\partial \theta} + \frac{v_\phi}{r \sin \theta} \frac{\partial T}{\partial \phi} \right)$$

$$= k \left[\frac{1}{r^2} \frac{\partial (r^2 \, \partial T / \partial r)}{\partial r} + \frac{1}{r^2 \sin \theta} \frac{\partial (\sin \theta \, \partial T / \partial \theta)}{\partial \theta} + \frac{1}{r^2 \sin \theta} \frac{\partial^2 T}{\partial \phi^2} \right] + q'''$$

$$+ \beta T \left(\frac{\partial p}{\partial t} + v_r \frac{\partial T}{\partial r} + \frac{v_\theta}{r} \frac{\partial T}{\partial \theta} + \frac{v_\phi}{r \sin \theta} \frac{\partial T}{\partial \phi} \right) + \mu \Phi$$

VISCOUS DISSIPATION FUNCTION

Rectangular

$$\Phi = 2 \left[\left(\frac{\partial u}{\partial x} \right)^2 + \left(\frac{\partial v}{\partial y} \right)^2 + \left(\frac{\partial w}{\partial z} \right)^2 \right]$$

$$+ \left[\frac{\partial v}{\partial x} + \frac{\partial u}{\partial y} \right]^2 + \left[\frac{\partial w}{\partial y} + \frac{\partial v}{\partial z} \right]^2 + \left[\frac{\partial u}{\partial z} + \frac{\partial w}{\partial x} \right]^2$$

$$- \frac{2}{3} \left[\frac{\partial u}{\partial x} + \frac{\partial v}{\partial y} + \frac{\partial w}{\partial z} \right]^2$$

Cylindrical

$$\Phi = 2 \left[\left(\frac{\partial v_r}{\partial r} \right)^2 + \left(\frac{1}{r} \frac{\partial v_\theta}{\partial \theta} + \frac{v_r}{r} \right)^2 + \left(\frac{\partial v_z}{\partial z} \right)^2 \right]$$

$$+ \left[r \frac{\partial}{\partial r} \left(\frac{v_\theta}{r} \right) + \frac{1}{r} \frac{\partial v_r}{\partial \theta} \right]^2 + \left[\frac{1}{r} \frac{\partial v_z}{\partial \theta} + \frac{\partial v_\theta}{\partial z} \right]^2$$

$$+ \left[\frac{\partial v_r}{\partial z} + \frac{\partial v_z}{\partial r} \right]^2 - \frac{2}{3} \left[\frac{1}{r} \frac{\partial}{\partial r} (r v_r) + \frac{1}{r} \frac{\partial v_\theta}{\partial \theta} + \frac{\partial v_z}{\partial z} \right]^2$$

Spherical

$$\Phi = 2\left[\left(\frac{\partial v_r}{\partial r}\right)^2 + \left(\frac{1}{r}\frac{\partial v_\theta}{\partial \theta} + \frac{v_r}{r}\right)^2 + \left(\frac{1}{r \sin \theta}\frac{\partial v_\phi}{\partial \phi} + \frac{v_r}{r} + \frac{v_\theta \cot \theta}{r}\right)^2\right]$$

$$+ \left[r\frac{\partial}{\partial r}\left(\frac{v_\theta}{r}\right) + \frac{1}{r}\frac{\partial v_r}{\partial \theta}\right]^2 + \left[\frac{\sin \theta}{r}\frac{\partial}{\partial \theta}\left(\frac{v_\phi}{\sin \theta}\right) + \frac{1}{r \sin \theta}\frac{\partial v_\theta}{\partial \phi}\right]^2$$

$$+ \left[\frac{1}{r \sin \theta}\frac{\partial v_r}{\partial \phi} + r\frac{\partial}{\partial r}\left(\frac{v_\phi}{r}\right)\right]^2$$

$$- \frac{2}{3}\left[\frac{1}{r^2}\frac{\partial}{\partial r}(r^2 v_r) + \frac{1}{r \sin \theta}\frac{\partial}{\partial \theta}(v_\theta \sin \theta) + \frac{1}{r \sin \theta}\frac{\partial v_\phi}{\partial \phi}\right]^2$$

HEAT-FLUX COMPONENTS

Rectangular

$$q_x = -k\frac{\partial T}{\partial x}, \qquad q_y = -k\frac{\partial T}{\partial y}, \qquad q_z = -k\frac{\partial T}{\partial z}$$

Cylindrical

$$q_r = -k\frac{\partial T}{\partial r}, \qquad q_\theta = -k\frac{1}{r}\frac{\partial T}{\partial \theta}, \qquad q_z = -k\frac{\partial T}{\partial z}$$

Spherical

$$q_r = -k\frac{\partial T}{\partial r}, \qquad q_\theta = -k\frac{1}{r}\frac{\partial T}{\partial \theta}, \qquad q_\phi = -k\frac{1}{r \sin \theta}\frac{\partial T}{\partial \phi}$$

BINARY DIFFUSION

In terms of mass fluxes

Rectangular

$$\rho\left(\frac{\partial \omega_1}{\partial t} + u\frac{\partial \omega_1}{\partial x} + v\frac{\partial \omega_1}{\partial y} + w\frac{\partial \omega_1}{\partial z}\right)$$

$$= r_1''' - \left(\frac{\partial \dot{m}_{1x}}{\partial x} + \frac{\partial \dot{m}_{1y}}{\partial y} + \frac{\partial \dot{m}_{1z}}{\partial z}\right)$$

Cylindrical

$$\rho\left(\frac{\partial \omega_1}{\partial t} + v_r\frac{\partial \omega_1}{\partial r} + \frac{v_\theta}{r}\frac{\partial \omega_1}{\partial \theta} + v_z\frac{\partial \omega_1}{\partial z}\right)$$

$$= r_1''' - \left[\frac{1}{r}\frac{\partial(r\dot{m}_{1r})}{\partial r} + \frac{1}{r}\frac{\partial \dot{m}_{1\theta}}{\partial \theta} + \frac{\partial \dot{m}_{1z}}{\partial z}\right]$$

Spherical

$$\rho\left(\frac{\partial \omega_1}{\partial t} + v_r\frac{\partial \omega_1}{\partial r} + \frac{v_\theta}{r}\frac{\partial \omega_1}{\partial \theta} + \frac{v_\phi}{r\sin\theta}\frac{\partial \omega_1}{\partial \phi}\right)$$

$$= r_1''' - \left[\frac{1}{r^2}\frac{\partial(r^2\dot{m}_{1r})}{\partial r} + \frac{1}{r\sin\theta}\frac{\partial(\dot{m}_{1\theta}\sin\theta)}{\partial \theta}\right.$$

$$\left. + \frac{1}{r\sin\theta}\frac{\partial(\dot{m}_{1\phi})}{\partial \phi}\right]$$

In terms of mass fraction gradients for ρ, D_{12} = const

Rectangular

$$\frac{\partial \omega_1}{\partial t} + u\frac{\partial \omega_1}{\partial x} + v\frac{\partial \omega_1}{\partial y} + w\frac{\partial \omega_1}{\partial z}$$

$$= \frac{r_1'''}{\rho} + D_{12}\left[\frac{\partial^2 \omega_1}{\partial x^2} + \frac{\partial^2 \omega_1}{\partial y^2} + \frac{\partial^2 \omega_1}{\partial z^2}\right]$$

Cylindrical

$$\frac{\partial \omega_1}{\partial t} + v_r\frac{\partial \omega_1}{\partial r} + \frac{v_\theta}{r}\frac{\partial \omega_1}{\partial \theta} + v_z\frac{\partial \omega_1}{\partial z}$$

$$= \frac{r_1'''}{\rho} + D_{12}\left[\frac{1}{r}\frac{\partial(r\,\partial \omega_1/\partial r)}{\partial r} + \frac{1}{r^2}\frac{\partial^2 \omega_1}{\partial \theta^2} + \frac{\partial^2 \omega_1}{\partial z^2}\right]$$

Spherical

$$\frac{\partial \omega_1}{\partial t} + v_r\frac{\partial \omega_1}{\partial r} + \frac{v_\theta}{r}\frac{\partial \omega_1}{\partial \theta} + \frac{v_\phi}{r\sin\theta}\frac{\partial \omega_1}{\partial \phi}$$

$$= \frac{r_1'''}{\rho} + D_{12}\left[\frac{1}{r^2}\frac{\partial(r^2\,\partial \omega_1/\partial r)}{\partial r} + \frac{1}{r^2\sin\theta}\frac{\partial(\sin\theta\,\partial \omega_1/\partial \theta)}{\partial \theta}\right.$$

$$\left. + \frac{1}{r^2\sin\theta}\frac{\partial^2 \omega_1}{\partial \phi^2}\right]$$

Appendix D

SIMILARITY TRANSFORMATION BY A SEPARATION OF VARIABLES

Methods by which similarity transformations can be discovered for reduction of the number of variables are treated by Hansen [1]. He discusses the free parameter, separation of variables (largely refined by Abbott and Kline [2]), group theory, and dimensional analysis methods. The method due to Moore [3] is particularly suited to unsteady problems. Similarity transformation of a boundary-layer motion equation by a separation of variables method is discussed in this appendix.

Generally speaking, two types of problems are encountered. One type is the *well-posed* problem, which has a completely prescribed set of boundary and initial conditions. For a general form of the similarity transformation, such a problem has either one similarity variable or none. A second type is the problem in which some, but not all, boundary and initial conditions are given. None, one, or many similarity variables may be found to exist. The laminar boundary-layer equations are usually of the second type inasmuch as the free-stream conditions are not specified precisely and a velocity distribution is not specified at one particular position.

The result of the similarity transformation usually, but not always, is a reduction from a partial differential to an ordinary differential equation that, because of the nonlinearity of the physical phenomena described, is nonlinear. The ordinary differential equation need only be solved once, albeit by numerical methods, to enable application to all situations encompassed by the original partial differential equation.

Because the method of separation of variables is familiar to many in the context of heat-conduction problem solution, it is used to show how a similarity form can be found for the x-motion laminar boundary-layer equation for steady flow of a constant-property fluid over a wedge.

The equation considered is

$$u\frac{\partial u}{\partial x} + v\frac{\partial u}{\partial y} = U\frac{dU}{dx} + v\frac{\partial^2 u}{\partial y^2} \tag{D-1}$$

$$u(x, y = 0) = 0 = v(x, y = 0)$$
$$u(x, y \to \infty) \to U(x) \tag{D-2}$$

The stream function, satisfying the continuity equation since $u = \partial\psi/\partial y$ and $v = -\partial\psi/\partial x$, is assumed to have the separation of variables form

$$\psi = H(\zeta)F(\eta) \tag{D-3}$$

where the coordinate transformation

$$x, y \to \zeta(x, y), \qquad \eta(x, y)$$

has been employed. For specific treatment let

$$\zeta = x \qquad \text{and} \qquad \eta = yg(x) \tag{D-4}$$

where $g(x)$ is a yet-unknown function of x.

The coordinate transformation specified by Eq. (D-4) gives

$$\frac{\partial\zeta}{\partial x} = 1, \qquad \frac{\partial\zeta}{\partial y} = 0, \qquad \frac{\partial\eta}{\partial x} = y\frac{dg}{dx} = \eta\frac{d\ln(g)}{dx}, \qquad \text{and} \qquad \frac{\partial\eta}{\partial y} = g$$

From this the velocities and their derivatives are found by the chain rule to be given in the transformed coordinates in terms of the stream function as

$$u = \frac{\partial\psi}{\partial y} = \frac{\partial\psi}{\partial\zeta}\frac{\partial\zeta}{\partial y} + \frac{\partial\psi}{\partial\eta}\frac{\partial\eta}{\partial y} = g\frac{\partial\psi}{\partial\eta} \tag{D-5}$$

$$-v = \frac{\partial\psi}{\partial x} = \frac{\partial\psi}{\partial\zeta}\frac{\partial\zeta}{\partial x} + \frac{\partial\psi}{\partial\eta}\frac{\partial\eta}{\partial x} = \frac{\partial\psi}{\partial\zeta} + \eta\frac{d\ln(g)}{dx}\frac{\partial\psi}{\partial\eta} \tag{D-6}$$

$$\frac{\partial u}{\partial x} = \frac{\partial}{\partial x}\frac{\partial\psi}{\partial y} = \frac{\partial}{\partial x}\left(g\frac{\partial\psi}{\partial\eta}\right) = \frac{dg}{dx}\frac{\partial\psi}{\partial\eta} + g\frac{\partial}{\partial x}\frac{\partial\psi}{\partial\eta}$$

$$= \frac{dg}{dx}\frac{\partial\psi}{\partial\eta} + g\left(\frac{\partial^2\psi}{\partial\zeta\partial\eta}\frac{\partial\zeta}{\partial x} + \frac{\partial^2\psi}{\partial\eta^2}\frac{\partial\eta}{\partial x}\right)$$

$$= \frac{dg}{dx}\frac{\partial\psi}{\partial\eta} + g\left[\frac{\partial^2\psi}{\partial\zeta\partial\eta} + \frac{\partial^2\psi}{\partial\eta^2}\eta\frac{d\ln(g)}{dx}\right] \tag{D-7}$$

$$\frac{\partial u}{\partial y} = \frac{\partial}{\partial y}\frac{\partial\psi}{\partial y} = \frac{\partial}{\partial y}\left(g\frac{\partial\psi}{\partial\eta}\right) = g\left[\frac{\partial^2\psi}{\partial\zeta\partial\eta}\frac{\partial\zeta}{\partial y} + \frac{\partial^2\psi}{\partial\eta^2}\frac{\partial\eta}{\partial y}\right] = g^2\frac{\partial^2\psi}{\partial\eta^2} \tag{D-8}$$

$$\frac{\partial^2 u}{\partial y^2} = \frac{\partial}{\partial y}\frac{\partial^2\psi}{\partial y^2} = \frac{\partial}{\partial y}\left(g^2\frac{\partial^2\psi}{\partial\eta^2}\right) = g^2\left[\frac{\partial^3\psi}{\partial\zeta\partial\eta^2}\frac{\partial\zeta}{\partial y} + \frac{\partial^3\psi}{\partial\eta^3}\frac{\partial\eta}{\partial y}\right] = g^3\frac{\partial^3\psi}{\partial\eta^3}$$

$$\tag{D-9}$$

Substitution of Eqs. (D-5)–(D-9) into Eq. (D-1) with $\psi = H(\zeta)F(\eta)$ according to Eq. (D-3) gives, after rearrangement and elimination of the distinction between ζ and x,

$$\frac{d^3F}{d\eta^3} + \left[\frac{1}{\nu g}\frac{dH}{dx}\right]F\frac{d^2F}{d\eta^2} - \frac{1}{\nu g}\left[\frac{dH}{dx} + \frac{d\ln(g)}{dx}H\right]\left(\frac{dF}{d\eta}\right)^2 = -\frac{U\,dU/dx}{\nu g^3 H}$$

$$\text{(D-10)}$$

A separation of variables will be possible in Eq. (D-10) if

$$\frac{1}{\nu g}\frac{dH}{dx} = c_1 \tag{D-11}$$

and

$$\frac{1}{\nu g}\left[\frac{dH}{dx} + \frac{d\ln(g)}{dx}H\right] = c_2 \tag{D-12}$$

Then Eq. (D-10) has the form

$$F''' + c_1 FF'' - c_2(F')^2 = \lambda = -\frac{U\,dU/dx}{\nu g^3 H} \tag{D-13}$$

where λ is a separation constant of yet unknown value as required by the fact that the left-hand side is solely a function of η whereas the right-hand side solely depends on x.

To determine allowable terms for $H(x)$ and $g(x)$, first consider Eq. (D-11) which, when differentiated once, shows that

$$\frac{d^2H}{dx^2} = c_1\nu\frac{dg}{dx}$$

which, together with Eq. (D-11), used in Eq. (D-12) to eliminate $g(x)$ gives

$$H\frac{d^2H}{dx^2} = \left(\frac{c_2}{c_1} - 1\right)\left(\frac{dH}{dx}\right)^2$$

On letting $n = c_2/c_1 - 1$, one finds the solution, for $n \neq 1$, to be

$$H = [(1-n)(c_3 x + c_4)]^{1/(1-n)}, \qquad g = \frac{c_3}{c_1\nu}[(1-n)(c_3 x + c_4)]^{n/(1-n)}$$

$$\text{(D-14)}$$

and, for $n = 1$,

$$H = c_4 e^{c_3 x}, \qquad g = \frac{c_3 c_4}{c_1\nu}e^{c_3 x} \tag{D-15}$$

From Eq. (D-13) the corresponding free-stream velocity is then, for $n \neq 1$,

$$U^2 = -\lambda \nu \int g^3 H \, dx + c_6$$

$$= -\frac{\lambda}{\nu^2} \frac{c_3^2}{c_1^3} \frac{(1-n)^{2(n+1)/(1-n)}}{1+n} (c_3 x + c_4)^{2(n+1)/(1-n)} + c_6$$

If $c_6 = 0$ arbitrarily, then

$$U = \left[-\frac{\lambda}{\nu^2} \frac{c_3^2}{c_1^3} \frac{1}{n+1} \right]^{1/2} [(1-n)(c_3 x + c_4)]^{(n+1)/(1-n)} \quad \text{(D-16)}$$

with only a slight loss of generality. For $n = 1$ this procedure gives

$$U^2 = -\frac{\lambda}{2\nu^2} \frac{c_3^2 c_4^4}{c_1^3} e^{4c_3 x} + c_6$$

which becomes, with $c_6 = 0$ arbitrarily,

$$U = \left[-\frac{\lambda}{2\nu^2} \frac{c_3^2 c_4^4}{c_1^3} \right]^{1/2} e^{2c_3 x} \quad \text{(D-17)}$$

Major interest is focused on the case in which $n \neq 1$.

Additional information can be deduced by considering the boundary condition of Eq. (D-2) that

$$u(x, y \to \infty) \to U(x)$$

In transformed variables this condition is expressed as

$$gH \frac{dF(\infty)}{d\eta} = U$$

Now $dF(\infty)/d\eta$ is a number that is here arbitrarily set equal to unity. Then

$$gH = U$$

Elimination of g by use of Eq. (D-11) puts this relationship into the form

$$\frac{dH^2}{dx} = 2c_1 \nu U$$

If $U = c(x + K)^m$ as suggested by Eq. (D-16), this relationship requires that

$$H = \left[\frac{2c_1\nu}{m + 1} c(x + K)^{m+1} \right]^{1/2} \tag{D-18}$$

and the corresponding form for g is then available from Eq. (D-11) as

$$g = \left[\frac{m + 1}{2} \frac{c}{c_1\nu} (x + K)^{m-1} \right]^{1/2} \tag{D-19}$$

The constant c_2 can now be evaluated in terms of the m of the assumed free-stream velocity variation $U = c(x + K)^m$. From Eq. (D-12) it is found with the help of Eqs. (D-18) and (D-19) that

$$c_2 = \frac{2c_1 m}{1 + m} \tag{D-20}$$

Similarly, the separation constant λ is found from Eq. (D-13)

$$\lambda = - \frac{dU^2/dx}{2\nu g^3 H}$$

with the aid of Eqs. (D-18) and (D-19) to be

$$\lambda = - \frac{2c_1 m}{1 + m} \tag{D-21}$$

The similarity equation [Eq. (D-13)] has been found to have the form

$$F''' + c_1 FF'' + \frac{2c_1 m}{1 + m} \left[1 - (F')^2 \right] = 0$$

where

$$\eta = y \left[\frac{m + 1}{2} \frac{c}{c_1\nu} (x + K)^{m-1} \right]^{1/2} = \frac{y}{x + K} \left(\frac{m + 1}{2c_1} \right) \text{Re}^{1/2}$$

with

$$\text{Re} = \frac{U(x + K)}{\nu}$$

There are no other conditions by which to evaluate constants, so c_1 could have any value. Although a variety of choices have been adopted in the literature, c_1 is here taken to equal $(m + 1)/2$. Then

$$F''' + \frac{m + 1}{2} FF'' + m\left[1 - (F')^2\right] = 0 \qquad \text{(D-22)}$$

where

$$\eta = y\left[\frac{c}{\nu}(x + K)^{m-1}\right]^{1/2}$$

The constant K cannot be evaluated without additional information. Measurements suggest that it is usually small. However, because it cannot be shown to identically equal zero, the similarity equation expressed by Eq. (D-22) is only accurate away from the leading edge if K is assumed to equal zero. In free convection it has been suggested [4] that use of $x + K$ rather than x (with K an experimentally determined constant) provides a better correlation of Nusselt number versus Rayleigh number. The effect of a nonzero K on the similarity solution is to displace downstream a distance K all solutions obtained on the assumption of $K = 0$.

REFERENCES

1. A. G. Hansen, *Similarity Analyses of Boundary Value Problems in Engineering*, Prentice-Hall, 1964.
2. D. E. Abbot and S. J. Kline, Simple methods for classification and construction of similarity solutions of partial differential equations, Report MD-6, Department of Mechanical Engineering, Stanford University, Stanford, CA. See also AFOSR-TN-60-1163.
3. F. K. Moore, Unsteady laminary boundary layer flow, NACA TN 2471 1951.
4. A. J. Edc, *Adv. Heat Transfer* **4**, 46 (1967).

Appendix E

INSTABILITY

In this appendix the presentations of the stability of a liquid–vapor interface set forth by Jordan [1] and Leppert and Pitts [2] are largely followed. The classical works of Milne-Thompson [3] and Lamb [4] can be consulted for additional reading on the subject.

TAYLOR INSTABILITY

Two fluids, one over the other, flow horizontally, as illustrated in Fig. E-1. The upper fluid lies in region R that is of indefinite x and y extent and that extends a distance a above the nominal interface. The lower fluid occupies region R' that is likewise of indefinite x and y extent and that extends a distance a' below the nominal interface. The fluids are incompressible, inviscid, and immiscible, and the flow is irrotational. The constant major fluid velocity is along the x axis of magnitude U and U' for the upper and lower fluid, respectively. Superimposed on this major velocity is a small perturbation due to the wavy character of the interface. Hence the velocity of the upper fluid is

$$\mathbf{V} = (U + u)\hat{\mathbf{i}} + v\hat{\mathbf{j}} + w\hat{\mathbf{k}} \tag{E-1a}$$

and the velocity of the lower liquid is

$$\mathbf{V}' = (U' + u')\hat{\mathbf{i}} + v'\hat{\mathbf{j}} + w'\hat{\mathbf{k}} \tag{E-1b}$$

where u, v, and w are velocity perturbations along the x, y, and z axes. The local instantaneous deviation from the nominal interface is $\eta(x, y, t)$.

597

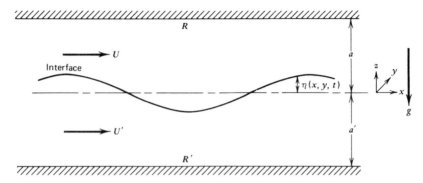

Figure E-1 Taylor instability in which one fluid overlays another less dense fluid.

Since the flow is irrotational, the fluid velocity is derivable from the gradient of a scalar. Let the negative of this scalar be termed the velocity potential Φ. Then $V = -\nabla\Phi$ in R and $V' = -\nabla\Phi'$ in R'. Or

$$U + u = -\frac{\partial\Phi}{\partial x}, \quad v = -\frac{\partial\Phi}{\partial y}, \quad w = -\frac{\partial\Phi}{\partial z} \quad \text{in} \quad R \quad \text{(E-2a)}$$

$$U' + u' = -\frac{\partial\Phi'}{\partial x}, \quad v' = -\frac{\partial\Phi'}{\partial y}, \quad w' = -\frac{\partial\Phi'}{\partial z} \quad \text{in} \quad R' \quad \text{(E-2b)}$$

The continuity equations, $\nabla \cdot V = 0$ as shown in Eq. 2-7, for the upper and lower fluids then are

$$\nabla^2\Phi = 0 \quad \text{in} \quad R \quad \text{(E-3a)}$$

$$\nabla^2\Phi' = 0 \quad \text{in} \quad R' \quad \text{(E-3b)}$$

where $\nabla^2 = \partial^2/\partial x^2 + \partial^2/\partial y^2 + \partial^2/\partial z^2$ in rectangular coordinates. The inviscid equation of motion is $(\rho/g_c)[\partial V/\partial t + (V \cdot \nabla)V] = -\nabla p - \rho g \hat{k}/g_c$ as shown in Eqs. (2-20) and (2-30). Additionally,

$$\nabla(V \cdot V) = 2(V \cdot \nabla)V + 2V \times \text{curl } V$$

For irrotational flow (curl $V = 0$) and in view of the fact that $V \cdot V = V^2_{\text{magnitude}}$, it then follows that the inviscid equation of motion can be rewritten as

$$\frac{\rho}{g_c}\left[\frac{\partial V}{\partial t} + \nabla\left(\frac{V^2_{\text{mag}}}{2} + \frac{g_c p}{\rho} + gz\right)\right] = 0$$

Introducing Eqs. (E-3) into this result leads to

$$\frac{g_c p}{\rho} = \frac{\partial \Phi}{\partial t} - \frac{V_{mag}^2}{2} - gz + C \quad \text{in} \quad R \qquad \text{(E-4a)}$$

$$\frac{g_c p'}{\rho'} = \frac{\partial \Phi'}{\partial t} - \frac{V_{mag}'^2}{2} - gz + C' \quad \text{in} \quad R' \qquad \text{(E-4b)}$$

where C and C' are arbitrary constants.

The inviscid assumption prevents the no-slip condition from being enforced at either the interface or the upper and lower surfaces. However, the impermeability of the upper and lower surfaces can be satisfied by imposing the two boundary conditions that

$$w(x, y, z = a, t) = 0 = w'(x, y, z = -a', t) \qquad \text{(E-5)}$$

A third boundary condition is obtained by requiring the wavy interface between the two fluids to be in equilibrium with pressure forces balanced by surface tension forces. For small deviations η from the nominal interface position the difference between the two fluid pressures, σ is surface tension,

$$p - p' = \sigma \left(\frac{\partial^2 \eta}{\partial x^2} + \frac{\partial^2 \eta}{\partial y^2} \right) \quad \text{at} \quad z = \eta(x, y, t) \qquad \text{(E-6)}$$

A fourth boundary condition is achievable by recognizing that at the wavy interface

$$z = \eta(x, y, t)$$

or, defining a function as the difference between z and η,

$$F(x, y, z, t) = z - \eta(x, y, t) = 0 \qquad \text{(E-7)}$$

Expansion of F in series about the point x_0, y_0, t_0 gives

$$F(x, y, z, t) = F(x_0, y_0, t_0) + dx \frac{\partial F(x_0, y_0, t_0)}{\partial x} + dy \frac{\partial F(x_0, y_0, t_0)}{\partial y}$$

$$+ dz \frac{\partial F(x_0, y_0, t_0)}{\partial z} + dt \frac{\partial F(x_0, y_0, t_0)}{\partial t} + \cdots \qquad \text{(E-8)}$$

Realize that $F = 0$ always and that dx/dt, dy/dt, and dz/dt can be interpreted as the three velocity components of a point on the wavy interface (subject to the understanding of Section 2.2) that must be the velocity of the fluids on either side of the interface. Then $dx/dt = -\partial \Phi/\partial x$, for example, and Eq. (E-8) can be written as

$$0 = -\frac{\partial \Phi}{\partial x}\frac{\partial F}{\partial x} - \frac{\partial \Phi}{\partial y}\frac{\partial F}{\partial y} - \frac{\partial \Phi}{\partial z}\frac{\partial F}{\partial z} + \frac{\partial F}{\partial t} = -\nabla \Phi \cdot \nabla F$$

Emplacing Eq. (E-7) into this result then yields the fourth boundary condition

$$-\frac{\partial \Phi}{\partial z} = \frac{\partial \eta}{\partial t} - \frac{\partial \Phi}{\partial x}\frac{\partial \eta}{\partial x} - \frac{\partial \Phi}{\partial y}\frac{\partial \eta}{\partial y} \qquad \text{at} \quad z = \eta \qquad \text{(E-9a)}$$

Since $dx/dt = -\partial \Phi'/\partial x$ as well, it follows in parallel fashion that

$$-\frac{\partial \Phi'}{\partial z} = \frac{\partial \eta}{\partial t} - \frac{\partial \Phi'}{\partial x}\frac{\partial \eta}{\partial x} - \frac{\partial \Phi'}{\partial y}\frac{\partial \eta}{\partial y} \qquad \text{at} \quad z = \eta \qquad \text{(E-9b)}$$

A fifth boundary condition is obtained by combining Eqs. (E-4) and (E-6) to achieve

$$\frac{\rho}{g_c}\left(\frac{\partial \Phi}{\partial t} - \frac{V^2_{\text{mag}}}{2} - g\eta + C\right) - \frac{\rho'}{g_c}\left[\frac{\partial \Phi'}{\partial t} - \frac{V'^2_{\text{mag}}}{2} - g\eta + C'\right]$$

$$= \sigma\left(\frac{\partial^2 \eta}{\partial x^2} + \frac{\partial^2 \eta}{\partial y^2}\right) \qquad \text{(E-10)}$$

The velocity potentials are taken to be

$$\Phi = \phi - Ux \qquad \text{(E-11a)}$$
$$\Phi' = \phi' - U'x \qquad \text{(E-11b)}$$

in which ϕ and ϕ' are the small perturbations in velocity potential that give the small perturbations in velocity cited in Eq. (E-1). Introduction of Eq. (E-11) into Eqs. (E-3), (E-5), (E-9), and (E-10) gives the final formulation of the mathematical problem to be solved as

$$\nabla^2\phi = 0 \quad \text{in} \quad R, \qquad \nabla^2\phi' = 0 \quad \text{in} \quad R' \qquad \text{(E-12a)}$$

$$\frac{\partial\phi(z = a)}{\partial z} = 0, \qquad \frac{\partial\phi'(z = -a)}{\partial z} = 0 \qquad \text{(E-12b)}$$

$$-\frac{\partial\phi(z = 0)}{\partial z} = \frac{\partial\eta}{\partial t} + U\frac{\partial\eta}{\partial x}, \qquad -\frac{\partial\phi'(z = 0)}{\partial z} = \frac{\partial\eta}{\partial t} + U'\frac{\partial\eta}{\partial x} \qquad \text{(E-12c)}$$

—such higher-order terms as $(\partial\phi/\partial x)(\partial\eta/\partial x) + (\partial\phi/\partial y)(\partial\eta/\partial y) + (\partial\phi/$

$\partial z)(\partial \eta / \partial z)$ have been neglected—

$$\frac{\rho}{g_c}\left[\frac{\partial \phi(z = 0)}{\partial t} + U\frac{\partial \phi(z = 0)}{\partial x} - g\eta\right]$$

$$-\frac{\rho'}{g_c}\left[\frac{\partial \phi'(z = 0)}{\partial t} + U'\frac{\partial \phi'(z = 0)}{\partial x} - g\eta\right] = \sigma\left(\frac{\partial^2 \eta}{\partial x^2} + \frac{\partial^2 \eta}{\partial y^2}\right)$$

$$(\text{E-12d})$$

In Eq. (E-12d) the unperturbed condition, $\eta = 0$ and $\phi = 0$ and $\phi' = 0$, is satisfied by letting $C = U^2/2$ and $C' = U'^2/2$. It has also been assumed that $\eta \approx 0$ and that since $u, v, w \ll U$,

$$\left(U - \frac{\partial \phi}{\partial x}\right)^2 + \left(\frac{\partial \phi}{\partial y}\right)^2 + \left(\frac{\partial \phi}{\partial z}\right)^2 \approx 2U\frac{\partial \phi}{\partial x}$$

The boundary condition in Eq. (E-12) is satisfied by

$$\phi = Af(x, y, t)\frac{\cosh[L(z - a)]}{\sinh La} \qquad (\text{E-13a})$$

$$\phi' = A'f(x, y, t)\frac{\cosh[L(z + a')]}{\sinh La'} \qquad (\text{E-13b})$$

Substitution of Eq. (E-13) into the continuity equation for each fluid [Eq. (E-12)] reveals that it is necessary that $f(x, y, t)$ be a solution to

$$\frac{\partial^2 f}{\partial x^2} + \frac{\partial^2 f}{\partial y^2} + L^2 f = 0 \qquad (\text{E-14})$$

Inasmuch as it is expected that the interface will be wavy in both space and time, the interface deviation η is assumed to have the functional form

$$\eta = \eta_0 \exp[\pm i(\omega t + m_1 x + m_2 y)] \qquad (\text{E-15})$$

where η_0 is an amplitude of oscillation, ω is a frequency of oscillation, and $m_{1,2}$ is a wave number such that the wavelength λ is given by $\lambda = 2\pi/m$. Since the fluid velocities are expected to have a similar wavy characteristic in space and time, a similar functional form is assumed for f as

$$f(x, y, t) = \exp[\pm i(\omega t + m_1 x + m_2 y)] \qquad (\text{E-16})$$

Substitution of Eq. (E-16) into Eq. (E-14) gives

$$L^2 = m_1^2 + m_2^2 \tag{E-17a}$$

whereas substitution of Eq. (E-16) into Eq. (E-12c) gives

$$A = \pm \frac{i\eta_0(\omega + m_1 U)}{L}, \qquad A' = \mp \frac{i\eta_0(\omega + m_1 U')}{L} \tag{E-17b}$$

and substitution into Eq. (E-12d) gives

$$\pm i\rho A(\omega + Um_1)\frac{\coth aL}{g_c} \mp i\rho' A'(\omega + U'm_1)\frac{\coth(a'L)}{g_c}$$

$$= -\sigma\eta_0(m_1^2 + m_2^2) + \eta_0(\rho - \rho')\frac{g}{g_c} \tag{E-17c}$$

Combination of Eqs. (E-17) yields the single relation between frequency ω and the wave numbers m_1 and m_2 of

$$\rho(\omega + Um_1)^2 \frac{\coth aL}{g_c} + \rho'(\omega + U'm_1)^2 \frac{\coth a'L}{g_c}$$

$$= \sigma L^3 - (\rho - \rho')\frac{Lg}{g_c} \tag{E-18}$$

A general condition for the stability of the interface can be obtained from Eq. (E-18). For present purposes, the specialization of negligible gross motion, $U \approx 0 \approx U'$, and deep fluid layers, $a \to \infty$ and $a' \to \infty$, is admissible and puts Eq. (E-18) into the simpler form of

$$\omega^2 = g_c L \frac{\sigma L^2 - (\rho - \rho')g/g_c}{\rho + \rho'} \tag{E-19}$$

Note from Eq. (E-15) that an imaginary part to ω results in an interface that experiences increasingly large undulations. In other words, the interface is stable only if

$$\sigma L^2 \geq (\rho - \rho')\frac{g}{g_c} \tag{E-20}$$

Violation of the stability criterion of Eq. (E-20) is called *Taylor instability* and is illustrated by the air–water interface below an inverted tumbler of water. Although such an interface can be stable (e.g., through the influence of a gauze), it is usually metastable, and any small disturbance usually grows and results in a two-dimensional pattern of vapor jets flowing upward and liquid flowing downward. The unstable arrangement of dense liquid above lighter vapor is characteristic of the peak heat flux condition in boiling.

Taylor's [5] original analysis, experimentally confirmed by Lewis [6], neglected surface tension and the density of one liquid. Effects of surface tension and viscosity were later incorporated by Bellman and Pennington [7] —viscosity usually has little effect.

No unique specification of the boundaries of the interface has been made, so there is no unique solution for the wave pattern and spacing of nodes. An admissible assumption, however, is that there is a square pattern (x and y directions are interchangeable) and a coordinate axis passes through a node, an antinode, and so on. For this to occur, $m_1 = m_2$ and $L^2 = 2m^2$. Furthermore, in a boiling situation the stability criterion of Eq. (E-20) is violated so that Eq. (E-19) is more conveniently rearranged into

$$b^2 = (i\omega)^2 = \frac{g_c L\left[(\rho - \rho')g/g_c - \sigma L^2\right]}{\rho + \rho'}$$ (E-21a)

and it is seen then that Eq. (E-15) becomes

$$\eta = \eta_0 \exp(bt) \exp\left[\pm i(m_1 x + m_2 y)\right]$$ (E-21b)

In such a case b is appropriately termed the *growth-rate parameter*.

The largest value of b—which is $b*$—is seen in Eq. (E-21a) to occur when L achieves the "most dangerous" value L_D of

$$L_D = \left[(\rho - \rho')\frac{g}{g_c}3\sigma\right]^{1/2}$$ (E-22)

which gives

$$b* = \left[\frac{(2/3)(\rho - \rho')g}{\rho + \rho'}\right]^{1/2}\left[\frac{(\rho - \rho')(g/g_c)}{3\sigma}\right]^{1/4}$$ (E-23)

Next, note that the barely stable case of $b = 0$ is found from Eq. (E-2a) to occur at $L_0^2 = (\rho - \rho')(g/g_c)/\sigma$—which, since $L^2 = 2m^2$ and wavelength λ is related to wave number m by $\lambda = 2\pi/m$, gives the associated wavelength as

$$\lambda_0 = 2\pi\left[\frac{\sigma(g_c/g)}{\rho - \rho'}\right]^{1/2}$$ (E-24)

where the $2^{1/2}$ factor is deleted to account for measurement of distance between nodes along the side of the presumed square array rather than along a diagonal. Similarly, the "most dangerous" wavelength corresponding to L_D is found to be

$$\lambda_D = 2\pi \left[\frac{3\sigma(g_c/g)}{\rho - \rho'} \right]^{1/2} \tag{E-25}$$

Although the distance λ between nodes along the sides of a square is not known, it is clear that

$$\lambda_0 \leqslant \lambda \leqslant \lambda_D \tag{E-26}$$

Also, Eq. (E-23) can be recast into

$$b^* = \left[\frac{(4\pi/3\lambda_D)(\rho - \rho')g}{\rho + \rho'} \right]^{1/2} \tag{E-27}$$

As suggested by Eq. (E-26), there is likely to be a spectrum of unstable wavelengths and, therefore, a spectrum of growth rate parameters rather than the single one given by Eq. (E-27). Further details and photographs are available [8].

Because of Taylor instability a horizontal liquid-over-vapor interface (as occurs at peak heat flux boiling conditions) tends to break up in a square pattern of side length λ. From each unit cell of area λ^2 rises a circular vapor column of diameter $\lambda/2$ and area $\pi\lambda^2/16$. This diameter is reasonable, although arbitrary, since it is the distance between a node and an antinode. Consequently, the rate of vapor release from the interface depends only on the maximum frequency of vapor bubble emission from the interface that is limited by Helmholtz instability.

HELMHOLTZ INSTABILITY

The preceding analysis for the condition illustrated in Fig. E-1 that led to Eq. (E-18) can be directly applied to the variant condition illustrated in Fig. E-2. Here, the primary specialization adopted is that gravity acts parallel to the interface and plays no first-order role in its deformation—$g = 0$ in Eq. (E-18). Further simplifications are that the fluid layers are very wide, $a \to \infty$ and $a' \to \infty$, so that Eq. (E-18) becomes

$$\rho(\omega^2 + Um_1)^2 + \rho'(\omega + U'm_1)^2 - \sigma L^3 g_c = 0$$

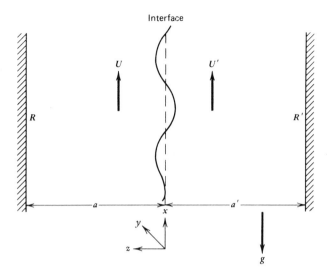

Figure E-2   Helmholtz instability in which two fluids flow parallel to a common wavy interface.

Solution of this relationship for ω yields

$$\omega = -m_1 U_{av} \pm m_1 \left[\frac{C_0^2 - \rho\rho'(U - U')^2}{(\rho + \rho')^2} \right]^{1/2} \tag{E-28}$$

where $U_{av} = (\rho U + \rho' U')/(\rho + \rho')$ is the average velocity of the two flows and $C_0 = [(\sigma g_c L^3/m_1^2)/(\rho + \rho')]^{1/2}$ is the velocity of a surface wave in the absence of bulk currents ($U_1 = 0 = U'$). It is seen that surface waves travel relative to U_{av} with speeds given by $\pm C$, where

$$C^2 = C_0^2 - \frac{\rho\rho'(U - U')^2}{(\rho + \rho')^2}$$

As before, a stable interface requires that the frequency of oscillation ω have no imaginary part. Hence the stability criterion is that

$$\frac{\sigma g_c L^3}{m_1^2} \geqslant \frac{\rho\rho'(U - U')^2}{\rho + \rho'} \tag{E-29}$$

If this criterion is not satisfied the interface oscillates with increasing amplitude until it disrupts entirely, a condition termed *Helmholtz instability* [9].

Application of these results to the peak heat-flux boiling condition leads to the identification of the vapor as being in region R' and flowing upward at velocity $U' = V_v$ and the liquid being in region R and flowing downward at velocity $U = -V_L$. The downflowing liquid must compete with the upflowing vapor; at a certain upward velocity of the vapor, the interface between the two counterflowing streams will be disrupted by Helmholtz instability. At incipient instability the equality of Eq. (E-29) holds, giving

$$g_c m_1 = \frac{\rho \rho'(V_v + V_L)^2}{\rho + \rho'} \tag{E-30}$$

where it has been assumed for simplicity that $L = m_1$—a one-dimensional assumption. A further condition is that, for the square array of vapor columns described at the end of the Taylor Instability section, equality of upward vapor mass flow and downward liquid mass flow requires that

$$\rho V_L \left(\lambda^2 - \frac{\pi \lambda^2}{16} \right) = \rho' V_v \frac{\pi \lambda^2}{16} \tag{E-31}$$

Equations (E-30) and (E-31) then yield the critical vapor velocity as

$$V_{v,\text{critical}} = \left(\frac{\sigma g_c m_1}{\rho'} \right)^{1/2} \frac{[(\rho + \rho')/\rho]^{1/2}}{1 + \pi(\rho'/\rho)/(16 - \pi)}$$

the last term of which is nearly unity. Rayleigh's analysis [10] of a circular gas jet in a liquid shows that axially symmetric disturbances with wavelengths larger than the jet circumference are unstable for all jet velocities. Recalling that wavelength λ is related to wave number m as

$$\lambda = \frac{2\pi}{m}$$

and that the vapor jet has diameter $\lambda/2$, one obtains $m = 4/\lambda$ and

$$V_{v,\text{critical}} = \left(4\sigma \frac{g_c}{\lambda \rho'} \right)^{1/2} \left(\frac{\rho + \rho'}{\rho} \right)^{1/2} \tag{E-32}$$

The heat flux carried away by the vapor (in saturated boiling) is given by

$$\frac{q}{A} = \rho' V_v \frac{A_v}{A} h_{fg}$$

which is further reduced at the critical conditions of peak heat flux, with the use of Eqs. (E-26) and (E-32), to

$$0.12 \leqslant \frac{(q/A)}{\rho_v h_{\text{fg}} \left[\sigma(\rho_L - \rho_v) g g_c / \rho_v^2 \right]^{1/4} \left[(\rho_L + \rho_v)/\rho_L \right]^{1/2}} \leqslant 0.16 \quad \text{(E-33)}$$

which is identical in form with the recommended Eq. (11-3) due to Zuber.

BOTTOM-HEATED HORIZONTAL FLUID LAYER

A layer of fluid, heated from below, saturates a porous medium contained between two infinite horizontal planes. If the temperature difference ΔT exceeds a critical value, the stagnant situation of denser (cooler) upper fluid overlaying lighter (warmer) fluid below becomes unstable and natural convection currents begin. Prior to the inception of motion the stagnant pressure p_0, temperature T_0, and velocity $\mathbf{V}_0$ distributions are

$$-\frac{\partial p_0}{\partial z} = p_0 g, \qquad T_0 = T_w + \left(1 - \frac{z}{L}\right)\Delta T, \qquad \text{and} \qquad \mathbf{V}_0 = 0$$

The motion is described by the continuity, motion, and energy equations from Chapter 2 in vector notation as

$$\nabla \cdot \mathbf{V} = 0 \qquad\qquad\qquad \text{(E-34a)}$$

$$\rho \frac{D\mathbf{V}}{Dt} = -\rho g - \nabla p - \frac{\mu}{K}\mathbf{V} + \mu \nabla^2 \mathbf{V} \qquad\qquad \text{(E-34b)}$$

$$\rho C_p \frac{DT}{Dt} = k \nabla^2 T \qquad\qquad\qquad \text{(E-34c)}$$

The pressure, temperature, and velocity are each expressed as the sum of a stagnant distribution and a small perturbation (denoted by a prime) as

$$p = p_0 + p', \qquad T = T_0 + T', \qquad \text{and} \qquad \mathbf{V} = \mathbf{V}_0 + \mathbf{V}'$$

with $\mathbf{V}' = u'\hat{\imath} + v'\hat{\jmath} + w'\hat{k}$.

Introduction of these expressions for p, T, and $\mathbf{V}$ into Eqs. (E-34) gives

$$\nabla \cdot \mathbf{V}' = 0$$

$$\rho\left[\frac{\partial \mathbf{V}'}{\partial T} + u\frac{\partial u}{\partial x}\hat{\imath} + \cdots \right] = -\rho g\hat{k} - \left[\frac{\partial p'}{\partial x}\hat{\imath} + \frac{\partial p'}{\partial y}\hat{\jmath} + \left(\frac{\partial p_0}{\partial z} + \frac{\partial p'}{\partial z}\right)\right]$$

$$\underbrace{\phantom{+ u\frac{\partial u}{\partial x}\hat{\imath}}}_{\text{negligible}} \qquad \qquad -\rho_0 g$$

$$-\frac{\mu}{K}\mathbf{V}' + \mu\nabla^2\mathbf{V}'$$

$$\rho C_p\left[\frac{\partial T'}{\partial t} + w'\frac{\partial T_0}{\partial z} + w'\frac{\partial T'}{\partial z} + \cdots\right] = k[\nabla^2 T_0 + \nabla^2 T']$$

$$-\Delta T/L \qquad \underbrace{\phantom{+ w'\frac{\partial T'}{\partial z}}}_{\text{negligible}} \qquad 0$$

The products of two perturbation quantities are neglected in favor of a linear description. And, because the onset of convection occurs when the departure from steady state is nonzero and independent of time as pointed out by Jeffreys [11], $\partial(\ )/\partial t = 0$ as well. The equations written above then are

$$\nabla \cdot \mathbf{V}' = 0 \tag{E-35a}$$

$$0 = g\rho\beta T'\hat{k} - \nabla p' - \frac{\mu}{K}\mathbf{V}' + \mu\nabla^2\mathbf{V}' \tag{E-35b}$$

$$-\frac{\Delta T}{L}w' = \alpha\nabla^2 T' \tag{E-35c}$$

The procedure is first to eliminate pressure and second to eliminate velocity from these equations, leaving only one equation of increased order for temperature. To eliminate pressure, apply the $\nabla \cdot$ operator to the motion equation [Eq. (E-35b)] to get

$$0 = g\rho\beta\frac{\partial(T' - T_w)}{\partial z}\hat{k} - \nabla \cdot \nabla p' - \frac{\mu}{K}\nabla \cdot \mathbf{V}' + \mu\nabla^2(\nabla \cdot \mathbf{V}')$$

which in view of Eq. (E-35a) is

$$\nabla^2 p' = g\rho\beta\frac{\partial(T' - T)}{\partial z}\hat{k} \tag{E-36}$$

Then apply the $\nabla^2 = \partial^2/\partial x^2 + \partial^2/\partial y^2 + \partial^2/\partial z^2$ operator to the z-motion equation to get

$$0 = g\rho\beta \nabla^2 T' - \frac{\partial \nabla^2 p'}{\partial z} - \frac{\mu}{K}\nabla^2 w' + \nabla^2\nabla^2 w' \tag{E-37}$$

Substitute Eq. (E-36) into Eq. (E-37) to get

$$0 = g\rho\beta \nabla^2 T' - g\rho\beta\frac{\partial^2(T' - T_w)}{\partial z^2} - \frac{\mu}{K}\nabla^2 w' + \nabla^2\nabla^2 w'$$

Letting $\nabla_1^2 = \partial^2/\partial x^2 + \partial^2/\partial y^2$, this equation is

$$0 = g\rho\beta \nabla_1^2 T' - \frac{\mu}{K}\nabla^2 w' + \nabla^4 w' \tag{E-38}$$

Next, use of the energy equation [Eq. (E-35c)] in Eq. (E-38) allows velocity to be eliminated resulting in

$$0 = g\rho\beta \nabla_1^2 T' + \frac{\mu}{K}\frac{\alpha L}{\Delta T}\nabla^4(T' - T_w) - \frac{\alpha L}{\Delta T}\nabla^6(T' - T_w) \tag{E-39}$$

Now assume a periodic solution for T', consistent with the expectation that natural convection will occur in an assembly of cells, as

$$T' - T_w = \Delta T\, \theta(Z) \sin(nX) \sin(pY)$$

where n and p are integers, $X = x/L$, $Y = y/L$, $Z = z/L$, $R = g\beta \Delta T L^3/\nu\alpha$, and $K' = K/L^2$. Substitution into Eq. (E-39) gives

$$a^2 R\theta = \frac{1}{K'}\left(\frac{d^2}{dZ^2} - a^2\right)^2 \theta - \left(\frac{d^2}{dZ^2} - a^2\right)^3 \theta \tag{E-40}$$

where $a^2 = n^2 + p^2$ is the wave number.

The boundary conditions for this problem with isothermal, impermeable, and no-slip surfaces are $T' = 0 = w'$ and $u' = 0 = v'$. Further since the latter, no-slip condition must be satisfied for all x and y on the surfaces, $\partial u'/\partial x = 0 = \partial v'/\partial y$ there; in view of the continuity equation, this requires $\partial w'/\partial z = 0$ there, too. Hence, in view of Eq. (E-35c), on the surfaces $T' = 0 = \nabla^2 T'$ and $\partial(\nabla^2 T')/\partial z = 0$. With the foregoing definitions for variables, these boundary conditions become, at $Z = 0$ and 1,

$$\theta = 0 = \frac{d^2\theta}{dZ^2} \tag{E-41a}$$

and

$$\frac{d^3\theta}{dZ^3} - a^2\frac{d\theta}{dZ} = 0 \qquad \text{(E-41b)}$$

This gives six conditions for the sixth-order Eq. (E-40).

The solution procedure is illustrated by considering the case of a porous medium of low permeability ($K' \to 0$). Then Eqs. (E-40) and (E-41) are

$$\frac{d^4\theta}{dZ^4} - 2a^2\frac{d^2\theta}{dZ^2} + a^4\theta = a^2K'R\theta \qquad \text{(E-42)}$$

subject to, since only four conditions can be satisfied,

$$\theta(0) = \frac{d^2\theta(0)}{dZ^2} \qquad \text{and} \qquad \theta(1) = \frac{d^2\theta(1)}{dZ^2}$$

The solution is assumed to be

$$\frac{d^4\theta}{dZ^4} = \sum_{m=1}^{\infty} A_m \sin(m\pi Z)$$

Successive integrations give

$$\theta = C_1 + C_2 Z + C_3 Z^2 + C_4 Z^3 + \sum_{m=1}^{\infty} \frac{A_m \sin(m\pi Z)}{(m\pi)^4}$$

Satisfaction of the boundary conditions results in $C_{1,2,3,4} = 0$. Substitution of this solution for θ into Eq. (E-42) gives

$$\sum_{m=1}^{\infty} \left\{ \frac{\left[(m\pi)^2 + a^2\right]^2 - a^2K'R}{(m\pi)^4} \right\} A_m \sin(m\pi Z) = 0$$

The only nontrivial solution with motion occurs when the coefficient in braces is zero so that

$$\text{Ra} = \frac{g\beta\,\Delta T LK}{\nu\alpha} = \left[\frac{(m\pi)^2 + a^2}{a}\right]^2 \qquad \text{(E-43)}$$

Ra depends on wave number as shown in Fig. E-3. The critical Rayleigh number Ra is the smallest for any of the allowed wave numbers m and a. Setting the derivative of Eq. (E-43) with respect to a for $m = 1$ equal to zero

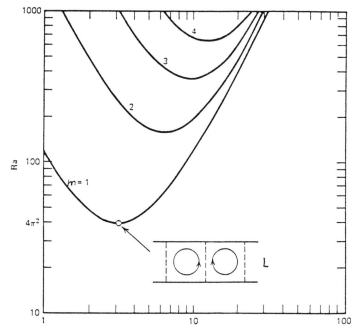

Figure E-3 The minimum Rayleigh number for neutrally stable convection in a porous layer heated from below. [Reprinted by permission from A. Bejan, *Convection Heat Transfer*, © 1984 John Wiley & Sons.]

gives, as first found by Lapwood [12],

$$Ra = 4\pi^2$$

and $a_{critical} = \pi$, showing that the convection cell width nearly equals the plate separation L. After the first instability, a second instability occurs at a higher critical Rayleigh number as temperature difference increases, to be followed by a third later, and so on.

Cases with distributed internal heat release and externally imposed downward flow have been treated [13] by similar methods. Other methods for determining stability criteria are available as discussed by Catton [14].

REFERENCES

1. D. P. Jordan, *Adv. Heat Transfer* **5**, 117–122, (1968).

2. G. Leppert and C. C. Pitts, *Adv. Heat Transfer* **1**, 234–238 (1964).

3. L. M. Milne-Thompson, *Theoretical Hydrodynamics*, 3rd ed., Macmillan, 1955, pp. 374–431.

4. H. Lamb, *Hydrodynamics*, 6th ed., Dover, 1945, p. 370–375.

5. G. Taylor, *Proc. Roy. Soc. London, Ser. A* **201**, 192 (1950).

6. D. J. Lewis, *Proc. Roy. Soc. London, Ser. A* **201**, 81 (1950).

7. R. Bellman and R. H. Pennington, *Quart. Appl. Math.* **12**, 151 (1954).

8. K. Taghavi-Tafreshi and V. K. Dhir, *Int. J. Heat Mass Transfer* **23**, 1433–1445 (1980).

9. H. von Helmholtz, *Berl. Monatsber* (April 1868); *Philos. Mag.* (November 1868); *Wissenschaftliche Abhandlungen*, Leipzig, **14b** (1882–1883).

10. Lord Rayleigh, *Theory of Sound*, Dover, 1945.

11. H. Jeffreys, *Philos. Mag.* **2**, 833–844 (1926).

12. E. R. Lapwood, *Proc. Cambridge Philo. Soc.* **44**, 508–521 (1948).

13. A. Hadim and L. Burmeister, *J. Thermophys. Heat Transfer* **2**, 343–351 (1988).

14. I. Catton, in S. Kakac, W. Aung, and R. Viskanta, Eds., *National Convection Fundamentals and Applications*, Hemisphere, 1985, p. 97–134.

INDEX